Progress in Brain Research
Volume 265

Nanomedicine and Neuroprotection in Brain Diseases

Serial Editor

Vincent Walsh
Institute of Cognitive Neuroscience
University College London
17 Queen Square
London WC1N 3AR UK

Editorial Board

Progress in Brain Research
Volume 265

Nanomedicine and Neuroprotection in Brain Diseases

Edited by

Hari Shanker Sharma

*International Experimental Central Nervous System Injury &
Repair (IECNSIR), Department of Surgical Sciences,
Anesthesiology & Intensive Care Medicine,
Uppsala University Hospital, Uppsala University,
Uppsala, Sweden*

Aruna Sharma

*International Experimental Central Nervous System Injury &
Repair (IECNSIR), Department of Surgical Sciences,
Anesthesiology & Intensive Care Medicine,
Uppsala University Hospital, Uppsala University,
Uppsala, Sweden*

ELSEVIER

Elsevier
Radarweg 29, PO Box 211, 1000 AE Amsterdam, Netherlands
The Boulevard, Langford Lane, Kidlington, Oxford, OX5 1GB, United Kingdom
50 Hampshire Street, 5th Floor, Cambridge, MA 02139, United States

First edition 2021

ISBN: 978-0-323-90162-8
ISSN: 0079-6123

For information on all Elsevier publications
visit our website at https://www.elsevier.com/books-and-journals

Publisher: Zoe Kruze
Acquisitions Editor: Sam Mahfoudh
Developmental Editor: Federico Paulo Mendoza
Production Project Manager: Abdulla Sait
Cover Designer: Alan Studholme

Typeset by STRAIVE, India

Working together
to grow libraries in
developing countries

www.elsevier.com • www.bookaid.org

Contributors

Igor Bryukhovetskiy
Department of Fundamental Medicine, School of Biomedicine, Far Eastern Federal University; Laboratory of Pharmacology, National Scientific Center of Marine Biology, Far East Branch of the Russian Academy of Sciences; School of Life Science & Biomedicine, Medical Center, Far Eastern Federal University (FEFU), Vladivostok, Russia

Anca D. Buzoianu
Department of Clinical Pharmacology and Toxicology, "Iuliu Hatieganu" University of Medicine and Pharmacy, Cluj-Napoca, Romania

Rudy J. Castellani
Department of Pathology, University of Maryland, Baltimore, MD, United States

Huijing Chen
Guangzhou University of Chinese Medicine, Guangzhou, China

Lin Chen
Department of Neurosurgery, Dongzhimen Hospital of Beijing University of Traditional Chinese Medicine, Beijing, China

Shuang Dai
Department of Oncology, The Seventh Affiliated Hospital, Sun Yat-sen University, Shenzhen, China

Lianyuan Feng
Department of Neurology, Bethune International Peace Hospital, Shijiazhuang, China

Hanyu Guo
The Second Affiliated Hospital of Guangzhou University of Chinese Medicine, Guangdong Province Hospital of Chinese Medical, Guangzhou, China

Hongyun Huang
Beijing Hongtianji Neuroscience Academy; Institute of Neurorestoratology, Third Medical Center of General Hospital of PLA, Beijing, People Republic of China

Aleksandra Kosianova
School of Life Science & Biomedicine, Medical Center, Far Eastern Federal University (FEFU), Vladivostok, Russia

José Vicente Lafuente
LaNCE, Department of Neuroscience, University of the Basque Country (UPV/EHU), Leioa, Bizkaia, Spain

Cong Li
The Second Affiliated Hospital of Guangzhou University of Chinese Medicine, Guangdong Province Hospital of Chinese Medical; Department of Neurosurgery, Chinese Medicine Hospital of Guangdong Province, The Second Affiliated Hospital, Guangzhou University of Chinese Medicine, Yuexiu, Guangzhou, China

Wenyu Li
Department of Oncology, The First Affiliated Hospital, Sun Yat-sen University, Guangzhou, China

Yanwen Luo
Qionghai Hospital of traditional Chinese Medicine, Qionghai, Hainan Province, China

Igor Manzhulo
Department of Fundamental Medicine, School of Biomedicine, Far Eastern Federal University; Laboratory of Pharmacology, National Scientific Center of Marine Biology, Far East Branch of the Russian Academy of Sciences, Vladivostok, Russia

Gengsheng Mao
Beijing Hongtianji Neuroscience Academy, Beijing, People Republic of China

Preeti K. Menon
Department of Biochemistry and Biophysics, Stockholm University, Stockholm, Sweden

Dafin F. Muresanu
Department of Clinical Neurosciences, University of Medicine & Pharmacy; "RoNeuro" Institute for Neurological Research and Diagnostic, Cluj-Napoca, Romania

Feng Niu
CSPC NBP Pharmaceutical Medicine, Shijiazhuang, China

Ala Nozari
Anesthesiology & Intensive Care, Massachusetts General Hospital, Boston, MA, United States

Oleg Pak
School of Life Science & Biomedicine, Medical Center, Far Eastern Federal University (FEFU), Vladivostok, Russia

Ranjana Patnaik
Department of Biomaterials, School of Biomedical Engineering, Indian Institute of Technology, Banaras Hindu University, Varanasi, India

Seaab Sahib
Department of Chemistry & Biochemistry, University of Arkansas, Fayetteville, AR, United States

Aruna Sharma
International Experimental Central Nervous System Injury & Repair (IECNSIR), Department of Surgical Sciences, Anesthesiology & Intensive Care Medicine, Uppsala University Hospital, Uppsala University, Uppsala, Sweden

Hari Shanker Sharma
International Experimental Central Nervous System Injury & Repair (IECNSIR), Department of Surgical Sciences, Anesthesiology & Intensive Care Medicine, Uppsala University Hospital, Uppsala University, Uppsala, Sweden

Qijia Tan
The Second Affiliated Hospital of Guangzhou University of Chinese Medicine, Guangdong Province Hospital of Chinese Medical, Guangzhou, China

Z. Ryan Tian
Department of Chemistry & Biochemistry, University of Arkansas, Fayetteville, AR, United States

Chao Wang
The Second Affiliated Hospital of Guangzhou University of Chinese Medicine, Guangdong Province Hospital of Chinese Medical, Guangzhou, China

Zhenguo Wang
CSPC NBP Pharmaceutical Medicine, Shijiazhuang, China

Lars Wiklund
International Experimental Central Nervous System Injury & Repair (IECNSIR), Department of Surgical Sciences, Anesthesiology & Intensive Care Medicine, Uppsala University Hospital, Uppsala University, Uppsala, Sweden

Youliang Wu
The Second Affiliated Hospital of Guangzhou University of Chinese Medicine, Guangdong Province Hospital of Chinese Medical, Guangzhou, China

Caijun Xie
The Second Affiliated Hospital of Guangzhou University of Chinese Medicine, Guangdong Province Hospital of Chinese Medical, Guangzhou, China

Sergeis Zaitsev
School of Life Science & Biomedicine, Medical Center, Far Eastern Federal University (FEFU), Vladivostok, Russia

Wengang Zhan
The Second Affiliated Hospital of Guangzhou University of Chinese Medicine, Guangdong Province Hospital of Chinese Medical, Guangzhou, China

Zhiqiang Zhang
The Second Affiliated Hospital of Guangzhou University of Chinese Medicine, Guangdong Province Hospital of Chinese Medical; Department of Neurosurgery, Chinese Medicine Hospital of Guangdong Province, The Second Affiliated Hospital, Guangzhou University of Chinese Medicine, Yuexiu, Guangzhou, China

Contents

CHAPTER 1 **Alzheimer's disease neuropathology is exacerbated following traumatic brain injury. Neuroprotection by co-administration of nanowired mesenchymal stem cells and cerebrolysin with monoclonal antibodies to amyloid beta peptide** 1

Hari Shanker Sharma, Dafin F. Muresanu,
Rudy J. Castellani, Ala Nozari, José Vicente Lafuente,
Anca D. Buzoianu, Seaab Sahib, Z. Ryan Tian,
Igor Bryukhovetskiy, Igor Manzhulo, Preeti K. Menon,
Ranjana Patnaik, Lars Wiklund, and Aruna Sharma

**CHAPTER 4 Nanodelivery of oxiracetam enhances memory,
 functional recovery and induces neuroprotection
 following concussive head injury** **139**
Feng Niu, Aruna Sharma, Zhenguo Wang, Lianyuan Feng,
Dafin F. Muresanu, Seaab Sahib, Z. Ryan Tian,
José Vicente Lafuente, Anca D. Buzoianu,
Rudy J. Castellani, Ala Nozari, Preeti K. Menon,
Ranjana Patnaik, Lars Wiklund, and Hari Shanker Sharma

CHAPTER 7 **Upregulation of hemeoxygenase enzymes HO-1 and HO-2 following ischemia-reperfusion injury in connection with experimental cardiac arrest and cardiopulmonary resuscitation: Neuroprotective effects of methylene blue**............**317**

Lars Wiklund, Aruna Sharma, Ranjana Patnaik, Dafin F. Muresanu, Seaab Sahib, Z. Ryan Tian, Rudy J. Castellani, Ala Nozari, José Vicente Lafuente, and Hari Shanker Sharma

CHAPTER 9   Manganese nanoparticles induce blood-brain barrier disruption, cerebral blood flow reduction, edema formation and brain pathology associated with cognitive and motor dysfunctions 385

Aruna Sharma, Lianyuan Feng, Dafin F. Muresanu,
Seaab Sahib, Z. Ryan Tian, José Vicente Lafuente,
Anca D. Buzoianu, Rudy J. Castellani, Ala Nozari,
Lars Wiklund, and Hari Shanker Sharma

Acknowledgments

We are greatly indebted to Sam Mahfoudh, Acquisitions Editor at Elsevier Science, Oxford, United Kingdom, for his excellent and constant help, encouragement, and support since the inception of this volume and through various stages of compilation and communication leading to its realization. Without his support, the volume in its current form would have been difficult to imagine. We thank Federico Paulo S. Mendoza, Editorial Project Manager, Elsevier, Manila, Philippines, for helping us with the compilation and publication procedure. We are grateful to Abdulla Sait, Project Manager, Elsevier, Chennai, India, for his excellent help during the production of this volume. We also thank Suraj Sharma, Blekinge Institute of Technology, Karlskrona, Sweden, and Dr. Saja Alshafeay, University of Arkansas, Fayetteville, AR, United States, for providing computer and graphics support.

HARI SHANKER SHARMA
ARUNA SHARMA
Uppsala, Sweden

Preface

"Nanomedicine uses nano-sized tools for the diagnosis, prevention and treatment of disease and to gain increased understanding of the complex underlying patho-physiology of disease. The ultimate goal is to improve quality of life."

European Science Foundation.
Forward Look Nanomedicine: An EMRC Consensus Opinion 2005.
http://www.esf.org (accessed 15 December 2005)

The history of nanomedicine could be traced back to the ancient times of *Bhasma* in *Ayurveda* (ca. 1500 BC) described in the *Charak Samhita* (Kaviratna and Sharma, 1913; Sharma, 2003; Sharma and Dash, 2000) and using colloidal gold preparation for electron microscopy or biomedicine (Anconi, 1618; Macker, 1766). Modern nanomedicine was first described by Ehrlich (1913) for work on cell-specific diagnostic and therapeutic uses (Astruc, 2016). Although nanoparticles related to therapy and toxicity have been mentioned in the scientific literature since 1970, the term "nanomedicine" was first described in early 2000 for treatment opportunities in modern medicine (Sharma, 2007a,b, 2009). In the last 50 years, the use of nanoparticles for nanomedicine has rapidly developed using Ag, Au, liposomes, antibodies, quantum dots, DNA, albumin and other proteins, micelles, and polymers conjugated with drugs for delivery either for diagnostic purposes or for therapeutic uses (Astruc, 2016; Ventola, 2012a,b,c). Recently, nanowires are used for effective delivery of drugs for superior neuroprotection in brain diseases (Sharma, 2007a,b; Sharma et al., 2007, 2009a,b,c Tian et al., 2012). These potential nanowires are able to deliver therapeutic drugs effectively to induce neuroprotection in various brain insults (Ozkizilcik et al., 2018; Sharma et al., 2016a,b).

However, nanomaterials have always been a matter of great concern since ancient times with regard to human health effects. The concept of nanomedicine is mentioned in the oldest medical literature available in the world *"Ayurveda,"* considered more than 5000 years old (Mishra, 2019; Sharma and Dash, 2000; Sharma et al., 2019). In the first treatise of medicine in the *Sanskrit* literature, *Atharvaveda* (the science of toxicology) is referred to as *Agada Tantra* (toxicological sciences). In *Ayurveda* as mentioned in the *Charak Samhita*, the toxicological aspects of science deal with the toxic effects of natural poisons that are divided into inanimate (*Sthaavara*) and animate (*Jangama*) disciplines (Kaviratna and Sharma, 1913; Sharma and Dash, 2000; Sharma and Clark, 1998). Toxicity caused by inanimate poisons could result from minerals, metals, or metal ores embedded in the earth, whereas animate toxicity includes venoms from reptiles, scorpions, worms, and other insects causing brain dysfunction or death (Wujastyk, 2003).

Apart from the knowledge of toxicity, in Ayurveda the use of nanomedicine was very common in the forms of *Bhasma* (fine powder) and other *Ayurvedic* drugs from the 7th to 15th century BC (Dash and Ramaswami, 1998; Pal, 2015; Prakash et al., 2010; Sarkar and Chaudhary, 2010). These *Ayurveda Bhasma* or drugs contain fine particles or powders prepared through special procedures and mixed with other herbal elements to enhance the properties of the medicine for effective therapy. This mixing of metal powders with herbs is known as *Rasa Shastra* and was developed by *Siddha Nagarjuna*, who is considered the father of Indian Alchemy (Mahdihassan, 1985; Sharma, 2003). These Ayurvedic drugs include gold (Au), silver (Ag), copper (Cu), lead (Pb), tin (Sn), iron (Fe), and some other alloys of mercury (Hg) and sulfur (S) (Sarkar and Chaudhary, 2010; Sharma et al., 2007a,b). To treat various diseases, different metal powders are prepared to cure the ailment Sharma, 2000.

Modern nanotechnology and nanomedicine equally possess these age-old problems even today. Thus, we still do not know the adverse effects of various nanoparticles on human health, which are used to deliver drugs for either more effective treatment of diseases or diagnostic purposes (Sharma, 2009; Sharma and Sharma, 2012, 2013, 2021). Moreover, our knowledge on the exposure of different nanoparticles present in air, food, water, and soil or those produced by industrial sources on human health is still rudimentary.

Nanomaterials (NMs) or nanoparticles (NPs) were earlier referred with different names, namely, microfine particles (MfPs) or ultrafine particles (UfPs), in atmospheric or environmental studies (Buzea et al., 2007; Handy and Shaw, 2007). They are also known as particulate matters (PMs) and represent all particles or a collection of them that are less than $10 \mu m$ (μm) in diameter. The normal range of these PMs in atmosphere could vary between 10 and 100 nm and could cause serious health hazards after condensation with other harmful matters or other NPs (Aitken et al., 2006). Sometimes NPs are also described as nanoparticulate matters (NPMs), which refer to a collection of NPs with their cumulative function or behavior. The UfPs in the environment are also NPs with their sizes smaller than 100 nm (Sharma, 2009; Sharma and Sharma, 2012; Sharma et al., 2009a,b,c). Dust, pollen, mold spores, and other particles from air are often referred to as UfPs (Buzea et al., 2007). Microfine particles, in the range of 1–2.5 μm in sizes, could come into the air as pollutants from combusted fuels and soot from diesel engines, coal burning, or from cigarette smoke (Sharma, 2009). The prominent sources of these microfine particles are poorly maintained diesel engine vehicles and some industrial sources. Soot particles are often clearly visible in the black smoke from vehicle's exhaust. These microfine and UfPs are potential health hazards (Sharma and Sharma 2007).

Military personnel are normally more susceptible to NP exposure either during combat operations or while engaging in peace-keeping activities across the world (Lewinski et al., 2008; Sharma et al., 2015, 2018a,b). They are also exposed to various other kinds of stress caused by the environment, e.g., heat, dust, and cold or gunpowder explosion (Chen et al., 2015; Sharma, 2007a,b; Sharma et al., 2017; Sharma and Hoopes, 2003; Sharma and Sharma, 2012, 2013, 2021; Sharma et al., 2007a,b,c, 2009a,b,c). Use of military equipment and missile explosion are good sources of various NPs to which military personnel are exposed during day-to-day activities. Dust exposure or dust storms in Middle Eastern countries expose military personnel to SiO_2 NPs, whereas gunpowder explosion or handling of explosive materials and other military equipment exposes them to C, S, Cu, Ag, or Ti NPs frequently (Sharma et al., 2009c). Moreover, the cities or areas where these personnel are engaged in combat activities are also heavily polluted with several other MfPs,

UfPs, and various NPs. They breathe the polluted and contaminated air for a long time. Accordingly, their health is most likely to be compromised after prolonged exposure to these ambient air NPs (Sharma et al., 2009b,c; Sharma and Sharma, 2012, 2013).

It is quite likely that when these NP-intoxicated military personnel get injured in their brain or spinal cord during combat operations, their brain pathology could be different from those with CNS injury without NP involvement (Sharma and Sharma, 2015; Sharma et al., 2009a,b,c,d). Furthermore, it is still unclear whether drugs that are able to thwart CNS injury and induce neuroprotection in normal healthy individuals after CNS trauma could also equally protect the brain or spinal cord damage in persons intoxicated with NPs following CNS insults (Sharma et al., 2016a,b, 2020). Thus, further studies on the involvement of NPs in CNS injury and their modification with drugs or a combination of them in achieving neuroprotection are needed in greater detail.

This volume is focused on the potential exposure of nanoparticles from various sources and their effects on brain dysfunction, including brain pathology. Moreover, the pathophysiological effects of neurodegeneration and use of nanomedicine either alone or in combination to induce superior neuroprotection is discussed. It appears that nanomedicine that can alleviate nanoneurotoxicity is the need of the hour to enhance the quality of life of patients either suffering from neurodegenerative diseases or after experiencing stroke, trauma, or other related brain diseases. This volume contains 10 peer-reviewed chapters by select invited leading experts in neurological diseases from across the world. We hope that this volume will further expand our knowledge with regard to nanomedicine in treating neurological diseases for achieving superior neuroprotection. The salient findings regarding brain disease in this volume include Alzheimer's disease (AD), glioblastoma, cardiac arrest, heat stroke and ischemic stroke, and concussive head injury using nanomedicine approaches such as cell therapy.

In Chapter 1, Hari Shanker Sharma (Uppsala, Sweden) and colleagues show exacerbation of brain pathology in Alzheimer's disease following concussive head injury. In such comorbidity situations in AD, single treatment strategies appear ineffective. Their research clearly suggests that nanowired delivery of cerebrolysin together with mesenchymal stem cells in combination with monoclonal antibodies to the amyloid beta protein is needed for superior neuroprotection. This suggests that if brain diseases like AD are associated with other comorbidity factors such as concussive head injury, then a select combination of agents using nanodelivery is beneficial.

In Chapter 2, Igor Bryukhovetskiy (Vladivostok, Russia) and colleagues discuss the current status and future strategies of patients with glioblastoma for superior beneficial effects and better therapeutic approaches.

Glioblastoma is a dreadful disease, and current therapeutic approaches so far are not enough and require further exploration in therapeutic strategies. In Chapter 3, Cong Li (Guangzhou, China) and colleagues present new research on the pharmacological characteristics of cinobufotalin in treating glioblastoma.

Feng Niu (Shijiazhuang, China) and colleagues in Chapter 4 explore new therapeutic strategies for concussive head injury-induced brain pathology and functional improvements using oxiracetam and compared its nanowired delivery using sodium or potassium titanate nanowires. The novel research shows that potassium titanate nanowires appear to have superior neuroprotective effects than sodium titanate nanowires under identical conditions. This novel research encourages further examining effects of different kinds of nanowired delivery in brain diseases to have better therapeutic effects in future.

In Chapter 5, Hongyun Huang (Beijing, China) and colleagues describe new approaches for cell therapy for better advancement of therapeutic strategies in ischemic stroke treatment.

In Chapter 6, Seaab Sahib (Fayetteville, AR, USA) and colleagues use nanodelivery of Chinese medicines from Gingko biloba extracts EGb-761 and BN-52021 in treating heat stroke-induced brain pathology. This novel research shows for the first time that nanowired delivery of EGb-761 or BN-52021 exerts superior neuroprotective effects in thwarting brain pathology in heat stroke.

Cardiac arrest is one of the key causes of brain damage in survivors associated with disabilities. In Chapter 7, Lars Wiklund (Uppsala, Sweden) and colleagues discuss the new role of heme oxygenase in cardiac arrest-induced brain pathology. They showed the significance of an old drug methylene blue in attenuating cardiac arrest-induced brain damage together with experimental hypothermia.

In Chapter 8, Huijing Chen (Guangzhou, China) and colleagues discuss multimodal imaging in glioblastoma and its recurrence, following treatment in light of the current literature.

In Chapter 9 Aruna Sharma (Uppsala, Sweden) and colleagues, the influence of manganese nanoparticles associated with Parkinson's disease is further evaluated in the brain pathology with reference to the blood–brain barrier breakdown, and edema formation including reduction in the cerebral blood flow.

In Chapter 10, Cong Li (Guangzhou, China) and colleagues describe the effects of anlotinib in glioblastoma recurrence in a patient with leptomeningeal spread in light of the current literature.

The new knowledge presented in this volume for novel treatment strategies in brain tumor, cardiac arrest, ischemic and heat strokes, concussive head injury, and AD expand our understanding about these brain diseases and shed light on better therapeutic efficacy in reducing brain pathology.

This volume is indispensable for educators, researchers, policy makers, clinicians, and in drug development and healthcare industries. Teachers, researchers, and students in several neuroscience disciplines including neuropathology, neuropharmacology, neuroradiology, neurology, neurorehabilitation, neuropsychiatry, and neurophysiology will benefit from the research and state-of-the art scientific literature presented in this volume. We sincerely hope that this volume will encourage further research in neuroscience and clinical practices with nanomedicine for the betterment of the quality of life of patients affected with brain diseases.

Hari Shanker Sharma

Aruna Sharma

Uppsala University, Uppsala, Sweden

References

Aitken, R.J., Chaudhry, M.Q., Boxall, A.B., Hull, M., 2006. Manufacture and use of nanomaterials: current status in the UK and global trends. Occup. Med. (Lond.) 56 (5), 300–306. https://doi.org/10.1093/occmed/kql051.

Anconi, F., 1618. Panacae Aurea-Auro Potabile. Bibliopolio Frobeniano, Hamburg, Germany.

Astruc, D., 2016. Editorial: introduction to nanomedicine. Molecules 21, 4. https://doi.org/10.3390/molecules21010004.

Buzea, C., Pacheco, I.I., Robbie, K., 2007. Nanomaterials and nanoparticles: sources and toxicity. Biointerphases 2 (4), MR17–71. https://doi.org/10.1116/1.2815690.

Chen, L., Ao, Q., Sharma, H.S., Wang, A., Feng, S., 2015. Neurorestoratologic strategies and mechanisms in the nervous system. Biomed. Res. Int. 2015, 163170. https://doi.org/10.1155/2015/163170 (Epub 2015 Sep 29).

Dash, B., Ramaswami, S., 1998. Ayurveda the Science of Traditional Indian Medicine. Published December 1st 1998, Lustre Press, Varanasi, India, ISBN: 8174360441, pp. 1–84 (ISBN13: 9788174360441).

Ehrlich, P., 1913. Address in pathology. On chemiotherapy. Delivered before the 17th international congress of medicine. Br. Med. J. 16, 353–359.

Handy, R.D., Shaw, B.J., 2007. Ecotoxicity of nanomaterials to fish: challenges for ecotoxicity testing. Integr. Environ. Assess. Manag. 3 (3), 458–460. https://doi.org/10.1002/ieam.5630030316.

Kaviratna, A.C., Sharma, P., 1913. The Charaka Samhita. 5 vols Sri Satguru Publications, ISBN: 978-81-7030-471-5.

Lewinski, N., Colvin, V., Drezek, R., 2008. Cytotoxicity of nanoparticles. Small 4 (1), 26–49. https://doi.org/10.1002/smll.200700595.

Macker, P.-J., 1766. Dictionnaire de Chymie. Lacombe, Paris, France.

Mahdihassan, S., 1985. Cinnabar-gold as the best alchemical drug of longevity, called Makaradhwaja in India. Am. J. Chin. Med. 13 (1–4), 93–108. https://doi.org/10.1142/S0192415X85000149.

Mishra, R.K., 2019. Ayurveda – 5,000 Years Young. https://www.maharishi.co.uk/blog/ayurveda-5000-years-young/. accessed January 10, 2020.

Ozkizilcik, A., Williams, R., Tian, Z.R., Muresanu, D.F., Sharma, A., Sharma, H.S., 2018. Synthesis of biocompatible titanate nanofibers for effective delivery of neuroprotective agents. Methods Mol. Biol. 1727, 433–442. https://doi.org/10.1007/978-1-4939-7571-6_35.

Pal, S.K., 2015. The *Ayurvedic Bhasma*: the ancient science of nanomedicine. Recent Pat. Nanomed. 2015 (5), 12–18.

Prakash, V.B., Prakash, S., Sharma, R., Pal, S.K., 2010. Sustainable effect of *Ayurvedic* formulations in the treatment of nutritional anemia in adolescent students. J. Altern. Complement. Med. 16 (2), 205–211.

Sarkar, P.K., Chaudhary, A.K., 2010. Ayurvedic Bhasma: the most ancient application of nanomedicine. J. Sci. Ind. Res. 69, 901–905.

Sharma, D.C., 2000. India raises standards for traditional drugs. Lancet 356 (9225), 231. https://doi.org/10.1016/S0140-6736(05)74488-3.

Sharma, H.S., 2003. Nagarjuna's Rasendra Mangala. Chaukhamba Orientalia Publications, Varanasi, India, pp. 1–250.

Sharma, H.S., 2007a. Nanoneuroscience: emerging concepts on nanoneurotoxicity and nanoneuroprotection. Nanomedicine (London) 2 (6), 753–758. https://doi.org/10.2217/17435889.2.6.753.

Sharma, H.S., 2007b. Neurotrophic factors in combination: a possible new therapeutic strategy to influence pathophysiology of spinal cord injury and repair mechanisms. Curr. Pharm. Des. 13 (18), 1841–1874. https://doi.org/10.2174/138161207780858410.

Sharma, H.S., 2009. Nanoneuroscience and nanoneuropharmacology. Prog. Brain Res. 180, 1–264. https://doi.org/10.1016/S0079-6123(08)80016-7. Epub 2009 Dec 8.S ISBN: 9780444534316; eBook ISBN: 9780080962245.

Sharma, H., Clark, C., 1998. Contemporary Ayurveda: Medicine and Research in Maharishi Ayurveda. Churchill Livingstone Press, London, pp. 1–184.

Sharma, R.K., Dash, B., 2000. Agnivesh's Charaka Samhita. vol. III Chowkhamba Sanskrit Series Publications, Varanasi, India, ISBN: 978-81-7080-014-9, pp. 43–44.

Sharma, H.S., Hoopes, P.J., 2003. Hyperthermia induced pathophysiology of the central nervous system. Int. J. Hyperth. 19 (3), 325–354. https://doi.org/10.1080/026567302 1000054621.

Sharma, H.S., Sharma, A., 2012. Nanowired drug delivery for neuroprotection in central nervous system injuries: modulation by environmental temperature, intoxication of nano-particles, and comorbidity factors. Wiley Interdiscip. Rev. Nanomed. Nanobiotechnol. 4 (2), 184–203. https://doi.org/10.1002/wnan.172 (Epub 2011 Dec 8).

Sharma, H.S., Sharma, A., 2013. New perspectives of nanoneuroprotection, nanoneurophar-macology and nanoneurotoxicity: modulatory role of amino acid neurotransmitters, stress, trauma, and co-morbidity factors in nanomedicine. Amino Acids 45 (5), 1055–1071. https://doi.org/10.1007/s00726-013-1584-z.

Sharma, H.S., Sharma, A., 2015. Neurotoxicity of nanomaterials, Chapter 14. In: Fadeel, B. (Ed.), Handbook of Safety Assessment of Nanomaterials: From Toxicological Testing to Personalized Medicine (Jenny Stanford Series on Biomedical Nanotechnology) 1st Edi-tion, first ed. Jenny Stanford Publishing, pp. 407–438. (December 10, 2014) ISBN-13: 978-9814463362.

Sharma, H.S., Sharma, A., 2021. Amine precursors in depressive disorders and the blood-brain barrier. In: Riederer, P., Laux, G., Nagatsu, T., Le, W., Riederer, C. (Eds.), NeuroPsycho-pharmacotherapy. Springer, Cham. https://doi.org/10.1007/978-3-319-56015-1_423-1.

Sharma, H.S., Ali, S.F., Dong, W., Tian, Z.R., Patnaik, R., Patnaik, S., Sharma, A., Boman, A., Lek, P., Seifert, E., Lundstedt, T., 2007a. Drug delivery to the spinal cord tagged with nanowire enhances neuroprotective efficacy and functional recovery following trauma to the rat spinal cord. Ann. N. Y. Acad. Sci. 1122, 197–218. https://doi.org/10.1196/annals.1403.014.

Sharma, H., Chandola, H.M., Singh, G., Basisht, G., 2007b. Utilization of Ayurveda in health care: an approach for prevention, health promotion, and treatment of disease. Part 1—ayurveda, the science of life. J. Altern. Complement. Med. 13 (9), 1011–1019. https://doi.org/10.1089/acm.2007.7017-A.

Sharma, H., Chandola, H.M., Singh, G., Basisht, G., 2007c. Utilization of Ayurveda in health care: an approach for prevention, health promotion, and treatment of disease. Part 2—ayurveda in primary health care. J. Altern. Complement. Med. 13 (10), 1135–1150. https://doi.org/10.1089/acm.2007.7017-B.

Sharma A., Menon P.K., Patnaik R., Muresanu D.F., Lafuente J.V., Tian Z.R., Ozkizilcik A., Castellani R.J., Mössler H., Sharma H.S., 2017. Novel Treatment Strategies Using TiO2-Nanowired Delivery of Histaminergic Drugs and Antibodies to Tau With Cerebrolysin for Superior Neuroprotection in the Pathophysiology of Alzheimer's Disease. Int. Rev. Neurobiol. 137, 123–165. https://doi.org/10.1016/bs.irn.2017.09.002 (Epub 2017 Nov 6).

Sharma, H.S., Muresanu, D.F., Sharma, A., Patnaik, R., Lafuente, J.V., 2009a. Chapter 9—Nanoparticles influence pathophysiology of spinal cord injury and repair. Prog. Brain Res. 180, 154–180. https://doi.org/10.1016/S0079-6123(08)80009-X (Epub 2009 Dec 8).

Sharma, H.S., Patnaik, R., Sharma, A., Sjöquist, P.O., Lafuente, J.V., 2009b. Silicon dioxide nanoparticles (SiO2, 40-50 nm) exacerbate pathophysiology of traumatic spinal cord injury and deteriorate functional outcome in the rat. An experimental study using pharmacological and morphological approaches. J. Nanosci. Nanotechnol. 9 (8), 4970–4980. https://doi.org/10.1166/jnn.2009.1717.

Sharma, H.S., Ali, S.F., Hussain, S.M., Schlager, J.J., Sharma, A., 2009c. Influence of engineered nanoparticles from metals on the blood-brain barrier permeability, cerebral blood flow, brain edema and neurotoxicity. An experimental study in the rat and mice using biochemical and morphological approaches. J. Nanosci. Nanotechnol. 9 (8), 5055–5072. https://doi.org/10.1166/jnn.2009.gr09.

Sharma, H.S., Ali, S.F., Tian, Z.R., Hussain, S.M., Schlager, J.J., Sjöquist, P.O., Sharma, A., Muresanu, D.F., 2009d. Chronic treatment with nanoparticles exacerbate hyperthermia induced blood-brain barrier breakdown, cognitive dysfunction and brain pathology in the rat. Neuroprotective effects of nanowired-antioxidant compound H-290/51. J. Nanosci. Nanotechnol. 9 (8), 5073–5090. https://doi.org/10.1166/jnn.2009.gr10.

Sharma, A., Muresanu, D.F., Lafuente, J.V., Patnaik, R., Tian, Z.R., Buzoianu, A.D., Sharma, H.S., 2015. Sleep deprivation-induced blood-brain barrier breakdown and brain dysfunction are exacerbated by size-related exposure to ag and cu nanoparticles. Neuroprotective effects of a 5-HT3 receptor antagonist ondansetron. Mol. Neurobiol. 52 (2), 867–881. https://doi.org/10.1007/s12035-015-9236-9 (Epub 2015 Jul 2).

Sharma, H.S., Muresanu, D.F., Sharma, A., 2016a. Alzheimer's disease: cerebrolysin and nanotechnology as a therapeutic strategy. Neurodegener. Dis. Manag. 6 (6), 453–456. https://doi.org/10.2217/nmt-2016-0037 (Epub 2016 Nov 9).

Sharma, A., Menon, P., Muresanu, D.F., Ozkizilcik, A., Tian, Z.R., Lafuente, J.V., Sharma, H.S., 2016b. Nanowired drug delivery across the blood-brain barrier in central nervous system injury and repair. CNS Neurol. Disord. Drug Targets 15 (9), 1092–1117. https://doi.org/10.2174/1871527315666160819123059.

Sharma, A., Muresanu, D.F., Lafuente, J.V., Sjöquist, P.O., Patnaik, R., Ryan Tian, Z., Ozkizilcik, A., Sharma, H.S., 2018a. Cold environment exacerbates brain pathology and oxidative stress following traumatic brain injuries: potential therapeutic effects of nanowired antioxidant compound H-290/51. Mol. Neurobiol. 55 (1), 276–285. https://doi.org/10.1007/s12035-017-0740-y.

Sharma, H.S., Muresanu, D.F., Lafuente, J.V., Patnaik, R., Tian, Z.R., Ozkizilcik, A., Castellani, R.J., Mössler, H., Sharma, A., 2018b. Co-administration of TiO_2 nanowired mesenchymal stem cells with cerebrolysin potentiates neprilysin level and reduces brain pathology in Alzheimer's disease. Mol. Neurobiol. 55 (1), 300–311. https://doi.org/10.1007/s12035-017-0742-9.

Sharma, H.S., Muresanu, D.F., Castellani, R.J., Nozari, A., Lafuente, J.V., Tian, Z.R., Sahib, S., Bryukhovetskiy, I., Bryukhovetskiy, A., Buzoianu, A.D., Patnaik, R., Wiklund, L., Sharma, A., 2020. Pathophysiology of blood-brain barrier in brain tumor. Novel therapeutic advances using nanomedicine. Int. Rev. Neurobiol. 151, 1–66. https://doi.org/10.1016/bs.irn.2020.03.001 (Epub 2020 May 13).

Sharma, A., Shukla, U., Mishra, R., 2019. Ayurveda concept of *Bhasma*, specific role in diseases management and precautionary measurements related to *Bhasma*: a review. J. Drug Deliv. Ther. 9 (2-s), 649–651. https://doi.org/10.22270/jddt.v9i2-s.2525.

Tian, Z.R., Sharma, A., Nozari, A., Subramaniam, R., Lundstedt, T., Sharma, H.S., 2012. Nanowired drug delivery to enhance neuroprotection in spinal cord injury. CNS Neurol. Disord. Drug Targets 11 (1), 86–95. https://doi.org/10.2174/187152712799960727.

Ventola, C.L., 2012a. The nanomedicine revolution. Part 1. Emerging concepts. P&T 37 (9), 512–525.

Ventola, C.L., 2012b. The nanomedicine revolution. Part 2. Current and future clinical applications. PT 37 (10), 582–591.

Ventola, C.L., 2012c. The nanomedicine revolution. Part 3. Regulatory and safety challenges. PT 37 (11), 631–639.

Wujastyk, D., 2003. The Roots of Ayurveda: Selections From Sanskrit Medical Writings. Penguin Books, New York, USA, pp. 1–408.

Alzheimer's disease neuropathology is exacerbated following traumatic brain injury. Neuroprotection by co-administration of nanowired mesenchymal stem cells and cerebrolysin with monoclonal antibodies to amyloid beta peptide

Hari Shanker Sharma[a],*, Dafin F. Muresanu[b,c], Rudy J. Castellani[d], Ala Nozari[e], José Vicente Lafuente[f], Anca D. Buzoianu[g], Seaab Sahib[h], Z. Ryan Tian[h], Igor Bryukhovetskiy[i,j], Igor Manzhulo[i,j], Preeti K. Menon[k], Ranjana Patnaik[l], Lars Wiklund[a], and Aruna Sharma[a],*

[a]International Experimental Central Nervous System Injury & Repair (IECNSIR), Department of Surgical Sciences, Anesthesiology & Intensive Care Medicine, Uppsala University Hospital, Uppsala University, Uppsala, Sweden
[b]Department of Clinical Neurosciences, University of Medicine & Pharmacy, Cluj-Napoca, Romania
[c]"RoNeuro" Institute for Neurological Research and Diagnostic, Cluj-Napoca, Romania
[d]Department of Pathology, University of Maryland, Baltimore, MD, United States
[e]Anesthesiology & Intensive Care, Massachusetts General Hospital, Boston, MA, United States
[f]LaNCE, Department of Neuroscience, University of the Basque Country (UPV/EHU), Leioa, Bizkaia, Spain
[g]Department of Clinical Pharmacology and Toxicology, "Iuliu Hatieganu" University of Medicine and Pharmacy, Cluj-Napoca, Romania
[h]Department of Chemistry & Biochemistry, University of Arkansas, Fayetteville, AR, United States
[i]Department of Fundamental Medicine, School of Biomedicine, Far Eastern Federal University, Vladivostok, Russia

Progress in Brain Research, Volume 265, ISSN 0079-6123, https://doi.org/10.1016/bs.pbr.2021.04.008

^{j}*Laboratory of Pharmacology, National Scientific Center of Marine Biology, Far East Branch of the Russian Academy of Sciences, Vladivostok, Russia*
^{k}*Department of Biochemistry and Biophysics, Stockholm University, Stockholm, Sweden*
^{l}*Department of Biomaterials, School of Biomedical Engineering, Indian Institute of Technology, Banaras Hindu University, Varanasi, India*
*Corresponding authors: Tel./Fax: +46-18-24-38-99, e-mail address: sharma@surgsci.uu.se; harishanker_sharma55@icloud.com; hssharma@aol.com; aruna.sharma@surgsci.uu.se

Abstract

Military personnel are prone to traumatic brain injury (TBI) that is one of the risk factors in developing Alzheimer's disease (AD) at a later stage. TBI induces breakdown of the blood-brain barrier (BBB) to serum proteins into the brain and leads to extravasation of plasma amyloid beta peptide (AβP) into the brain fluid compartments causing AD brain pathology. Thus, there is a need to expand our knowledge on the role of TBI in AD. In addition, exploration of the novel roles of nanomedicine in AD and TBI for neuroprotection is the need of the hour. Since stem cells and neurotrophic factors play important roles in TBI and in AD, it is likely that nanodelivery of these agents exert superior neuroprotection in TBI induced exacerbation of AD brain pathology. In this review, these aspects are examined in details based on our own investigations in the light of current scientific literature in the field. Our observations show that TBI exacerbates AD brain pathology and TiO$_2$ nanowired delivery of mesenchymal stem cells together with cerebrolysin—a balanced composition of several neurotrophic factors and active peptide fragments, and monoclonal antibodies to amyloid beta protein thwarted the development of neuropathology following TBI in AD, not reported earlier.

Keywords

Alzheimer's disease, Traumatic brain injury, Brain pathology, Mesenchymal stem cells, Cerebrolysin, Nanowired delivery, Nanomedicine

1 Introduction

Alzheimer's disease (AD) is a slowly progressing brain disease resulting in memory loss and other intellectual functions (Bondi et al., 2017; Kirova et al., 2015; Leli and Scott, 1982; Lyketsos et al., 2011; Shimada et al., 2001). The AD may develop over 20 years from early signs of inception resulting in gradually interfering with normal daily life (Besser et al., 2018; O'Bryant et al., 2015; Weiner et al., 2015). AD is not a disease of aging as it can appears in patients of 40 or 50 years of age known as younger onset (Juan and Adlard, 2019; Lane et al., 2018; Soria Lopez et al., 2019;

Tromp et al., 2015). In 2016, AD represents the sixth leading causes of death in the United States of America (USA) (Alzheimer's Association, 2016; Calamia et al., 2016; Yeh et al., 2017). In the beginning memory loss in AD is mild that become severe with advancing age and disease (Jahn, 2013; Wang and Holtzman, 2020). People with AD normally live 8–10 years after initial diagnosis depending on the age and other co-morbidity factors (Alzheimer's Association, 2015; Mueller et al., 2019; Zhang et al., 2020a,b). Till date there are no cure for AD but available treatments could slow down the progression of disease and symptoms (Cortes-Canteli and Iadecola, 2020; Cummings et al., 2019; Shao et al., 2020). Thus, there is an urgent need to explore novel treatment strategies to alleviate brain pathology in AD or thwart the developmental processes of the disease.

The normal healthy human brain contains about 100 billions of neurons that are connected with each other with synapses for normal communication to maintain brain functions (Pfisterer and Khodosevich, 2017; Sporns et al., 2005; von Bartheld et al., 2016). About 100 trillions synapses allow flow of information through the neuronal circuits forming the cellular basis of memory, thoughts, emotion and other intellectual functions (Mesulam, 1998; Stampanoni Bassi et al., 2019; Zimmer, 2011). The number of neurons and their connections through synapses severely deteriorate during AD resulting in loss of memory and higher bran functions (Jackson et al., 2019; Meyer et al., 2019; Osborn et al., 2016). Deposits of the peptide fragment known as amyloid beta peptide (AβP) outside of the neurons and accumulation of the abnormal form of tau protein inside the neurons are the hallmark of AD brain (Blennow et al., 2006; Busche and Hyman, 2020; van der Kant et al., 2020). The AβP plaques contribute to cell death by interfering with neuronal communication at the synapses in AD (Leong et al., 2020; Malishev et al., 2019; Picone et al., 2020). The abnormal tau proteins known as tau tangles inhibit the transport of essential molecules and nutrients within the neurons resulting in cell death (Blurton-Jones and Laferla, 2006; Mata, 2018; Rodríguez-Martín et al., 2013). When the level of AβP deposits increases the tau tangles spread throughout the brain resulting in profound symptoms of AD and brain damage (Bassil et al., 2020; Bloom, 2014; Lowe et al., 2018).

The progress of AD with time results in shrinkage of the brain tissues together with the gradual enlargement of all the cerebral ventricles (Ferreira et al., 2015; Kuroda et al., 2020; Mentis et al., 2020; Ye et al., 2016). The hippocampus cells shrinking causes neurodegeneration resulting in memory decline (Jaroudi et al., 2017; Khanal et al., 2014; Mu and Gage, 2011; Tabatabaei-Jafari et al., 2017). With time AD spreads throughout the cerebral cortex and brain resulting in behavioral changes, higher mental and vital function deterioration affecting language, judgment, memory, daily life and other body systems until death within 9–10 years after diagnosis (Brookmeyer and Abdalla, 2019; Skaper et al., 2017; van Engelen et al., 2020).

The trigger of the molecular mechanisms leading to AβP deposits and/or abnormal or phosphorylated tau (p-tau) accumulation leading to AD pathology is still unclear (Gatti et al., 2020; Paroni et al., 2019; Wang and Zhang, 2018). In healthy conditions, very low levels of AβP and tau is present in the plasma that appears to be increased in patients with AD (Karikari et al., 2020; Mattsson et al., 2016, 2017;

Schultz et al., 2019; Wei et al., 2017; Yun et al., 2020). It appears that in AD plasma AβP and tau are transported to the brain and in the cerebrospinal fluid (CSF) probably due to a compromise of the blood-brain barrier (BBB) and the blood-CSF-barrier (BCSFB) function (Sharma, 2004, 2009; Sharma and Sharma, 2010; Sharma and Westman, 2004; Sharma et al., 2012). The healthy brain normally removes abnormal proteins from the brain via efflux transporters located at the cerebrovascular unit to maintain homeostasis (Bao et al., 2020; Iliff et al., 2012; Sharma, 2004). However, accumulation of the AβP and p-tau in AD brain suggests that these transporters or brain-blood barrier (bbb) is also affected (Hladky and Barrand, 2018; Jadiya et al., 2019).

One of the earliest precipitating factors in inducing AD appears to be traumatic brain injury (TBI) (Armstrong, 2019; Kempuraj et al., 2020; Wu et al., 2020). The linkage between brain trauma and neurodegenerative diseases is responsible for late onset of several chronic brain disorders (Bertogliat et al., 2020; Gardner and Yaffe, 2015; Ramos-Cejudo et al., 2018). Increasing evidences show that even a single mild TBI sustained in early life may lead to a cascade that could manifest in late development of AD-like disorders (Becker et al., 2018; Griesbach et al., 2018; Gupta and Sen, 2016). Several cases of boxers that get repetitive mild head trauma leads to neurodegenerative symptoms in some cases as early as in their 20s and 30s years of life (Alosco and Stern, 2019; McKee et al., 2018; Ossenkoppele et al., 2020). Repetitive head trauma causing chronic traumatic encephalopathy (CTE) results in several neurological disorders including AD (Iverson et al., 2019; McKee et al., 2009; Simom et al., 2017). Neurotrauma could induce cerebrovascular diseases (CBD) leading to the disruption of the BBB and BCSFB (Johanson et al., 2011; O'Keeffe et al., 2020; Sahyouni et al., 2017; Sharma and Johanson, 2007a,b). These cerebrovascular disturbances may trigger AβP and p-tau accumulation leading to AD (Patnaik et al., 2018; Sharma et al., 2012, 2016a,b). Military personnel are often prone to TBI resulting in post-traumatic stress disorders (PTSD) that are prevalent in soldiers associated with combat experiences (Adrian et al., 2018; Groer et al., 2015; Kritikos et al., 2019). Thus, it is quite likely that TBI could exacerbate AD brain pathology (Moore et al., 2020; Proessl et al., 2020; Silverberg et al., 2020). Accordingly, treatment modalities for improving or exploring patient care is needed to enhance clinical efficacy in situations of AD with TBI.

In this review we discuss novel therapeutic strategies to treat exacerbation of brain pathology in AD with TBI using nanomedicine based on our own observations. New evidences suggest that nanowired delivery of neurotrophic factors and stem cells could be the potential novel therapeutic agents in treating AD cases with TBI, not reported earlier.

1.1 Blood-brain barrier leakage in AD patients

Several lines of evidences suggest that impairment of cerebrovascular function is the prominent contributing factor in the pathophysiology of AD (Bell and Zlokovic, 2009; Nelson et al., 2016; Zlokovic, 2011). Pathological evaluation of albumin leakage within the AD brain suggests that breakdown of the BBB plays a key role (Montagne et al., 2017; Sweeney et al., 2018, 2019). The BBB resides in the cerebral capillaries that are connected with tight junctions (Abbott et al., 2010; Sharma, 2009; Sharma and Westman, 2004; Zlokovic, 2008). Thus, the cellular and subcellular structures comprising endothelial cell covering of basal lamina, glial end feet and neuronal contacts with the microvessels are actively separating the blood from brain parenchyma (Abbott et al., 2006; Liebner et al., 2018; Profaci et al., 2020). The BBB is thus maintaining strict microenvironment of the brain by controlling passage of essential nutrients through active and passive transport mechanisms (Banks, 2016; Erdő et al., 2017; Sharif et al., 2018). The BBB also prevents neurotoxins from entering into the brain (Alexander, 2018; Obermeier et al., 2013, 2016). The clearance function of the BBB using active transport and selective transporters maintains the healthy environment of the brain in normal conditions (Bell et al., 2007; Deane et al., 2009; Shibata et al., 2000). Deterioration of the BBB active maintenance of the brain function or reduction in the clearance capacity of toxins and unwanted substances results in microvascular abnormalities, microbleeds, protein deposits, neuroinflammation and neuronal cell death (Cai et al., 2018; Ransohoff, 2016; Sulhan et al., 2020). These pathological phenomena within the brain lead to decline of intellectual and cognitive functions promoting dementia and eventually AD (Eratne et al., 2018; Masters et al., 2015; Sweeney et al., 2019).

This idea is well substantiated by albumin ratio measurement and contrast material enhancement using magnetic resonance imaging (MRI) to detect BBB leakage in neurodegenerative diseases (van de Haar et al., 2016a,b, 2017). Using these techniques neuroradiologists in the Netherlands examined BBB leakage in patients with early AD and compared with age-matched healthy controls. The authors used a dynamic contrast enhanced MRI with dual-time resolution that separates the leakage from the microvascular fillings (van de Haar et al., 2016a,b). In addition, local blood plasma volume and its relationship with the BBB permeability and global cognition were also investigated.

The authors find increased BBB leakage in patients with early AD that was globally distributed within the cerebrum and correlates well with the declined cognitive performances (van de Haar et al., 2016a,b). Using dynamic contrast-enhanced MRI for dual resolution the BBB leakage strength or rate is seen in the gray matter as well as in the white matter within the cerebrum (van de Haar et al., 2017). However, the leakage volume was greater than the leakage rate in early AD cases. Thus, BBB breakdown appears to be global in early AD cases (van de Haar et al., 2016a,b, 2017).

Apart from BBB leakage the local plasma volume was also decreased in the areas showing BBB breakdown indicating ischemic hypoperfusion of the brain (Dong et al., 2018; Feng et al., 2019; Love and Miners, 2016). Ischemia induced cerebrovascular perfusion and endothelial cell dysfunction including leakage results in microbleed and small vessel diseases quite common in dementia and in AD (Arvanitakis et al., 2016; Blumenau et al., 2020; Hase et al., 2020). Another interesting point from this study showed a close correspondence between the strength of BBB leakage and decline in cognitive function (Kamintsky et al., 2020;

Raja et al., 2018; Uemura et al., 2020). Although the mechanism of the BBB leakage in AD is still uncertain, it appears that endothelial cell permeability along with loss or damage of the tight junctions are both could be responsible for such leakage (Sharma, 2004, 2009; Sharma et al., 2012; Sweeney et al., 2019). These observations indicate that both influx and efflux transport systems across the BBB is compromised in AD (Jeynes and Provias, 2011; Kim et al., 2020a,b,c; Zhou et al., 2017).

1.2 Blood-borne factors influence AD brain pathology

There are reasons to believe that BBB is a crucial determining factor for AβP and tau deposition in the AD brain (Kent et al., 2020; Laurent et al., 2018; Liu et al., 2019). Obviously, a breakdown of the influx and efflux transport system at the BBB will allow greater deposits and accumulation of these neurotoxins responsible for AD brain pathology (Sweeney et al., 2019; Xin et al., 2018; Zenaro et al., 2017). This is further evident from the findings that there is a close association between vascular inflammation and neuroinflammation in AD brain pathology (Klohs, 2019; Loera-Valencia et al., 2019; Sung et al., 2020). In AD, two key factors namely AβP deposition and neuroinflammation play predominant roles in brain pathology (Calsolaro and Edison, 2016). Although the precise mechanisms between the interaction of these two factors are not well known, it is amply clear that neuroinflammation predates the AβP deposition (González-Reyes et al., 2017; Hoozemans et al., 2006; Regen et al., 2017).

Several lines of evidences suggest that innate immune system plays pivotal role in inflammation and subsequent progression of AD (Gate et al., 2020; Park et al., 2020a,b; Webers et al., 2020). Activation of microglia and macrophages together with AβP and p-tau are responsible for AD brain pathology (Bartels et al., 2020; Heneka, 2020; Yuan et al., 2020). Activation of neuroimmune system allows clearance of AβP or other neurotoxic materials from the brain and leads to cell death of damaged neurons (Bedoui et al., 2018; Griffiths et al., 2007). However, in chronic activation of immune system leads to production of neurotoxic cytokines, chemokines and oxidative stress resulting in slow neuronal death in AD (Friker et al., 2020; Hoarau et al., 2011; Welikovitch et al., 2020).

Recent evidences show that one of the peripheral blood enzymes plasmin (PLS) is an important regulatory factor of neuroinflammation and brain pathology in AD (Baker et al., 2018, 2019; Paul et al., 2007). PLS enzyme is formed from cleavage of the primary blood protein plasminogen (PLG) that is synthesized in the liver (Huebner et al., 2018). The PLS is involved in the cell signaling, inflammatory responses, fibrinolysis and wound healing (Castellino and Ploplis, 2005). PLG plays a key role in activating inflammation and immune responses (Heissig et al., 2020). This is a chemoattractant for macrophages and monocytes have binding sites for PLG (Syrovets et al., 2012). Depletion of PLS pharmacologically results in increased survival rate due to less inflammation caused by excessive immunoactivation (Cuzner and Opdenakker, 1999; Gur-Wahnon et al., 2013). This suggests that PL is involved in production of proinflammatory cytokines and chemokines (Kang et al., 2020).

Since PLG and PLS are involved in neuroinflammatory behavior a possibility exists that pharmacological depletion of these enzymes may have some neuroprotective effects in AD.

1.2.1 Neuroinflammation and AD

The mechanisms behind the progression of AD brain pathology are still not well known. However, it appears that neuroinflammation is one of the key factors in contributing AD brain pathology (Newcombe et al., 2018; Regen et al., 2017; Wilkins and Swerdlow, 2016). The hallmark of AD includes AβP extracellular deposition and p-tau intracellular accumulation causing brain damage is preceded with activation of neuroimmune system and inflammatory responses of glial cells (LaFerla et al., 2007; Šimić et al., 2016; Villemagne et al., 2018). Activation of microglia could lead to clear AβP deposition through phagocytes up to some extent (Süß and Schlachetzki, 2020). However, chronic activation of immune system causes misfolding to AβP aggregation that is difficult to clear or dissolve in AD (Griciuc et al., 2013). In addition, AβP could activate proinflammatory cytokines through activating the glial cells by binding on the receptors present on them (Minter et al., 2016; Wang et al., 2015). The AβP can also activate proinflammatory cell signaling system in plasma and in cerebrospinal fluid (CSF) to induce neuroinflammation (Pillai et al., 2019; Sutton et al., 1999). This idea is further supported by the fact that depletion of AβP proinflammatory contact system results in decrease in neuroinflammation, brain pathology and cognitive dysfunction in AD (Kinney et al., 2018).

1.2.2 Plasminogen and plasmin in AD

Several plasma factors are known activators of PLG and PLS affecting AD brain pathology (Baker et al., 2018; Simão et al., 2017). Depletion of blood coagulation factor XII (CFXII) responsible for intrinsic pathways for coagulation and immune contact activation is able to reduce brain pathology and neuroinflammation in AD (Chen et al., 2017). In addition CFXIIa, CFXI1, kallikrein and bradykinin that are key molecules for immune contact pathways are known activators of PLG and PLS (Göbel et al., 2019). When PLG and PLS are activated neuroinflammation is enhanced leading to brain pathology (Reuland and Church, 2020). CFXII is also activated by PLS that causes release of bradykinin—a proinflammatory agent and mediator of vascular permeability causing edema and neuroinflammation (Mugisho et al., 2019). In addition CFXII also induces thrombin generation another potent proinflammatory factor induces neuroinflammation through activation of protease receptors in the CNS (Iannucci et al., 2020).

One of the key factors in neuroinflammation is the deposition of fibrinogen within the CNS (Cortes-Canteli et al., 2012; Sulimai and Lominadze, 2020). When the BBB is compromised fibrin enters into the brain and deposited into several brain areas causing neuroinflammation (Cortes-Canteli et al., 2015). Deposition of fibrinogen is associated with depletion of PLG or PL in the CNS (Shaw et al., 2017). This idea is further supported by the fact that inhibition of PLG by tranexamic acid (TXA) enhances neuroinflammatory response in the CNS (Atsev and Tomov, 2020). Interestingly TXA is able to inhibit only free PLG or PL but could not affect cell bound PL activity (Baker et al., 2018).

Evidences show that AβP plaques are significantly decreased in PLG depletion in mice model of AD using antisense oligonucleotide (ASO) treatment (Baker et al., 2018). However, this decrease of AβP plaques with PLG depletion is unclear. There are reasons to believe that PL could enhance the cleavage of both a- and b-amyloid precursor protein (APP) (Baranello et al., 2015). There are reports that PLG activation (PA) is induced by AβP that could degrade oligomeric and fibrillar AβP (Tucker et al., 2000). Interestingly, tissue PLG activation (tPA) results in inhibition of AβP aggregation and neurotoxicity (Yang et al., 2020). Both PLG and tPA decreases with advancing age are associated with AD but their expression is increased around AβP plaques (Cacquevel et al., 2007). This indicates that PL is associated with tPA-induced proteolysis in AD (Nalivaeva et al., 2008). This aspect is further confirmed in experiments with genetic depletion of protein PAl-1, the major inhibitor of tPA that resulted in decrease in AβP deposition in mouse model of AD (Kutz et al., 2012). This observation clearly suggests an important role of PL in AβP clearance (Baker et al., 2018). PLG depletion thus reduces that ability to induce prominent neuroinflammation and therefore less AβP deposition in the AD brains (Baker et al., 2018, 2019).

Increased activation of PL is often seen in chronic inflammatory or autoimmune disorders (Cañas et al., 2015). In addition, PLG and tPA are found localized around AβP plaques that could further enhance local inflammatory responses (Riemenschneider et al., 2006). However, further studies are needed to examine the therapeutic effects of PLG in AD.

1.3 Specificity of amyloid beta peptide deposition in AD brain

The signature characteristics of AD include extracellular deposition of AβP plaques, intracellular tangles of p-tau proteins, loss of synapses and neurons leading to brain pathology (Brun et al., 1995; Liu et al., 1999). However, apart from classical AD cases similar deposition of AβP is seen in TBI as well as PTSD cases (Emmerling et al., 2000; Mohamed et al., 2018; Raby et al., 1998). However, in each case the intensity and magnitude together with regional variation of AD deposition occurs. These specificities led to several functional imaging studies resulting in novel discoveries and fingerprinting of AβP deposits in several diseases (Veitch et al., 2019).

There exists a close relationship between AβP deposits and cognitive impairments in several imaging studies of TBI and AD brains of short or long-term survival (Abrahamson and Ikonomovic, 2020; Ayubcha et al., 2021). Increased AβP deposits found in TBI patients with PET studies that were largely confined into the posterior cingulate cortex, cerebellum and striatum very similar to that seen in AD cases (Hong et al., 2014; Stern et al., 2019; Ubukata et al., 2020; Yang et al., 2015). AβP plaques are seen in more than one third of all TBI patients irrespective of their age (Mohamed et al., 2018). This indicates that TBI is an important risk factor for AD development.

Interestingly, AβP deposits in TBI is seen following short term injury and could not be found in many long-term survivor of brain trauma (Mohamed et al., 2020; Scott et al., 2016). This indicates that enhanced AβP clearance is active in eliminating AβP from the brain fluid microenvironment. However, when the clearance mechanisms are compromised AβP accumulation persists and spread through different parts of brain leading to full development of AD cases (Reddy and Oliver, 2019; Suzuki et al., 2015). Taken together these studies clearly suggest that TBI induce deposition of AβP is a risk factor for later development of AD brain pathology (Djordjevic et al., 2016; Jellinger, 2004; LoBue and Cullum, 2020).

Military personnel while engaged in combat operations are susceptible to TBI followed by PTSD (Chin and Zeber, 2020; Iljazi et al., 2020; Mac Donald et al., 2021; Moore et al., 2020). There are several reports suggesting that these military personnel are highly vulnerable to later development of AD brain pathology (Khachaturian and Khachaturian, 2014; Veitch et al., 2013; Weiner et al., 2013). Interestingly, like TBI, PTSD also shows deposition of AβP in specific brain areas (Mohamed et al., 2018, 2019). Thus, it is suggested that veterans with PTSD could have a twofold increase in the risk of developing AD (Chan et al., 2017; Desmarais et al., 2020; Günak et al., 2020; Rafferty et al., 2018). Recently, imaging studies of veterans with PTSD showed AβP deposition in the precuneus, frontal, temporal, parietal and anterior and posterior cingulate cortices associated with cognitive functional decline very similar to that of AD cases (Mohamed et al., 2018, 2020). These observations suggest that AβP could be a possible link between PTSD and AD.

It appears that stress related changes in hypothalamic-pituitary-adrenal (HPA) axis could play an important role between PTSD and AD. This is further evident from the fact that loss of the volume of hippocampus and associated cognitive decline is common in both the PTSD and AD (Bonne et al., 2008; Scheff et al., 2006; Schuff et al., 2009; Wang and Xiao, 2010). Several MRI studies in PTSD show decrease in volume of hippocampus, anterior cingulate cortex and prefrontal structures in patients and in AD (Boutet et al., 2014; Murphy et al., 2003; Parker et al., 2019; Schuff et al., 2009). In addition, PTSD and AD are associated with several neuroinflammatory factors including cytokines, chemokines and other neurodegenerative elements including AβP accumulation (Cai et al., 2014; Gill et al., 2018; Webers et al., 2020; Webster et al., 2015).

Studies showing good correlation in CSF markers of AβP and p-tau in TBI, PTSD and AD further suggest a close link between them (Blennow et al., 2010; Mohamed et al., 2018, 2019; Rabbito et al., 2020; Zou et al., 2020). This suggests an increased risk of TBI and or PTSD in development of AD brain pathology in later periods of life. However, slight and subtle differences in AβP accumulation occur in all three groups of patients with TBI, TBI and PTSD or AD (Mohamed et al., 2018). Thus, patients with TBI alone increased accumulation of AβP are seen in the cerebellum and precuneus (Hortobágyi et al., 2007; Mohamed et al., 2018). On the other hand, TBI and PTSD group showed a substantial increase in accumulation of AβP in the white matter besides the other brain regions (Mohamed et al., 2018, 2019). On the other hand PTSD alone exhibited AβP deposits in the frontal, occipital and temporal cortices (Mohamed et al., 2018). These differences in accumulation of AβP distribution show differential relationships between TBI, TBI and PTSD and PTSD alone in relation to development of AD brain pathology.

1.4 Mesenchymal stem cells and AD brain pathology

Mesenchymal stem cells (MSCs) are one of the prominent therapeutic tools in the treatment strategies of AD (Chakari-Khiavi et al., 2019; Kim et al., 2020a,b,c; Wang et al., 2019; Zhang et al., 2020a,b). Several studies have shown the efficacy of MSCs treatment in AD because of their immune modulatory and neurotrophic functions (Elia et al., 2019; Lo Furno et al., 2018; Mehrabadi et al., 2020; Zhang et al., 2020a,b). Some studies have also investigated the use of AD patients CSF samples as formulations of MSCs for the treatment of AD therapy (Benhamron et al., 2020; Lee et al., 2019).

The most important aspect of stem cell therapy in neurodegenerative disease like AD is to choose appropriate cell sources (Han et al., 2020). There are evidences that embryonic stem cells (ESCs), MSCs, brain-derived neural stem cells (NSCs) and induced pluripotent stem cells (iPSCs) are used in AD therapy (Farahzadi et al., 2020; Kim et al., 2020a,b,c; Liu et al., 2020; Lin et al., 2018; Penney et al., 2020). The MSCs induce development of mesenchymal tissue and can be harvested from umbilical cord blood (UCB-MSCs) or Wharton's jelly (Kim et al., 2015; Petukhova et al., 2019). In addition MSCs are present in adult stem cells niches, e.g., bone marrow and adipose tissues (Nakano et al., 2020; Nasiri et al., 2019; Qin et al., 2020; Reza-Zaldivar et al., 2018).

The MSCs are originally discovered from stroma of the bone marrow are the multipotent progenitor cells that are extensively used for neurodegenerative therapy due to their immune modulatory and neurotrophic properties (Esmaeilzade et al., 2019; Wang et al., 2019; Yang et al., 2016). AD is one of the most important neurodegenerative diseases for which no suitable therapeutic strategies are developed until today (Sharma et al., 2012, 2016a,b). The MSCs are used in AD therapy largely because of their intrinsic ability to reduce AβP levels in the brain and attenuate neuroinflammation (Ding et al., 2018; Habisch et al., 2010; Yokokawa et al., 2019). In addition MSCs are also capable to enhance endogenous neurogenesis and improve behavioral performances in AD. Based on these evidences several clinical trials are underway to explore the safety and efficacy of MSCs in AD patients (Mohamed et al., 2018, 2019). Regarding the most efficient routes of MSCs administration in AD preclinical results suggests that intracerebroventricular (i.c.v.) or intraparenchymal injections yield superior effects as compared to intravenous (i.v.) or intra-arterial (i.a.) injections (Elia et al., 2019; Kim et al., 2015; Mohammadi et al., 2019).

Another important point is to use clinical grade of MSCs preparation as drug in AD therapy (Samsonraj et al., 2017). However, further studies are needed to further explore this aspect in greater details. Normally, the MSCs are used currently include various supplements to promote the survival of MSCs. This increases the possibilities of unavoidable interactions among cellular system of the recipients. Thus, this is important to enhance survival of MSCs in the recipient brain using other safe techniques such as nanodelivery of MSCs in AD (Andrzejewska et al., 2020; Sarnowska et al., 2013; Sharma et al., 2018).

Another way of optimizing MSCs formulation for AD therapy is the CSF samples of AD patients (Joerger-Messerli et al., 2016; Lee et al., 2019). This has several advantages. Since MSCs are normally administered into the CSF of patients using intracerebroventricular routes that have higher chances to penetrate into the brain parenchyma (Johanson et al., 2011; Sharma and Johanson, 2007a,b). Since, the CSF flow is responsible for AβP clearance strategy in AD this route of administration helps MSCs to enhance positive effects (Sharma and Westman, 2004). CSF from AD patients represents brain fluid microenvironment and thus MSCs given through CSF of individual patients may better survive with the disease environment as compared to naive MSCs (Dehghanian et al., 2020).

Recent observations using administration of Wharton's jelly MSCs formulations with AD patients CSF resulted in gene expression that are known to block apoptosis, cell proliferation and neurogenesis (Bodart-Santos et al., 2019; Lee et al., 2019, 2020). Furthermore, this treatment enhances extracellular transport of AβP and exhibited neuroprotective and neurotrophic activity (Kim et al., 2020a,b,c). In addition, MSCs therapy in patients CSF increased expression of genes responsible for cell migration and cell adhesion indicating potential beneficial effects on cell survival (Benvenuti et al., 2006). These observations indicate that patients CSF could be used as optimal formulation for MSCs for effective AD therapy (Sharma et al., 2012, 2016a,b, 2018). However, further studies using other techniques such as nanodelivery of MSCs are needed to expand our knowledge in effective AD therapy.

2 Neurotrophic factors in AD

Neurotrophic factors (NF) are secreted proteins affecting several neural and non-neural tissues involved in development, survival and maintenance of the central nervous system (CNS) and the peripheral nervous system (PNS) (Huang and Reichardt, 2001; Lewin and Barde, 1996; Sharma et al., 1998a,b,c; Skaper, 2018). In AD and other neurodegenerative disease changes in NF and their receptors occur that are critical for cell death and neuronal degeneration (Amidfar et al., 2020; Bharani et al., 2020; Schindowski et al., 2008; Sopova et al., 2014; Tanila, 2017). The NF prevents cell death (Diener and Bregman, 1994; Lu and Hagg, 1997) and enhance growth (Markus et al., 2002; Vanmali et al., 2003), survival and function of affected neurons in AD or neurodegenerative diseases (Cui, 2006; Granholm, 2000; Hock et al., 2000; Sharma, 2007; Sharma et al., 2019a,b,c; Thoenen et al., 1987; Zhao et al., 2014). In AD several biochemical and cellular and/or molecular mechanisms are altered causing cell death and neurodegeneration. These include synaptic dysfunction, impairment of NF activity or energy deficit due to mitochondrial disorders, enhanced oxidative stress and neuroinflammation (Cai et al., 2014; Calsolaro and Edison, 2016). All these factors result in a progressive decline of cognitive functions in AD triggered by pathological changes in the brain affecting the basal forebrain cholinergic neurons (Fernández-Cabello et al., 2020; Hampel et al., 2018, 2019). In this cascade NF paly prominent roles indicating exogenous delivery of them to alleviate AD brain pathology (Sharma et al., 2019a,b,c).

2.1 Neurotrophic factors changes in human brain with AD

Postmortem studies in human brains of AD patients show widespread alterations in NF levels (Du et al., 2018; Durany et al., 2000; Murer et al., 1999). Decreased levels of brain derived NF (BDNF) occurred in hippocampus and in parietal cerebral cortex in AD that showed histopathological degeneration (Connor et al., 1997; Durany et al., 2000). On the other hand nerve growth factor (NGF) showed increased levels in hippocampus and in frontal cortex (Hefti and Mash, 1989). Measurement of NF/neurotensin-3 (NT-3) did not show significant changes in these brain areas. The ratio of BDNF protein/NT-3 ratio decreased in the parietal cortex and frontal cortex in AD (Zhao et al., 2014). This decrease was also seen in the hippocampus of AD brains. On the other hand, measurement of NT-4/NT-3 ratios indicated significant decrease in NT-4 in the hippocampus of AD brains (Hock et al., 2000). Interestingly, a decrease in NT-4/5 ratio was seen in the cerebellum of AD brains (Braun et al., 2017; Hock et al., 2000).

NGF is known to regulate amyloidogenic pathways and AβP production (Iulita et al., 2016). NGF maintains that amyloidogenic route through decrease in APP phosphorylation that results in diminishing in APP-b-secretase interaction by AβP formation (He et al., 2007; Jovanovic et al., 2014).

These observations strongly point out a therapeutic value of exogenous supplement with NF as AD therapy (Sharma et al., 2012, 2016a,b, 2019a,b,c).

2.1.1 Neurotrophic factors therapy in AD

Several lines of evidences suggest that treatment with neurotrophic factors and their select combination may reduce the pathophysiology of AD. Thus, administration of ciliary neurotrophic factor (CNTF) and BDNF were used for the treatment of AD using their subcutaneous or intrathecal administration (Maruyama and Naoi, 2013; Sampaio et al., 2017). A dose related effects on AD was observed associated with side effects (Cui et al., 2016; Kazim and Iqbal, 2016; Windisch et al., 1998). Later on intracerebroventricular (i.c.v.) administration of nerve growth factor (NGF) and glial derived neurotrophic factor (GDNF) were used in AD patients with some success and side effects that dependent on the dosage used (Straten et al., 2011, 2009). After that intracerebral injection of NF were used to treat AD cases with some success and less side effects (Cuello et al., 2007; Krames et al., 1999; Sopova et al., 2014). These observations suggest that further research is needed to deliver NF using nanodelivery for effective treatment. Also, a good combination of NFs is needed to induce better therapeutic approaches in AD (see Sharma et al., 2012, 2019a,b,c).

2.2 Serotonin and oxidative stress in AD

In addition to degeneration of basal forebrain cholinergic system serotonergic neurons also showed decrease in 5-hydroxytryptamine (5-HT, serotonin) content in AD (Bostancıklıoğlu, 2020; Rodríguez et al., 2012). The 5-HT decrease is associated with depression and behavioral symptoms in AD (Halliday et al., 1992; Marner et al., 2012; Meltzer et al., 1998; Palmer and DeKosky, 1993; Reinikainen et al., 1988; Šimić et al., 2017; Vakalopoulos, 2017; Yun et al., 2015). Deposition of AβP in AD brain is related to neurotoxicity, formation of neurofibrillary tangles, neuroinflammation and oxidative damages (Bonda et al., 2010; Chen and Zhong, 2014; Kamat et al., 2016; Su et al., 2008; Tönnies and Trushina, 2017; Wang et al., 2014; Zhu et al., 2005, 2007).

Oxidative stress is associated with AβP induced brain toxicity, tau pathology, mitochondrial dysfunctions and alterations in metals, i.e., copper, zinc and iron homeostasis in AD (Huat et al., 2019; Jiang et al., 2019). Amyloidosis and tau protein accumulation in AD brain induce formation of reactive oxygen species (ROS) and promoted redox imbalance (Ahmad et al., 2017; Chun and Lee, 2018; Manoharan et al., 2016).

2.3 Antibodies to AβP is neuroprotective in AD

Several studies show that therapy with monoclonal antibodies to AβP in clinical or pre-clinical cases reduces the deposition of AβP in brain tissue resulting in neuroprotection in AD with or without improvement of cognitive functions (Geylis and Steinitz, 2006; Panza et al., 2011, 2019a,b; Sevigny et al., 2016; Stern et al., 1989; van Dyck, 2018; Zayas-Santiago et al., 2020). This is partially due to blockade of some AβP oligomers where other AβP types are not affected (Loureiro et al., 2020; Lozupone et al., 2020). AβP is ubiquitously present in the brain and plasma of healthy person and help in multiple aspects of brain function (Chen et al., 2019; Dawkins and Small, 2014; Morley and Farr, 2014; Morley et al., 2019; Zhang et al., 2019). Presynaptic localization of AβP in neurons appears to play key roles in regulating neural transmission, long-term potentiation, learning and memory, synaptic plasticity, electrophysiology and neuronal survival (Honcharenko et al., 2019; Jiang et al., 2019; Morley and Farr, 2014).

Increased production of AβP occurs following several neurological or psychiatric conditions apart from AD (Morgese et al., 2017; Shishido et al., 2019; Walsh et al., 2002). Thus, traumatic brain injury (TBI), chronic encephalopathy, sleep deprivation, cerebral ischemia, stroke, cardiac arrest, general anesthesia, major depression and amyotrophic lateral sclerosis (ALS) increase AβP deposition in the CSF (Kang et al., 2009; Lee et al., 2018; Nguyen, 2019; Pluta et al., 2019; Selkoe and Hardy, 2016; Shishido et al., 2019; Shokri-Kojori et al., 2018; Stein et al., 2015; Wiśniewski and Maślińska, 1996; Xie and Xu, 2013; Yasuno et al., 2019). In such situations, increased AβP production is likely to be responsible for elevation in the CSF rather that decreased clearance of the peptide (Chen et al., 2018; Iliff et al., 2012; Xin et al., 2018).

Based on these observations, it appears that increased AβP production possibly represent adaptive response to ameliorate neuronal dysfunction caused by chronic inflammation, metabolic failure and/or tau associated disruption of network, microglial activation, oxidative stress and related disorders (Sharma et al., 2012, 2017, 2018, 2019a,b,c, 2020a,b,c). Prolonged elevation of AβP with continued neuronal dysfunction results in AβP aggregates formation and AβP oligomers that induce neurotoxicity and other pathophysiological cascades (Mucke and Selkoe, 2012).

Accordingly several clinical and preclinical studies demonstrated use of monoclonal antibodies (mAbs) to AβP or its oligomers resulting in mixed outcome (van Dyck, 2018). In clinical study several agents are used against AβP oligomers that are highly neurotoxic. One such mAb solanezumab was able to stabilize AβP monomers and prevented neurotoxic AβP oligomers (Honig et al., 2018). Crenezumab binds specifically to AβP oligomers and fibrils in high affinity (Yang et al., 2019). The other mAb gantenerumab exhibited about 20-time high affinity to AβP oligomers than AβP monomers (Tolar et al., 2020). Intravenous administration of human immunoglobulin (IVIg) that contains several types of anti-AβP antibody that recognizes AβP oligomers and fibrils resulting neuroprotection against AβP induced neurotoxicity (Manolopoulos et al., 2019). Aducanumab a potent anti-AβP oligomer mAb induces superior clinical efficacy in AD (Sevigny et al., 2016). Likewise BAN2401 mAb that selectively binds soluble AβP protofibrils exhibited promising results with prodromal or mild AD in more than 856 patients (Logovinsky et al., 2016; Tolar et al., 2020). A higher dose of BAN2401 (10 mg) biweekly resulted in significantly slowing of both clinical deterioration and AβP accumulation in the brain (Tolar et al., 2020). These observations encourage further research in mAbs therapy using nanowired delivery in AD to understand disease progression and superior neuroprotection.

3 Our observation of nanowired delivery of agents in AD

In order to explore novel strategies in AD therapy we investigated combination of neuroprotective agents delivered through nanowired technology in AD models showing promising results. A brief description of the methods and observations are given below.

3.1 Methodological consideration

3.1.1 Animals

Experiments were carried out on inbred Male Sprague-Dawley rats (age 10–15 weeks, body weight 300–400 g) housed at control ambient temperature ($21 \pm 1\,°C$) with 12 h light and 12 h dark schedule. Food and water ere supplied ad libitum.

All experiments and handling of animals were carried out according National Institute of Health (NIH) Guidelines for the Care and use of laboratory animals and approved by Local Ethics Committee (National Research Council, 2011).

3.1.2 Surgery and cannulation
The animals were implanted with lateral cerebral ventricular cannula as well as the femoral artery and right jugular vein cannula aseptically 1 week before the experiment as described earlier (Sharma, 1987; Sharma and Dey, 1986, 1987).

3.1.2.1 Intracerebroventricular cannulation
Animals were anesthetized with Equithesin (3 mL/kg, i.p.) and their heads were fixed in the rat Stereotaxic instrument (Harvard apparatus, Holliston, MA, USA). After making a sterile incision on the skin of the skull the underlying skull was freed from muscle to expose parietal bone. A burr hole was made on the left parietal bone using dental drill bit (MF-5176 BASI, West Lafayette IN, USA) making 1.28 mm circle and a stainless steel screw (Hagerstown, MD, USA) was inserted. Another hole was made over the parietal bone with co-ordinates for lateral ventricle cannulation 0.8 m posterior to Bregma and 1.5 mm lateral to midline. A 21 gauge Scorex stainless steel cannula (Fischer Scientific, Landsmeer, Netherlands) was inserted ascetically into the brain 3.5 mm ventral to the skull and fixed with dental cement (BDH, London, UK) with the adjacent skull screw. A stainless steel wire stellate was inserted into the cannula of same length to avoid blockage of the needle and sealed the opening with a rubber plug. The skin was then sutured and sprinkled with antiseptic boric acid powder (Sigma Chemical Co. St. Louis, USA) and the suture animals were places individually in the cage after the operation.

The intracerebroventricular (i.c.v.) cannula was flushed out daily with artificial 30 μL CSF for maintenance of its patency.

3.1.2.2 Arterial and venous cannula implantation
For this purpose, the right femoral vein was exposed and a sterile polythene catheter PE 10 (Instech, Plymouth Meeting, PA, USA) was inserted filled with heparinized saline was inserted about 1 cm into the artery and sutured with black nylon surgical thread (Micromed, Wurmlingen, Germany). The other end of the catheter was taken out from back skin and sealed with heat.

Another catheter PE 10 was placed in the right jugular vein filled with heparinized saline and sutured with the surgical nylon thread. The other end of the catheter was taken out from the neck skin and sealed with heat. These catheters were also flushed daily with 100 μL of heparinized saline to keep them patent throughout the experimental protocol (Sharma, 1987; Sharma and Dey, 1986).

All cannulated animals are placed individually in cage with food and water supplied ad libitum. The body weight and rectal temperature were monitored daily for their well being until recovery for further experiments after 1 week of surgery (Sharma, 1987).

3.2 **Alzheimer's disease model**
Alzheimer's disease (AD) like symptoms were produced by intracerebroventricular (i.c.v.) administration of water soluble form of AβP 200 ng/30 μL in artificial CSF or in 0.9% saline once per day into the left lateral ventricle for 4 weeks through the

implanted cannula as described earlier (Sharma et al., 2012, 2015a,b, 2019a,b,c). In brief, AβP (Amyloid β-Peptide (1–40) (human), Cat. No. 1190, Tocris, Bio-Techne R&D Systems, S.LU, Madrid, Spain) was dissolved in commercial artificial CSF (1 mg/mL, #59-7316, Harvard Apparatus, Holliston, MA, USA) and filled in 100 μL Hamilton syringe (Salt Lake City, UT, USA) and connected with the micro-infusion pump (#70-4501 Pump II, Elite Infusion Dual syringe, Harvard Apparatus, Holliston, MA; USA). The syringe needle and cannula in the rat were connected with a PE10 sterile Polythene tubing filled with AβP solution. The rats were partially restrained without having any stress and the infusion was made at the rate of 10 μL/min. After the infusion, the i.c.v. cannula the stellate steel wire was inserted into the cannula and sealed with a plug. These procedures were repeated once daily for 4 weeks (Sharma et al., 2012, 2015a,b, 2019a,b,c, 2020a,b,c). All i.c.v. injections were commenced between 8:00 to 9:00 AM everyday to avoid diurnal variation and stress response among the animals.

Control group of animals were injected artificial CSF or 0.9% saline at room temperature (30 μL) into the left lateral cerebral ventricle daily in identical manner for 4 weeks. Some naïve group of animals was also used for normal control. There was no difference between injection of AβP in artificial CSF or in 0.9% saline of the results in AD or in control animal models (Sharma HS, unpublished observations).

3.3 Concussive head injury

Concussive head injury (CHI) was inflicted in Equithesin (3 mL/kg, i.p.) anesthetized animals to induce mild traumatic brain injury (mTBI) as described earlier in details (Sharma et al., 2016a,b, 2019a,b,c, 2020a,b,c). In brief, a tapered stainless steel rod weighing 114.6 g was dropped over the right exposed parietal bone of rats head fixed in the stereotaxic apparatus on a cotton wool cushion to avoid rebound movement from a 20 cm height through an aluminum guide tube. This weight and height adjustment delivers an impact injury of 0.224 N on the right parietal skull bone leading to counter-coup brain injury (Dey and Sharma, 1984; Sharma et al., 2007, 2016a,b, 2019a,b,c, 2020a,b,c). Immediately after CHI, the animals were taken out from the stereotaxic instrument and the skin was sutured and placed them on a head pad to maintain their body temperature until they recover from the anesthesia. The animals were checked for any pain and the wound is closed with boric acid powder and Xylocaine spray (10 mg, London, UK).

In some groups CHI was dome after AβP infusion and in some animal groups, CHI was done before the onset of AβP infusion. The results show that whether CHI is done before or after AβP infusion after 4 weeks the outcome is not very statistically different in terms of brain pathophysiology (Sharma HS, unpublished observations).

3.4 Treatment strategies in AD

The following therapeutic agents were delivered either through i.c.v. cannula or through intravenous route using standard procedures in control and AD groups.

3.4.1 Nanowired delivery of therapeutic agents

Cerebrolysin (Ever NeuroPharma, Austria), melanocyte stimulating hormone (MSH, M4135, Sigma-Aldrich, St. Louis, USA) and beta Amyloid Monoclonal Antibody (BAmAb, DE2B4; ab11132, abcam, Cambridge, MA, USA) was processed for TiO_2 nanowired delivery as described earlier (Sharma et al., 2016a,b).

In brief, Titanium Dioxide nanopowder (Degussa P25, 0.375 g, Nanoshel, Wilmington, DE, USA) was mixed with 50 mL of 10 M NaOH solution and stirred for 24 h. After that the solution was sonicated for 15 min transferred to a stainless steel Teflon linear container and placed in an oven at 240 °C for 72 h. The suspension was washed with double distilled water with alternate sonication and vertexing 5 min between the washes until the material gets neutralized (pH 7.2).

Equal amounts of cerebrolysin, MSH or BAmAb were mixed nanowires (TiO_2) with a w/w ratio of 1:10 and the mixtures were rotated gently for 48 h at room temperature. Amount of therapeutic agents or nanomaterials alone were also rotated at the same temperature, using identical amount of time to use as controls (Sharma et al., 2018, 2019a,b,c, 2020a,b,c).

3.4.2 TiO_2 nanowired delivery of cerebrolysin

Animals subjected to AβP infusion with or without CHI were administered cerebrolysin 50 µL i.c.v. daily starting from 1 week after AβP infusion once daily for 3 weeks. Nanowired cerebrolysin (NWCBL) was administered 1 h after AβP infusion under identical condition.

In another set of experiments, NWCBL was administered 5 mL/kg intravenously into the right jugular vein 1 h after AβP i.c.v. infusion. Intravenous NWCBL infusion was also done once daily for 3 weeks in AD groups of rats with or without CHI. For comparison conventional cerebrolysin was also administered in identical manner.

Control rats received artificial CSF or 0.9% saline in equal amount given either i.c.v. or i.v. for comparison.

3.4.3 TiO_2 nanowired mesenchymal stem cells

Nanowired MSC about 1×10^6 cells was administered intravenously 1 week after AβP infusion. This dose of MSC was repeated after 2 weeks from the onset of AβP infusion with our without CHI (Sharma et al., 2019a,b,c, 2020a,b,c). To compare the effects of nanodelivery and conventional delivery normal MSC was also given in identical manner.

In control animals, 0.9% saline was administered under identical conditions.

3.4.4 TiO_2 nanowired amyloid beta monoclonal antibodies

Nanowired Amyloid beta monoclonal antibody (AβPmAb) 1.20 dilution was administered i.c.v. through intracerebroventricular cannula in a dose of 50 µL in AD group of animals with or without CHI (Sharma and Sharma, 2012a,b). The AβPmAb was given 3 h after AβP infusion following 1 week after the onset in AD group once daily for 3 weeks.

To compare normal AβPmAb was also given in identical manner to understand the effects of nanowired delivery and conventional delivery of antibodies.

The control group received artificial CSF or 0.9% saline instead of BAmAb under identical conditions.

3.5 Parameters measured

The following parameters were evaluated to explore AD pathophysiology and neuroprotection using standard protocol.

3.5.1 Blood-brain barrier permeability

The BBB was measured using two exogenous tracers Evans blue albumin (EBA) and radioiodine ([131]-Iodine) as described earlier (Sharma, 1987; Sharma et al., 2018, 2019a,b,c, 2020a,b,c). In brief, EBA (2%, 3 mL/kg, i.v.) and radioiodine (100 μCi/kg, i.v.) were injected into the right jugular vein 5–10 min before termination of the experiments. The intravascular tracers were washed out by cold 0.9% saline perfusion in situ using a peristaltic pump (P-70, Harvard Apparatus, Holliston, MA, USA) maintaining a pressure of 90 Torr (Sharma and Dey, 1986). For this purpose, under deep anesthesia, the chest was rapidly opened and the heart was exposed and the right auricle was cut. A 21 G butterfly needle inserted into the left cardiac ventricle that is connected to the pump attached with 0.9% saline bottle. The intravascular blood was washed out by 50–60 mL of 0.9 and saline (Sharma, 1987).

Immediately before perfusion about 1 mL of whole blood was withdrawn from the left cardiac ventricle for measurement of whole blood radioactivity or EBA concentration (Sharma and Dey, 1987; Sharma et al., 1990a,b).

After the end of the perfusion, the brain was dissected out and examined for the EBA leakage on the dorsal and ventral sides of the brain. After that a mid sagittal section was done to examine leakage of EBA dye into the ventricles and other subcortical areas of the brain. After that small tissue pieces were dissected out from the desired brain areas weighed immediately and counted in a 3-in Gamma Counter (Cobra II Model 5003, Packard, USA) for radioactivity. After counting the radioactivity, the tissue pieces were homogenized in a mixture of sodium sulfate (0.5%) and pure acetone to extract EBA dye from the brain tissues. The samples were homogenized and centrifuged at $900 \times g$ and the supernatant was measured calorimetrically in a spectrophotometer (Thermo Fisher Scientific, UK) at 620 nm for blue dye concentration. The dye concentration was calculated according to the standard curve of EBA dye (Sharma and Dey, 1986).

Extravasation of radioactivity in the brain was expressed as percentage increase from the whole blood radioactivity as described earlier (Sharma, 1987; Sharma and Dey, 1986, 1987).

3.5.2 Brain edema formation

Brain edema formation was measured using brain water content as described earlier (Sharma and Cervós-Navarro, 1990; Sharma and Olsson, 1990; Sharma et al., 1991). In brief, after the experiments, the brain was dissected out and placed on a cold 0.9%

saline wetted filter paper (Whatman Nr. 1, USA). The large superficial blood vessels and blood clots, if any were removed. The desired brain regions were dissected out and weighed on a preweighed filter paper immediately. After that the brain samples were placed in an incubator (BD115, Fisher Scientific, Pittsburgh, PA, USA) maintained at 90 °C for 72 h to evaporate the water content of the tissue. When the three consecutive dry weight of the samples became constant the brain water content was calculated from the differences between wet and dry weight as described earlier (Sharma and Olsson, 1990).

The volume swelling (%f) was calculated from the differences of brain water content between control and experimental group using the formula of Elliott and Jasper (1949) as described earlier (Sharma and Cervós-Navarro, 1990). In general about 1% difference in brain water represent about 4% of volume swelling (Sharma et al., 1991).

3.5.3 Brain pathology

Brain pathology was examined using standard histopathology, immunohistochemistry and transmission electron microscopy of AD group and compared with the treatment strategies according to the standard protocol (Sharma et al., 2016a,b, 2018, 2019a,b,c, 2020a,b,c).

3.5.3.1 Perfusion and fixation

After the end of the experiments animals were deeply anesthetized with Equithesin (3 mL/kg, i.p.) and the chest was rapidly opened and the hearty was exposed. The right auricle was cut and a 21 G Butterfly needle was inserted into the left cardiac ventricle. Cold phosphate buffer saline (0.1 M, pH 7.0) was perfused though wash out the remaining blood from the vessels using the peristaltic pump (P 70, Harvard Apparatus, Holliston, MA; USA). After about 150 mL PBS was washed out the animals were perfused in situ with cold 4% buffered paraformaldehyde solution (pH 6.9, Sigma, Aldrich, St. Louis, MO, USA) (Sharma and Olsson, 1990; Sharma et al., 2011a,b). About 250 mL fixative was used to perfuse the animals in situ. The perfusion pressure was maintained at 90 Torr for both PBS or paraformaldehyde perfusion. In some cases Somogyi Fixative containing 0.25% picric acid was used under identical condition (Sharma and Cervós-Navarro, 1990).

3.5.3.1.1 Light microscopy. After perfusion fixation in situ the animals were wrapped in aluminum foil and placed overnight at 4 °C in a refrigerator. On the next day, the brains were dissected out and coronal sections at the desired level were cut and processed automatically and embedded in paraffin (Sakura FineTek, USA Inc., Torrance, CA, USA). About 3-μm serial thick sections were cut and stained with Nissl, Hematoxylin and Eosin (H&E), or Toluidine Blue according to standard protocol.

The sections were examined under a Carl Zeiss Bright Field Light Microscope (AxioVert 200 M, Oberkochen, Germany) attached with a digital camera (Zeiss Axiocam 500 color, Thornwood, NY, USA) and images were taken at 20× or 40× magnification (Sharma and Sjöquist, 2002). The images were stored in an

Apple McIntosh Power Book (System Mc Oss, El Capitan, 10.11.6) and processed using commercial software Adobe Photoshop CSS extended version 12.0.4 × 84 (San Jose, CA, USA) using identical contrast and color filters from control and experimental animals (Sharma and Kiyatkin, 2009).

3.5.3.1.2 Immunohistochemistry. Random serial sections fixed with either Somogyi or paraformaldehyde fixatives were used for immunohistochemistry for Albumin, glial fibrillary acidic protein (GFAP), myelin basic protein (MBP) and amyloid beta peptide (AβP) using commercial protocol as described earlier (Sharma et al., 1992a,b, 1993, 2015a,b, 2016a,b, 2018, 2019a,b,c, 2020a,b,c). For myelin examination Luxol Fast Blue (LFB) stain was also used and compared with the MBP immunostaining (Sharma et al., 2011a,b).

In brief, recombinant rabbit anti-bovine monoclonal albumin antibody (1500, Abcam, EPR12774, Cambridge, UK); recombinant rabbit monoclonal anti-GFAP-antibody GFAP (1500, ab68428, EPR1034Y, Cambridge, MA, USA); recombinant rabbit monoclonal anti-myelin basic protein antibody (1200, ab 133620, EPR 6652, Cambridge, UK); recombinant rabbit monoclonal anti-beta amyloid antibody (16000 ab 205340, mOC23, Cambridge, MA, USA) was employed on paraffin sections to visualize immunoreaction product using Avidin-Biotin-complex (ABC technique, Burlingame, CA, USA) according to commercial protocol (Sharma et al., 1993, 2016a,b, 2018, 2019a,b,c, 2020a,b,c). Myelin was also visualized using LFB staining as described earlier (Sharma et al., 1992a,b, Sharma et al., 2011a,b). No significant differences between immunostaining of albumin, AβP, GFAP or MBP were detected in tissue sections either produced after Somogyi or paraformaldehyde fixatives (Sharma HS, unpublished observation).

3.5.3.1.3 Transmission electron microscopy. Brain tissues fixed with either Somogyi or paraformaldehyde was processed for transmission electron microscopy (TEM) as described earlier (Sharma and Olsson, 1990; Sharma et al., 1991, 1992a,b, 1993, 1995, 1998a,b,c). In brief, small tissue pieces from the brains were cut and post fixed in osmium tetroxide (OsO_4) and embedded in plastic (Eon 812). Semithin sections (1-μm thick) were cut and stained with Toluidine blue to identify the areas to be investigated under TEM. The desired areas were then trimmed and ultrathin sections (50 nm thick) were cut on an Ultramicrotome (Ultracut E Reichert-Jung, Bayreuth, Germany) using diamond knife (Diatome, Hatfield, PA, USA). The serial sections were cut and placed on single hole (600 μm), copper TEM grids (Stansted, Essex, UK).

Some sections were counterstained with lead citrate and uranyl acetate for contrast and some sections were viewed under Philipp's 400 TEM (Eindhoven, Netherlands) and images are recorded at 4 k or 6 k magnifications on the attached digital camera system (Gatan K3 IS camera, Pleasanton, CA, USA). The images were processed using Adobe Photoshop commercial software using identical contrast and image filters (Sharma and Sjöquist, 2002).

3.6 Biochemical measurement

Several key elements involved in the pathophysiology of AD were measured in the CSF of animals in control, experimental and drug treated groups as follows.

3.6.1 Sampling of CSF

For this purpose, at the end of the experiment, about 100 μL of CSF was withdrawn from the cisterna magna by puncturing the atlanto-occipital membrane and the samples were stored at −80 °C till further biochemical measurements as described earlier (Sharma et al., 2018, 2019a,b,c, 2020a,b,c). Any CSF sample (less than 2%) that is contaminated with blood is discarded (Sharma and Dey, 1981).

3.6.1.1 AβP measurement

For AβP measurement in the CSF rat amyloid beta peptide 1–40 (Ab1–40) ELISA kit from Biomatik (EKU02329, Wilmington, DE, USA) sensitivity 4.84 pg/mL in biological fluid was used according to commercial protocol.

3.6.1.2 Albumin measurement

Albumin in CSF was measured using rat albumin ELISA kit (ab108790, Cambridge, MA, USA) sensitivity 0.7 μg/mL in biological fluids according to commercial protocol (Sharma et al., 2019a,b,c).

3.6.1.3 Plasminogen measurement

Plasminogen was measured in CSF using Plasminogen ELISA kit sensitivity 20 ng/mL (P00747, Thermo Fischer Scientific, Waltham, MA, USA) according to commercial protocol.

3.6.1.4 Statistical analyses of the data

Statistical analyses of the data were dome using commercial software StatView5 (Abacus Concept Ltd., Piscataway, NJ, USA) on a Macintosh Power Book computer (Classical environment Os 9.6.2) using ANOVA followed by Dunnett's test for multiple group comparison from one control for the quantitative data obtained. For semiquantitative data the non-parametric Chi Square test was performed. A P-value less than 0.05 was considered significant.

4 Results

4.1 Rat model of AD induced by AβP infusion

Our observations show that once daily intracerebroventricular (i.c.v.) administration of soluble form of AβP 200 ng/30 μL per day into the left lateral ventricle for 4 weeks resulted into brain pathology and behavioral symptoms (Sharma HS, results not shown) quite similar to clinical cases of AD (Fig. 1). Thus, deposition of AβP seen using immunohistochemistry is very similar to that of seen in human cases of AD. Biochemical measurement CSF exhibited significant increase in AβP as compared to

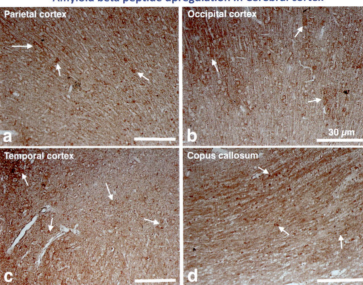

FIG. 1

Representative example of Amyloid beta peptide immunochemistry in the AβP infusion induced Alzheimer's disease in rat model. Low power light microscopic images show AβP immunoreactivity (arrows) in several brain areas including parietal cortex (A), occipital cortex (B), temporal cortex (C) and corpus callosum (D) areas. Bar = 30 μm.

the control group (Fig. 2). Morris water maze test regarding search of hidden plat-form shows significant delay in AβP infused rats that started after 3 weeks of daily infusion and further continued to increase after 4 weeks of AβP administration (Sharma HS and Sharma A, unpublished observation, Samaey et al., 2019).

These observations suggest that our AD model could be used for drug treatment strategies as described earlier (Sharma et al., 2012, 2016a,b, 2017, 2018, 2019a,b,c).

4.1.1 AD pathology exacerbated by infliction of concussive head injury

Infliction of concussive head injury (CHI) without AβP infusion resulted in enhanced AβP levels in the CSF and exhibited increased immunostaining in the brain after 4 weeks survival (Figs. 2 and 3). The CHI rats also showed memory disturbances as seen on Morris water maze enhanced time for platform search (Results not shown; Sharma HS and Sharma A, unpublished observation, Samaey et al., 2019). However, the deposition of AβP in the brain and/or elevation in the CSF is significantly much smaller in CHI after 4 weeks as compared to 4 weeks of AβP infusion without injury (Figs. 2 and 3).

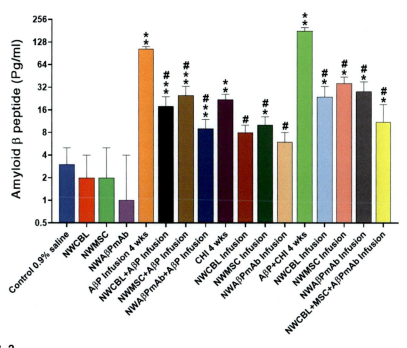

FIG. 2

Cerebrospinal fluid (CSF) Amyloid beta peptide measurement in Alzheimer's disease model of rats in combination with concussive head injury (CHI) and treatments. Values are Mean ± SD of 6–8 rats at each point. * $P < 0.05$; ** $P < 0.01$ from control; # $P < 0.05$ from untreated AD or CHI group. ANOVA followed by Dunnett's test for multiple comparisons using one control. AβP, amyloid beta peptide; CHI, concussive head injury; LH, left hemisphere; mAb, monoclonal antibody; NWAβPmAb, TiO_2 nanowired monoclonal antibody against amyloid beta peptide; NWCBL, TiO_2 nanowired cerebrolysin; NWMSC, TiO_2 nanowired melanocyte stimulating hormone; RH, right hemisphere. CHI was inflicted on the right parietal bone with an impact of 0.224 N as described earlier (Sharma et al., 2016a,b). AβP infusion (200 ng, i.c.v.) was dome once daily for 4 weeks. CHI was inflicted either 24 h before commencement of AβP infusion or 24 h after the onset of AβP infusion. NWCBL, NWMSH or NWAβPmAb were administered 1 week after AβP infusion with or without CHI for 3 weeks. All the parameters were examined after 4 weeks of AβP infusion with or without CHI. For details see text.

However, when AβP infusion was carried out in animals after CHI for 4 weeks the pathological outcome, deposition of AβP in the brain and elevation in the level of AβP in CSF all are exacerbated as compared to either AβP infusion or CHI alone (Figs. 2 and 3). The memory impairment as measured using Morris water maze test in finding hidden platform in opaque water was significantly enhanced as compared to AβP Infusion or CHI alone (Sharma HS and Sharma A unpublished observations, Results not shown). These observations further suggests that our model of AD using

CHI exacerbates Amyloid beta peptide upregulation

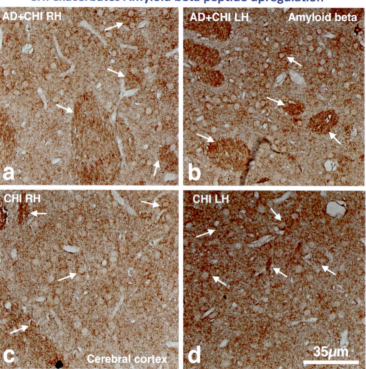

FIG. 3

Representative example of Amyloid beat peptide (AβP) immunohistochemistry in the cerebral cortex of Alzheimer's disease model in rats with concussive head injury (CHI) in the right hemisphere (RH) and in left hemisphere (LH). The intensity of AβP immunostaining (arrows) appears to be greater in the LH in AD with CHI (B) or CHI alone (D) as compared to the RH in AD with CHI (A) or CHI alone (C). Bar = 35 μm.

AβP infusion represents significant advances in AD research in line with earlier published studies from our laboratory (Sharma et al., 2012, 2016a,b, 2017, 2018, 2019a,b,c).

4.2 Blood-brain barrier breakdown in AD

Alzheimer's disease results in the BBB breakdown (Zenaro et al., 2017). In our model of AβP infusion, the BBB to Evans blue albumin (EBA) and radioiodine [131]-Iodine was significantly increased in the various brain regions after 4 weeks (Fig. 4). As shown in Fig. 4, the right hemisphere and left hemispheres both shows almost similar increase in EBA and radioiodine extravasation although the AβP infusion was given into the left cerebral ventricle. Thus, as compared to saline infusion

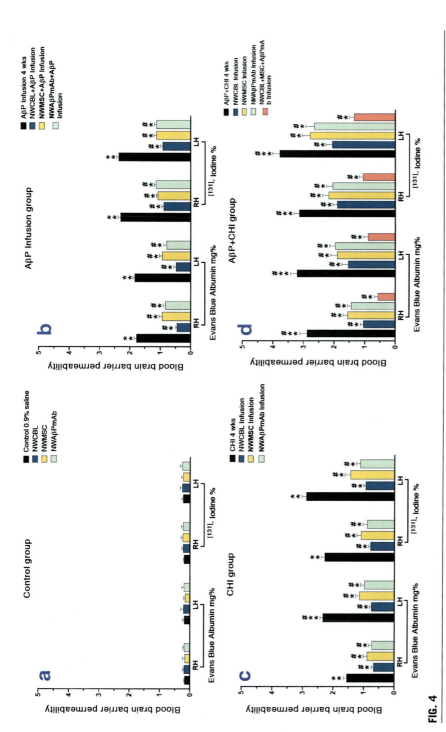

FIG. 4

Blood-brain barrier (BBB) permeability to EBA and Radioiodine ([131]-Iodine) in right injured hemisphere (RH) and left uninjured hemisphere (LH) in amyloid beta peptide (AβP) infused Alzheimer's disease (AD) and their modification with closed head injury (CHI) on RH and therapeutic agents in control group (A); AβP infusion group (B); CHI group (C) and CHI+AβP group (D). Values are Mean ±SD of 6–8 rats at each point. * P<0.05; ** P<0.01 from control; # P<0.05 from untreated AD or CHI group. ANOVA followed by Dunnett's test for multiple comparisons using one control. AβP, amyloid beta peptide; CHI, concussive head injury; LH, left hemisphere; mAb, monoclonal antibody; NWAβPmAb, TiO₂ nanowired monoclonal antibody against amyloid beta peptide; NWCBL, TiO₂ nanowired cerebrolysin; NWMSC, TiO₂ nanowired melanocyte stimulating hormone; RH, right hemisphere. CHI was inflicted on the right parietal bone with an impact of 0.224 N. AβP infusion (200 ng, i.c.v.) was dome once daily for 4 weeks. CHI was inflicted either 24 h before commencement of AβP infusion or 24 h after the onset of AβP infusion. NWCBL, NWMSH or NWAβPmAb were administered 1 week after AβP infusion with or without CHI for 3 weeks. All the parameters were examined after 4 weeks of AβP infusion with or without CHI. For details see text.

the EBA content was in left half (0.19 ± 0.05 mg %) and right half (0.18 ± 0.02 mg %). After 4 weeks of AβP infusion in the left cerebral ventricles the EBA extravasation in left half significantly increased to 1.84 ± 0.06 mg % ($P < 0.01$ from left half saline control) and the right half shows 1.78 ± 0.08 mg % EBA ($P < 0.01$ from right half saline control). Likewise the leakage of radioiodine in saline control in the left half was 0.23 ± 0.04% and in right half this was 0.20 ± 0.02% that increased to 2.37 ± 0.07% in left half ($P < 0.01$ compared from left half saline control) and in right half the value was 2.31 ± 0.09% ($P < 0.01$ compared from right half saline control) after 4 weeks of AβP infusion.

Visual examination of EBA leakage showed mild to deep blue staining in several cerebral cortical and sub cortical regions. Thus, cingulate, parietal and piriform cortices showed moderate blue staining while fontal, occipital and temporal cortices exhibited deep blue staining on both halves of the brain after AβP infusion. In subcortical areas, hippocampus, thalamus, hypothalamus exhibited deep blue staining and caudate nucleus, putamen, colliculi and brain stem showed mild to moderate staining. Interestingly the vermis and lateral cerebellar cortices showed moderate to deep blue staining. The walls of the ventricular system (lateral and fourth ventricles) also showed mild blue staining indicating breakdown of the blood-CSF barrier along with the BBB (results not shown).

Leakage of radioactivity in AD brain following infusion of AβP was also significantly higher than in saline infusion control group at 4 weeks. Thus, saline infusion shows 0.23 ± 0.04% radioiodine leakage in left half and 0.20 ± 0.02% in right half. After 4 weeks of AβP Infusion extravasation of radioiodine was significantly enhanced in the left half by 2.37 ± 0.07% ($P < 0.01$ as compared to left half from saline) and the right half showed 2.31 ± 0.09% ($P < 0.01$ compared to saline infusion in right half) extravasation of radioiodine.

The regional variation in radioiodine leakage was highest in the cingulate, frontal and temporal cortices followed by parietal, occipital and piriform cortices. The hippocampus showed high radioiodine leakage followed by caudate nucleus, putamen, thalamus, hypothalamus, colliculi and brainstem. The cerebellar vermis and lateral cortices exhibited radioiodine extravasation very similar to the occipital and frontal cortices (results not shown, Sharma HS and Sharma A, unpublished observation). Small CSF sample (40 μL) obtained from cisterna magna also showed significant higher radioiodine leakage as compared to saline infused identical CSF samples (results not shown). This suggests that AβP infusion model induces breakdown of both the BBB and BCSFB to radioiodine (Sharma HS and Sharma A, unpublished observations).

4.2.1 Concussive head injury exacerbates BBB breakdown in AD

Concussive head injury (CHI) was inflicted on the right parietal skull with an impact of 0.224 N resulted in significant increase in the BBB permeability to EBA and radioiodine tracers after 4 weeks of primary injury that was the most pronounced in the left half of the brain as compared to the right injured half (Fig. 4). Thus, extravasation of EBA after CHI in the right injured half was 1.57 ± 0.08 mg % ($P < 0.01$ from control) and the left uninjured half showed 2.34 ± 0.09 mg % ($P < 0.01$ from control) leakage of EBA, Likewise the extravasation of radioiodine in the right half was $2.18 \pm 0.11\%$ ($P < 0.01$ from control) and in the left half was $2.89 \pm 0.14\%$ ($P < 0.01$ from control). This increase in the BBB breakdown to EBA and radioiodine was slightly higher in CHI as compared to AβP infusion after 4 weeks,

Visual inspection of EBA leakage in CHI brains showed mild to moderate blue staining on the occipital, parietal, temporal cortices in the whereas the cingulate and frontal cortices exhibited mild blue staining (Results not shown). The walls of the cerebral ventricles in lateral and the fourth ventricles were moderately stained after CHI indicating breakdown of the BCSFB permeability as well. The subcortical structures showed mild to moderate blue staining. Thus, hippocampus, caudate nucleus, putamen, thalamus and hypothalamus exhibited moderate staining. The colliculi and brainstem showed light blue staining. Interestingly, the cerebellar vermis showed moderate blue staining while the cerebellar cortices showed mild EBA staining after 4 weeks of CHI (Results not shown). In general, the intensity on the blue staining was higher in most brain areas after CHI in the left half as compared to the right injured half (Sharma HS and Sharma A, unpublished observations).

Extravasation of the radioiodine was also prominent in the left in injured half after 4 weeks of CHI as compared to the right injured half (Fig. 4). Thus, the right half showed $2.38 \pm 0.11\%$ leakage of radioactivity ($P < 0.01$ from control) whereas the left half exhibited $2.89 \pm 0.14\%$ ($P < 0.01$ from control) extravasation of radioiodine after 4 weeks of primary injury (Fig. 4).

Regional analyses of the radioactivity leakage exhibited high concentration in the parietal, temporal, occipital cortices and cerebellar vermis followed by cingulate, frontal and cerebellar cortices. The hippocampus, caudate nucleus, putamen, thalamus and hypothalamus showed moderate radioactivity leakage followed by colliculi, and brain stem. The left uninjured half exhibited higher intensity of radioiodine leakage as compared to the identical brain areas in the right injured half.

In the CSF samples (40 μL) taken from cisterna magna after 4 weeks of CHI exhibited about 250% increase in the radioiodine leakage as compared to the control group CSF (Results not shown). This indicates breakdown of the BCSFB to radioiodine in CHI after 4 weeks of primary injury (Sharma HS and Sharma A, unpublished observation).

When AβP infusion was made after CHI, a synergistic higher increase in the BBB breakdown in the right injured and left uninjured halves were observed (Fig. 4). Thus, in CHI with AβP infusion of 4 weeks resulted in leakage of EBA in the right half was 2.89 ± 0.23 mg % ($P < 0.01$ from control) and in the left half was 3.21 ± 0.022 mg % ($P < 0.01$). The EBA extravasation in this group was significantly higher than the AβP or CHI group alone after 4 weeks.

The regional blue staining was observed in identical brain areas as seen in CHI or AβP infusion alone but the intensity of blue staining was much more deeper in combined CHI with AβP infusion group (Results nor shown).

The radioactivity leakage was also greater in AβP infusion with CHI as compared to CHI or AβP infusion alone (Fig. 4). Thus, the right injured half showed

3.14 ± 0.17 mg % ($P < 001$ from control) EBA leakage and the left uninjured half revealed 3.78 ± 0.23 mg % leakage of radioactivity (Fig. 4). These values of radio-iodine leakage in CHI with AβP infusion are higher than the values obtained in CHI or AβP infusion alone after 4 weeks of survival (Fig. 4).

The magnitude and intensity of radioiodine leakage in CHI with AβP infusion was higher in identical brain regions and in the cisterna magna CSF as compared to the similar brain regions and in CSF in CHI or AβP Infusion alone after 4 weeks of survival (Sharma HS and Sharma A, unpublished observations). These observations clearly suggest that combination of CHI with AβP infusion exacerbates BBB and BCSFB breakdown to EBA and radioiodine tracers.

4.3 Brain edema formation in AβP infusion induced AD

Breakdown of the BBB is an important factor for vasogenic brain edema formation (Sharma and Sharma, 2019; Sharma et al., 1998a,b,c). Accordingly, measurement of brain edema after 4 weeks of AβP infusion shows a significant increase in the brain water content in right half to $75.16 \pm 0.13\%$ from control value of $74.34 \pm 0.056\%$ ($P < 0.01$) equal to about 4% increase in volume swelling (Fig. 5). Similar increase in volume swelling of 4% was also seen in the left half in AβP infusion as the brain water content $75.21 \pm 0.20\%$ increased from control value of 74.53 ± 0.61 ($P < 0.01$). This observation indicates that AβP infusion induced brain edema formation in AD model.

4.3.1 Concussive head injury exacerbated brain edema formation in AD

Concussive head injury (CHI) significantly enhanced brain edema formation that was most prominent in the uninjured half of the brain as compared to the injured half (Fig. 5). Thus, the after 4 weeks of CHI the right half sowed an increase in the brain water content from control ($74.34 \pm 0.56\%$) to $76.48 \pm 0.23\%$ ($P < 0.01$) that equals to about 8% volume swelling (Fig. 6). Whereas, the uninjured left half exhibited an increase of brain content to $77.67 \pm 0.31\%$ from control value of $74.53 \pm 0.61\%$ ($P < 0.01$) amounting to 12% volume swelling (Fig. 6).

When AβP infusion was made for 4 weeks after CHI the brain edema formation showed synergistic higher values as command to the CHI or AβP infusion alone (Fig. 4). Thus the combined CHI and AβP infusion resulted in brain water content of the right inured half after 4 weeks of survival to $77.67 \pm 0.12\%$ ($P < 0.01$ from control) corresponding to an increase in volume swelling by 13%. On the other hand, the left half resulted in an increase in water content by $79.48 \pm 0.21\%$ ($P < 0.01$) from the control values equal to a volume swelling of 20% (Fig. 6). These observations are in line with the idea that a combination of CHI and AβP infusion exacerbates brain edema formation as compared to AβP infusion or CHI alone after 4 weeks of survival.

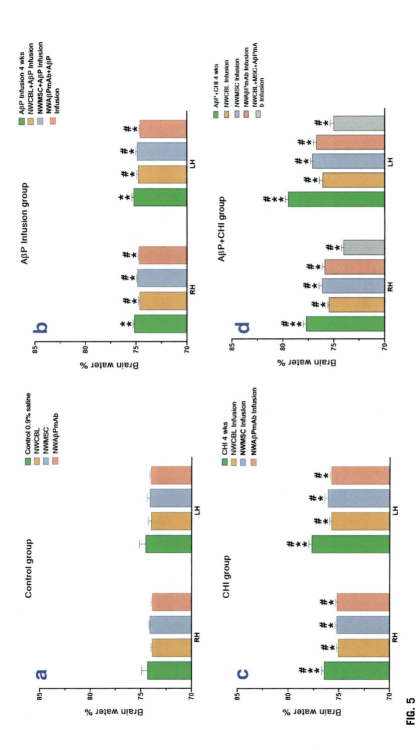

FIG. 5

Brain water content in right injured hemisphere (RH) and left uninjured hemisphere (LH) in amyloid beta peptide (AβP) infused Alzheimer's disease (AD) and their modification with closed head injury (CHI) on RH and therapeutic agents in control group (A); AβP infusion group (B); CHI group (C) and CHI+AβP group (D). Values are Mean±SD of 6–8 rats at each point. * $P < 0.05$; ** $P < 0.01$ from control; # $P < 0.05$ from untreated AD or CHI group. ANOVA followed by Dunnett's test for multiple comparisons using one control. AβP, amyloid beta peptide; CHI, concussive head injury; LH, left hemisphere; mAb, monoclonal antibody; NWAβPmAb, TiO$_2$ nanowired monoclonal antibody against amyloid beta peptide; NWCBL, TiO$_2$ nanowired cerebrolysin; NWMSC, TiO$_2$ nanowired melanocyte stimulating hormone; RH, right hemisphere. CHI was inflicted on the right parietal bone with an impact of 0.224N. AβP infusion (200 ng, i.c.v.) was dome once daily for 4 weeks. CHI was inflicted either 24 h before commencement of AβP infusion or 24 h after the onset of AβP infusion. NWCBL, NWMSH or NWAβPmAb were administered 1 week after AβP infusion with or without CHI for 3 weeks. All the parameters were examined after 4 weeks of AβP infusion with or without CHI. For details see text.

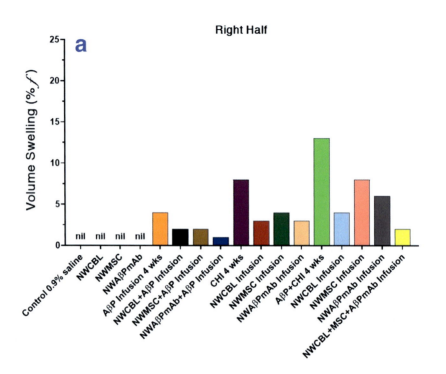

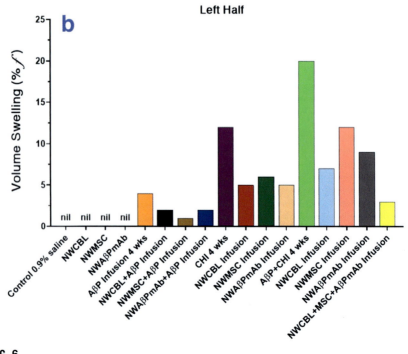

FIG. 6

See figure legend on opposite page.

4.4 Brain pathology in AβP infusion induced AD

Breakdown of the BBB and vasogenic edema results in alterations in the fluid microenvironment of the central nervous system (CNS) leading to brain pathology (Sharma, 2009; Sharma and Westman, 2004; Sharma et al., 1998a,b,c).

4.4.1 Structural changes in AβP induced AD brain at light microscopy

We examined neural and glial and myelin changes at light microscopy.

4.4.1.1 Neuronal changes

AβP infusion induced profound neuronal injuries starting from 3 weeks after infusion and reached to its maximum after 4 weeks of survival as compared to saline infusion (Fig. 7). Thus, about 484 ± 21 number of neurons are distorted and damaged in the cerebral cortex as compared to saline infusion where only 3 ± 2 neurons show distortion (Fig. 7). Many neurons were shrunken or swollen with perineuronal edema are evident in the neuropil (Fig. 8). These neurons are located in the areas of EBA stained regions. A clear sponginess and expansion of the neuropil is evident that exhibited BBB breakdown to EBA and radiotracer (Fig. 4). These neuronal damages are present in almost all the cerebral cortical region in cerebellum. In cerebellum both Purkinje cells and granule cells are distorted following AβP infusion (Fig. 8). The other areas of brain where neuronal distortion is evident include hippocampus, caudate nucleus, putamen, thalamus and hypothalamus. In the hippocampus A-3 and dentate gyrus exhibited neuronal degeneration, dark and distorted neurons within the expanded neuropil (Sharma HS and Sharma A, unpublished observation). In several neurons a distinct nucleus is not seen and in some cases if the neuronal nucleus is present it shows the eccentric nucleolus (Results not shown).

FIG. 6—Cont'd

Volume swelling (% f) calculated from the differences between brain water content in control and experimental cases in right injured hemisphere (RH) and left uninjured hemisphere (LH) in amyloid beta peptide (AβP) infused Alzheimer's disease (AD) and their modification with closed head injury (CHI) on RH and therapeutic agents in control group. (A) RH; (B) LH. AβP, amyloid beta peptide; CHI, concussive head injury; LH, left hemisphere; mAb, monoclonal antibody; NWAβPmAb, TiO_2 nanowired monoclonal antibody against amyloid beta peptide; NWCBL, TiO_2 nanowired cerebrolysin; NWMSC, TiO_2 nanowired melanocyte stimulating hormone; RH, right hemisphere. CHI was inflicted on the right parietal bone with an impact of 0.224 N. AβP infusion (200 ng, i.c.v.) was dome once daily for 4 weeks. CHI was inflicted either 24 h before commencement of AβP infusion or 24 h after the onset of AβP infusion. NWCBL, NWMSH or NWAβPmAb were administered 1 week after AβP infusion with or without CHI for 3 weeks. All the parameters were examined after 4 weeks of AβP infusion with or without CHI. For details see text.

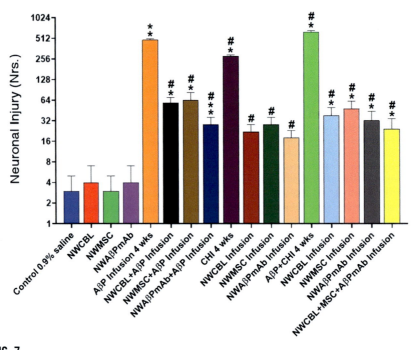

FIG. 7

Left parietal cerebral cortex semiquantitative data on Neuronal injury at light microscopy in Alzheimer's disease model in rat with concussive head injury (CHI) and treatments. Values are Mean $\pm$ SD of 6–8 rats at each point. * $P < 0.05$; ** $P < 0.01$ from control; # $P < 0.05$ from untreated AD or CHI group. ANOVA followed by Dunnett's test for multiple comparisons using one control. AβP, amyloid beta peptide; CHI, concussive head injury; LH, left hemisphere; mAb, monoclonal antibody; NWAβPmAb, TiO$_2$ nanowired monoclonal antibody against amyloid beta peptide; NWCBL, TiO$_2$ nanowired cerebrolysin; NWMSC, TiO$_2$ nanowired melanocyte stimulating hormone; RH, right hemisphere. CHI was inflicted on the right parietal bone with an impact of 0.224 N as described earlier (Sharma et al., 2016a,b). AβP infusion (200 ng, i.c.v.) was dome once daily for 4 weeks. CHI was inflicted either 24 h before commencement of AβP infusion or 24 h after the onset of AβP infusion. NWCBL, NWMSH or NWAβPmAb were administered 1 week after AβP infusion with or without CHI for 3 weeks. All the parameters were examined after 4 weeks of AβP infusion with or without CHI. For details see text.

4.4.1.2 Glial changes

Changes in glial cells were examined using GFAP immunostaining of astrocytes. Astrocytes are activated in the cerebral cortex, hippocampus, thalamus and hypothalamus after AβP infusion seen after 4 weeks of infusion. These star showed activated glia are largely located around the perivascular and perineuronal regions.

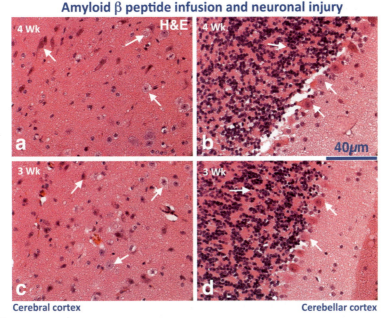

FIG. 8

Representative example of Amyloid beta peptide infusion induced brain pathology in the cerebral cortex (A and C) and in cerebellar cortex (B and D) after 3 weeks (C and D) and 4 weeks (A and B) infusion. Both Purkinje cell and granule cells show degenerative changes following AβP infusion into the lateral cerebral ventricle (arrows) after 3 weeks (D) and further enhanced after 4 weeks (C). A similar change of neuronal degeneration in cerebral cortex is also seen at 3 weeks (B, arrows) and further increased at 4 weeks after AβP infusion (A). H&E staining on Paraffin sections 3-μm thick, Bar=40 μm.

GFAP activation is evident in the areas of BBB disruption in AD model (Fig. 9). Semiquantitative data shows that AβP infusion resulted in about 284 ± 21 GFAP positive sells located in the cerebral cortex after 4 weeks as compared to saline infusion where only 6 ± 3 GFAP positive cells were seen in the cortex (Fig. 10).

4.4.1.3 Myelin changes

Myelin changes were examined using LFB staining (Fig. 11). Significant reduction in the LFB staining is seen in AD in several areas of the brain including cerebral cortex, thalamus, hypothalamus and the brain stem (results not shown). These LFB changes seen in areas are further confirmed by myelin basic protein immunoreactivity (Sharma HS and Sharma A, unpublished observations) and transmission electron microscopy (Fig. 12).

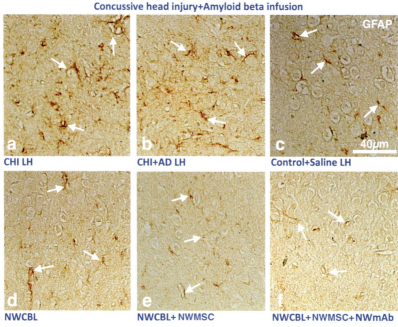

FIG. 9

Representing example of glial fibrillary acidic protein (GFAP) immunoreactivity (arrows) in the left hemisphere (LH) of cerebral cortex after concussive head injury at right hemisphere (A) and amyloid beta peptide (AβP) infusion in CHI (B) as compared to saline treated control rat (C). Treatment with NWCBL (D), NWCBL + NWMSC (E) and NWCBL + NWMSC + NWmAb (F) attenuated GFAP positive cells in the LH effectively. It appears that NWMSC and NWmAb of AβP combine has superior effect on reduction in GFAP positive cells in LH. Bar = 40 μm. AβP, amyloid beta peptide; CHI, concussive head injury; LH, left hemisphere; mAb, monoclonal antibody; NWAβPmAb, TiO$_2$ nanowired monoclonal antibody against amyloid beta peptide; NWCBL, TiO$_2$ nanowired cerebrolysin; NWMSC, TiO$_2$ nanowired melanocyte stimulating hormone; RH, right hemisphere. CHI was inflicted on the right parietal bone with an impact of 0.224 N as described earlier (Sharma et al., 2016a,b). AβP infusion (200 ng, i.c.v.) was dome once daily for 4 weeks. CHI was inflicted either 24 h before commencement of AβP infusion or 24 h after the onset of AβP infusion. NWCBL, NWMSH or NWAβPmAb were administered 1 week after AβP infusion with or without CHI for 3 weeks. All the parameters were examined after 4 weeks of AβP infusion with or without CHI. For details see text.

4.4.1.4 Albumin immunoreactivity

Immunohistochemistry of albumin clearly shows staining of neurons and the neuropil and often located in the perivascular areas exhibiting leakage from these cerebral microvessels (Sharma HS and Sharma A, unpublished observation) and within the neuropil (Fig. 13). These observations further conform the breakdown of the BBB permeability as seen using EBA and radiotracers and biochemical measurements in the CSF (Fig. 14).

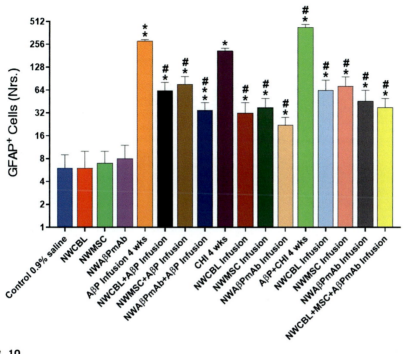

FIG. 10

Left hemisphere Glial Fibrillary Acidic Protein (GFAP) positive immunoreactive cells at light microscopy semiquantitative data in Alzheimer's disease model in rat with concussive head injury (CHI) and treatments. Values are Mean $\pm$ SD of 6–8 rats at each point. * $P < 0.05$; ** $P < 0.01$ from control; # $P < 0.05$ from untreated AD or CHI group. ANOVA followed by Dunnett's test for multiple comparisons using one control. AβP, amyloid beta peptide; CHI, concussive head injury; LH, left hemisphere; mAb, monoclonal antibody; NWAβPmAb, TiO$_2$ nanowired monoclonal antibody against amyloid beta peptide; NWCBL, TiO$_2$ nanowired cerebrolysin; NWMSC, TiO$_2$ nanowired melanocyte stimulating hormone; RH, right hemisphere. CHI was inflicted on the right parietal bone with an impact of 0.224 N as described earlier (Sharma et al., 2016a,b). AβP infusion (200 ng, i.c.v.) was dome once daily for 4 weeks. CHI was inflicted either 24 h before commencement of AβP infusion or 24 h after the onset of AβP infusion. NWCBL, NWMSH or NWAβPmAb were administered 1 week after AβP infusion with or without CHI for 3 weeks. All the parameters were examined after 4 weeks of AβP infusion with or without CHI. For details see text.

4.4.2 Concussive head injury exacerbates brain pathology in AβP induced AD

AβP infusion in CHI group further exacerbated the neuronal, glial and myelin damage. This damage is much more pronounced in the uninjured left half as compared to the injured right half.

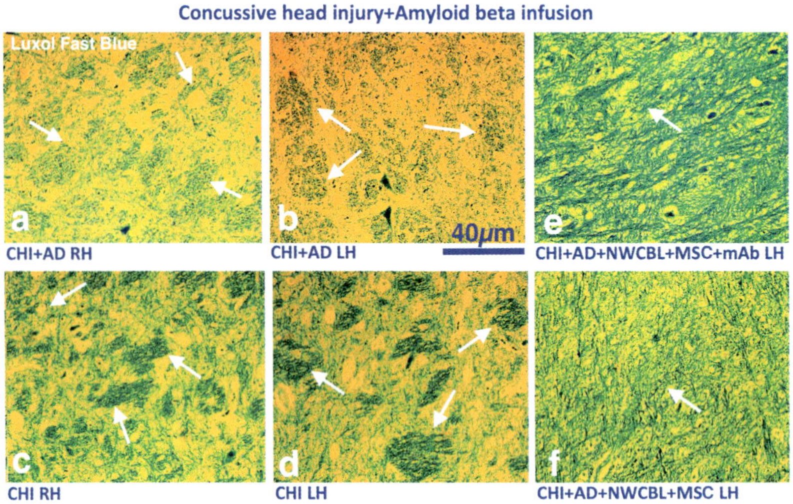

FIG. 11

Representative examples of Luxol fast blue (LFB) staining in the cerebral cortex following concussive head injury (CHI) in Alzheimer's disease (AD) model in the rat. Degeneration of myelin is evident by loss of LFB staining (arrows) in CHI + AD in RH (A) and LH (B). CHI alone also induced myelin degeneration as evidenced with loss of LFB staining (arrows) in the cerebral cortex of RH (C) and LH (D). CHI + AD exacerbates loss of LFB (A and B) that is most prominent in LH as compared to the RH. Treatment with combination of NWCBL with NWMSC and NWmAb has superior effects in preventing loss of LFB in CHI with AD in LH (E) as compared to NWCBL + NWMSC alone in identical condition (F). Bar = 40 μm. CHI, concussive head injury; LH, left hemisphere; NWCBL, TiO$_2$ nanowired cerebrolysin; NWMSC, TiO$_2$ nanowired mesenchymal stem cells; NWmAb, TiO$_2$ nanowired monoclonal antibody against amyloid beta peptide (AβP); RH, right hemisphere. CHI was inflicted on the right parietal bone with an impact of 0.224 N as described earlier (Sharma et al., 2016a,b). AβP infusion (200 ng, i.c.v.) was dome once daily for 4 weeks. CHI was inflicted either 24 h before commencement of AβP infusion or 24 h after the onset of AβP infusion. NWCBL or NWAβPmAb were administered 1 week after AβP infusion with or without CHI for 3 weeks. All the parameters were examined after 4 weeks of AβP infusion with or without CHI. For details see text.

4.4.2.1 Neuronal injury

In CHI alone about 282 ± 12 nerve cells ($P < 0.05$ from control) showed distortion and damage in the left temporal cortex (Fig. 7) whereas the right half showed only 156 ± 12 ($P < 0.05$ from control) injured neuronal cells. The neurons show dark and distorted and some of them are swollen and some are shrunken. Perineuronal edema along with perivascular damage is apparent in many parts of the neuropil (Fig. 15). These changes are present in the cerebral cortex, hippocampus, caudate nucleus, putamen, thalamus and hypothalamus (results not shown).

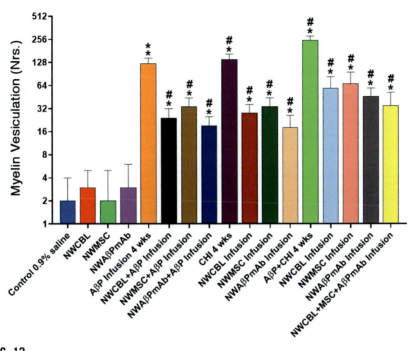

FIG. 12

Left hemisphere myelin vesiculation at transmission electron microscopy (TEM) semiquantitative data in Alzheimer's disease model in rat with concussive head injury (CHI) and treatments. Values are Mean ± SD of 6–8 rats at each point. * $P < 0.05$; ** $P < 0.01$ from control; # $P < 0.05$ from untreated AD or CHI group. ANOVA followed by Dunnett's test for multiple comparisons using one control. AβP, amyloid beta peptide; CHI, concussive head injury; LH, left hemisphere; mAb, monoclonal antibody; NWAβPmAb, TiO$_2$ nanowired monoclonal antibody against amyloid beta peptide; NWCBL, TiO$_2$ nanowired cerebrolysin; NWMSC, TiO$_2$ nanowired melanocyte stimulating hormone; RH, right hemisphere. CHI was inflicted on the right parietal bone with an impact of 0.224 N as described earlier (Sharma et al., 2016a,b). AβP infusion (200 ng, i.c.v.) was dome once daily for 4 weeks. CHI was inflicted either 24 h before commencement of AβP infusion or 24 h after the onset of AβP infusion. NWCBL, NWMSH or NWAβPmAb were administered 1 week after AβP infusion with or without CHI for 3 weeks. All the parameters were examined after 4 weeks of AβP infusion with or without CHI. For details see text.

The magnitude and intensity of these neuronal changes are much more pronounced in the uninjured left half as compared to the right injured half (Sharma HS and Sharma A, unpublished observation).

The neuronal changes in CHI alone are significantly less than AβP infusion alone after 4 weeks of survival (Fig. 7).

However, when AD was produced by AβP infusion in CHI rats the magnitude and intensity of neuronal injuries are synergistically exacerbated (Fig. 15). Thus, about

Concussive head injury+Amyloid beta infusion

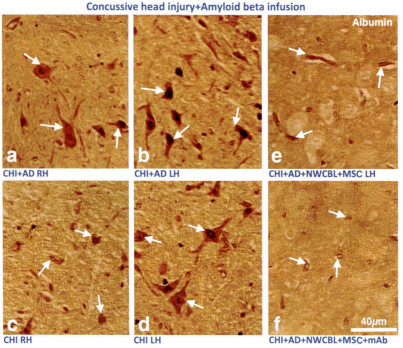

FIG. 13

Representative example of albumin immunoreactivity following Amyloid beta peptide infusion in concussive head injury (CHI) induced in the injured right hemisphere (RH, A) and uninjured left hemisphere (LH, B) as compared to CHI (RH, C) and CHI (LH, D). Combined treatment of NWCBL+NWMSC+NWmAb (F), NWCBL+NWMSC (E) attenuated albumin positive cells in CHI+AD in LH as compared to untreated CHI+AD (A and B). Albumin positive cells are more pronounced in LH as compared to the RH in CHI or CHI+AD. AD+CHI exacerbated albumin positive cells in the neuropil (A and B). AβP, amyloid beta peptide; CHI, concussive head injury; LH, left hemisphere; mAb, monoclonal antibody; NWAβPmAb, TiO$_2$ nanowired monoclonal antibody against amyloid beta peptide; NWCBL, TiO$_2$ nanowired cerebrolysin; NWMSC, TiO$_2$ nanowired melanocyte stimulating hormone; RH, right hemisphere. CHI was inflicted on the right parietal bone with an impact of 0.224 N as described earlier (Sharma et al., 2016a,b). AβP infusion (200 ng, i.c.v.) was dome once daily for 4 weeks. CHI was inflicted either 24 h before commencement of AβP infusion or 24 h after the onset of AβP infusion. NWCBL, NWMSH or NWAβPmAb were administered 1 week after AβP infusion with or without CHI for 3 weeks. All the parameters were examined after 4 weeks of AβP infusion with or without CHI. For details see text.

638 ± 38 neurons showed cell injuries in CHI with AD combination (Fig. 7). These damaged neurons are present in several brain areas including cerebral cortex, cerebellum, hippocampus, caudate nucleus, putamen, thalamus, hypothalamus and brain stem. The left uninjured half showed extensive neuronal damages as compared to the

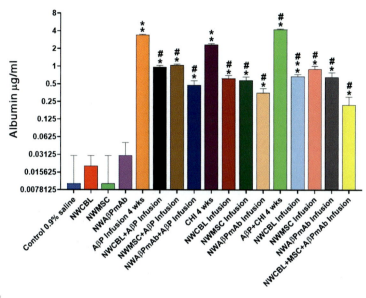

FIG. 14

Cerebrospinal fluid (CSF) Albumin measurement in Alzheimer's disease model in rats with concussive head injury (CHI) and treatments. Values are Mean ± SD of 6–8 rats at each point. * $P<0.05$; ** $P<0.01$ from control; # $P<0.05$ from untreated AD or CHI group. ANOVA followed by Dunnett's test for multiple comparisons using one control. AβP, amyloid beta peptide; CHI, concussive head injury; LH, left hemisphere; mAb, monoclonal antibody; NWAβPmAb, TiO_2 nanowired monoclonal antibody against amyloid beta peptide; NWCBL, TiO_2 nanowired cerebrolysin; NWMSC, TiO_2 nanowired melanocyte stimulating hormone; RH, right hemisphere. CHI was inflicted on the right parietal bone with an impact of 0.224 N as described earlier (Sharma et al., 2016a,b). AβP infusion (200 ng, i.c.v.) was dome once daily for 4 weeks. CHI was inflicted either 24 h before commencement of AβP infusion or 24 h after the onset of AβP infusion. NWCBL, NWMSH or NWAβPmAb were administered 1 week after AβP infusion with or without CHI for 3 weeks. All the parameters were examined after 4 weeks of AβP infusion with or without CHI. For details see text.

right injured half in CHI in combination with AβP infusion (results not shown). These observations suggest that AD induced brain damage is exacerbated by prior CHI.

4.4.2.2 Glial changes

Concussive head injury alone induced significant induced gliosis in the brain that was the most prominent in the left uninjured half as compared to the right injured half (Fig. 9). Distribution of GFAP positive astrocytes is seen in the areas of the BBB breakdown and around perineuronal and perivascular regions (Fig. 9). Semi-quantitative data show that about 210 ± 18 GFAP positive astrocytes are seen in the temporal cortex of the left half (Fig. 10). The right half showed 154 ± 12 GFAP positive

CHI exacerbates Brain Pathology in AD

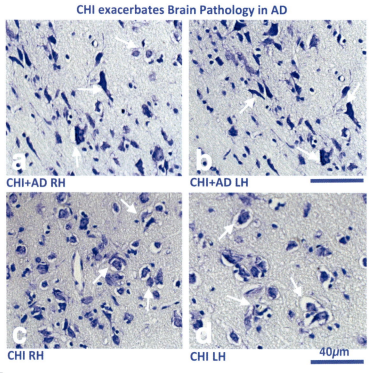

FIG. 15

Representative example of Neuronal cell injuries following Amyloid beta peptide infusion with concussive head Injury (CHI) in the cerebral cortex in Alzheimer's disease model in the rat in the right hemisphere (RH, injured) and in left hemisphere (LH). For details see text. Many nerve cells were shrunken (arrows) with perineuronal edema is quite prominent in both hemispheres RH and LH in CHI (C and D). These nerve cell injuries appears to be exacerbated by CHI+AD group (A and B). LH exhibited greater neuronal injuries after CHI and in AD+CHI cases. Toluidine stain on Paraffin section 3-μm thick; Bar=40 μm.

cells (Fig. 10). GFAP positive cells are also present in the hippocampus, cerebellum, thalamus, hypothalamus and the brainstem. However, the magnitude and intensity of GFAP immunoreactivity is much higher in the left half as compared to the right half (Fig. 9).

When AβP infusion was given after CHI there was a synergistic higher GFAP activation in the brain after 4 weeks of survival (Figs. 9 and 10). Thus, the number of GFAP positive astrocytes rose to about two times from CHI alone (Fig. 10). About 628 ± 38 GFAP positive astrocytes are seen in the left temporal cortex after 4 weeks of AβP infusion in CHI combination. This indicates that prior CHI exacerbates AD induces astrogliosis.

4.4.2.3 Myelin changes

CHI alone induces profound reduction in LFB staining after 4 weeks indicating myelin degeneration (Fig. 11). LFB reduction is seen following CHI in several brain regions indicating widespread myelin damage in CHI brain. However, Left half exhibited greater myelin degeneration as compared to the right injured half as evident with decrease in LFB staining (Fig. 11).

When AβP infusion was done in CHI rats this degeneration in myelin was further aggravated (Fig. 12). This indicates that AD caused by AβP infusion in prior CHI results in greater degeneration of myelinated axons. These LFB changes are also supported by immunohistochemistry of MBP in identical brain areas (Sharma HS and Sharma A, unpublished observation). The intensity of MBP reduction is far greater in the left half following AβP infusion in CHI rats after 4 weeks as compared to the right injured half (Results not shown). Taken together these observations show that AβP infusion in CHI exacerbates brain pathology and axonal damages.

4.5 Ultrastructural brain pathology in AD induced by AβP infusion

Brain pathology was examined at the ultrastructural level in AβP induced AD of neuropil using transmission electron microscope (TEM). Focus was given on myelin vesiculation, edema formation and neuronal injuries (Fig. 16). In the cerebral cortex we found several shrunken neurons with dark cytoplasm, irregular nuclear membrane and electron dense karyoplasm (results not shown). Several myelinated axons show myelin vesiculation (Fig. 16), membrane vacuolation, and synaptic damages after 4 weeks of AβP infusion (Fig. 12). About 140 ± 24 myelinated axons exhibited myelin vesiculation, distortion in shape of axons and damage to neuropil as compared to saline infusion group of controls (2 ± 2, $P < 0.01$). Intracellular swelling, membrane disruption exhibiting edema formation and electron dense cytoplasm of neurons with distorted shape are the common findings in AD at TEM (Sharma HS and Sharma A, unpublished observations). These changes at the ultrastructural level are frequent in the cerebellum, hippocampus and thalamus that were examined by TEM from AD group (Results not shown).

4.5.1 Concussive head injury exacerbates ultrastructural brain pathology in AD induced by AβP infusion

Concussive head injury alone resulted in ultrastructural changes in brain pathology in neurons, astrocytes, microglia, endothelial cells as well as myelinated axons, vesiculation of myelin and synaptic damages (Sharma HS and Sharma A, unpublished observation). The magnitude and intensity of these changes were much more higher in the left uninjured half as compared to the injured right half after 4 weeks of CHI. Thus the incidences of myelin vesiculation in the left half of temporal cortex showed 140 ± 24 cases after CHI whereas only 88 ± 12 ($P < 0.01$ from control) incidences of myelin vesiculation was seen in the right injured half in the cerebral cortex (Fig. 7).

Concussive head injury+Amyloid beta infusion

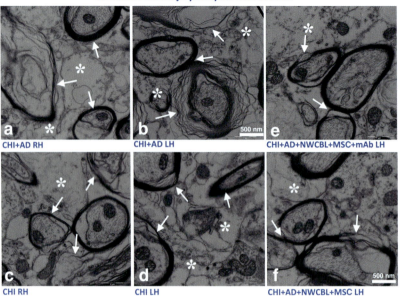

FIG. 16

Representative examples Transmission electron microscopy (TEM) exhibiting myelin loss and vesiculation following concussive head injury (CHI) in Alzheimer's disease (AD) model in the rat in the injured right hemisphere (RH, A) and left hemisphere (LH, B) and CHI alone in RH (C) and LH (D). Degeneration of myelin (arrows) and vacuolation (*) is prominent in CHI+AD and in CHI alone cases. Combined treatment with NWCBL with NWMSC and NWmAb induces slightly better preservation of myelin in CHI+AD cases in LH (E) as compared to the treatment with NWCBL+NWMSC in LH under identical conditions (F). Bar = 500 nm. CHI, concussive head injury; LH, left hemisphere; NWCBL, TiO_2 nanowired cerebrolysin; NWMSC, TiO_2 nanowired mesenchymal stem cells; NWmAb, TiO_2 nanowired monoclonal antibody against amyloid beta peptide (AβP); RH, right hemisphere. CHI was inflicted on the right parietal bone with an impact of 0.224 N as described earlier (Sharma et al., 2016a,b). AβP infusion (200 ng, i.c.v.) was dome once daily for 4 weeks. CHI was inflicted either 24 h before commencement of AβP infusion or 24 h after the onset of AβP infusion. NWCBL or NWAβPmAb were administered 1 week after AβP infusion with or without CHI for 3 weeks. All the parameters were examined after 4 weeks of AβP infusion with or without CHI. For details see text.

When AβP infusion was given before CHI these ultrastructural brain pathology was synergistically exacerbated after 4 weeks of survival. About 250 ± 32 incidences of myelin vesiculation was noted in in AβP infusion in CHI in the left half as compared to 124 ± 18 cases of demyelination at TEM studies (Fig. 7). Membrane disruption, vacuolation and edema associated with distortion of neurons, endothelial cells were significantly higher in numbers in AβP infusion in CHI group as compared to either AβP infusion or CHO alone after 4 weeks of survival (Fig. 7) (other results not shown).

These ultrastructural studies further support the light microscopic events and suggest that AβP infusion in CHI aggravates this brain pathology at TEM.

4.6 Biochemical changes in AβP infusion induced AD model

In order to understand the pathological mechanisms of AβP infusion biochemical measurement cisterna magna samples of CSF of agents involved in AD mechanisms are examined using standard protocol.

4.6.1 Amyloid beta peptide

Measurement of AβP in cisterna magna samples of CSF showed significant elevation following 4 weeks of AβP infusion as compared to control group. Thus, in AD model 104 ± 8 pg/mL AβP was found after 4 weeks of 200 ng AβP infusion in the third lateral cerebral ventricle daily for 4 weeks as compared to saline infusion control (3 ± 2 pg/mL, $P < 0.01$) (Fig. 2).

4.6.2 Plasminogen

Measurement of plasminogen in cisterna magna CSF revealed significant decrease after 4 weeks of AβP infusion as compared to saline control. Thus, the control group CSF from cisterna magna showed 19 ± 4 μg/mL plasminogen level that is significantly decreased in the CSF to 5 ± 3 μg/mL after 4 weeks of AβP infusion as compared to saline control (Fig. 17). Thus, the control group CSF from cisterna magna showed 19 ± 4 μg/mL plasminogen level that is significantly decreased in the CSF to 5 ± 3 μg/mL after 4 weeks of AβP infusion ($P < 0.01$). This suggests that decrease of plasminogen in CSF is crucial for AD brain pathology.

4.6.3 Albumin

Measurement of albumin was dome in the CSF obtained from Cisterna magna in AβP infusion induced AD model. We found significant increase in CSF albumin of 3.42 ± 0.09 μg/mL after 4 weeks of AβP infusion as compared to saline infusion control that showed albumin level of 0.01 ± 0.02 μg/mL (Fig. 14). An increase in CSF albumin from the negligible control group suggests breakdown of the BBB and BCSFB in AD model that is one of the key factors in the development of brain pathology.

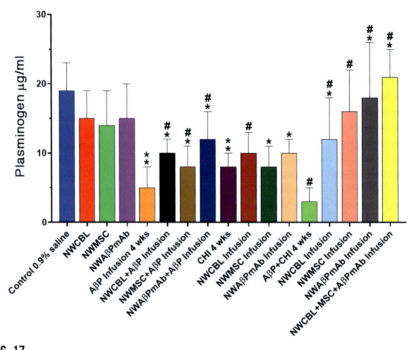

FIG. 17

Cerebrospinal fluid (CSF) Plasminogen measurement in Alzheimer's disease model in rat with concussive head injury (CHI) and treatments. Values are Mean ± SD of 6–8 rats at each point. * $P < 0.05$; ** $P < 0.01$ from control; # $P < 0.05$ from untreated AD or CHI group. ANOVA followed by Dunnett's test for multiple comparisons using one control. AβP, amyloid beta peptide; CHI, concussive head injury; LH, left hemisphere; mAb, monoclonal antibody; NWAβPmAb, TiO_2 nanowired monoclonal antibody against amyloid beta peptide; RH, right hemisphere; NWCBL, TiO_2 nanowired cerebrolysin; NWMSC, TiO_2 nanowired melanocyte stimulating hormone. CHI was inflicted on the right parietal bone with an impact of 0.224 N as described earlier (Sharma et al., 2016a,b). AβP infusion (200 ng, i.c.v.) was dome once daily for 4 weeks. CHI was inflicted either 24 h before commencement of AβP infusion or 24 h after the onset of AβP infusion. NWCBL, NWMSH or NWAβPmAb were administered 1 week after AβP infusion with or without CHI for 3 weeks. All the parameters were examined after 4 weeks of AβP infusion with or without CHI. For details see text.

4.7 Concussive head injury exacerbated biochemical changes in CSF of AβP infusion induced AD

Concussive head injury alone resulted in increase of several biochemical factors in the CSF capable to induce brain pathology.

4.7.1 Amyloid beta peptide

CHI alone increased AβP level in the cisterna magna CSF after 4 weeks to 22 ± 4 pg/mL ($P < 0.01$) from the control level (3 ± 2 pg/mL). This suggests that brain injury itself is capable to enhance AβP levels causing brain pathology. However, when

AβP infusion was induced in CHI groups the increase in AβP was synergistically enhanced in the cisterna magna CSF after 4 weeks to the level of 180 ± 21 pg/mL ($P < 0.01$) (Fig. 2). This increase was significantly higher than AβP infusion or CHI alone. This suggests that CHI aggravates AD induced AβP levels in the brain.

4.7.2 Plasminogen

There are reasons to believe that plasminogen and its product plasmin are also affected by brain injury apart from AD (Baker et al., 2018). We observed a significant decline in plasminogen in the cisterna magna CSF to 8 ± 2 μg/mL ($P < 0.01$) after 4 weeks of primary injury as compared to control group (19 ± 4 μg/mL) (Fig. 17). This observation clearly indicates that brain injury induces reduction in plasminogen.

Interestingly, when AβP infusion is made in CHI animals this reduction was synergistically exacerbated after 4 weeks of survival (Fig. 17). As a result the combination of CHI and AβP infusion resulted in severe reduction of plasminogen in the CSF to the value of 3 ± 2 μg/mL ($P < 0.01$) as compared to the control group after 4 weeks of survival. This suggests that CHI exacerbates plasminogen decline in AD induced by AβP infusion.

4.7.3 Albumin

Measurement of albumin in the cisterna magna SF following CHI alone after 4 weeks showed significant increase to the value of 2.34 ± 0.10 μg/mL as compared to the negligible amount of albumin in the control CSF (0.01 ± 0.02 μg/mL, $P < 0.01$) (Fig. 14). This indicates breakdown of the BCSFB as well in CHI.

On the other hand when AβP infusion was made in CHI rats this increase in the CSF albumin was synergistically enhanced up to a level of 4.23 ± 0.12 μg/mL ($P < 0.01$) as compared to the control values (Fig. 14). Increased albumin level into CSF is far higher in CHI with AβP infusion as compared to the CHI or AβP infusion alone after 4 week of survival. This indicates that CHI aggravates AD pathology induced by AβP infusion model.

5 Neuroprotective strategies in AD induced by AβP infusion

The pathophysiology in AD model induced by infusion of AβP and their exacerbation by CHI requires modification of neuroprotective strategies to counteract the synergistically enhanced pathological responses (Sharma et al., 2012, 2016a,b, 2017, 2018, 2019a,b,c). Thus, we have combined the potential neuroprotective effects of several powerful agents to achieve neuroprotection in AD model with CHI as co-morbidity factor (Sharma et al., 2017, 2018, 2019a,b,c). As a result we used three key neuroprotective agents to achieve this target by using cerebrolysin (Sharma et al., 2012, 2016a,b), mesenchymal stem cells (MSCs) (Sharma et al., 2018) and monoclonal antibodies to AβP (AβPmAb) (Sharma et al., 2017). To enhance their neuroprotective effects all the agents were administered using TiO_2 nanowired delivery in AD with CHI (Sharma et al., 2018, 2019a,b,c). A brief decryption of the results is summarized below.

5.1 Nanowired cerebrolysin with MSCs and AβPmAb reduces blood-brain barrier breakdown

Treatment with TiO_2 nanowired cerebrolysin (NWCBL), nanowired MSCs (NWMSC) or nanowired monoclonal AβP antibodies (NWAβPmAb) in control animals did not influence BBB to EBA or radioiodine in any brain region as compared to the control group (Fig. 4).

5.1.1 AβP infusion group

In AβP infusion for 4 weeks, treatment with NWCBL was very effective in reducing BBB breakdown to EBA and radioiodine (Fig. 4). Thus, the BBB breakdown to EBA in NWCBL treated group was 0.45 ± 0.05 mg % for right half and 0.48 ± 0.08 mg % for left half ($P < 0.05$) from untreated AβP infused group (Fig. 4). The radioiodine values were also considerably reduced to $0.89 \pm 0.06\%$ for right half and $0.92 \pm 0.07\%$ for left half ($P < 0.05$). Treatment with NWMSC in AD group resulted in slight but significant reduction ($P < 0.05$) in the BBB breakdown to EBA in right half (0.94 ± 0.010 mg %) and in left half (0.94 ± 0.10 mg %) from the AD group. While radioiodine also showed significant reduction ($P < 0.05$) by NWMSC in Ad group right half ($1.08 \pm 0.05\%$) and in left half ($1.12 \pm 0.09\%$) as compared to the untreated AD group (Fig. 4). On the other hand treatment with NWAβPmAb exhibited significantly moderate effects ($P < 0.05$) in reducing BBB breakdown to EBA in right half (0.84 ± 0.07 mg %) and in Left half (0.80 ± 0.09 mg %) as compared to untreated AD group. The reductions in radioiodine leakage by NWAβPmAb were also considerably moderate significant reduction ($P < 0.05$) in the right half ($1.14 \pm 0.06\%$) and in left half ($1.12 \pm 0.07\%$) from the untreated AD group (Fig. 4). This suggests that NWCBL has the most prominent effects in reducing BBB breakdown in AD.

5.1.2 CHI group

In CHI group of 4 weeks NWCBL was the most effective treatment in reducing BBB breakdown to EBA and radioiodine followed by NWAβPmAb and NWMSC (Fig. 3). Thus, in NWCBL treated CHI group significant reduction in the EBA leakage ($P < 0.05$) was seen in the right half (0.67 ± 0.08 mg %) and in left half (0.75 ± 0.06 mg %) as compared to the untreated CHI rats. Likewise the radioiodine extravasation in NWCBL treated CHI group was reduced in right half to $0.78 \pm 0.06\%$ and in the left half to $0.94 \pm 0.07\%$ significantly ($P < 0.05$) from the untreated CHI group (Fig. 4).

Treatment with NWMSC in CHI group was also slightly but significantly ($P < 0.05$) reduced EBA leakage in the right half (0.89 ± 0.12 mg %) and in left half (1.14 ± 0.13 mg %). Likewise, radioiodine extravasation in CHI group treated with NWMSC was also mildly but significantly ($P < 0.05$) thwarted in the right half ($1.09 \pm 0.14\%$) and in the left half ($1.45 \pm 0.18\%$) as compared to the untreated CHI rats (Fig. 4).

Interestingly, NWAβPmAb treatment in CHI resulted in significant moderate reduction ($P < 0.05$) in the BBB to EBA in right half (0.74 ± 0.10 mg %) and in the left half (0.98 ± 0.12 mg %) from the untreated CHI rats. Similarly, radioiodine extravasation in NWAβPmAb treated group thwarted radioiodine leakage in the right half ($0.98 \pm 0.14\%$) and in the left half ($1.12 \pm 0.14\%$) in untreated CHI group (Fig. 4).

5.1.3 AβP infusion in CHI group

In this group the most effective treatment strategies was the combination of all the agents together in AD with CHI in reducing the BBB breakdown of EBA and radioiodine after 4 weeks of survival (Fig. 4). Thus, the combination of NWCBL with NWMSC and NWAβPmAb resulted in the most efficient reduction in the BBB to EBA in the right half (0.56 ± 0.12 mg %) and in the left half (0.89 ± 0.14 mg %) compared to the untreated AD in combination with CHI ($P < 0.05$). Likewise radioiodine extravasation was also most effectively reduced by triple combination of NWCBL with NWMSC and NWAβPmAb in AD with CHI in the right half ($1.05 \pm 0.12\%$) and in the left half ($1.34 \pm 0.16\%$) as compared to the untreated AD with CHI rats ($P < 0.05$). NWCBL treatment alone was also reduced the BBB leakage to EBA and radioiodine in AD with CHI with significantly higher degree as compared to the NWMSC and NWAβPmAb treatment (Fig. 4). Thus, NWCBL treatment resulted in reduction of EBA in right half to 1.04 ± 0.10 mg % and in the kef half to 1.54 ± 0.21 mg % as compared to the untreated AD with CHI ($P < 0.05$) (Fig. 4). The radioiodine extravasation was reduced to $1.05 \pm 0.12\%$ in the right half and $1.34 \pm 0.16\%$ in the left half in AD with CHI group treated with NWCBL in combination with NWMSC and NWAβPmAb treatment as compared to the untreated AD in CHI (Fig. 4). These observations suggest that the exacerbation of BBB breakdown by CHI in AD require combination of nanowired MSCs and AβPmAb together with cerebrolysin to enhance superior neuroprotective effects.

5.2 Nanowired cerebrolysin with MSCs and AβPmAb reduces brain edema formation

Treatment with NWCBL, NWMSC or NWAβPmAb in control group did not induce any alterations in the brain water content or volume swelling (Fig. 5). However when these agents were given alone or in combination they exerted powerful anti edema effects either in AβP infusion or CHI alone and in combination (Fig. 5).

However, NWCBL appears to be the most potent in anti edema effects in AD or in CHI either alone or in combination.

Treatment with NWCBL, NWMSC or NWAβPmAb either alone or in combination significantly reduced brain edema formation and volume swelling in AD or CHI either alone or in combination (Fig. 5). NWCBL appears to me most effective in reducing brain edema and volume swelling in all groups of AD, CHI and in combination. A brief description is given below.

5.2.1 Amyloid beta group

Infusion of NWCBL, NWMSC or NWAβPmAb in normal animals did not influence brain water content or volume swelling (Fig. 5). However, when these agents were infused alone following AβP infusion for 4 weeks significant reduction in edema formation and volume swelling was seen (Fig. 5). NWCBL appears to be the most potent in reducing brain edema formation and volume swelling following AβP infusion.

Thus, NWCBL treatment results in significant higher reduction in brain water in the left half to $74.65 \pm 0.12\%$ and in the right half $75.21 \pm 0.20\%$ compared to the untreated AβP infusion model after 4 weeks ($P < 0.01$). This indicates a reduction in volume swelling by 2% from the untreated group in the right half and in 3% in left half from the untreated AD model by AβP infusion (Fig. 5).

Treatment with NWMSC in AβP infusion group resulted in mild but significant reduction in brain edema in right half ($74.85 \pm 0.12\%$) and in left half ($74.89 \pm 0.12\%$) from the untreated group of AβP infusion for 4 weeks (Fig. 5). This amount to a reduction on 2% volumes swelling the tight half and about 1% in left half in the treated group (Fig. 6).

When NWAβPmAb was given to AβP infusion induced AD group reduction of brain edema was also mild but significant in the right half ($74.74 \pm 0.06\%$) and in left half ($74.60 \pm 0.09\%$) amounting to 3% reduction in volume swelling in the right half and 2% in left half as compared to the untreated AD group (Fig. 6).

5.2.2 CHI group

In CHI group NWCBL is the most effective in reducing brain edema and volume swelling after 4 weeks of survival followed by NWAβPmAb and NWMSC (Fig. 6).

Thus, when NWCBL treatment was done in CHI group significant higher reduction in the brain water content was observed in the right half ($75.07 \pm 0.17\%$) and in left half ($74.75 \pm 0.18\%$) after 4 weeks of survival as compared to the untreated CHI group ($P < 0.05$) (Fig. 5). This amounts to 5% reduction in the volume swelling in right half and 7% reduction in left half from the untreated injured group (Fig. 6).

NWMSC treatment resulted in a reduction in the water content in right half ($75.21 \pm 0.16\%$) and in the left half ($76.10 \pm 0.32\%$) after 4 weeks of CHI as compared to the untreated CHI group. This amounts to about 4% reduction in volume swelling in the right half and about 6% reduction on the left half (Fig. 5).

Treatment with NWAβPmAb in 4 weeks CHI resulted in moderate reduction in the brain water content in the right half ($75.18 \pm 0.16\%$) and in the left half ($75.76 \pm 0.14\%$) significantly ($P < 0.05$) as compared to the untreated CHI rats. This amount to a reduction in the volume swelling of 5% in the right half and 7% in the left half (Figs. 5 and 6).

This suggests that NWCBL and NWAβPmAb are both equally effective in reducing brain edema formation in CHI after 4 weeks of survival.

5.2.3 AβP infusion and CHI group

In combined cases of CHI with AβP infusion brain edema formation was exacerbated as compared to the AD or CHI alone (Fig. 5). This in this situation a combination of all the three agents was administered to achieve superior neuroprotection in AD with comorbidity of CHI factor.

When NWAβPmAb was administered in CHI with AβP infusion significantly higher reduction in the brain edema formation and volume swelling was observed after 4 weeks of survival (Fig. 6). Thus, the right half in NWAβPmAb treated group shows $74.89 \pm 0.21\%$ brain water whereas the left half exhibited $75.01 \pm 0.32\%$ brain water as compared to the untreated right half ($77.67 \pm 0.32\%$) and in left half ($79.480.28\%$) ($P < 0.05$). This amounts to about 11% reduction in volume swelling in the right half and 17% reduction in volume swelling in the left half (Figs. 5 and 6).

Treatment with NWCBL alone resulted in significant reduction in the brain water in right half ($75.45 \pm 0.14\%$) and in the left half ($76.08 \pm 0.32\%$) in AD with CHI group (Fig. 5). This amount to about 9% reduction in the volume swelling in right half and about 13% reduction in the left half (Fig. 6).

NWMSC infusion in AD with CHI group resulted in reduction in brain water in right half to $76.12 \pm 0.32\%$ and in left half to $77.06 \pm 0.21\%$ ($P < 0.05$) that amounts to about 5% reduction in volume swelling in the right half and about 8% reduction in in left half (Figs. 5 and 6).

Treatment with NWAβPmAb also resulted in significant reduction in brain edema in AD with CHI after 4 weeks of survival. Thus, the right half of brain water was reduced to $75.87 \pm 0.23\%$ and in the left half the values was $76.67 \pm 0.31\%$ in AD with CHI ($P < 0.05$) as compared to the untreated AD in CHI group (Fig. 5). Thus amounts to a reduction in 7% volume swelling in the right half and about 11% reduction in the left half (Fig. 6).

These observations suggests that in combination with AD and CHI a combination of NWCBL with NWMSC and NWAβPmAb induces superior anti edema effects.

5.3 Nanowired cerebrolysin with MSCs and AβPmAb reduces brain pathology

Treatment with NWCBL, NWMSC and NWAβPmAb in AβP infusion induced AD and CHI either alone or in combination markedly reduced brain pathology after 4 weeks of survival. In case of AD with CHI the combined treatment with all the agents induced superior neuroprotection on brain pathology. On the other hand NWCBL appears to be one of key neuroprotective agents in reducing brain pathology in AD and CHI. A brief description is given below.

5.3.1 Amyloid beta group

Normal animals received NWCBL, NWMSC or NWAβPmAb did not affect brain pathology in any region (Fig. 7). However, when NWCBL is administered in AβP infusion group, significant neuroprotection is evident after 4 weeks of survival. As a result only 58 ± 12 neurons were found mild to moderately distorted in AD group (Fig. 7). The number of GFAP positive astrocytes was significantly reduced in this group to 58 ± 12 cells ($P < 005$ from untreated AD group) (Figs. 9 and 10). Myelin vesiculation seen at the ultrastructural level in NWCBL treated AD group also showed marked decrease (24 ± 8 as compared to untreated 124 ± 22, $P < 0.05$) (Figs. 11 and 12). This effect of myelin preservation is also seen using LFB staining and MBP immunostaining in the AD brain treated with NWCBL (Fig. 11, results not shown). Albumin immunoreactivity indicating breakdown of the BBB to endogenous serum protein is significantly thwarted in AD by NWCBL treatment (Figs. 13 and 14). Also, the AβP immunoreactivity in the brain was considerable reduced by NWCBL treatment in AD model (Fig. 18).

Similarly, treatment with NWMSC and NWAβPmAb treatment reduced the number of neuroma injury in AD brain by 64 ± 19 and 28 ± 8, respectively by NWMSC and NWAβPmAb (Fig. 7). The number of GFAP positive astrocytes is also reduced to 76 ± 21 cells by NWMSC and 35 ± 9 cells by NWAβPmAb treatment in AD brains (Figs. 9 and 10). Reduction in albumin, AβP and MBP immunoreactivity was also seen in AD after NWMSC or NWAβPmAb treatment (results not shown). LFB staining recovered markedly in AD brain after treatment with NWMSC or NWAβPmAb therapy (results not shown).

5.3.2 CHI group

In CHI, treatment with NWCBL markedly attenuated brain pathology after 4 weeks of survival. Thus, the number of distorted neurons was significant down 22 ± 6 cells as compared to the untreated CHI (282 ± 12 cells, $P < 0.05$). The GFAP positive astrocytes were also reduced significantly to 32 ± 12 cells in treated group as compared to the untreated CHI (210 ± 18, $P < 0.05$) (Figs. 9 and 10). The number of axons showing myelin vesiculation was also reduced to 28 ± 8 in treated group as compared to 140 ± 24 in untreated CHI (Figs. 12 and 16). These ultrastructural observations on myelin vesiculation was also confirmed using LFB staining (Fig. 11) and MBP immunoreactivity (Results not shown). Immunoreactivity to albumin in the neuropil exhibiting BBB breakdown was also markedly reduced in NWCBL treated CHI (Fig. 13) as compared to the untreated injured group (results not shown). Interestingly, NWCBL treatment also reduced the AβP immunoreactivity in the brain (Fig. 18) as compared to the untreated CHI (Sharma HS and Sharma A, unpublished observation).

Treatment with NWMSC was also able to thwart brain pathology in CHI. Thus, the number of neurons exhibiting distortion was limited to 28 ± 8 cells and GFAP positive cells were also reduced to 38 ± 12 cells in treated groups as compared to the untreated CHI after 4 weeks of survival (Fig. 7). The number of axons showing myelin vesiculation in NWMSC was 34 ± 10 as seen at TEM investigation (Fig. 7) and this observation was also supported at light microscopy using LFB staining and AβP immunoreactivity (Fig. 12; Results not shown). The AβP immunoreactivity was also considerable reduced (Figs. 13 and 14) by NWMSC treatment in CHI (Sharma HS and Sharma A, unpublished observations).

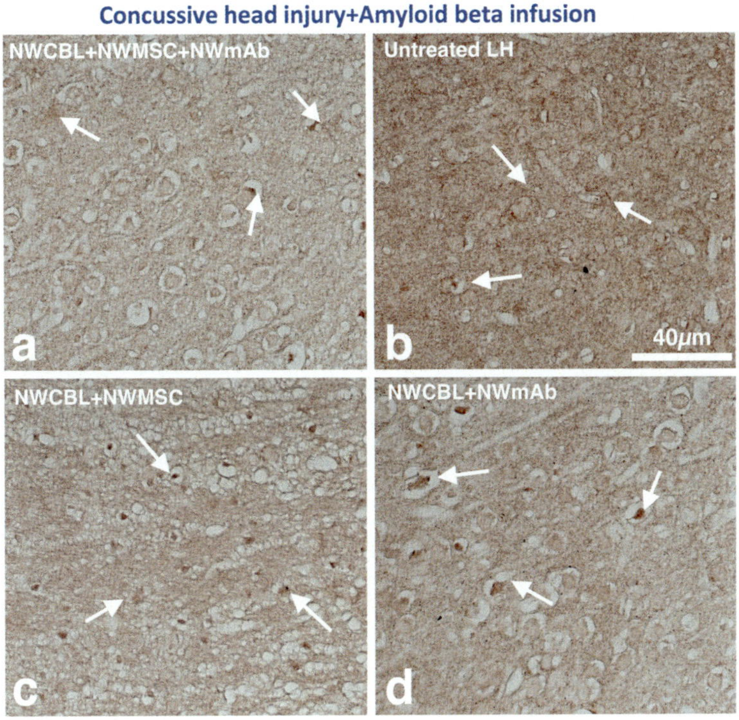

FIG. 18

Representative example of Amyloid beta peptide infusion in concussive head injury (CHI) induced AβP immunoreactivity in untreated (B) its modification with combined treatment of NWCBL + NWMSC + NWmAb (A), NWCBL + NWMSC (C) and NWCBL + NWmAb (D). It appears that treatment with NWCBL + NWMSC + NWmAb has the superior effects in reducing AβP immunoreactivity in AD with CHI as compared to the other treatments. AβP, amyloid beta peptide; CHI, concussive head injury; LH, left hemisphere; mAb, monoclonal antibody; NWAβPmAb, TiO$_2$ nanowired monoclonal antibody against amyloid beta peptide; NWCBL, TiO$_2$ nanowired cerebrolysin; NWMSC, TiO$_2$ nanowired melanocyte stimulating hormone; RH, right hemisphere. CHI was inflicted on the right parietal bone with an impact of 0.224 N as described earlier (Sharma et al., 2016a,b). AβP infusion (200 ng, i.c.v.) was dome once daily for 4 weeks. CHI was inflicted either 24 h before commencement of AβP infusion or 24 h after the onset of AβP infusion. NWCBL, NWMSH or NWAβPmAb were administered 1 week after AβP infusion with or without CHI for 3 weeks. All the parameters were examined after 4 weeks of AβP infusion with or without CHI. For details see text.

Interestingly, when NWAβPmAb was administered in CHI, the brain pathology is considerable reduced as compared to untreated CHI after 4 weeks of survival. Thus, only 18 ± 6 neurons show distortion together with 22 ± 6 GFAP positive cells in this treatment as compared to the untreated CHI ($P < 0.05$) (Fig. 7). The number of axons showing myelin vesiculation were rescued to 18 ± 8 as compared to the untreated CHI (Fig. 13; $P < 0.05$) that is also evident with LFB staining and MBP immunoreactivity (results not shown). The AβP and albumin immunoreactivity was also greatly reduced in CHI with NWAβPmAb therapy (Fig. 14) as compared to the untreated CHI after 4 weeks of survival (Sharma HS and Sharma A, unpublished observation).

5.3.3 AβP infusion and CHI group

Exacerbation of brain pathology in AD in combination with CHI was significantly reduced by the combination of NWCBL with NWMSC and NWAβPmAb treatment (Fig. 7). Thus this triple agents treatment resulted in significant reduction on the number of damaged or distorted neurons (24 ± 10 calls) as compared to the untreated CHI with AD ($P < 0.05$). The number of GFAP positive astrocytes was also declined to 38 ± 12 ($P < 005$) as compared to the untreated CHI after 4 weeks of survival. Only 35 ± 5 axons exhibited myelin vesiculation in combined treatment group of CHI with AD as compared to the untreated group ($P < 0.05$). In this treatment group the upregulation of AβP immunoreactivity was also thwarted (Fig. 2). At light microcopy preservation of LFB staining (Fig. 13) and MBP immunoreactivity (results not shown) was clearly seen with this combined treatment in AD with CHI. The albumin immunoreactivity was also markedly reduced in AD with CHI combination by the triple agents treatment (Fig. 14).

On the other hand, individual treatment with NWCBL, NWMSC or NWAβPmAb given alone was also able to reduce brain pathology significantly (Figs. 15, 19 and 20). Thus NWCBL educed the neuronal damage to 38 ± 12 cells with GFAP positive cells down to 64 ± 23 cells in AD with CHI as compared to the untreated group ($P < 0.05$). The number of axons showing myelin vesiculation were 59 ± 24 in NWCBL treated AD with CHI (Fig. 7) that is also supported by preservation of LFB staining (Fig. 13) and MBP immunoreactivity (results not shown). The immunoreactivity to Albumin and AβP was also considerably reduced by NWCBL in AD with CHI group (Fig. 14).

Likewise NWMSC was able to reduce the neuronal injury limited to 48 ± 12 cells and GFAP positive astrocytes to 72 ± 24 cells in AD with CHI (Figs. 9 and 10) while myelin vesiculation was seen in only 67 ± 28 cases (Fig. 12) as compared to the untreated CHI with AD. LFB staining and MBP immunoreactivity appears to be well preserved after 4 weeks of combined injury of AD and CHI and the albumin and AβP immunoreactivity was considerably reduced as compared to the untreated CHI with AD group (Results not shown).

NWAβPmAb treatment in AD with CHI was successful in reducing brain pathology as evident with 32 ± 12 distorted neurons and 46 ± 18 GFAP positive astrocytes (Fig. 7). The myelin vesiculation of the axons was also reduced to 46 ± 13 cases as seen at TEM and is supported with LFB staining (Fig. 16) and MBP immunoreactivity (Results not shown). The AβP and albumin immunoreactivities were also markedly reduced in AD with CH after 24 h by NWAβPmAb treatment as compared to the untreated AD with CHI after 4 weeks of survival (Sharma HS and Sharma A, unpublished observation).

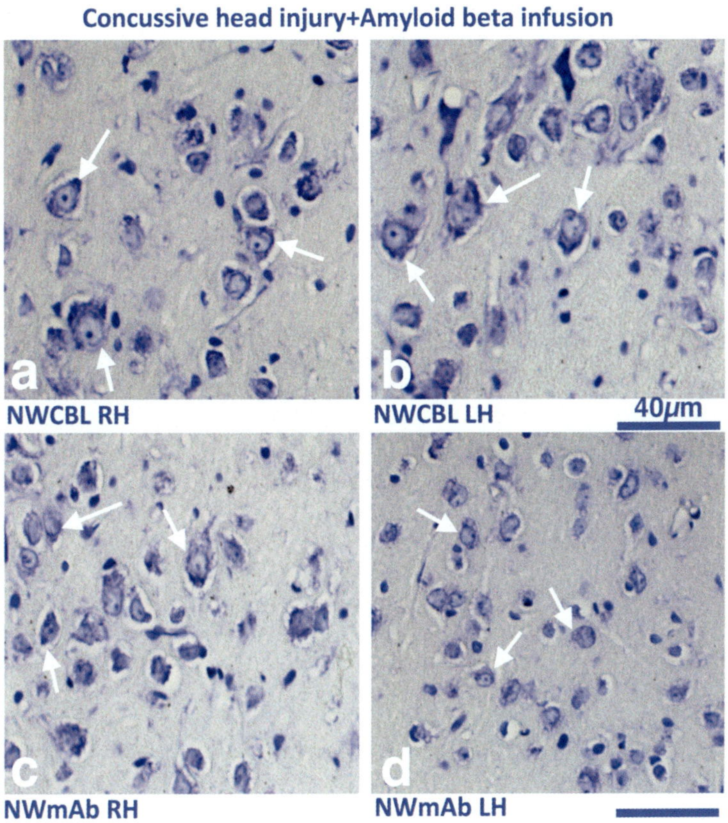

FIG. 19

Representative example of Toluidine stained nerve cell morphology on paraffin sections (3-μm thick) following amyloid beta peptide infusion in concussive head injury (CHI) in right hemisphere (RH, A) and left hemisphere (LH, B) after treatment with NWCBL (A and B) and NWmAb (RH, C) and LH (D). Treatment with NWCBL and NWmAb alone is able to rescue majority of nerve cells from damage after CHI with AD (arrows). Bar = 40 μm. CHI, concussive head injury; LH, left hemisphere; NWCBL, TiO$_2$ nanowired cerebrolysin; NWmAb, TiO$_2$ nanowired monoclonal antibody against amyloid beta peptide (AβP); RH, right hemisphere. CHI was inflicted on the right parietal bone with an impact of 0.224 N as described earlier (Sharma et al., 2016a,b). AβP infusion (200 ng, i.c.v.) was dome once daily for 4 weeks. CHI was inflicted either 24 h before commencement of AβP infusion or 24 h after the onset of AβP infusion. NWCBL or NWAβPmAb were administered 1 week after AβP infusion with or without CHI for 3 weeks. All the parameters were examined after 4 weeks of AβP infusion with or without CHI. For details see text.

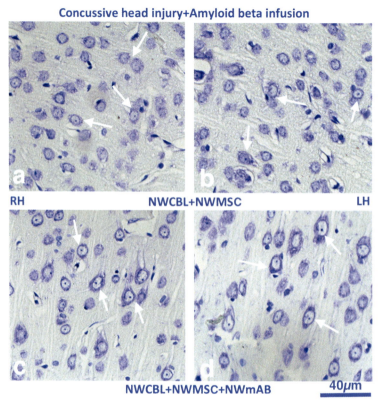

Concussive head injury+Amyloid beta infusion

FIG. 20

Representative example of Toluidine stained nerve cell morphology on paraffin sections (3-μm thick) following amyloid beta peptide infusion in concussive head injury (CHI) in right hemisphere (RH, A) and left hemisphere (LH, B) after treatment with NWCBL+NWMSC (A and B) and NWCBL+NWMSC+NWmAb (RH, C) and LH (D). Combined treatment with NWCBL, NWNSC and NWmAb has superior neuroprotective effects in rescuing nerve cells from damage after CHI+AD (arrows). Bar = 40 μm. CHI, concussive head injury; LH, left hemisphere; NWCBL, TiO_2 nanowired cerebrolysin; NWMSC, TiO_2 nanowired mesenchymal stem cells; NWmAb, TiO_2 nanowired monoclonal antibody against amyloid beta peptide (AβP); RH, right hemisphere. CHI was inflicted on the right parietal bone with an impact of 0.224 N as described earlier (Sharma et al., 2016a,b). AβP infusion (200 ng, i.c.v.) was dome once daily for 4 weeks. CHI was inflicted either 24 h before commencement of AβP infusion or 24 h after the onset of AβP infusion. NWCBL or NWAβPmAb were administered 1 week after AβP infusion with or without CHI for 3 weeks. All the parameters were examined after 4 weeks of AβP infusion with or without CHI. For details see text.

5.4 **Nanowired cerebrolysin with MSCs and AβPmAb reduces biochemical changes in CSF**

Biochemical changes in the CSF that are responsible for AD and/or CHI pathology was evaluated following treatment with NWCBL, NWMSC, and NWAβPmAb to examine whether they are also modified. Administration of these agents in normal control group did not modify CSF contents of AβP, Plasminogen or albumin after 4 weeks of survival (Figs. 2, 14 and 17).

5.4.1 Amyloid beta group

In AβP infusion model of AD NWCBL treatment resulted in significant reduction in CSF AβP (18 ± 6 pg/mL) as compared to untreated group (104 ± 8 pg/mL, $P < 001$). On the other hand, plasminogen level in NWCBL treated group showed significant increase (10 ± 2 μg/mL) in AD CSF as compared to the untreated group (5 ± 3 μg/mL, $P < 0.01$) (Fig. 2). Whereas, albumin level in the CSF of AβP infused AD group was significantly reduced in NWCBL treated animals (0.96 ± 0.07 μg/mL) as compared to untreated AD group (3.42 ± 0.09, $P < 0.01$). These observations show that the CSF composition following AβP infusion model in NWCBL treated group is significantly restored leading to marked neuroprotection.

Treatment with NWMSC mildly but significantly reduced AβP level in CSF of AD (25 ± 8 pg/mL) and also reduced albumin level in CSF (1.04 ± 0.05 μg/mL) as well as enhanced the plasminogen level in the CSF (8 ± 3 μg/mL) as compared to respective control in untreated AD group ($P < 0.05$) (Fig. 2).

On the other hand, NWAβPmAb therapy in AβP infused AD model exhibited significant downregulation of CSF AβP (25 ± 8 pg/mL), CSF albumin (0.48 ± 0.09 μg/mL) and enhanced plasminogen level in the CSF (10 ± 2 μg/mL) as compared to the untreated AD controls ($P < 0.05$). This suggests that NWAβPmAb also effectively ameliorates CSF disturbance of harmful agents in AD.

5.4.2 CHI group

In CHI group also NWCBL treatment showed protective effects in CSF microenvironment leading to significant reduction in AβP level (8 ± 2 pg/mL) as compared to the untreated traumatized group (22 ± 4 pg/mL, $P < 0.01$). The albumin level is also reduced significantly in treated AD group (0.62 ± 0.08 μg/mL) as compared to the untreated AD animals (2.34 ± 0.10 pg/mL, $P < 0.01$). NWCBL was also able to enhance plasminogen level in the CSF (10 ± 3 μg/mL) as compared to the untreated AD group (8 ± 2 μg/mL, $P < 0.05$).

On the other hand NWMSC treatment significantly reduced the AβP (10 ± 3 pg/mL) and albumin (0.57 ± 0.09 μg/mL) as compared to the untreated CHI group ($P < 0.05$). Interestingly, NWMSC was unable to enhance plasminogen level in CHI treated group (Figs. 2, 14 and 17).

However, NWAβPmAb therapy significantly restored CSF microenvironment in CHI resulting in reduction of AβP (6 ± 2 pg/mL) an albumin (0.35 ± 0.07 μg/mL) from the untreated CHI group ($P < 0.05$). This treatment was successful in elevating plasminogen level in CSF of CHI treated group ($10 \pm 2 \mu g/mL$) significantly as compared to the untreated CHI group after 4 weeks of survival ($P < 0.05$). This suggests that NWCBL and NWAβPmAb therapy is highly successful in restoring CSF microenvironment in CHI (Figs. 2, 14 and 17).

5.4.3 AβP infusion in CHI group

Our observations show that combined treatment with NWCBL, NWMSC and NWAβPmAb was the most efficient combination to restore CNS microenvironment in AD that is exacerbated by CHI (Fig. 2). Thus, this triple agent treatment resulted in significant reduction in CSF AβP (11 ± 8 pg/mL, untreated control 180 ± 21 pg/mL, $P < 0.01$) and CSF albumin ($0.22 \pm 0.08 \mu g/mL$, untreated control 4.23 ± 0.12, $P < 0.01$) and enhanced the plasminogen level in the CSF ($21 \pm 4 \mu g/mL$, untreated control $3 \pm 2 \mu g/mL$) (Figs. 2, 14 and 17).

NWCBL treatment in AβP infusion in CHI group alone resulted in significant reduction in CSF AβP (24 ± 9 pg/mL, $P < 0.01$) and albumin ($0.67 \pm 0.06 \mu g/mL$, $P < 0.01$) and also enhanced the plasminogen level significantly in the CSF ($12 \pm 6 \mu g/mL$, $P < 0.05$).

NWMSC treatment in AD with CHI resulted in mild but significant reduction in CSF AβP (36 ± 8 pg/mL, $P < 0.01$) and albumin ($0.89 \pm 0.12 \mu g/mL$, $P < 0.01$) (Fig. 2). This treatment also enhanced the CSF level of plasminogen significantly ($16 \pm 6 \mu g/mL$, $P < 0.01$) in AD with CHI after 4 weeks of survival (Fig. 2).

Therapy with NWAβPmAb caused significant reduction in CSF AβP (28 ± 10 pg/mL, $P < 0.01$) and CSF albumin ($0.65 \pm 0.13 \mu g/mL$) and enhanced the plasminogen level in the CSF ($18 \pm 8 \mu g/mL$, $P < 001$) in AD with CHI after 4 weeks of survival (Figs. 2, 14 and 17).

These observations suggests that nanodelivery of AβP antibodies, MSC and cerebrolysin are effective alone in AD with CHI. However, their combination in AD with CHI results in superior restoration of CSF microenvironment leading to exacerbation of their neuroprotective ability.

6 Discussion

The salient features of this investigation show that AβP infusion into the lateral cerebral ventricle of rats once daily for 4 weeks leads to the development of AD like disease symptoms and pathophysiology (Frautschy et al., 1996; Nitta et al., 1994, 1997). We have used 200 ng soluble AβP infusions into the lateral cerebral ventricle once daily for 4 weeks resulting in profound AD like behavioral and pathophysiological symptoms (Sharma et al., 2012, 2016a,b, 2018, 2019a,b,c). This observation suggests that this AD model in our hands could be used for drug discovery in AD for neuroprotection.

In this model of AD biobehavioral effects are considerably prominent (Sharma HS, unpublished observations). Thus, after 4 weeks the animals show significant reduction in latency of Rota-rod treadmill performances (Results not shown). Furthermore a significant increase in latency of hidden platform search in Morris water maze was also seen (Results not shown). These behavioral deficits are similar to that seen in several other experimental models in AD (Karthick et al., 2019; Morales et al., 2020; Muñoz-Moreno et al., 2018; Shallie et al., 2020; Sturchler-Pierrat et al., 1997).

The hallmark of AD is deposition of amyloid beta peptide in the brain (DeTure and Dickson, 2019; Gouras et al., 2015; Thal and Fändrich, 2015). In our model of AD profound deposition of AβP is seen using immunohistochemistry in the cerebral cortex and other brain structures (Fig. 4). Furthermore measurement of AβP in CSF exhibited profound elevation of the peptide in our AD model. These observations strongly validate our model of AD for drug discovery and study on the pathophysiology of neuroprotection.

AβP is a neurotoxic agent (Darling and Shorter, 2020; Mucke and Selkoe, 2012; Rauk, 2008; Walsh et al., 2002). Thus after deposition of AβP in the brain leads to alterations in brain dysfunction by inducing oxidative stress and free radical production causing membrane damage and perturbation in the fluid microenvironment of the brain (Elangovan and Holsinger, 2020). In AD soluble aggregated form AβP is associated with divalent copper (Coskuner, 2016; Teng et al., 2019; Tõugu et al., 2008). The AβP with copper ion adheres to plasma or intracellular membranes and cause lesions by a combination of radical-initiated lipid peroxidation and formation of the cerebral endothelial cells leading disruption of the integrity of the BBB (Carrano et al., 2011; Cuevas et al., 2019). Binding of AβP and copper ion complex to the intracellular membranes results in loss of calcium homeostasis (Green, 2009; Tong et al., 2018). AβP after binding to tau, and several kinases results in hyperphosphorylation of tau and wreckage of microtubules system that supports and maintain axonal structure (Zempel and Mandelkow, 2012). This will lead to synaptic pathology and neuronal death (Bloom, 2014; Merlo et al., 2018).

The AβP deposition induce breakdown of the BBB is clearly evident from our present investigation of significant leakage of exogenous protein tracers EBA and radioiodine in several brain regions. The EBA and radioiodine when administered into the systemic circulation they binds to serum proteins largely to albumin and thus their leakage into the brain compartment represents tracer-protein complex (Sharma, 1987; Sharma and Dey, 1986, 1987). EBA binds to about 68% of serum albumin and radioiodine binds to 52% of serum albumin (Sharma and Dey, 1986). This makes EBA in circulation large molecular tracer than radioiodine. This could be the basic region that radioiodine penetration in the brain tissues is higher than EBA (Sharma and Dey, 1986). Apart from the endogenous tracers EBA and radioiodine, our observation further confirms the breakdown of the BBB by analyzing immunohistochemistry of endogenous serum albumin (Fig. 2). Several cells and perivascular regions show albumin positive immunoreaction in the neuropil of AD model. This indicates that AβP deposition induces breakdown of the BBB (Rauk, 2008). This observation is in line with our previous findings in AD (Sharma et al., 2012, 2016a,b, 2017, 2018, 2019a,b,c).

We infused AβP into the lateral cerebral ventricle and observed 4 weeks after deposition into the cerebral cortex and other brain structures such as hippocampus, caudate nucleus, putamen, thalamus, hypothalamus and in cerebellum (Sharma HS, unpublished observation). This observation suggests that during the 4 weeks period the AβP reached in different parts of the brain due to disruption of the BBB caused by AβP. However, another possibility could be due to transport of plasma AβP into the brain after BBB breakdown. AβP infusion into the brain could damage the BBB and as a results AβP could be transported into the systemic circulation and from there it reaches to various parts of the brain (Sharma, 2004, 2009). However, further studies are needed to find our passage of AβP into brain in AD after their infusion. For this, measurement of plasma AβP in this model is needed. Increased AβP in the cisterna magna CSF clearly support the idea that lateral ventricular infusion of AβP is distributed to the various brain regions. Since the barrier between CSF and brain is not very tight (Sharma, 2009; Sharma and Sharma, 2019) the AβP could easily enter into the brain fluid microenvironment (Sharma and Westman, 2004).

Leakage of serum proteins across the BBB into the brain fluid microenvironment results in vasogenic brain edema formation. In our investigations in AD model significant edema formation was observed following AβP infusion. This observation is in line with several reports of pre-clinical and clinical cases of brain edema in AD (Carlson et al., 2011; Ly et al., 2021; Zhou et al., 2014). Brain edema in AβP infusion was seen in several areas including cerebral cortex, cerebellum, hippocampus, thalamus and hypothalamus. This indicates that like AβP deposits in the brain areas associated with BBB breakdown are associated with brain edema formation (Sharma et al., 1998a,b,c, 2016a,b, 2019a,b,c).

Breakdown of the BBB and vasogenic edema formation are crucial for structural changes in the brain (Sharma and Sharma, 2019). Spread of edema fluid into the brain fluid environment allows several prohibited biochemical, immunological and neurotoxic agents into the brain compartments (Kiyatkin and Sharma, 2009; Muresanu et al., 2019; Sharma et al., 1991, 1992a,b, 1993, 1995; Stokum et al., 2016; Timaru-Kast et al., 2012). As a result, AβP infusion model of AD shows profound neuronal, glial and axonal changes in various parts of the brain (Figs. 4–7). Damage to neurons located into the cerebral cortex, cerebellum and hippocampus are well known to cause functional disturbances and memory impairment (Sharma and Sharma, 2010; Sharma et al., 1998a,b,c). These observations are in line with the idea that AβP infusion model of AD are very useful to explore novel therapeutic strategies to treat brain pathology in AD.

The potential trigger apart from genetic factors about precipitation and development of AD is not well known (Montagne et al., 2020; Serý et al., 2013; Sharma et al., 2017, 2019a,b,c; Sweeney et al., 2019; Wardlaw et al., 2020). However, several lines of evidences suggest that traumatic brain injury is a predisposing factor in the development of AD (Armstrong, 2019; Gardner and Yaffe, 2015; Jellinger, 2004; Sharma et al., 2018). Early TBI could lead to the development of AD within 10–15 years (Abrahamson and Ikonomovic, 2020; Dams-O'Connor et al., 2016). Thus, we examined the effects of CHI on AβP induced AD model in this investigation. A mild concussive head injury (Sharma et al., 2020a,b,c) before the onset of AβP infusion significantly exacerbated AD induced brain pathology suggest that our animal model is highly accurate and reproducible in nature (Sharma et al., 2017, 2018, 2019a,b,c).

Mild CHI in itself enhanced AβP deposition in the brain and CSF in our investigation (Ramos-Cejudo et al., 2018; Shishido et al., 2019). Breakdown of the BBB and development of profound brain edema and cell injuries are quite prominent in CHI alone after 4 weeks of survival. Since impact was delivered on the right parietal bone of the skull the left half showing more pathological injuries as compared to the right injured side of the brain (Sharma et al., 2016a). This indicates that our model of CHI represents clinical feature of counter coup injury (Sharma et al., 2020a,b,c). As a result AβP infusion after CHI also resulted in greater pathophysiological changes in the uninjured left half than the right injured half in this investigation. These observations are in line with the idea that CHI exacerbates brain pathology in AD.

Interestingly, CHI alone enhanced a mild but significant increase in AβP level in the CSF that could be responsible for breakdown of the BBB, edema formation and cell injuries (Sharma et al., 2016a,b). These pathological changes are greater in the left half than the right injured half in CHI. It is likely that AβP infusion after CHI the passage of AβP in left half could be greater due to greater BBB breakdown as compared to the right half. This could result in greater brain pathology following AβP infusion in CHI dominated within the left side. As a result, greater neuropathological damage associated with the BBB breakdown and edema formation is seen in CHI with AD. These observations clearly support the crucial role of BBB breakdown in the development of brain pathology and edema in CHI or in AD. Obviously, a combination of both CHI and AD could synergistically enhance BBB breakdown, edema formation and brain pathology (Sharma et al., 2017, 2018, 2019a,b,c, 2020a,b,c). A greater accumulation of AβP in CSF from CHI with AD cases further strengthens this hypothesis.

AD brain pathology is significantly exacerbated in CHI in association with greater AβP accumulation. This suggests that AβP efflux and/or clearance mechanisms both are downregulated. The efflux of AβP is compromised because of the BBB breakdown. This is also possible that due to malfunction of the barrier between brain and blood plasma AβP is also elevated. This has a rebound effect on cerebrovascular barriers in AD. There are reports that the plasminogen deficiency in AD could be one of the mechanisms of retarded clearance of AβP from brain or CSF (Baker et al., 2018, 2019; Barker et al., 2010; Yepes, 2021).

The plasminogen is a fibrin-degrading enzyme for plasmin formation that is needed for cleavage of AβP in healthy situation (Baker et al., 2018, 2019). However, in AD enough plasminogen is not available to covert to plasmin resulting in accumulation of AβP in the AD brain (Yepes, 2021). In our model of AβP infusion induced AD the plasminogen level in CSF is significantly decreased while AβP levels are increased profoundly. This indicates that a decrease in plasminogen enzyme is crucial for AD induced brain pathology. Interestingly, plasminogen also decreased in CHI associated with significant increase in CSF AβP level. When AβP was infused in CHI the value of plasminogen decrease in CSF was further exacerbated and the increase in AβP was aggravated. This increase in AβP correlates well with increased albumin levels in the CSF of AD brain or following CHI alone or in combination. These observations suggests that drugs that enhance plasminogen level in AD or CHI could thwart AβP increase and reduce brain pathology (Baker et al., 2018, 2019; Barker et al., 2010).

Recent evidences show that neurotrophins could be one of the important agents in reducing AD brain pathology (Amidfar et al., 2020; Bharani et al., 2020). In addition, mesenchymal stem cells are also beneficial in reducing AD induced neurodegeneration (Park et al., 2020a,b; Qin et al., 2020; Zhang et al., 2020a,b). Treatment with monoclonal AβP antibodies are also beneficial in AD in clinical and in preclinical cases. Immunotherapy is able to neutralize the toxic effect of amyloid pathology (Dodel et al., 2002, 2004; Panza et al., 2010, 2011; van Dyck, 2018).

Accordingly, we used cerebrolysin, a select combination of neurotrophins and peptide fragments, MSCs and monoclonal antibodies to AβP for the treatment of AD and CHI cases including their combinations. We used nanodelivery of these agents in AD, CHI or their combination for their superior effects on neuroprotection (Sharma and Sharma, 2012a,b, 2013).

Nanodelivery of drugs induces superior neuroprotection in brain pathology as compared to their conventional delivery (Sharma et al., 2012, 2016a,b, 2017, 2018, 2019a,b,c, 2020a,b,c). Thus, we used nanowired technology to treat AD and CHI cases in our animal model for achieving superior neuroprotection in AD with CHI. Our observation is the first to show that combination of NWCBL with NWMSH and NWAβPmAb together induced significant neuroprotection in AD cases with CHI, not reported earlier. These triple agents's treatment acted in synergy to thwart exacerbation of brain pathology by CHI in AD. When these agents are given individually in AD with CHI each were able to reduce brain pathology significantly with mild to moderate degree. Thus, a combined treatment profoundly reduced the pathological changes in the brain in animals with AD and CHI. This suggests that to treat AD cases with CHI supplement of neurotrophins, stem cells and antibody therapy to neutralize AβP effects are needed to achieve superior neuroprotection and enhanced the plasminogen levels in the CSF. These individual agents are able to induce marked neuroprotection in AD and CHI models alone. This indicates that each type of agents is needed in neuroprotection in AD and CHI. When these three forces are combined they are able to thwart combined pathology of AD with CHI, a feature not reported earlier.

7 Conclusion and future perspectives

In conclusion, our study clearly points out that the model of AD and CHI induces clinical settings of the disease and thus could be used to explore novel drug development for therapy in clinical cases. Our results strongly indicate that AβP is toxic and is crucial in inducing BBB breakdown resulting in edema formation and cell injury. When AD symptoms were induced in CHI cases the brain pathology was exacerbated. Ti induces neuroprotection in AD with CHI combination of therapy is needed to achieve superior neuroprotection. Our study also suggests that enhancing plasminogen in AD or CHI is neuroprotective in nature. Thus, drugs or agents alone or in combination enhancing plasminogen levels could be suitable in clinics to induce neuroprotection in AD or CHI. It would be interesting to administer nanowired plasminogen or plasmin in AD and CHI to evaluate their effects in reducing brain pathology in clinics. The salient points of this investigation highlight the importance of BBB as a crucial factor in brain pathology in AD and CHI. It remains to be seen whether other co-morbidity factors such hypertension, diabetes, emotional stress and nanoparticles exposure could also exacerbate AD induced brain pathology and if so, whether these drug treatments can also induce neuroprotection in such combination of cases. Our laboratory is currently examining these possibilities in greater details.

Acknowledgments

This investigation is supported by grants from the Air Force Office of Scientific Research (EOARD, London, UK), and Air Force Material Command, USAF, under grant number FA8655-05-1-3065; Grants from the Alzheimer's Association (IIRG-09-132087), the National Institutes of Health (R01 AG028679) and the Dr. Robert M. Kohrman Memorial Fund (RJC); Swedish Medical Research Council (Nr 2710-HSS), the Ministry of Science & Technology, People Republic of China, Göran Gustafsson Foundation, Stockholm, Sweden (HSS), Astra Zeneca, Mölndal, Sweden (HSS/AS), The University Grants Commission, New Delhi, India (HSS/AS), Ministry of Science & Technology, Govt. of India (HSS/AS), Indian Medical Research Council, New Delhi, India (HSS/AS) and India-EU Co-operation Program (RP/AS/HSS) and IT-901/16 (JVL), Government of Basque Country and PPG 17/51 (JVL), JVL thanks to the support of the University of the Basque Country (UPV/EHU) PPG 17/51 and 14/08, the Basque Government (IT-901/16 and CS-2203) Basque Country, Spain; and Foundation for Nanoneuroscience and Nanoneuroprotection (FSNN), Romania. Technical and human support provided by Dr. Ricardo Andrade from SGIker (UPV/EHU) is gratefully acknowledged. Dr. Seaab Sahib is supported by Research Fellowship at the University of Arkansas Fayetteville AR by Department of Community Health; Middle Technical University; Wassit; Iraq, and The Higher Committee for Education Development in Iraq; Baghdad; Iraq. We thank Suraj Sharma, Blekinge Inst. Technology, Karlskrona, Sweden and Dr. Saja Alshafeay, University of Arkansas Fayetteville, Fayetteville AR, USA for computer and graphic support. The U.S. Government is authorized to reproduce and distribute reprints for Government purpose notwithstanding any copyright notation thereon. The views and conclusions contained herein are those of the authors and should not be interpreted as necessarily representing the official policies or endorsements, either expressed or implied, of the Air Force Office of Scientific Research or the U.S. Government.

Conflict of interest

There is no conflict of interest between any entity and/or organization mentioned here.

References

Abbott, N.J., Rönnbäck, L., Hansson, E., 2006. Astrocyte-endothelial interactions at the blood-brain barrier. Nat. Rev. Neurosci. 7 (1), 41–53. https://doi.org/10.1038/nrn1824.

Abbott, N.J., Patabendige, A.A., Dolman, D.E., Yusof, S.R., Begley, D.J., 2010. Structure and function of the blood-brain barrier. Neurobiol. Dis. 37 (1), 13–25. https://doi.org/10.1016/j.nbd.2009.07.030 (Epub 2009 Aug 5).

Abrahamson, E.E., Ikonomovic, M.D., 2020. Brain injury-induced dysfunction of the blood brain barrier as a risk for dementia. Exp. Neurol. 328, 113257. https://doi.org/10.1016/j.expneurol.2020.113257 (Epub 2020 Feb 21).

Adrian, A.L., Thomas, J.L., Adler, A.B., 2018. Soldiers and leaders with combat experience: unit health and climate. Psychiatry 81 (4), 391–407. https://doi.org/10.1080/00332747.2018.1485372 (Epub 2018 Aug 17).

Ahmad, W., Ijaz, B., Shabbiri, K., Ahmed, F., Rehman, S., 2017. Oxidative toxicity in diabetes and Alzheimer's disease: mechanisms behind ROS/RNS generation. J. Biomed. Sci. 24 (1), 76. https://doi.org/10.1186/s12929-017-0379-z.

Alexander, J.J., 2018. Blood-brain barrier (BBB) and the complement landscape. Mol. Immunol. 102, 26–31. https://doi.org/10.1016/j.molimm.2018.06.267 (Epub 2018 Jul 12).

Alosco, M.L., Stern, R.A., 2019. The long-term consequences of repetitive head impacts: chronic traumatic encephalopathy. Handb. Clin. Neurol. 167, 337–355. https://doi.org/10.1016/B978-0-12-804766-8.00018-2.

Alzheimer's Association, 2015. 2015 Alzheimer's disease facts and figures. Alzheimers Dement. 11 (3), 332–384. https://doi.org/10.1016/j.jalz.2015.02.003.

Alzheimer's Association, 2016. 2016 Alzheimer's disease facts and figures. Alzheimers Dement. 12 (4), 459–509. https://doi.org/10.1016/j.jalz.2016.03.001.

Amidfar, M., de Oliveira, J., Kucharska, E., Budni, J., Kim, Y.K., 2020. The role of CREB and BDNF in neurobiology and treatment of Alzheimer's disease. Life Sci. 257, 118020. https://doi.org/10.1016/j.lfs.2020.118020 (Epub 2020 Jun 27).

Andrzejewska, A., Dabrowska, S., Nowak, B., Walczak, P., Lukomska, B., Janowski, M., 2020. Mesenchymal stem cells injected into carotid artery to target focal brain injury home to perivascular space. Theranostics 10 (15), 6615–6628. https://doi.org/10.7150/thno.43169 (eCollection 2020).

Armstrong, R., 2019. Risk factors for Alzheimer's disease. Folia Neuropathol. 57 (2), 87–105. https://doi.org/10.5114/fn.2019.85929.

Arvanitakis, Z., Capuano, A.W., Leurgans, S.E., Bennett, D.A., Schneider, J.A., 2016. Relation of cerebral vessel disease to Alzheimer's disease dementia and cognitive function in elderly people: a cross-sectional study. Lancet Neurol. 15 (9), 934–943. https://doi.org/10.1016/S1474-4422(16)30029-1 (Epub 2016 Jun 14).

Atsev, S., Tomov, N., 2020. Using antifibrinolytics to tackle neuroinflammation. Neural Regen. Res. 15 (12), 2203–2206. https://doi.org/10.4103/1673-5374.284979.

Ayubcha, C., Revheim, M.E., Newberg, A., Moghbel, M., Rojulpote, C., Werner, T.J., Alavi, A., 2021. A critical review of radiotracers in the positron emission tomography imaging of traumatic brain injury: FDG, tau, and amyloid imaging in mild traumatic brain injury and chronic traumatic encephalopathy. Eur. J. Nucl. Med. Mol. Imaging 48 (2), 623–641. https://doi.org/10.1007/s00259-020-04926-4 (Epub 2020 Jul 21).

Baker, S.K., Chen, Z.L., Norris, E.H., Revenko, A.S., MacLeod, A.R., Strickland, S., 2018. Blood-derived plasminogen drives brain inflammation and plaque deposition in a mouse model of Alzheimer's disease. Proc. Natl. Acad. Sci. U. S. A. 115 (41), E9687–E9696. https://doi.org/10.1073/pnas.1811172115. Epub.

Baker, S.K., Chen, Z.L., Norris, E.H., Strickland, S., 2019. Plasminogen mediates communication between the peripheral and central immune systems during systemic immune challenge with lipopolysaccharide. J. Neuroinflammation 16 (1), 172. https://doi.org/10.1186/s12974-019-1560-y.

Banks, W.A., 2016. From blood-brain barrier to blood-brain interface: new opportunities for CNS drug delivery. Nat. Rev. Drug Discov. 15 (4), 275–292. https://doi.org/10.1038/nrd.2015.21 (Epub 2016 Jan 22).

Bao, X., Wu, J., Xie, Y., Kim, S., Michelhaugh, S., Jiang, J., Mittal, S., Sanai, N., Li, J., 2020. Protein expression and functional relevance of efflux and uptake drug transporters at the blood-brain barrier of human brain and glioblastoma. Clin. Pharmacol. Ther. 107 (5), 1116–1127. https://doi.org/10.1002/cpt.1710 (Epub 2019 Dec 10).

Baranello, R.J., Bharani, K.L., Padmaraju, V., Chopra, N., Lahiri, D.K., Greig, N.H., Pappolla, M.A., Sambamurti, K., 2015. Amyloid-beta protein clearance and degradation (ABCD) pathways and their role in Alzheimer's disease. Curr. Alzheimer Res. 12 (1), 32–46. https://doi.org/10.2174/1567205012666141218140953.

Barker, R., Love, S., Kehoe, P.G., 2010. Plasminogen and plasmin in Alzheimer's disease. Brain Res. 1355, 7–15. https://doi.org/10.1016/j.brainres.2010.08.025 (Epub 2010 Aug 13).

Bartels, T., De Schepper, S., Hong, S., 2020. Microglia modulate neurodegeneration in Alzheimer's and Parkinson's diseases. Science 370 (6512), 66–69. https://doi.org/10.1126/science.abb8587.

von Bartheld, C.S., Bahney, J., Herculano-Houzel, S., 2016. The search for true numbers of neurons and glial cells in the human brain: a review of 150 years of cell counting. J. Comp. Neurol. 524 (18), 3865–3895. https://doi.org/10.1002/cne.24040 (Epub 2016 Jun 16).

Bassil, F., Brown, H.J., Pattabhiraman, S., Iwasyk, J.E., Maghames, C.M., Meymand, E.S., Cox, T.O., Riddle, D.M., Zhang, B., Trojanowski, J.Q., Lee, V.M., 2020. Amyloid-beta (abeta) plaques promote seeding and spreading of alpha-synuclein and tau in a mouse model of Lewy body disorders with abeta pathology. Neuron 105 (2), 260–275.e6. https://doi.org/10.1016/j.neuron.2019.10.010 (Epub 2019 Nov 20).

Becker, R.E., Kapogiannis, D., Greig, N.H., 2018. Does traumatic brain injury hold the key to the Alzheimer's disease puzzle? Alzheimers Dement. 14 (4), 431–443. https://doi.org/10.1016/j.jalz.2017.11.007 (Epub 2017 Dec 12).

Bedoui, Y., Neal, J.W., Gasque, P., 2018. The neuro-immune-regulators (NIREGs) promote tissue resilience; a vital component of the host's defense strategy against neuroinflammation. J. NeuroImmune Pharmacol. 13 (3), 309–329. https://doi.org/10.1007/s11481-018-9793-6 (Epub 2018 Jun 16).

Bell, R.D., Zlokovic, B.V., 2009. Neurovascular mechanisms and blood-brain barrier disorder in Alzheimer's disease. Acta Neuropathol. 118 (1), 103–113. https://doi.org/10.1007/s00401-009-0522-3 (Epub 2009 Mar 25).

Bell, R.D., Sagare, A.P., Friedman, A.E., Bedi, G.S., Holtzman, D.M., Deane, R., Zlokovic, B.V., 2007. Transport pathways for clearance of human Alzheimer's amyloid beta-peptide and apolipoproteins E and J in the mouse central nervous system. J. Cereb. Blood Flow Metab. 27 (5), 909–918. https://doi.org/10.1038/sj.jcbfm.9600419 (Epub 2006 Nov 1).

Benhamron, S., Nitzan, K., Valitsky, M., Lax, N., Karussis, D., Kassis, I., Rosenmann, H., 2020. Cerebrospinal fluid (CSF) exchange therapy with artificial CSF enriched with mesenchymal stem cell secretions ameliorates cognitive deficits and brain pathology in Alzheimer's disease mice. J. Alzheimers Dis. 76 (1), 369–385. https://doi.org/10.3233/JAD-191219.

Benvenuti, S., Saccardi, R., Luciani, P., Urbani, S., Deledda, C., Cellai, I., Francini, F., Squecco, R., Rosati, F., Danza, G., Gelmini, S., Greeve, I., Rossi, M., Maggi, R., Serio, M., Peri, A., 2006. Neuronal differentiation of human mesenchymal stem cells: changes in the expression of the Alzheimer's disease-related gene seladin-1. Exp. Cell Res. 312 (13), 2592–2604. https://doi.org/10.1016/j.yexcr.2006.04.016 (Epub 2006 May 6).

Bertogliat, M.J., Morris-Blanco, K.C., Vemuganti, R., 2020. Epigenetic mechanisms of neurodegenerative diseases and acute brain injury. Neurochem. Int. 133, 104642. https://doi.org/10.1016/j.neuint.2019.104642 (Epub 2019 Dec 12).

Besser, L.M., Kukull, W.A., Teylan, M.A., Bigio, E.H., Cairns, N.J., Kofler, J.K., Montine, T.-J., Schneider, J.A., Nelson, P.T., 2018. The revised National Alzheimer's coordinating center's neuropathology form-available data and new analyses. J. Neuropathol. Exp. Neurol. 77 (8), 717–726. https://doi.org/10.1093/jnen/nly049.

Bharani, K.L., Ledreux, A., Gilmore, A., Carroll, S.L., Granholm, A.C., 2020. Serum pro-BDNF levels correlate with phospho-tau staining in Alzheimer's disease. Neurobiol. Aging 87, 49–59. https://doi.org/10.1016/j.neurobiolaging.2019.11.010 (Epub 2019 Nov 22).

Blennow, K., de Leon, M.J., Zetterberg, H., 2006. Alzheimer's disease. Lancet 368 (9533), 387–403. https://doi.org/10.1016/S0140-6736(06)69113-7.

Blennow, K., Hampel, H., Weiner, M., Zetterberg, H., 2010. Cerebrospinal fluid and plasma biomarkers in Alzheimer disease. Nat. Rev. Neurol. 6 (3), 131–144. https://doi.org/10.1038/nrneurol.2010.4 (Epub 2010 Feb 16).

Bloom, G.S., 2014. Amyloid-beta and tau: the trigger and bullet in Alzheimer disease pathogenesis. JAMA Neurol. 71 (4), 505–508. https://doi.org/10.1001/jamaneurol.2013.5847.

Blumenau, S., Foddis, M., Müller, S., Holtgrewe, M., Bentele, K., Berchtold, D., Beule, D., Dirnagl, U., Sassi, C., 2020. Investigating APOE, APP-Abeta metabolism genes and Alzheimer's disease GWAS hits in brain small vessel ischemic disease. Sci. Rep. 10 (1), 7103. https://doi.org/10.1038/s41598-020-63183-5.

Blurton-Jones, M., Laferla, F.M., 2006. Pathways by which Abeta facilitates tau pathology. Curr. Alzheimer Res. 3 (5), 437–448. https://doi.org/10.2174/156720506779025242.

Bodart-Santos, V., de Carvalho, L.R.P., de Godoy, M.A., Batista, A.F., Saraiva, L.M., Lima, L.G., Abreu, C.A., De Felice, F.G., Galina, A., Mendez-Otero, R., Ferreira, S.T., 2019. Extracellular vesicles derived from human Wharton's jelly mesenchymal stem cells protect hippocampal neurons from oxidative stress and synapse damage induced by amyloid-beta oligomers. Stem Cell Res Ther 10 (1), 332. https://doi.org/10.1186/s13287-019-1432-5.

Bonda, D.J., Wang, X., Perry, G., Nunomura, A., Tabaton, M., Zhu, X., Smith, M.A., 2010. Oxidative stress in Alzheimer disease: a possibility for prevention. Neuropharmacology 59 (4–5), 290–294. https://doi.org/10.1016/j.neuropharm.2010.04.005. Epub.

Bondi, M.W., Edmonds, E.C., Salmon, D.P., 2017. Alzheimer's disease: past, present, and future. J. Int. Neuropsychol. Soc. 23 (9–10), 818–831. https://doi.org/10.1017/S135561771700100X.

Bonne, O., Vythilingam, M., Inagaki, M., Wood, S., Neumeister, A., Nugent, A.C., Snow, J., Luckenbaugh, D.A., Bain, E.E., Drevets, W.C., Charney, D.S., 2008. Reduced posterior hippocampal volume in posttraumatic stress disorder. J. Clin. Psychiatry 69 (7), 1087–1091. https://doi.org/10.4088/jcp.v69n0707.

Bostancıklıoğlu, M., 2020. Optogenetic stimulation of serotonin nuclei retrieve the lost memory in Alzheimer's disease. J. Cell. Physiol. 235 (2), 836–847. https://doi.org/10.1002/jcp.29077 (Epub 2019 Jul 22).

Boutet, C., Chupin, M., Lehéricy, S., Marrakchi-Kacem, L., Epelbaum, S., Poupon, C., Wiggins, C., Vignaud, A., Hasboun, D., Defontaines, B., Hanon, O., Dubois, B., Sarazin, M., Hertz-Pannier, L., Colliot, O., 2014. Detection of volume loss in hippocampal layers in Alzheimer's disease using 7 T MRI: a feasibility study. Neuroimage Clin. 5, 341–348. https://doi.org/10.1016/j.nicl.2014.07.011 (eCollection 2014).

Braun, D.J., Kalinin, S., Feinstein, D.L., 2017. Conditional depletion of hippocampal brain-derived neurotrophic factor exacerbates neuropathology in a mouse model of Alzheimer's disease. ASN Neuro 9 (2). https://doi.org/10.1177/1759091417696161, 1759091417696161.

Brookmeyer, R., Abdalla, N., 2019. Multistate models and lifetime risk estimation: application to Alzheimer's disease. Stat. Med. 38 (9), 1558–1565. https://doi.org/10.1002/sim.8056 (Epub 2018 Dec 3).

Brun, A., Liu, X., Erikson, C., 1995. Synapse loss and gliosis in the molecular layer of the cerebral cortex in Alzheimer's disease and in frontal lobe degeneration. Neurodegeneration 4 (2), 171–177. https://doi.org/10.1006/neur.1995.0021.

Busche, M.A., Hyman, B.T., 2020. Synergy between amyloid-beta and tau in Alzheimer's disease. Nat. Neurosci. 23 (10), 1183–1193. https://doi.org/10.1038/s41593-020-0687-6 (Epub 2020 Aug 10).

Cacquevel, M., Launay, S., Castel, H., Benchenane, K., Chéenne, S., Buée, L., Moons, L., Delacourte, A., Carmeliet, P., Vivien, D., 2007. Ageing and amyloid-beta peptide deposition contribute to an impaired brain tissue plasminogen activator activity by different mechanisms. Neurobiol. Dis. 27 (2), 164–173. https://doi.org/10.1016/j.nbd.2007.04.004 (Epub 2007 Apr 24).

Cai, Z., Hussain, M.D., Yan, L.J., 2014. Microglia, neuroinflammation, and beta-amyloid protein in Alzheimer's disease. Int. J. Neurosci. 124 (5), 307–321. https://doi.org/10.3109/00207454.2013.833510 (Epub 2013 Sep 12).

Cai, Z., Qiao, P.F., Wan, C.Q., Cai, M., Zhou, N.K., Li, Q., 2018. Role of blood-brain barrier in Alzheimer's disease. J. Alzheimers Dis. 63 (4), 1223–1234. https://doi.org/10.3233/JAD-180098.

Calamia, M., Bernstein, J.P., Keller, J.N., 2016. I'd do anything for research, but I won't do that: interest in pharmacological interventions in older adults enrolled in a longitudinal aging study. PLoS One 11 (7). https://doi.org/10.1371/journal.pone.0159664, e0159664 (eCollection 2016).

Calsolaro, V., Edison, P., 2016. Neuroinflammation in Alzheimer's disease: current evidence and future directions. Alzheimers Dement. 12 (6), 719–732. https://doi.org/10.1016/j.jalz.2016.02.010 (Epub 2016 May 11).

Cañas, F., Simonin, L., Couturaud, F., Renaudineau, Y., 2015. Annexin A2 autoantibodies in thrombosis and autoimmune diseases. Thromb. Res. 135 (2), 226–230. https://doi.org/10.1016/j.thromres.2014.11.034 (Epub 2014 Dec 13).

Carlson, C., Estergard, W., Oh, J., Suhy, J., Jack Jr., C.R., Siemers, E., Barakos, J., 2011. Prevalence of asymptomatic vasogenic edema in pretreatment Alzheimer's disease study cohorts from phase 3 trials of semagacestat and solanezumab. Alzheimers Dement. 7 (4), 396–401. https://doi.org/10.1016/j.jalz.2011.05.2353.

Carrano, A., Hoozemans, J.J., van der Vies, S.M., Rozemuller, A.J., van Horssen, J., de Vries, H.E., 2011. Amyloid Beta induces oxidative stress-mediated blood-brain barrier changes in capillary amyloid angiopathy. Antioxid. Redox Signal. 15 (5), 1167–1178. https://doi.org/10.1089/ars.2011.3895 (Epub 2011 Mar 23).

Castellino, F.J., Ploplis, V.A., 2005. Structure and function of the plasminogen/plasmin system. Thromb. Haemost. 93 (4), 647–654. https://doi.org/10.1160/TH04-12-0842.

Chakari-Khiavi, F., Dolati, S., Chakari-Khiavi, A., Abbaszadeh, H., Aghebati-Maleki, L., Pourlak, T., Mehdizadeh, A., Yousefi, M., 2019. Prospects for the application of mesenchymal stem cells in Alzheimer's disease treatment. Life Sci. 231, 116564. https://doi.org/10.1016/j.lfs.2019.116564 (Epub 2019 Jun 13).

Chan, Y.E., Bai, Y.M., Hsu, J.W., Huang, K.L., Su, T.P., Li, C.T., Lin, W.C., Pan, T.L., Chen, T.J., Tsai, S.J., Chen, M.H., 2017. Post-traumatic stress disorder and risk of Parkinson disease: a nationwide longitudinal study. Am. J. Geriatr. Psychiatry 25 (8), 917–923. https://doi.org/10.1016/j.jagp.2017.03.012 (Epub 2017 Mar 23).

Chen, Z., Zhong, C., 2014. Oxidative stress in Alzheimer's disease. Neurosci. Bull. 30 (2), 271–281. https://doi.org/10.1007/s12264-013-1423-y (Epub 2014 Mar 24).

Chen, Z.L., Revenko, A.S., Singh, P., MacLeod, A.R., Norris, E.H., Strickland, S., 2017. Depletion of coagulation factor XII ameliorates brain pathology and cognitive impairment in Alzheimer disease mice. Blood 129 (18), 2547–2556. https://doi.org/10.1182/blood-2016-11-753202 (Epub 2017 Feb 27).

Chen, D.W., Wang, J., Zhang, L.L., Wang, Y.J., Gao, C.Y., 2018. Cerebrospinal fluid amyloid-beta levels are increased in patients with insomnia. J. Alzheimers Dis. 61 (2), 645–651. https://doi.org/10.3233/JAD-170032.

Chen, T.B., Lai, Y.H., Ke, T.L., Chen, J.P., Lee, Y.J., Lin, S.Y., Lin, P.C., Wang, P.N., Cheng, I.H., 2019. Changes in plasma amyloid and tau in a longitudinal study of normal aging, mild cognitive impairment, and Alzheimer's disease. Dement. Geriatr. Cogn. Disord. 48 (3–4), 180–195. https://doi.org/10.1159/000505435 (Epub 2020 Jan 28).

Chin, D.L., Zeber, J.E., 2020. Mental health outcomes among military service members after severe injury in combat and TBI. Mil. Med. 185 (5–6), e711–e718. https://doi.org/10.1093/milmed/usz440.

Chun, H., Lee, C.J., 2018. Reactive astrocytes in Alzheimer's disease: a double-edged sword. Neurosci. Res. 126, 44–52. https://doi.org/10.1016/j.neures.2017.11.012 (Epub 2017 Dec 7).

Connor, B., Young, D., Yan, Q., Faull, R.L., Synek, B., Dragunow, M., 1997. Brain-derived neurotrophic factor is reduced in Alzheimer's disease. Brain Res. Mol. Brain Res. 49 (1–2), 71–81. https://doi.org/10.1016/s0169-328x(97)00125-3.

Cortes-Canteli, M., Iadecola, C., 2020. Alzheimer's disease and vascular aging: JACC focus seminar. J. Am. Coll. Cardiol. 75 (8), 942–951. https://doi.org/10.1016/j.jacc.2019.10.062.

Cortes-Canteli, M., Zamolodchikov, D., Ahn, H.J., Strickland, S., Norris, E.H., 2012. Fibrinogen and altered hemostasis in Alzheimer's disease. J. Alzheimers Dis. 32 (3), 599–608. https://doi.org/10.3233/JAD-2012-120820.

Cortes-Canteli, M., Mattei, L., Richards, A.T., Norris, E.H., Strickland, S., 2015. Fibrin deposited in the Alzheimer's disease brain promotes neuronal degeneration. Neurobiol. Aging 36 (2), 608–617. https://doi.org/10.1016/j.neurobiolaging.2014.10.030. Epub 2014 Oct.

Coskuner, O., 2016. Divalent copper ion bound amyloid-beta(40) and amyloid-beta(42) alloforms are less preferred than divalent zinc ion bound amyloid-beta(40) and amyloid-beta(42) alloforms. J. Biol. Inorg. Chem. 21 (8), 957–973. https://doi.org/10.1007/s00775-016-1392-5 (Epub 2016 Sep 22).

Cuello, A.C., Bruno, M.A., Bell, K.F., 2007. NGF-cholinergic dependency in brain aging, MCI and Alzheimer's disease. Curr. Alzheimer Res. 4 (4), 351–358. https://doi.org/10.2174/156720507781788774.

Cuevas, E., Rosas-Hernandez, H., Burks, S.M., Ramirez-Lee, M.A., Guzman, A., Imam, S.Z., Ali, S.F., Sarkar, S., 2019. Amyloid beta 25-35 induces blood-brain barrier disruption in vitro. Metab. Brain Dis. 34 (5), 1365–1374. https://doi.org/10.1007/s11011-019-00447-8 (Epub 2019 Jul 2).

Cui, Q., 2006. Actions of neurotrophic factors and their signaling pathways in neuronal survival and axonal regeneration. Mol. Neurobiol. 33 (2), 155–179. https://doi.org/10.1385/MN:33:2:155.

Cui, G.H., Shao, S.J., Yang, J.J., Liu, J.R., Guo, H.D., 2016. Designer self-assemble peptides maximize the therapeutic benefits of neural stem cell transplantation for Alzheimer's disease via enhancing neuron differentiation and paracrine action. Mol. Neurobiol. 53 (2), 1108–1123. https://doi.org/10.1007/s12035-014-9069-y (Epub 2015 Jan 14).

Cummings, J.L., Tong, G., Ballard, C., 2019. Treatment combinations for Alzheimer's disease: current and future pharmacotherapy options. J. Alzheimers Dis. 67 (3), 779–794. https://doi.org/10.3233/JAD-180766.

Cuzner, M.L., Opdenakker, G., 1999. Plasminogen activators and matrix metalloproteases, mediators of extracellular proteolysis in inflammatory demyelination of the central nervous system. J. Neuroimmunol. 94 (1–2), 1–14. https://doi.org/10.1016/s0165-5728(98)00241-0.

Dams-O'Connor, K., Guetta, G., Hahn-Ketter, A.E., Fedor, A., 2016. Traumatic brain injury as a risk factor for Alzheimer's disease: current knowledge and future directions. Neurodegener. Dis. Manag. 6 (5), 417–429. https://doi.org/10.2217/nmt-2016-0017 (Epub 2016 Sep 7).

Darling, A.L., Shorter, J., 2020. Atomic structures of amyloid-beta oligomers illuminate a neurotoxic mechanism. Trends Neurosci. 43 (10), 740–743. https://doi.org/10.1016/j.tins.2020.07.006 (Epub 2020 Aug 10).

Dawkins, E., Small, D.H., 2014. Insights into the physiological function of the beta-amyloid precursor protein: beyond Alzheimer's disease. J. Neurochem. 129 (5), 756–769. https://doi.org/10.1111/jnc.12675 (Epub 2014 Mar 7).

Deane, R., Bell, R.D., Sagare, A., Zlokovic, B.V., 2009. Clearance of amyloid-beta peptide across the blood-brain barrier: implication for therapies in Alzheimer's disease. CNS Neurol. Disord. Drug Targets 8 (1), 16–30. https://doi.org/10.2174/187152709787601867.

Dehghanian, F., Soltani, Z., Khaksari, M., 2020. Can mesenchymal stem cells act multipotential in traumatic brain injury? J. Mol. Neurosci. 70 (5), 677–688. https://doi.org/10.1007/s12031-019-01475-w (Epub 2020 Jan 2).

Desmarais, P., Weidman, D., Wassef, A., Bruneau, M.A., Friedland, J., Bajsarowicz, P., Thibodeau, M.P., Herrmann, N., Nguyen, Q.D., 2020. The interplay between posttraumatic stress disorder and dementia: a systematic review. Am. J. Geriatr. Psychiatry 28 (1), 48–60. https://doi.org/10.1016/j.jagp.2019.08.006 (Epub 2019 Aug 9).

DeTure, M.A., Dickson, D.W., 2019. The neuropathological diagnosis of Alzheimer's disease. Mol. Neurodegener. 14 (1), 32. https://doi.org/10.1186/s13024-019-0333-5.

Dey, P.K., Sharma, H.S., 1984. Influence of ambient temperature and drug treatments on brain oedema induced by impact injury on skull in rats. Indian J. Physiol. Pharmacol. 28 (3), 177–186.

Diener, P.S., Bregman, B.S., 1994. Neurotrophic factors prevent the death of CNS neurons after spinal cord lesions in newborn rats. Neuroreport 5 (15), 1913–1917. https://doi.org/10.1097/00001756-199410000-00018.

Ding, M., Shen, Y., Wang, P., Xie, Z., Xu, S., Zhu, Z., Wang, Y., Lyu, Y., Wang, D., Xu, L., Bi, J., Yang, H., 2018. Exosomes isolated from human umbilical cord mesenchymal stem cells alleviate neuroinflammation and reduce amyloid-beta deposition by modulating microglial activation in Alzheimer's disease. Neurochem. Res. 43 (11), 2165–2177. https://doi.org/10.1007/s11064-018-2641-5 (Epub 2018 Sep 26).

Djordjevic, J., Sabbir, M.G., Albensi, B.C., 2016. Traumatic brain injury as a risk factor for Alzheimer's disease: is inflammatory signaling a key player? Curr. Alzheimer Res. 13 (7), 730–738. https://doi.org/10.2174/1567205013666160222110320.

Dodel, R., Hampel, H., Depboylu, C., Lin, S., Gao, F., Schock, S., Jäckel, S., Wei, X., Buerger, K., Höft, C., Hemmer, B., Möller, H.J., Farlow, M., Oertel, W.H., Sommer, N., Du, Y., 2002. Human antibodies against amyloid beta peptide: a potential treatment for Alzheimer's disease. Ann. Neurol. 52 (2), 253–256. https://doi.org/10.1002/ana.10253.

Dodel, R.C., Du, Y., Depboylu, C., Hampel, H., Frölich, L., Haag, A., Hemmeter, U., Paulsen, S., Teipel, S.J., Brettschneider, S., Spottke, A., Nölker, C., Möller, H.J., Wei, X., Farlow, M., Sommer, N., Oertel, W.H., 2004. Intravenous immunoglobulins containing antibodies against beta-amyloid for the treatment of Alzheimer's disease. J. Neurol. Neurosurg. Psychiatry 75 (10), 1472–1474. https://doi.org/10.1136/jnnp.2003.033399.

Dong, S., Maniar, S., Manole, M.D., Sun, D., 2018. Cerebral hypoperfusion and other shared brain pathologies in ischemic stroke and Alzheimer's disease. Transl. Stroke Res. 9 (3), 238–250. https://doi.org/10.1007/s12975-017-0570-2 (Epub 2017 Oct 2).

Du, Y., Wu, H.T., Qin, X.Y., Cao, C., Liu, Y., Cao, Z.Z., Cheng, Y., 2018. Postmortem brain, cerebrospinal fluid, and blood neurotrophic factor levels in Alzheimer's disease: a systematic review and meta-analysis. J. Mol. Neurosci. 65 (3), 289–300. https://doi.org/10.1007/s12031-018-1100-8 (Epub 2018 Jun 28).

Durany, N., Michel, T., Kurt, J., Cruz-Sánchez, F.F., Cervós-Navarro, J., Riederer, P., 2000. Brain-derived neurotrophic factor and neurotrophin-3 levels in Alzheimer's disease brains. Int. J. Dev. Neurosci. 18 (8), 807–813.

Elangovan, S., Holsinger, R.M.D., 2020. Cyclical amyloid beta-astrocyte activity induces oxidative stress in Alzheimer's disease. Biochimie 171–172, 38–42. https://doi.org/10.1016/j.biochi.2020.02.003 (Epub 2020 Feb 13).

Elia, C.A., Tamborini, M., Rasile, M., Desiato, G., Marchetti, S., Swuec, P., Mazzitelli, S., Clemente, F., Anselmo, A., Matteoli, M., Malosio, M.L., Coco, S., 2019. Intracerebral injection of extracellular vesicles from mesenchymal stem cells exerts reduced Abeta plaque burden in early stages of a preclinical model of Alzheimer's disease. Cell 8 (9), 1059. https://doi.org/10.3390/cells8091059.

Elliott, K.A., Jasper, H., 1949. Measurement of experimentally induced brain swelling and shrinkage. Am. J. Phys. 157 (1), 122–129. https://doi.org/10.1152/ajplegacy.1949.157.1.122.

Emmerling, M.R., Morganti-Kossmann, M.C., Kossmann, T., Stahel, P.F., Watson, M.D., Evans, L.M., Mehta, P.D., Spiegel, K., Kuo, Y.M., Roher, A.E., Raby, C.A., 2000. Traumatic brain injury elevates the Alzheimer's amyloid peptide A beta 42 in human CSF. A possible role for nerve cell injury. Ann. N. Y. Acad. Sci. 903, 118–122. https://doi.org/10.1111/j.1749-6632.2000.tb06357.x.

Eratne, D., Loi, S.M., Farrand, S., Kelso, W., Velakoulis, D., Looi, J.C., 2018. Alzheimer's disease: clinical update on epidemiology, pathophysiology and diagnosis. Australas. Psychiatry 26 (4), 347–357. https://doi.org/10.1177/1039856218762308 (Epub 2018 Apr 3).

Erdő, F., Denes, L., de Lange, E., 2017. Age-associated physiological and pathological changes at the blood-brain barrier: a review. J. Cereb. Blood Flow Metab. 37 (1), 4–24. https://doi.org/10.1177/0271678X16679420 (Epub 2016 Nov).

Esmaeilzade, B., Artimani, T., Amiri, I., Najafi, R., Shahidi, S., Sabec, M., Farzadinia, P., Zare, M., Zahiri, M., Soleimani, A.S., 2019. Dimethyloxalylglycine preconditioning enhances protective effects of bone marrow-derived mesenchymal stem cells in Abeta-induced Alzheimer disease. Physiol. Behav. 199, 265–272. https://doi.org/10.1016/j.physbeh.2018.11.034 (Epub 2018 Nov 27).

Farahzadi, R., Fathi, E., Vietor, I., 2020. Mesenchymal stem cells could be considered as a candidate for further studies in cell-based therapy of Alzheimer's disease via targeting the signaling pathways. ACS Chem. Neurosci. 11 (10), 1424–1435. https://doi.org/10.1021/acschemneuro.0c00052 (Epub 2020 Apr 29).

Feng, T., Yamashita, T., Shang, J., Shi, X., Nakano, Y., Morihara, R., Tsunoda, K., Nomura, E., Sasaki, R., Tadokoro, K., Matsumoto, N., Hishikawa, N., Ohta, Y., Abe, K., 2019. Clinical and pathological benefits of edaravone for Alzheimer's disease with chronic cerebral hypoperfusion in a novel mouse model. J. Alzheimers Dis. 71 (1), 327–339. https://doi.org/10.3233/JAD-190369.

Fernández-Cabello, S., Kronbichler, M., Van Dijk, K.R.A., Goodman, J.A., Spreng, R.N., Schmitz, T.W., Alzheimer's Disease Neuroimaging Initiative, 2020. Basal forebrain volume reliably predicts the cortical spread of Alzheimer's degeneration. Brain 143 (3), 993–1009. https://doi.org/10.1093/brain/awaa012.

Ferreira, D., Westman, E., Eyjolfsdottir, H., Almqvist, P., Lind, G., Linderoth, B., Seiger, A., Blennow, K., Karami, A., Darreh-Shori, T., Wiberg, M., Simmons, A., Wahlund, L.O., Wahlberg, L., Eriksdotter, M., 2015. Brain changes in Alzheimer's disease patients with implanted encapsulated cells releasing nerve growth factor. J. Alzheimers Dis. 43 (3), 1059–1072. https://doi.org/10.3233/JAD-141068.

Frautschy, S.A., Yang, F., Calderón, L., Cole, G.M., 1996. Rodent models of Alzheimer's disease: rat A beta infusion approaches to amyloid deposits. Neurobiol. Aging 17 (2), 311–321. https://doi.org/10.1016/0197-4580(95)02073-x.

Friker, L.L., Scheiblich, H., Hochheiser, I.V., Brinkschulte, R., Riedel, D., Latz, E., Geyer, M., Heneka, M.T., 2020. Beta-amyloid clustering around ASC fibrils boosts its toxicity in microglia. Cell Rep. 30 (11), 3743–3754.e6. https://doi.org/10.1016/j.celrep.2020.02.025.

Gardner, R.C., Yaffe, K., 2015. Epidemiology of mild traumatic brain injury and neurodegenerative disease. Mol. Cell. Neurosci. 66 (Pt. B), 75–80. https://doi.org/10.1016/j.mcn.2015.03.001 (Epub 2015 Mar 5).

Gate, D., Saligrama, N., Leventhal, O., Yang, A.C., Unger, M.S., Middeldorp, J., Chen, K., Lehallier, B., Channappa, D., De Los Santos, M.B., McBride, A., Pluvinage, J., Elahi, F., Tam, G.K., Kim, Y., Greicius, M., Wagner, A.D., Aigner, L., Galasko, D.R., Davis, M.M., Wyss-Coray, T., 2020. Clonally expanded CD8 T cells patrol the cerebrospinal fluid in Alzheimer's disease. Nature 577 (7790), 399–404. https://doi.org/10.1038/s41586-019-1895-7 (Epub 2020 Jan 8).

Gatti, L., Tinelli, F., Scelzo, E., Arioli, F., Di Fede, G., Obici, L., Pantoni, L., Giaccone, G., Caroppo, P., Parati, E.A., Bersano, A., 2020. Understanding the pathophysiology of cerebral amyloid angiopathy. Int. J. Mol. Sci. 21 (10), 3435. https://doi.org/10.3390/ijms21103435.

Geylis, V., Steinitz, M., 2006. Immunotherapy of Alzheimer's disease (AD): from murine models to anti-amyloid beta (Abeta) human monoclonal antibodies. Autoimmun. Rev. 5 (1), 33–39. https://doi.org/10.1016/j.autrev.2005.06.007 (Epub 2005 Aug 1).

Gill, J., Mustapic, M., Diaz-Arrastia, R., Lange, R., Gulyani, S., Diehl, T., Motamedi, V., Osier, N., Stern, R.A., Kapogiannis, D., 2018. Higher exosomal tau, amyloid-beta 42 and IL-10 are associated with mild TBIs and chronic symptoms in military personnel. Brain Inj. 32 (10), 1277–1284. https://doi.org/10.1080/02699052.2018.1471738 (Epub 2018 Jun 18).

Göbel, K., Asaridou, C.M., Merker, M., Eichler, S., Herrmann, A.M., Geuß, E., Ruck, T., Schüngel, L., Groeneweg, L., Narayanan, V., Schneider-Hohendorf, T., Gross, C.C., Wiendl, H., Kehrel, B.E., Kleinschnitz, C., Meuth, S.G., 2019. Plasma kallikrein modulates immune cell trafficking during neuroinflammation via PAR2 and bradykinin release. Proc. Natl. Acad. Sci. U. S. A. 116 (1), 271–276. https://doi.org/10.1073/pnas.1810020116 (Epub 2018 Dec 17).

González-Reyes, R.E., Nava-Mesa, M.O., Vargas-Sánchez, K., Ariza-Salamanca, D., Mora-Muñoz, L., 2017. Involvement of astrocytes in Alzheimer's disease from a neuroinflammatory and oxidative stress perspective. Front. Mol. Neurosci. 10, 427. https://doi.org/10.3389/fnmol.2017.00427 (eCollection 2017).

Gouras, G.K., Olsson, T.T., Hansson, O., 2015. Beta-amyloid peptides and amyloid plaques in Alzheimer's disease. Neurotherapeutics 12 (1), 3–11. https://doi.org/10.1007/s13311-014-0313-y.

Granholm, A.C., 2000. Oestrogen and nerve growth factor—neuroprotection and repair in Alzheimer's disease. Expert Opin. Investig. Drugs 9 (4), 685–694. https://doi.org/10.1517/13543784.9.4.685.

Green, K.N., 2009. Calcium in the initiation, progression and as an effector of Alzheimer's disease pathology. J. Cell. Mol. Med. 13 (9A), 2787–2799. https://doi.org/10.1111/j.1582-4934.2009.00861.x (Epub 2009 Jul).

Griciuc, A., Serrano-Pozo, A., Parrado, A.R., Lesinski, A.N., Asselin, C.N., Mullin, K., Hooli, B., Choi, S.H., Hyman, B.T., Tanzi, R.E., 2013. Alzheimer's disease risk gene CD33 inhibits microglial uptake of amyloid beta. Neuron 78 (4), 631–643. https://doi.org/10.1016/j.neuron.2013.04.014 (Epub 2013 Apr 25).

Griesbach, G.S., Masel, B.E., Helvie, R.E., Ashley, M.J., 2018. The impact of traumatic brain injury on later life: effects on normal aging and neurodegenerative diseases. J. Neurotrauma 35 (1), 17–24. https://doi.org/10.1089/neu.2017.5103 (Epub 2017 Oct 27).

Griffiths, M., Neal, J.W., Gasque, P., 2007. Innate immunity and protective neuroinflammation: new emphasis on the role of neuroimmune regulatory proteins. Int. Rev. Neurobiol. 82, 29–55. https://doi.org/10.1016/S0074-7742(07)82002-2.

Groer, M.W., Kane, B., Williams, S.N., Duffy, A., 2015. Relationship of PTSD symptoms with combat exposure, stress, and inflammation in American soldiers. Biol. Res. Nurs. 17 (3), 303–310. https://doi.org/10.1177/1099800414544949 (Epub 2014 Sep 7).

Günak, M.M., Billings, J., Carratu, E., Marchant, N.L., Favarato, G., Orgeta, V., 2020. Posttraumatic stress disorder as a risk factor for dementia: systematic review and meta-analysis. Br. J. Psychiatry 217 (5), 600–608. https://doi.org/10.1192/bjp.2020.150.

Gupta, R., Sen, N., 2016. Traumatic brain injury: a risk factor for neurodegenerative diseases. Rev. Neurosci. 27 (1), 93–100.

Gur-Wahnon, D., Mizrachi, T., Maaravi-Pinto, F.Y., Lourbopoulos, A., Grigoriadis, N., Higazi, A.A., Brenner, T., 2013. The plasminogen activator system: involvement in central nervous system inflammation and a potential site for therapeutic intervention. J. Neuroinflammation 10, 124. https://doi.org/10.1186/1742-2094-10-124.

Habisch, H.J., Schmid, B., von Arnim, C.A., Ludolph, A.C., Brenner, R., Storch, A., 2010. Efficient processing of Alzheimer's disease amyloid-beta peptides by neuroectodermally converted mesenchymal stem cells. Stem Cells Dev. 19 (5), 629–633. https://doi.org/10.1089/scd.2009.0045.

Halliday, G.M., McCann, H.L., Pamphlett, R., Brooks, W.S., Creasey, H., McCusker, E., Cotton, R.G., Broe, G.A., Harper, C.G., 1992. Brain stem serotonin-synthesizing neurons in Alzheimer's disease: a clinicopathological correlation. Acta Neuropathol. 84 (6), 638–650. https://doi.org/10.1007/BF00227741.

Hampel, H., Mesulam, M.M., Cuello, A.C., Farlow, M.R., Giacobini, E., Grossberg, G.T., Khachaturian, A.S., Vergallo, A., Cavedo, E., Snyder, P.J., Khachaturian, Z.S., 2018. The cholinergic system in the pathophysiology and treatment of Alzheimer's disease. Brain 141 (7), 1917–1933. https://doi.org/10.1093/brain/awy132.

Hampel, H., Mesulam, M.M., Cuello, A.C., Khachaturian, A.S., Vergallo, A., Farlow, M.R., Snyder, P.J., Giacobini, E., Khachaturian, Z.S., 2019. Revisiting the cholinergic hypothesis in Alzheimer's disease: emerging evidence from translational and clinical research. J. Prev. Alzheimers Dis. 6 (1), 2–15. https://doi.org/10.14283/jpad.2018.43.

Han, F., Bi, J., Qiao, L., Arancio, O., 2020. Stem cell therapy for Alzheimer's disease. Adv. Exp. Med. Biol. 1266, 39–55. https://doi.org/10.1007/978-981-15-4370-8_4.

Hase, Y., Polvikoski, T.M., Firbank, M.J., Craggs, L.J.L., Hawthorne, E., Platten, C., Stevenson, W., Deramecourt, V., Ballard, C., Kenny, R.A., Perry, R.H., Ince, P., Carare, R.O., Allan, L.M., Horsburgh, K., Kalaria, R.N., 2020. Small vessel disease pathological changes in neurodegenerative and vascular dementias concomitant with autonomic dysfunction. Brain Pathol. 30 (1), 191–202. https://doi.org/10.1111/bpa.12769 (Epub 2019 Aug 16).

He, X., Cooley, K., Chung, C.H., Dashti, N., Tang, J., 2007. Apolipoprotein receptor 2 and X11 alpha/beta mediate apolipoprotein E-induced endocytosis of amyloid-beta precursor protein and beta-secretase, leading to amyloid-beta production. J. Neurosci. 27 (15), 4052–4060. https://doi.org/10.1523/JNEUROSCI.3993-06.2007.

Hefti, F., Mash, D.C., 1989. Localization of nerve growth factor receptors in the normal human brain and in Alzheimer's disease. Neurobiol. Aging 10 (1), 75–87. https://doi.org/10.1016/s0197-4580(89)80014-4.

Heissig, B., Salama, Y., Takahashi, S., Osada, T., Hattori, K., 2020. The multifaceted role of plasminogen in inflammation. Cell. Signal. 75, 109761. https://doi.org/10.1016/j.cellsig.2020.109761 (Epub 2020 Aug 28).

Heneka, M.T., 2020. An immune-cell signature marks the brain in Alzheimer's disease. Nature 577 (7790), 322–323. https://doi.org/10.1038/d41586-019-03892-8.

Hladky, S.B., Barrand, M.A., 2018. Elimination of substances from the brain parenchyma: efflux via perivascular pathways and via the blood-brain barrier. Fluids Barriers CNS 15 (1), 30. https://doi.org/10.1186/s12987-018-0113-6.

Hoarau, J.J., Krejbich-Trotot, P., Jaffar-Bandjee, M.C., Das, T., Thon-Hon, G.V., Kumar, S., Neal, J.W., Gasque, P., 2011. Activation and control of CNS innate immune responses in health and diseases: a balancing act finely tuned by neuroimmune regulators (NIReg). CNS Neurol. Disord. Drug Targets 10 (1), 25–43. https://doi.org/10.2174/187152711794488601.

Hock, C., Heese, K., Hulette, C., Rosenberg, C., Otten, U., 2000. Region-specific neurotrophin imbalances in Alzheimer disease: decreased levels of brain-derived neurotrophic factor and increased levels of nerve growth factor in hippocampus and cortical areas. Arch. Neurol. 57 (6), 846–851. https://doi.org/10.1001/archneur.57.6.846.

Honcharenko, D., Juneja, A., Roshan, F., Maity, J., Galán-Acosta, L., Biverstål, H., Hjorth, E., Johansson, J., Fisahn, A., Nilsson, L., Strömberg, R., 2019. Amyloid-beta peptide targeting peptidomimetics for prevention of neurotoxicity. ACS Chem. Neurosci. 10 (3), 1462–1477. https://doi.org/10.1021/acschemneuro.8b00485 (Epub 2019 Feb 6).

Hong, Y.T., Veenith, T., Dewar, D., Outtrim, J.G., Mani, V., Williams, C., Pimlott, S., Hutchinson, P.J., Tavares, A., Canales, R., Mathis, C.A., Klunk, W.E., Aigbirhio, F.I., Coles, J.P., Baron, J.C., Pickard, J.D., Fryer, T.D., Stewart, W., Menon, D.K., 2014.

Amyloid imaging with carbon 11-labeled Pittsburgh compound B for traumatic brain injury. JAMA Neurol. 71 (1), 23–31. https://doi.org/10.1001/jamaneurol.2013.4847.

Honig, L.S., Vellas, B., Woodward, M., Boada, M., Bullock, R., Borrie, M., Hager, K., Andreasen, N., Scarpini, E., Liu-Seifert, H., Case, M., Dean, R.A., Hake, A., Sundell, K., Poole Hoffmann, V., Carlson, C., Khanna, R., Mintun, M., DeMattos, R., Selzler, K.J., Siemers, E., 2018. Trial of solanezumab for mild dementia due to Alzheimer's disease. N. Engl. J. Med. 378 (4), 321–330. https://doi.org/10.1056/NEJMoa1705971.

Hoozemans, J.J., Veerhuis, R., Rozemuller, J.M., Eikelenboom, P., 2006. Neuroinflammation and regeneration in the early stages of Alzheimer's disease pathology. Int. J. Dev. Neurosci. 24 (2–3), 157–165. https://doi.org/10.1016/j.ijdevneu.2005.11.001 (Epub 2005 Dec 27).

Hortobágyi, T., Wise, S., Hunt, N., Cary, N., Djurovic, V., Fegan-Earl, A., Shorrock, K., Rouse, D., Al-Sarraj, S., 2007. Traumatic axonal damage in the brain can be detected using beta-APP immunohistochemistry within 35 min after head injury to human adults. Neuropathol. Appl. Neurobiol. 33 (2), 226–237. https://doi.org/10.1111/j.1365-2990.2006.00794.x.

Huang, E.J., Reichardt, L.F., 2001. Neurotrophins: roles in neuronal development and function. Annu. Rev. Neurosci. 24, 677–736. https://doi.org/10.1146/annurev.neuro.24.1.677.

Huat, T.J., Camats-Perna, J., Newcombe, E.A., Valmas, N., Kitazawa, M., Medeiros, R., 2019. Metal toxicity links to Alzheimer's disease and neuroinflammation. J. Mol. Biol. 431 (9), 1843–1868. https://doi.org/10.1016/j.jmb.2019.01.018 (Epub 2019 Jan 18).

Huebner, B.R., Moore, E.E., Moore, H.B., Gonzalez, E., Kelher, M.R., Sauaia, A., Banerjee, A., Silliman, C.C., 2018. Thrombin stimulates increased plasminogen activator inhibitor-1 release from liver compared to lung endothelium. J. Surg. Res. 225, 1–5. https://doi.org/10.1016/j.jss.2017.12.017 (Epub 2018 Jan 5).

Iannucci, J., Renehan, W., Grammas, P., 2020. Thrombin, a mediator of coagulation, inflammation, and neurotoxicity at the neurovascular interface: implications for Alzheimer's disease. Front. Neurosci. 14, 762. https://doi.org/10.3389/fnins.2020.00762 (eCollection 2020).

Iliff, J.J., Wang, M., Liao, Y., Plogg, B.A., Peng, W., Gundersen, G.A., Benveniste, H., Vates, G.E., Deane, R., Goldman, S.A., Nagelhus, E.A., Nedergaard, M., 2012. A paravascular pathway facilitates CSF flow through the brain parenchyma and the clearance of interstitial solutes, including amyloid beta. Sci. Transl. Med. 4 (147). https://doi.org/10.1126/scitranslmed.3003748, 147ra111.

Iljazi, A., Ashina, H., Al-Khazali, H.M., Lipton, R.B., Ashina, M., Schytz, H.W., Ashina, S., 2020. Post-traumatic stress disorder after traumatic brain injury—a systematic review and meta-analysis. Neurol. Sci. 41 (10), 2737–2746. https://doi.org/10.1007/s10072-020-04458-7 (Epub 2020 May 15).

Iulita, M.F., Caraci, F., Cuello, A.C., 2016. A link between nerve growth factor metabolic deregulation and amyloid-beta-driven inflammation in down syndrome. CNS Neurol. Disord. Drug Targets 15 (4), 434–447. https://doi.org/10.2174/1871527315666160321104916.

Iverson, G.L., Gardner, A.J., Shultz, S.R., Solomon, G.S., McCrory, P., Zafonte, R., Perry, G., Hazrati, L.N., Keene, C.D., Castellani, R.J., 2019. Chronic traumatic encephalopathy neuropathology might not be inexorably progressive or unique to repetitive neurotrauma. Brain 142 (12), 3672–3693. https://doi.org/10.1093/brain/awz286.

Jackson, J., Jambrina, E., Li, J., Marston, H., Menzies, F., Phillips, K., Gilmour, G., 2019. Targeting the synapse in Alzheimer's disease. Front. Neurosci. 13, 735. https://doi.org/10.3389/fnins.2019.00735 (eCollection 2019).

Jadiya, P., Kolmetzky, D.W., Tomar, D., Di Meco, A., Lombardi, A.A., Lambert, J.P., Luongo, T.S., Ludtmann, M.H., Praticò, D., Elrod, J.W., 2019. Impaired mitochondrial calcium efflux contributes to disease progression in models of Alzheimer's disease. Nat. Commun. 10 (1), 3885. https://doi.org/10.1038/s41467-019-11813-6.

Jahn, H., 2013. Memory loss in Alzheimer's disease. Dialogues Clin. Neurosci. 15 (4), 445–454. https://doi.org/10.31887/DCNS.2013.15.4/hjahn.

Jaroudi, W., Garami, J., Garrido, S., Hornberger, M., Keri, S., Moustafa, A.A., 2017. Factors underlying cognitive decline in old age and Alzheimer's disease: the role of the hippocampus. Rev. Neurosci. 28 (7), 705–714. https://doi.org/10.1515/revneuro-2016-0086.

Jellinger, K.A., 2004. Traumatic brain injury as a risk factor for Alzheimer's disease. J. Neurol. Neurosurg. Psychiatry 75 (3), 511–512.

Jeynes, B., Provias, J., 2011. The case for blood-brain barrier dysfunction in the pathogenesis of Alzheimer's disease. J. Neurosci. Res. 89 (1), 22–28. https://doi.org/10.1002/jnr.22527.

Jiang, Z., Shen, B., Xiang, J., 2019. Metal-dependent interactions of metallothionein-3 beta-domain with amyloid-beta peptide and related physiological implications. J. Inorg. Biochem. 196, 110693. https://doi.org/10.1016/j.jinorgbio.2019.110693 (Epub 2019 Apr 13).

Joerger-Messerli, M.S., Marx, C., Oppliger, B., Mueller, M., Surbek, D.V., Schoeberlein, A., 2016. Mesenchymal stem cells from Wharton's jelly and amniotic fluid. Best Pract. Res. Clin. Obstet. Gynaecol. 31, 30–44. https://doi.org/10.1016/j.bpobgyn.2015.07.006 (Epub 2015 Sep 10).

Johanson, C., Stopa, E., Baird, A., Sharma, H., 2011. Traumatic brain injury and recovery mechanisms: peptide modulation of periventricular neurogenic regions by the choroid plexus-CSF nexus. J. Neural Transm. (Vienna) 118 (1), 115–133. https://doi.org/10.1007/s00702-010-0498-0. Epub 2010 Oct.

Jovanovic, K., Loos, B., Da Costa, D.B., Penny, C., Weiss, S.F., 2014. High resolution imaging study of interactions between the 37 kDa/67 kDa laminin receptor and APP, beta-secretase and gamma-secretase in Alzheimer's disease. PLoS One 9 (6). https://doi.org/10.1371/journal.pone.0100373, e100373 (eCollection 2014).

Juan, S.M.A., Adlard, P.A., 2019. Ageing and cognition. Subcell. Biochem. 91, 107–122. https://doi.org/10.1007/978-981-13-3681-2_5.

Kamat, P.K., Kalani, A., Rai, S., Swarnkar, S., Tota, S., Nath, C., Tyagi, N., 2016. Mechanism of oxidative stress and synapse dysfunction in the pathogenesis of Alzheimer's disease: understanding the therapeutics strategies. Mol. Neurobiol. 53 (1), 648–661. https://doi.org/10.1007/s12035-014-9053-6 (Epub 2014 Dec 17).

Kamintsky, L., Beyea, S.D., Fisk, J.D., Hashmi, J.A., Omisade, A., Calkin, C., Bardouille, T., Bowen, C., Quraan, M., Mitnitski, A., Matheson, K., Friedman, A., Hanly, J.G., 2020. Blood-brain barrier leakage in systemic lupus erythematosus is associated with gray matter loss and cognitive impairment. Ann. Rheum. Dis. 79 (12), 1580–1587. https://doi.org/10.1136/annrheumdis-2020-218004 (Epub 2020).

Kang, J.E., Lim, M.M., Bateman, R.J., Lee, J.J., Smyth, L.P., Cirrito, J.R., Fujiki, N., Nishino, S., Holtzman, D.M., 2009. Amyloid-beta dynamics are regulated by orexin and the sleep-wake cycle. Science 326 (5955), 1005–1007. https://doi.org/10.1126/science.1180962 (Epub 2009 Sep 24).

Kang, S., Tanaka, T., Inoue, H., Ono, C., Hashimoto, S., Kioi, Y., Matsumoto, H., Matsuura, H., Matsubara, T., Shimizu, K., Ogura, H., Matsuura, Y., Kishimoto, T., 2020. IL-6 trans-signaling induces plasminogen activator inhibitor-1 from vascular endothelial cells in cytokine release syndrome. Proc. Natl. Acad. Sci. U. S. A. 117 (36), 22351–22356. https://doi.org/10.1073/pnas.2010229117 (Epub 2020 Aug 21).

Karikari, T.K., Pascoal, T.A., Ashton, N.J., Janelidze, S., Benedet, A.L., Rodriguez, J.L., Chamoun, M., Savard, M., Kang, M.S., Therriault, J., Schöll, M., Massarweh, G., Soucy, J.P., Höglund, K., Brinkmalm, G., Mattsson, N., Palmqvist, S., Gauthier, S., Stomrud, E., Zetterberg, H., Hansson, O., Rosa-Neto, P., Blennow, K., 2020. Blood phosphorylated tau 181 as a biomarker for Alzheimer's disease: a diagnostic performance and prediction modelling study using data from four prospective cohorts. Lancet Neurol. 19 (5), 422–433. https://doi.org/10.1016/S1474-4422(20)30071-5.

Karthick, C., Nithiyanandan, S., Essa, M.M., Guillemin, G.J., Jayachandran, S.K., Anusuyadevi, M., 2019. Time-dependent effect of oligomeric amyloid-beta (1-42)-induced hippocampal neurodegeneration in rat model of Alzheimer's disease. Neurol. Res. 41 (2), 139–150. https://doi.org/10.1080/01616412.2018.1544745 (Epub 2018 Nov 19).

Kazim, S.F., Iqbal, K., 2016. Neurotrophic factor small-molecule mimetics mediated neuroregeneration and synaptic repair: emerging therapeutic modality for Alzheimer's disease. Mol. Neurodegener. 11 (1), 50. https://doi.org/10.1186/s13024-016-0119-y.

Kempuraj, D., Ahmed, M.E., Selvakumar, G.P., Thangavel, R., Dhaliwal, A.S., Dubova, I., Mentor, S., Premkumar, K., Saeed, D., Zahoor, H., Raikwar, S.P., Zaheer, S., Iyer, S.S., Zaheer, A., 2020. Brain injury-mediated neuroinflammatory response and Alzheimer's disease. Neuroscientist 26 (2), 134–155. https://doi.org/10.1177/1073858419848293 (Epub 2019 May 16).

Kent, S.A., Spires-Jones, T.L., Durrant, C.S., 2020. The physiological roles of tau and Abeta: implications for Alzheimer's disease pathology and therapeutics. Acta Neuropathol. 140 (4), 417–447. https://doi.org/10.1007/s00401-020-02196-w (Epub 2020 Jul 29).

Khachaturian, A.S., Khachaturian, Z.S., 2014. Military risk factors for Alzheimer's dementia and neurodegenerative disease. Alzheimers Dement. 10 (Suppl. 3), S90–S91. https://doi.org/10.1016/j.jalz.2014.05.1085.

Khanal, B., Lorenzi, M., Ayache, N., Pennec, X., 2014. A biophysical model of shape changes due to atrophy in the brain with Alzheimer's disease. Med. Image Comput. Comput. Assist. Interv. 17 (Pt. 2), 41–48. https://doi.org/10.1007/978-3-319-10470-6_6.

Kim, H.J., Seo, S.W., Chang, J.W., Lee, J.I., Kim, C.H., Chin, J., Choi, S.J., Kwon, H., Yun, H.J., Lee, J.M., Kim, S.T., Choe, Y.S., Lee, K.H., Na, D.L., 2015. Stereotactic brain injection of human umbilical cord blood mesenchymal stem cells in patients with Alzheimer's disease dementia: a phase 1 clinical trial. Alzheimers Dement. (N. Y.) 1 (2), 95–102. https://doi.org/10.1016/j.trci.2015.06.007 (eCollection 2015).

Kim, D.Y., Choi, S.H., Lee, J.S., Kim, H.J., Kim, H.N., Lee, J.E., Shin, J.Y., Lee, P.H., 2020a. Feasibility and efficacy of intra-arterial administration of embryonic stem cell derived-mesenchymal stem cells in animal model of Alzheimer's disease. J. Alzheimers Dis. 76 (4), 1281–1296. https://doi.org/10.3233/JAD-200026.

Kim, S.H., Ahn, J.H., Yang, H., Lee, P., Koh, G.Y., Jeong, Y., 2020b. Cerebral amyloid angiopathy aggravates perivascular clearance impairment in an Alzheimer's disease mouse model. Acta Neuropathol. Commun. 8 (1), 181. https://doi.org/10.1186/s40478-020-01042-0.

Kim, J., Lee, Y., Lee, S., Kim, K., Song, M., Lee, J., 2020c. Mesenchymal stem cell therapy and Alzheimer's disease: current status and future perspectives. J. Alzheimers Dis. 77 (1), 1–14. https://doi.org/10.3233/JAD-200219.

Kinney, J.W., Bemiller, S.M., Murtishaw, A.S., Leisgang, A.M., Salazar, A.M., Lamb, B.T., 2018. Inflammation as a central mechanism in Alzheimer's disease. Alzheimers Dement. (N. Y.) 4, 575–590. https://doi.org/10.1016/j.trci.2018.06.014 (eCollection 2018).

Kirova, A.M., Bays, R.B., Lagalwar, S., 2015. Working memory and executive function decline across normal aging, mild cognitive impairment, and Alzheimer's disease. Biomed. Res. Int. 2015, 748212. https://doi.org/10.1155/2015/748212 (Epub 2015 Oct 15).

Kiyatkin, E.A., Sharma, H.S., 2009. Acute methamphetamine intoxication: brain hyperthermia, blood-brain barrier, brain edema, and morphological cell abnormalities. Int. Rev. Neurobiol. 88, 65–100. https://doi.org/10.1016/S0074-7742(09)88004-5.

Klohs, J., 2019. An integrated view on vascular dysfunction in Alzheimer's disease. Neurodegener. Dis. 19 (3–4), 109–127. https://doi.org/10.1159/000505625 (Epub 2020 Feb 14).

Krames, E., Buchser, E., Hassenbusch, S.J., Levy, R., 1999. Future trends in the development of local drug delivery systems: intraspinal, intracerebral, and intraparenchymal therapies. Neuromodulation 2 (2), 133–148. https://doi.org/10.1046/j.1525-1403.1999.00133.x.

Kritikos, T.K., Comer, J.S., He, M., Curren, L.C., Tompson, M.C., 2019. Combat experience and posttraumatic stress symptoms among military-serving parents: a meta-analytic examination of associated offspring and family outcomes. J. Abnorm. Child Psychol. 47 (1), 131–148. https://doi.org/10.1007/s10802-018-0427-5.

Kuroda, T., Honma, M., Mori, Y., Futamura, A., Sugimoto, A., Yano, S., Kinno, R., Murakami, H., Ono, K., 2020. Increased presence of cerebral microbleeds correlates with ventricular enlargement and increased white matter hyperintensities in Alzheimer's disease. Front. Aging Neurosci. 12, 13. https://doi.org/10.3389/fnagi.2020.00013 (eCollection 2020).

Kutz, S.M., Higgins, C.E., Higgins, P.J., 2012. Novel combinatorial therapeutic targeting of PAI-1 (SERPINE1) gene expression in Alzheimer's disease. Mol. Med. Ther. 1 (2), 106. https://doi.org/10.4172/2324-8769.1000106.

LaFerla, F.M., Green, K.N., Oddo, S., 2007. Intracellular amyloid-beta in Alzheimer's disease. Nat. Rev. Neurosci. 8 (7), 499–509. https://doi.org/10.1038/nrn2168.

Lane, C.A., Hardy, J., Schott, J.M., 2018. Alzheimer's disease. Eur. J. Neurol. 25 (1), 59–70. https://doi.org/10.1111/ene.13439 (Epub 2017 Oct 19).

Laurent, C., Buée, L., Blum, D., 2018. Tau and neuroinflammation: what impact for Alzheimer's disease and tauopathies? Biom. J. 41 (1), 21–33. https://doi.org/10.1016/j.bj.2018.01.003 (Epub 2018 Mar 20).

Lee, B.G., Leavitt, M.J., Bernick, C.B., Leger, G.C., Rabinovici, G., Banks, S.J., 2018. A systematic review of positron emission tomography of tau, amyloid beta, and neuroinflammation in chronic traumatic encephalopathy: the evidence to date. J. Neurotrauma 35 (17), 2015–2024. https://doi.org/10.1089/neu.2017.5558 (Epub 2018 Jun 12).

Lee, J., Kwon, S.J., Kim, J.H., Jang, H., Lee, N.K., Hwang, J.W., Kim, J.H., Chang, J.W., Na, D.L., 2019. Cerebrospinal fluid from Alzheimer's disease patients as an optimal formulation for therapeutic application of mesenchymal stem cells in Alzheimer's disease. Sci. Rep. 9 (1), 564. https://doi.org/10.1038/s41598-018-37252-9.

Lee, N.H., Myeong, S.H., Son, H.J., Hwang, J.W., Lee, N.K., Chang, J.W., Na, D.L., 2020. Ethionamide preconditioning enhances the proliferation and migration of human Wharton's Jelly-derived mesenchymal stem cells. Int. J. Mol. Sci. 21 (19), 7013. https://doi.org/10.3390/ijms21197013.

Leli, D.A., Scott, L.H., 1982. Cross-validation of two indexes of intellectual deterioration on patients with Alzheimer's disease. J. Consult. Clin. Psychol. 50 (3), 468. https://doi.org/10.1037//0022-006x.50.3.468.

Leong, Y.Q., Ng, K.Y., Chye, S.M., Ling, A.P.K., Koh, R.Y., 2020. Mechanisms of action of amyloid-beta and its precursor protein in neuronal cell death. Metab. Brain Dis. 35 (1), 11–30. https://doi.org/10.1007/s11011-019-00516-y (Epub 2019 Dec 6).

Lewin, G.R., Barde, Y.A., 1996. Physiology of the neurotrophins. Annu. Rev. Neurosci. 19, 289–317. https://doi.org/10.1146/annurev.ne.19.030196.001445.

Liebner, S., Dijkhuizen, R.M., Reiss, Y., Plate, K.H., Agalliu, D., Constantin, G., 2018. Functional morphology of the blood-brain barrier in health and disease. Acta Neuropathol. 135 (3), 311–336. https://doi.org/10.1007/s00401-018-1815-1 (Epub 2018 Feb 6).

Lin, Y.T., Seo, J., Gao, F., Feldman, H.M., Wen, H.L., Penney, J., Cam, H.P., Gjoneska, E., Raja, W.K., Cheng, J., Rueda, R., Kritskiy, O., Abdurrob, F., Peng, Z., Milo, B., Yu, C.J., Elmsaouri, S., Dey, D., Ko, T., Yankner, B.A., Tsai, L.H., 2018. APOE4 causes widespread molecular and cellular alterations associated with Alzheimer's disease phenotypes in human iPSC-derived brain cell types. Neuron 98 (6), 1141–1154.e7. https://doi.org/10.1016/j.neuron.2018.05.008 (Epub 2018 May 31) 29861287.

Liu, X., Passant, U., Risberg, J., Warkentin, S., Brun, A., 1999. Synapse density related to cerebral blood flow and symptomatology in frontal lobe degeneration and Alzheimer's disease. Dement. Geriatr. Cogn. Disord. 10 (Suppl. 1), 64–70. https://doi.org/10.1159/000051216.

Liu, X., Hou, D., Lin, F., Luo, J., Xie, J., Wang, Y., Tian, Y., 2019. The role of neurovascular unit damage in the occurrence and development of Alzheimer's disease. Rev. Neurosci. 30 (5), 477–484. https://doi.org/10.1515/revneuro-2018-0056.

Liu, Y., Huber, C.C., Wang, H., 2020. Disrupted blood-brain barrier in 5FAD mouse model of Alzheimer's disease can be mimicked and repaired in vitro with neural stem cell-derived exosomes. Biochem. Biophys. Res. Commun. 525 (1), 192–196. https://doi.org/10.1016/j.bbrc.2020.02.074.

Lo Furno, D., Mannino, G., Giuffrida, R., 2018. Functional role of mesenchymal stem cells in the treatment of chronic neurodegenerative diseases. J. Cell. Physiol. 233 (5), 3982–3999. https://doi.org/10.1002/jcp.26192 (Epub 2017 Oct 20).

LoBue, C., Cullum, C.M., 2020. POINT/COUNTER-POINT-beyond the headlines: the actual evidence that traumatic brain injury is a risk factor for later-in-life dementia. Arch. Clin. Neuropsychol. 35 (2), 123–127. https://doi.org/10.1093/arclin/acz074.

Loera-Valencia, R., Cedazo-Minguez, A., Kenigsberg, P.A., Page, G., Duarte, A.I., Giusti, P., Zusso, M., Robert, P., Frisoni, G.B., Cattaneo, A., Zille, M., Boltze, J., Cartier, N., Buee, L., Johansson, G., Winblad, B., 2019. Current and emerging avenues for Alzheimer's disease drug targets. J. Intern. Med. 286 (4), 398–437. https://doi.org/10.1111/joim.12959 (Epub 2019 Aug 29).

Logovinsky, V., Satlin, A., Lai, R., Swanson, C., Kaplow, J., Osswald, G., Basun, H., Lannfelt, L., 2016. Safety and tolerability of BAN2401—a clinical study in Alzheimer's disease with a protofibril selective Abeta antibody. Alzheimers Res. Ther. 8 (1), 14. https://doi.org/10.1186/s13195-016-0181-2.

Loureiro, J.C., Pais, M.V., Stella, F., Radanovic, M., Teixeira, A.L., Forlenza, O.V., de Souza, L.C., 2020. Passive antiamyloid immunotherapy for Alzheimer's disease. Curr. Opin. Psychiatry 33 (3), 284–291. https://doi.org/10.1097/YCO.0000000000000587.

Love, S., Miners, J.S., 2016. Cerebral hypoperfusion and the energy deficit in Alzheimer's disease. Brain Pathol. 26 (5), 607–617. https://doi.org/10.1111/bpa.12401.

Lowe, V.J., Wiste, H.J., Senjem, M.L., Weigand, S.D., Therneau, T.M., Boeve, B.F., Josephs, K.A., Fang, P., Pandey, M.K., Murray, M.E., Kantarci, K., Jones, D.T., Vemuri, P., Graff-Radford, J., Schwarz, C.G., Machulda, M.M., Mielke, M.M., Roberts, R.O., Knopman, D.S., Petersen, R.C., Jack Jr., C.R., 2018. Widespread brain tau and its association with ageing, Braak stage and Alzheimer's dementia. Brain 141 (1), 271–287. https://doi.org/10.1093/brain/awx320.

Lozupone, M., Solfrizzi, V., D'Urso, F., Di Gioia, I., Sardone, R., Dibello, V., Stallone, R., Liguori, A., Ciritella, C., Daniele, A., Bellomo, A., Seripa, D., Panza, F., 2020. Anti-amyloid-beta protein agents for the treatment of Alzheimer's disease: an update on emerging drugs. Expert Opin. Emerg. Drugs 25 (3), 319–335. https://doi.org/10.1080/14728214.2020.1808621. Epub.

Lu, X., Hagg, T., 1997. Glial cell line-derived neurotrophic factor prevents death, but not reductions in tyrosine hydroxylase, of injured nigrostriatal neurons in adult rats. J. Comp. Neurol. 388 (3), 484–494.

Ly, M., Raji, C.A., Yu, G.Z., Wang, Q., Wang, Y., Schindler, S.E., An, H., Samara, A., Eisenstein, S.A., Hershey, T., Smith, G., Klein, S., Liu, J., Xiong, C., Ances, B.M., Morris, J.C., Benzinger, T.L.S., 2021. Obesity and white matter neuroinflammation related edema in Alzheimer's disease dementia biomarker negative cognitively normal individuals. J. Alzheimers Dis. 79 (4), 1801–1811. https://doi.org/10.3233/JAD-201242.

Lyketsos, C.G., Carrillo, M.C., Ryan, J.M., Khachaturian, A.S., Trzepacz, P., Amatniek, J., Cedarbaum, J., Brashear, R., Miller, D.S., 2011. Neuropsychiatric symptoms in Alzheimer's disease. Alzheimers Dement. 7 (5), 532–539. https://doi.org/10.1016/j.jalz.2011.05.2410.

Mac Donald, C.L., Barber, J., Patterson, J., Johnson, A.M., Parsey, C., Scott, B., Fann, J.R., Temkin, N.R., 2021. Comparison of clinical outcomes 1 and 5 years post-injury following combat concussion. Neurology 96 (3), e387–e398. https://doi.org/10.1212/WNL.0000000000011089 (Epub 2020 Nov 11).

Malishev, R., Nandi, S., Śmiłowicz, D., Bakavayev, S., Engel, S., Bujanover, N., Gazit, R., Metzler-Nolte, N., Jelinek, R., 2019. Interactions between BIM protein and beta-amyloid may reveal a crucial missing link between Alzheimer's disease and neuronal cell death. ACS Chem. Neurosci. 10 (8), 3555–3564. https://doi.org/10.1021/acschemneuro.9b00177 (Epub 2019 Jun 12).

Manoharan, S., Guillemin, G.J., Abiramasundari, R.S., Essa, M.M., Akbar, M., Akbar, M.D., 2016. The role of reactive oxygen species in the pathogenesis of Alzheimer's disease, Parkinson's disease, and Huntington's disease: a mini review. Oxidative Med. Cell. Longev. 2016, 8590578. https://doi.org/10.1155/2016/8590578 (Epub 2016 Dec 27).

Manolopoulos, A., Andreadis, P., Malandris, K., Avgerinos, I., Karagiannis, T., Kapogiannis, D., Tsolaki, M., Tsapas, A., Bekiari, E., 2019. Intravenous immunoglobulin for patients with Alzheimer's disease: a systematic review and meta-analysis. Am. J. Alzheimers Dis. Other Dement. 34 (5), 281–289. https://doi.org/10.1177/1533317519843720 (Epub 2019 Apr 15).

Markus, A., Patel, T.D., Snider, W.D., 2002. Neurotrophic factors and axonal growth. Curr. Opin. Neurobiol. 12 (5), 523–531. https://doi.org/10.1016/s0959-4388(02)00372-0.

Marner, L., Frokjaer, V.G., Kalbitzer, J., Lehel, S., Madsen, K., Baaré, W.F., Knudsen, G.M., Hasselbalch, S.G., 2012. Loss of serotonin 2A receptors exceeds loss of serotonergic projections in early Alzheimer's disease: a combined [11C]DASB and [18F]altanserin-PET study. Neurobiol. Aging 33 (3), 479–487. https://doi.org/10.1016/j.neurobiolaging.2010.03.023 (Epub 2010 May 26).

Maruyama, W., Naoi, M., 2013. "70th Birthday Professor Riederer" induction of glial cell line-derived and brain-derived neurotrophic factors by rasagiline and (−)deprenyl: a way to a disease-modifying therapy? J. Neural Transm. (Vienna) 120 (1), 83–89. https://doi.org/10.1007/s00702-012-0876-x (Epub 2012 Aug 15).

Masters, C.L., Bateman, R., Blennow, K., Rowe, C.C., Sperling, R.A., Cummings, J.L., 2015. Alzheimer's disease. Nat. Rev. Dis. Primers. 1, 15056. https://doi.org/10.1038/nrdp.2015.56.

Mata, A.M., 2018. Functional interplay between plasma membrane Ca(2+)-ATPase, amyloid beta-peptide and tau. Neurosci. Lett. 663, 55–59. https://doi.org/10.1016/j.neulet.2017.08.004 (Epub 2017 Aug 2).

Mattsson, N., Zetterberg, H., Janelidze, S., Insel, P.S., Andreasson, U., Stomrud, E., Palmqvist, S., Baker, D., Tan Hehir, C.A., Jeromin, A., Hanlon, D., Song, L., Shaw, L.M., Trojanowski, J.Q., Weiner, M.W., Hansson, O., Blennow, K., ADNI Investigators, 2016. Plasma tau in Alzheimer disease. Neurology 87 (17), 1827–1835. https://doi.org/10.1212/WNL.0000000000003246 (Epub 2016 Sep 30).

Mattsson, N., Andreasson, U., Zetterberg, H., Blennow, K., Alzheimer's Disease Neuroimaging Initiative, 2017. Association of plasma neurofilament light with neurodegeneration in patients with Alzheimer disease. JAMA Neurol. 74 (5), 557–566. https://doi.org/10.1001/jamaneurol.2016.6117.

McKee, A.C., Cantu, R.C., Nowinski, C.J., Hedley-Whyte, E.T., Gavett, B.E., Budson, A.E., Santini, V.E., Lee, H.S., Kubilus, C.A., Stern, R.A., 2009. Chronic traumatic encephalopathy in athletes: progressive tauopathy after repetitive head injury. J. Neuropathol. Exp. Neurol. 68 (7), 709–735. https://doi.org/10.1097/NEN.0b013e3181a9d503.

Mckee, A.C., Abdolmohammadi, B., Stein, T.D., 2018. The neuropathology of chronic traumatic encephalopathy. Handb. Clin. Neurol. 158, 297–307. https://doi.org/10.1016/B978-0-444-63954-7.00028-8.

Mehrabadi, S., Motevaseli, E., Sadr, S.S., Moradbeygi, K., 2020. Hypoxic-conditioned medium from adipose tissue mesenchymal stem cells improved neuroinflammation through alternation of toll like receptor (TLR) 2 and TLR4 expression in model of Alzheimer's disease rats. Behav. Brain Res. 379, 112362. https://doi.org/10.1016/j.bbr.2019.112362 (Epub 2019 Nov 15).

Meltzer, C.C., Smith, G., DeKosky, S.T., Pollock, B.G., Mathis, C.A., Moore, R.Y., Kupfer, D.J., Reynolds 3rd., C.F., 1998. Serotonin in aging, late-life depression, and Alzheimer's disease: the emerging role of functional imaging. Neuropsychopharmacology 18 (6), 407–430. https://doi.org/10.1016/S0893-133X(97)00194-2.

Mentis, A.A., Dardiotis, E., Chrousos, G.P., 2020. Apolipoprotein E4 and meningeal lymphatics in Alzheimer disease: a conceptual framework. Mol. Psychiatry 26, 1075–1097. https://doi.org/10.1038/s41380-020-0731-7. Online ahead of print.

Merlo, S., Spampinato, S.F., Beneventano, M., Sortino, M.A., 2018. The contribution of microglia to early synaptic compensatory responses that precede beta-amyloid-induced neuronal death. Sci. Rep. 8 (1), 7297. https://doi.org/10.1038/s41598-018-25453-1.

Mesulam, M.M., 1998. From sensation to cognition. Brain 121 (Pt. 6), 1013–1052. https://doi.org/10.1093/brain/121.6.1013.

Meyer, K., Feldman, H.M., Lu, T., Drake, D., Lim, E.T., Ling, K.H., Bishop, N.A., Pan, Y., Seo, J., Lin, Y.T., Su, S.C., Church, G.M., Tsai, L.H., Yankner, B.A., 2019. REST and neural gene network dysregulation in iPSC models of Alzheimer's disease. Cell Rep. 26 (5), 1112–1127.e9. https://doi.org/10.1016/j.celrep.2019.01.023.

Minter, M.R., Taylor, J.M., Crack, P.J., 2016. The contribution of neuroinflammation to amyloid toxicity in Alzheimer's disease. J. Neurochem. 136 (3), 457–474. https://doi.org/10.1111/jnc.13411 (Epub 2015 Nov 18).

Mohamed, A.Z., Cumming, P., Srour, H., Gunasena, T., Uchida, A., Haller, C.N., Nasrallah, F., Department of Defense Alzheimer's Disease Neuroimaging Initiative, 2018. Amyloid pathology fingerprint differentiates post-traumatic stress disorder and traumatic brain injury. Neuroimage Clin. 19, 716–726. https://doi.org/10.1016/j.nicl.2018.05.016 (eCollection 2018).

Mohamed, A.Z., Cumming, P., Götz, J., Nasrallah, F., Department of Defense Alzheimer's Disease Neuroimaging Initiative, 2019. Tauopathy in veterans with long-term posttraumatic stress disorder and traumatic brain injury. Eur. J. Nucl. Med. Mol. Imaging 46 (5), 1139–1151. https://doi.org/10.1007/s00259-018-4241-7.

Mohamed, A.Z., Corrigan, F., Collins-Praino, L.E., Plummer, S.L., Soni, N., Nasrallah, F.A., 2020. Evaluating spatiotemporal microstructural alterations following diffuse traumatic brain injury. Neuroimage Clin. 25, 102136. https://doi.org/10.1016/j.nicl.2019.102136 (Epub 2019 Dec 14).

Mohammadi, A., Maleki-Jamshid, A., Milan, P.B., Ebrahimzadeh, K., Faghihi, F., Joghataei, M.T., 2019. Intrahippocampal transplantation of undifferentiated human chorionic-derived mesenchymal stem cells does not improve learning and memory in the rat model of sporadic Alzheimer disease. Curr. Stem Cell Res. Ther. 14 (2), 184–190. https://doi.org/10.2174/1574888X13666180723111249.

Montagne, A., Zhao, Z., Zlokovic, B.V., 2017. Alzheimer's disease: a matter of blood-brain barrier dysfunction? J. Exp. Med. 214 (11), 3151–3169. https://doi.org/10.1084/jem.20171406 (Epub 2017 Oct 23).

Montagne, A., Nation, D.A., Sagare, A.P., Barisano, G., Sweeney, M.D., Chakhoyan, A., Pachicano, M., Joe, E., Nelson, A.R., D'Orazio, L.M., Buennagel, D.P., Harrington, M.G., Benzinger, T.L.S., Fagan, A.M., Ringman, J.M., Schneider, L.S., Morris, J.C., Reiman, E.M., Caselli, R.J., Chui, H.C., Tcw, J., Chen, Y., Pa, J., Conti, P.S., Law, M., Toga, A.W., Zlokovic, B.V., 2020. APOE4 leads to blood-brain barrier dysfunction predicting cognitive decline. Nature 581 (7806), 71–76. https://doi.org/10.1038/s41586-020-2247-3 (Epub 2020 Apr 29).

Moore, B.A., Brock, M.S., Brager, A., Collen, J., LoPresti, M., Mysliwiec, V., 2020. Posttraumatic stress disorder, traumatic brain injury, sleep, and performance in military personnel. Sleep Med. Clin. 15 (1), 87–100. https://doi.org/10.1016/j.jsmc.2019.11.004 (Epub 2020 Jan 8).

Morales, R., Duran-Aniotz, C., Bravo-Alegria, J., Estrada, L.D., Shahnawaz, M., Hu, P.P., Kramm, C., Morales-Scheihing, D., Urayama, A., Soto, C., 2020. Infusion of blood from mice displaying cerebral amyloidosis accelerates amyloid pathology in animal models of Alzheimer's disease. Acta Neuropathol. Commun. 8 (1), 213. https://doi.org/10.1186/s40478-020-01087-1.

Morgese, M.G., Schiavone, S., Trabace, L., 2017. Emerging role of amyloid beta in stress response: implication for depression and diabetes. Eur. J. Pharmacol. 817, 22–29. https://doi.org/10.1016/j.ejphar.2017.08.031 (Epub 2017 Aug 24).

Morley, J.E., Farr, S.A., 2014. The role of amyloid-beta in the regulation of memory. Biochem. Pharmacol. 88 (4), 479–485. https://doi.org/10.1016/j.bcp.2013.12.018 (Epub 2014 Jan 4).

Morley, J.E., Farr, S.A., Nguyen, A.D., Xu, F., 2019. Editorial: what is the physiological function of amyloid-beta protein? J. Nutr. Health Aging 23 (3), 225–226. https://doi.org/10.1007/s12603-019-1162-5.

Mu, Y., Gage, F.H., 2011. Adult hippocampal neurogenesis and its role in Alzheimer's disease. Mol. Neurodegener. 6, 85. https://doi.org/10.1186/1750-1326-6-85.

Mucke, L., Selkoe, D.J., 2012. Neurotoxicity of amyloid beta-protein: synaptic and network dysfunction. Cold Spring Harb. Perspect. Med. 2 (7), a006338. https://doi.org/10.1101/cshperspect.a006338.

Mueller, C., Soysal, P., Rongve, A., Isik, A.T., Thompson, T., Maggi, S., Smith, L., Basso, C., Stewart, R., Ballard, C., O'Brien, J.T., Aarsland, D., Stubbs, B., Veronese, N., 2019. Survival time and differences between dementia with Lewy bodies and Alzheimer's disease following diagnosis: a meta-analysis of longitudinal studies. Ageing Res. Rev. 50, 72–80. https://doi.org/10.1016/j.arr.2019.01.005 (Epub 2019 Jan 6).

Mugisho, O.O., Robilliard, L.D., Nicholson, L.F.B., Graham, E.S., O'Carroll, S.J., 2019. Bradykinin receptor-1 activation induces inflammation and increases the permeability of human brain microvascular endothelial cells. Cell Biol. Int. 44, 343–351. https://doi.org/10.1002/cbin.11232.

Muñoz-Moreno, E., Tudela, R., López-Gil, X., Soria, G., 2018. Early brain connectivity alterations and cognitive impairment in a rat model of Alzheimer's disease. Alzheimers Res. Ther. 10 (1), 16. https://doi.org/10.1186/s13195-018-0346-2.

Murer, M.G., Boissiere, F., Yan, Q., Hunot, S., Villares, J., Faucheux, B., Agid, Y., Hirsch, E., Raisman-Vozari, R., 1999. An immunohistochemical study of the distribution of brain-derived neurotrophic factor in the adult human brain, with particular reference to Alzheimer's disease. Neuroscience 88 (4), 1015–1032. https://doi.org/10.1016/s0306-4522(98)00219-x.

Muresanu, D.F., Sharma, A., Patnaik, R., Menon, P.K., Mössler, H., Sharma, H.S., 2019. Exacerbation of blood-brain barrier breakdown, edema formation, nitric oxide synthase upregulation and brain pathology after heat stroke in diabetic and hypertensive rats. Potential neuroprotection with cerebrolysin treatment. Int. Rev. Neurobiol. 146, 83–102. https://doi.org/10.1016/bs.irn.2019.06.007 (Epub 2019 Jul 18).

Murphy, C., Jernigan, T.L., Fennema-Notestine, C., 2003. Left hippocampal volume loss in Alzheimer's disease is reflected in performance on odor identification: a structural MRI study. J. Int. Neuropsychol. Soc. 9 (3), 459–471. https://doi.org/10.1017/S1355617703930116.

Nakano, M., Kubota, K., Kobayashi, E., Chikenji, T.S., Saito, Y., Konari, N., Fujimiya, M., 2020. Bone marrow-derived mesenchymal stem cells improve cognitive impairment in an Alzheimer's disease model by increasing the expression of microRNA-146a in hippocampus. Sci. Rep. 10 (1), 10772. https://doi.org/10.1038/s41598-020-67460-1.

Nalivaeva, N.N., Fisk, L.R., Belyaev, N.D., Turner, A.J., 2008. Amyloid-degrading enzymes as therapeutic targets in Alzheimer's disease. Curr. Alzheimer Res. 5 (2), 212–224. https://doi.org/10.2174/156720508783954785.

Nasiri, E., Alizadeh, A., Roushandeh, A.M., Gazor, R., Hashemi-Firouzi, N., Golipoor, Z., 2019. Melatonin-pretreated adipose-derived mesenchymal stem cells effceintly improved learning, memory, and cognition in an animal model of Alzheimer's disease. Metab. Brain Dis. 34 (4), 1131–1143. https://doi.org/10.1007/s11011-019-00421-4 (Epub 2019 May 25).

National Research Council, 2011. Guide for the Care and Use of Laboratory Animals, eighth ed. The National Academic Press, Washington, DC. www.nap.edu.

Nelson, A.R., Sweeney, M.D., Sagare, A.P., Zlokovic, B.V., 2016. Neurovascular dysfunction and neurodegeneration in dementia and Alzheimer's disease. Biochim. Biophys. Acta 1862 (5), 887–900. https://doi.org/10.1016/j.bbadis.2015.12.016 (Epub 2015).

Newcombe, E.A., Camats-Perna, J., Silva, M.L., Valmas, N., Huat, T.J., Medeiros, R., 2018. Inflammation: the link between comorbidities, genetics, and Alzheimer's disease. J. Neuroinflammation 15 (1), 276. https://doi.org/10.1186/s12974-018-1313-3.

Nguyen, K.V., 2019. Beta-amyloid precursor protein (APP) and the human diseases. AIMS Neurosci. 6 (4), 273–281. https://doi.org/10.3934/Neuroscience.2019.4.273 (eCollection 2019).

Nitta, A., Itoh, A., Hasegawa, T., Nabeshima, T., 1994. Beta-amyloid protein-induced Alzheimer's disease animal model. Neurosci. Lett. 170 (1), 63–66. https://doi.org/10.1016/0304-3940(94)90239-9.

Nitta, A., Fukuta, T., Hasegawa, T., Nabeshima, T., 1997. Continuous infusion of beta-amyloid protein into the rat cerebral ventricle induces learning impairment and neuronal and morphological degeneration. Jpn. J. Pharmacol. 73 (1), 51–57. https://doi.org/10.1254/jjp.73.51.

O'Bryant, S.E., Gupta, V., Henriksen, K., Edwards, M., Jeromin, A., Lista, S., Bazenet, C., Soares, H., Lovestone, S., Hampel, H., Montine, T., Blennow, K., Foroud, T., Carrillo, M., Graff-Radford, N., Laske, C., Breteler, M., Shaw, L., Trojanowski, J.Q., Schupf, N., Rissman, R.A., Fagan, A.M., Oberoi, P., Umek, R., Weiner, M.W., Grammas, P., Posner, H., Martins, R., STAR-B and BBBIG working groups, 2015. Guidelines for the standardization of preanalytic variables for blood-based biomarker studies in Alzheimer's disease research. Alzheimers Dement. 11 (5), 549–560. https://doi.org/10.1016/j.jalz.2014.08.099 (Epub 2014 Oct 1).

O'Keeffe, E., Kelly, E., Liu, Y., Giordano, C., Wallace, E., Hynes, M., Tiernan, S., Meagher, A., Greene, C., Hughes, S., Burke, T., Kealy, J., Doyle, N., Hay, A., Farrell, M., Grant, G.A., Friedman, A., Veksler, R., Molloy, M.G., Meaney, J.F., Pender, N., Camarillo, D., Doherty, C.P., Campbell, M., 2020. Dynamic blood-brain barrier regulation in mild traumatic brain injury. J. Neurotrauma 37 (2), 347–356. https://doi.org/10.1089/neu.2019.6483 (Epub 2019 Nov 8).

Obermeier, B., Daneman, R., Ransohoff, R.M., 2013. Development, maintenance and disruption of the blood-brain barrier. Nat. Med. 19 (12), 1584–1596. https://doi.org/10.1038/nm.3407 (Epub 2013 Dec 5).

Obermeier, B., Verma, A., Ransohoff, R.M., 2016. The blood-brain barrier. Handb. Clin. Neurol. 133, 39–59. https://doi.org/10.1016/B978-0-444-63432-0.00003-7.

Osborn, L.M., Kamphuis, W., Wadman, W.J., Hol, E.M., 2016. Astrogliosis: an integral player in the pathogenesis of Alzheimer's disease. Prog. Neurobiol. 144, 121–141. https://doi.org/10.1016/j.pneurobio.2016.01.001 (Epub 2016 Jan 12).

Ossenkoppele, R., Lyoo, C.H., Jester-Broms, J., Sudre, C.H., Cho, H., Ryu, Y.H., Choi, J.Y., Smith, R., Strandberg, O., Palmqvist, S., Kramer, J., Boxer, A.L., Gorno-Tempini, M.L., Miller, B.L., La Joie, R., Rabinovici, G.D., Hansson, O., 2020. Assessment of demographic, genetic, and imaging variables associated with brain resilience and cognitive resilience to pathological tau in patients with Alzheimer disease. JAMA Neurol. 77 (5), 632–642. https://doi.org/10.1001/jamaneurol.2019.5154.

Palmer, A.M., DeKosky, S.T., 1993. Monoamine neurons in aging and Alzheimer's disease. J. Neural Transm. Gen. Sect. 91 (2–3), 135–159. https://doi.org/10.1007/BF01245229.

Panza, F., Frisardi, V., Imbimbo, B.P., D'Onofrio, G., Pietrarossa, G., Seripa, D., Pilotto, A., Solfrizzi, V., 2010. Bapineuzumab: anti-beta-amyloid monoclonal antibodies for the treatment of Alzheimer's disease. Immunotherapy 2 (6), 767–782. https://doi.org/10.2217/imt.10.80.

Panza, F., Frisardi, V., Imbimbo, B.P., Seripa, D., Solfrizzi, V., Pilotto, A., 2011. Monoclonal antibodies against beta-amyloid (Abeta) for the treatment of Alzheimer's disease: the Abeta target at a crossroads. Expert. Opin. Biol. Ther. 11 (6), 679–686. https://doi.org/10.1517/14712598.2011.579099 (Epub 2011 Apr 19).

Panza, F., Lozupone, M., Logroscino, G., Imbimbo, B.P., 2019a. A critical appraisal of amyloid-beta-targeting therapies for Alzheimer disease. Nat. Rev. Neurol. 15 (2), 73–88. https://doi.org/10.1038/s41582-018-0116-6.

Panza, F., Lozupone, M., Dibello, V., Greco, A., Daniele, A., Seripa, D., Logroscino, G., Imbimbo, B.P., 2019b. Are antibodies directed against amyloid-beta (Abeta) oligomers the last call for the Abeta hypothesis of Alzheimer's disease? Immunotherapy 11 (1), 3–6. https://doi.org/10.2217/imt-2018-0119.

Park, J.C., Han, S.H., Mook-Jung, I., 2020a. Peripheral inflammatory biomarkers in Alzheimer's disease: a brief review. BMB Rep. 53 (1), 10–19. https://doi.org/10.5483/BMBRep.2020.53.1.309.

Park, B.N., Kim, J.H., Lim, T.S., Park, S.H., Kim, T.G., Yoon, B.S., Son, K.S., Yoon, J.K., An, Y.S., 2020b. Therapeutic effect of mesenchymal stem cells in an animal model of Alzheimer's disease evaluated by beta-amyloid positron emission tomography imaging. Aust. N. Z. J. Psychiatry 54 (9), 883–891. https://doi.org/10.1177/0004867420917467. Epub 2020 May.

Parker, T.D., Slattery, C.F., Yong, K.X.X., Nicholas, J.M., Paterson, R.W., Foulkes, A.J.M., Malone, I.B., Thomas, D.L., Cash, D.M., Crutch, S.J., Fox, N.C., Schott, J.M., 2019. Differences in hippocampal subfield volume are seen in phenotypic variants of early onset Alzheimer's disease. Neuroimage Clin. 21, 101632. https://doi.org/10.1016/j.nicl.2018.101632 (Epub 2018 Dec 11).

Paroni, G., Bisceglia, P., Seripa, D., 2019. Understanding the amyloid hypothesis in Alzheimer's disease. J. Alzheimers Dis. 68 (2), 493–510. https://doi.org/10.3233/JAD-180802.

Patnaik, R., Sharma, A., Skaper, S.D., Muresanu, D.F., Lafuente, J.V., Castellani, R.J., Nozari, A., Sharma, H.S., 2018. Histamine H3 inverse agonist BF 2649 or antagonist with partial H4 agonist activity clobenpropit reduces amyloid beta peptide-induced brain pathology in Alzheimer's disease. Mol. Neurobiol. 55 (1), 312–321. https://doi.org/10.1007/s12035-017-0743-8. 28861757.

Paul, J., Strickland, S., Melchor, J.P., 2007. Fibrin deposition accelerates neurovascular damage and neuroinflammation in mouse models of Alzheimer's disease. J. Exp. Med. 204 (8), 1999–2008. https://doi.org/10.1084/jem.20070304 (Epub 2007 Jul 30).

Penney, J., Ralvenius, W.T., Tsai, L.H., 2020. Modeling Alzheimer's disease with iPSC-derived brain cells. Mol. Psychiatry 25 (1), 148–167. https://doi.org/10.1038/s41380-019-0468-3 (Epub 2019 Aug 7).

Petukhova, E.O., Mukhamedshina, Y.O., Salafutdinov, I.I., Garanina, E.E., Kaligin, M.S., Leushina, A.V., Rizvanov, A.A., Reis, H.J., Palotás, A., Zefirov, A.L., Mukhamedyarov, M.A., 2019. Effects of transplanted umbilical cord blood mononuclear cells overexpressing GDNF on spatial memory and hippocampal synaptic proteins in a mouse model of Alzheimer's disease. J. Alzheimers Dis. 69 (2), 443–453. https://doi.org/10.3233/JAD-190150.

Pfisterer, U., Khodosevich, K., 2017. Neuronal survival in the brain: neuron type-specific mechanisms. Cell Death Dis. 8 (3). https://doi.org/10.1038/cddis.2017.64, e2643.

Picone, P., Nuzzo, D., Giacomazza, D., Di Carlo, M., 2020. Beta-amyloid peptide: the cell compartment multi-faceted interaction in Alzheimer's disease. Neurotox. Res. 37 (2), 250–263. https://doi.org/10.1007/s12640-019-00116-9 (Epub 2019 Dec 6).

Pillai, J.A., Maxwell, S., Bena, J., Bekris, L.M., Rao, S.M., Chance, M., Lamb, B.T., Leverenz, J.B., Alzheimer's Disease Neuroimaging Initiative, 2019. Key inflammatory pathway activations in the MCI stage of Alzheimer's disease. Ann. Clin. Transl. Neurol. 6 (7), 1248–1262. https://doi.org/10.1002/acn3.50827 (Epub 2019 Jul 4).

Pluta, R., Ułamek-Kozioł, M., Januszewski, S., Czuczwar, S., 2019. Amyloid pathology in the brain after ischemia. Folia Neuropathol. 57 (3), 220–226. https://doi.org/10.5114/fn.2019.88450.

Proessl, F., Dretsch, M.N., Connaboy, C., Lovalekar, M., Dunn-Lewis, C., Canino, M.C., Sterczala, A.J., Deshpande, G., Katz, J.S., Denney, T.S., Flanagan, S.D., 2020. Structural connectome disruptions in military personnel with mild traumatic brain injury and post-traumatic stress disorder. J. Neurotrauma 37 (19), 2102–2112. https://doi.org/10.1089/neu.2020.6999 (Epub 2020 Jun 4).

Profaci, C.P., Munji, R.N., Pulido, R.S., Daneman, R., 2020. The blood-brain barrier in health and disease: important unanswered questions. J. Exp. Med. 217 (4). https://doi.org/10.1084/jem.20190062, e20190062.

Qin, C., Lu, Y., Wang, K., Bai, L., Shi, G., Huang, Y., Li, Y., 2020. Transplantation of bone marrow mesenchymal stem cells improves cognitive deficits and alleviates neuropathology in animal models of Alzheimer's disease: a meta-analytic review on potential mechanisms. Transl. Neurodegener. 9 (1), 20. https://doi.org/10.1186/s40035-020-00199-x.

Rabbito, A., Dulewicz, M., Kulczyńska-Przybik, A., Mroczko, B., 2020. Biochemical markers in Alzheimer's disease. Int. J. Mol. Sci. 21 (6), 1989. https://doi.org/10.3390/ijms21061989.

Raby, C.A., Morganti-Kossmann, M.C., Kossmann, T., Stahel, P.F., Watson, M.D., Evans, L.-M., Mehta, P.D., Spiegel, K., Kuo, Y.M., Roher, A.E., Emmerling, M.R., 1998. Traumatic brain injury increases beta-amyloid peptide 1-42 in cerebrospinal fluid. J. Neurochem. 71 (6), 2505–2509. https://doi.org/10.1046/j.1471-4159.1998.71062505.x.

Rafferty, L.A., Cawkill, P.E., Stevelink, S.A.M., Greenberg, K., Greenberg, N., 2018. Dementia, post-traumatic stress disorder and major depressive disorder: a review of the mental health risk factors for dementia in the military veteran population. Psychol. Med. 48 (9), 1400–1409. https://doi.org/10.1017/S0033291717001386 (Epub 2018 Mar 8).

Raja, R., Rosenberg, G.A., Caprihan, A., 2018. MRI measurements of Blood-Brain Barrier function in dementia: a review of recent studies. Neuropharmacology 134 (Pt. B), 259–271. https://doi.org/10.1016/j.neuropharm.2017.10.034 (Epub 2017 Oct 28).

Ramos-Cejudo, J., Wisniewski, T., Marmar, C., Zetterberg, H., Blennow, K., de Leon, M.J., Fossati, S., 2018. Traumatic brain injury and Alzheimer's disease: the cerebrovascular link. EBioMedicine 28, 21–30. https://doi.org/10.1016/j.ebiom.2018.01.021 (Epub 2018 Jan 31).

Ransohoff, R.M., 2016. How neuroinflammation contributes to neurodegeneration. Science 353 (6301), 777–783. https://doi.org/10.1126/science.aag2590.

Rauk, A., 2008. Why is the amyloid beta peptide of Alzheimer's disease neurotoxic? Dalton Trans. 10, 1273–1282. https://doi.org/10.1039/b718601k (Epub 2008 Feb 12).

Reddy, P.H., Oliver, D.M., 2019. Amyloid beta and phosphorylated tau-induced defective autophagy and mitophagy in Alzheimer's disease. Cell 8 (5), 488. https://doi.org/10.3390/cells8050488.

Regen, F., Hellmann-Regen, J., Costantini, E., Reale, M., 2017. Neuroinflammation and Alzheimer's disease: implications for microglial activation. Curr. Alzheimer Res. 14 (11), 1140–1148. https://doi.org/10.2174/1567205014666170203141717.

Reinikainen, K.J., Paljärvi, L., Huuskonen, M., Soininen, H., Laakso, M., Riekkinen, P.J., 1988. A post-mortem study of noradrenergic, serotonergic and GABAergic neurons in Alzheimer's disease. J. Neurol. Sci. 84 (1), 101–116. https://doi.org/10.1016/0022-510x(88)90179-7.

Reuland, C.J., Church, F.C., 2020. Synergy between plasminogen activator inhibitor-1, alpha-synuclein, and neuroinflammation in Parkinson's disease. Med. Hypotheses 138, 109602. https://doi.org/10.1016/j.mehy.2020.109602 (Epub 2020 Jan 28).

Reza-Zaldivar, E.E., Hernández-Sapiéns, M.A., Minjarez, B., Gutiérrez-Mercado, Y.K., Márquez-Aguirre, A.L., Canales-Aguirre, A.A., 2018. Potential effects of MSC-derived exosomes in neuroplasticity in Alzheimer's disease. Front. Cell. Neurosci. 12, 317. https://doi.org/10.3389/fncel.2018.00317 (eCollection 2018).

Riemenschneider, M., Konta, L., Friedrich, P., Schwarz, S., Taddei, K., Neff, F., Padovani, A., Kölsch, H., Laws, S.M., Klopp, N., Bickeböller, H., Wagenpfeil, S., Mueller, J.C., Rosenberger, A., Diehl-Schmid, J., Archetti, S., Lautenschlager, N., Borroni, B.,

Müller, U., Illig, T., Heun, R., Egensperger, R., Schlegel, J., Förstl, H., Martins, R.N., Kurz, A., 2006. A functional polymorphism within plasminogen activator urokinase (PLAU) is associated with Alzheimer's disease. Hum. Mol. Genet. 15 (16), 2446–2456. https://doi.org/10.1093/hmg/ddl167 (Epub 2006 Jul 6).

Rodríguez, J.J., Noristani, H.N., Verkhratsky, A., 2012. The serotonergic system in ageing and Alzheimer's disease. Prog. Neurobiol. 99 (1), 15–41. https://doi.org/10.1016/j.pneurobio.2012.06.010 (Epub 2012 Jul 2).

Rodríguez-Martín, T., Cuchillo-Ibáñez, I., Noble, W., Nyenya, F., Anderton, B.H., Hanger, D.P., 2013. Tau phosphorylation affects its axonal transport and degradation. Neurobiol. Aging 34 (9), 2146–2157. https://doi.org/10.1016/j.neurobiolaging.2013.03.015 (Epub 2013 Apr 17).

Sahyouni, R., Gutierrez, P., Gold, E., Robertson, R.T., Cummings, B.J., 2017. Effects of concussion on the blood-brain barrier in humans and rodents. J. Concussion 1, 1–15. https://doi.org/10.1177/2059700216684518.

Samaey, C., Schreurs, A., Stroobants, S., Balschun, D., 2019. Early cognitive and behavioral deficits in mouse models for tauopathy and Alzheimer's disease. Front. Aging Neurosci. 11, 335. https://doi.org/10.3389/fnagi.2019.00335 (eCollection 2019).

Sampaio, T.B., Savall, A.S., Gutierrez, M.E.Z., Pinton, S., 2017. Neurotrophic factors in Alzheimer's and Parkinson's diseases: implications for pathogenesis and therapy. Neural Regen. Res. 12 (4), 549–557. https://doi.org/10.4103/1673-5374.205084.

Samsonraj, R.M., Raghunath, M., Nurcombe, V., Hui, J.H., van Wijnen, A.J., Cool, S.M., 2017. Concise review: multifaceted characterization of human mesenchymal stem cells for use in regenerative medicine. Stem Cells Transl. Med. 6 (12), 2173–2185. https://doi.org/10.1002/sctm.17-0129 (Epub 2017 Oct 26).

Sarnowska, A., Jablonska, A., Jurga, M., Dainiak, M., Strojek, L., Drela, K., Wright, K., Tripathi, A., Kumar, A., Jungvid, H., Lukomska, B., Forraz, N., McGuckin, C., Domanska-Janik, K., 2013. Encapsulation of mesenchymal stem cells by bioscaffolds protects cell survival and attenuates neuroinflammatory reaction in injured brain tissue after transplantation. Cell Transplant. 22 (Suppl. 1), S67–S82. https://doi.org/10.3727/096368913X672172 (Epub 2013 Sep 10).

Scheff, S.W., Price, D.A., Schmitt, F.A., Mufson, E.J., 2006. Hippocampal synaptic loss in early Alzheimer's disease and mild cognitive impairment. Neurobiol. Aging 27 (10), 1372–1384. https://doi.org/10.1016/j.neurobiolaging.2005.09.012 (Epub 2005 Nov 9).

Schindowski, K., Belarbi, K., Buée, L., 2008. Neurotrophic factors in Alzheimer's disease: role of axonal transport. Genes Brain Behav. 7 (Suppl. 1(1)), 43–56. https://doi.org/10.1111/j.1601-183X.2007.00378.x.

Schuff, N., Woerner, N., Boreta, L., Kornfield, T., Shaw, L.M., Trojanowski, J.Q., Thompson, P.M., Jack Jr., C.R., Weiner, M.W., Alzheimer's Disease Neuroimaging initiative, 2009. MRI of hippocampal volume loss in early Alzheimer's disease in relation to ApoE genotype and biomarkers. Brain 132 (Pt. 4), 1067–1077. https://doi.org/10.1093/brain/awp007 (Epub 2009 Feb 27).

Schultz, N., Janelidze, S., Byman, E., Minthon, L., Nägga, K., Hansson, O., Wennström, M., 2019. Levels of islet amyloid polypeptide in cerebrospinal fluid and plasma from patients with Alzheimer's disease. PLoS One 14 (6), e0218561. https://doi.org/10.1371/journal.pone.0218561. (eCollection 2019).

Scott, G., Ramlackhansingh, A.F., Edison, P., Hellyer, P., Cole, J., Veronese, M., Leech, R., Greenwood, R.J., Turkheimer, F.E., Gentleman, S.M., Heckemann, R.A., Matthews, P.M.,

Brooks, D.J., Sharp, D.J., 2016. Amyloid pathology and axonal injury after brain trauma. Neurology 86 (9), 821–828. https://doi.org/10.1212/WNL.0000000000002413 (Epub 2016 Feb 3).

Selkoe, D.J., Hardy, J., 2016. The amyloid hypothesis of Alzheimer's disease at 25 years. EMBO Mol. Med. 8 (6), 595–608. https://doi.org/10.15252/emmm.201606210. Print 2016 Jun.

Serý, O., Povová, J., Míšek, I., Pešák, L., Janout, V., 2013. Molecular mechanisms of neuropathological changes in Alzheimer's disease: a review. Folia Neuropathol. 51 (1), 1–9. https://doi.org/10.5114/fn.2013.34190.

Sevigny, J., Chiao, P., Bussière, T., Weinreb, P.H., Williams, L., Maier, M., Dunstan, R., Salloway, S., Chen, T., Ling, Y., O'Gorman, J., Qian, F., Arastu, M., Li, M., Chollate, S., Brennan, M.S., Quintero-Monzon, O., Scannevin, R.H., Arnold, H.M., Engber, T., Rhodes, K., Ferrero, J., Hang, Y., Mikulskis, A., Grimm, J., Hock, C., Nitsch, R.M., Sandrock, A., 2016. The antibody aducanumab reduces Abeta plaques in Alzheimer's disease. Nature 537 (7618), 50–56. https://doi.org/10.1038/nature19323.

Shallie, O.F., Dalle, E., Mabandla, M.V., 2020. Memory decline correlates with increased plasma cytokines in amyloid-beta (1-42) rat model of Alzheimer's disease. Neurobiol. Learn. Mem. 169, 107187. https://doi.org/10.1016/j.nlm.2020.107187 (Epub 2020 Feb 12).

Shao, X., Cui, W., Xie, X., Ma, W., Zhan, Y., Lin, Y., 2020. Treatment of Alzheimer's disease with framework nucleic acids. Cell Prolif. 53 (4). https://doi.org/10.1111/cpr.12787, e12787 (Epub 2020 Mar 12).

Sharif, Y., Jumah, F., Coplan, L., Krosser, A., Sharif, K., Tubbs, R.S., 2018. Blood brain barrier: a review of its anatomy and physiology in health and disease. Clin. Anat. 31 (6), 812–823. https://doi.org/10.1002/ca.23083 (Epub 2018 Oct 18).

Sharma, H.S., 1987. Effect of captopril (a converting enzyme inhibitor) on blood-brain barrier permeability and cerebral blood flow in normotensive rats. Neuropharmacology 26 (1), 85–92. https://doi.org/10.1016/0028-3908(87)90049-9.

Sharma, H.S., 2004. Blood-brain and spinal cord barriers in stress. In: Sharma, H.S., Westman, J. (Eds.), The Blood-Spinal Cord and Brain Barriers in Health and Disease. Elsevier Academic Press, San Diego, pp. 231–298.

Sharma, H.S., 2007. Neurotrophic factors in combination: a possible new therapeutic strategy to influence pathophysiology of spinal cord injury and repair mechanisms. Curr. Pharm. Des. 13 (18), 1841–1874. https://doi.org/10.2174/138161207780858410.

Sharma, H.S., 2009. Blood–central nervous system barriers: the gateway to neurodegeneration, neuroprotection and neuroregeneration. In: Lajtha, A., Banik, N., Ray, S.K. (Eds.), Handbook of Neurochemistry and Molecular Neurobiology: Brain and Spinal Cord Trauma. Springer Verlag, Berlin, Heidelberg, New York, pp. 363–457.

Sharma, H.S., Cervós-Navarro, J., 1990. Brain oedema and cellular changes induced by acute heat stress in young rats. Acta Neurochir. Suppl. (Wien) 51, 383–386. https://doi.org/10.1007/978-3-7091-9115-6_129.

Sharma, H.S., Dey, P.K., 1981. Impairment of blood-brain barrier (BBB) in rat by immobilization stress: role of serotonin (5-HT). Indian J. Physiol. Pharmacol. 25 (2), 111–122.

Sharma, H.S., Dey, P.K., 1986. Influence of long-term immobilization stress on regional blood-brain barrier permeability, cerebral blood flow and 5-HT level in conscious normotensive young rats. J. Neurol. Sci. 72 (1), 61–76. https://doi.org/10.1016/0022-510x(86)90036-5.

Sharma, H.S., Dey, P.K., 1987. Influence of long-term acute heat exposure on regional blood-brain barrier permeability, cerebral blood flow and 5-HT level in conscious normotensive young rats. Brain Res. 424 (1), 153–162. https://doi.org/10.1016/0006-8993(87)91205-4.

Sharma, H.S., Johanson, C.E., 2007a. Blood-cerebrospinal fluid barrier in hyperthermia. Prog. Brain Res. 162, 459–478. https://doi.org/10.1016/S0079-6123(06)62023-2.

Sharma, H.S., Johanson, C.E., 2007b. Intracerebroventricularly administered neurotrophins attenuate blood cerebrospinal fluid barrier breakdown and brain pathology following whole-body hyperthermia: an experimental study in the rat using biochemical and morphological approaches. Ann. N. Y. Acad. Sci. 1122, 112–129. https://doi.org/10.1196/annals.1403.008.

Sharma, H.S., Kiyatkin, E.A., 2009. Rapid morphological brain abnormalities during acute methamphetamine intoxication in the rat: an experimental study using light and electron microscopy. J. Chem. Neuroanat. 37 (1), 18–32. https://doi.org/10.1016/j.jchemneu.2008.08.002 (Epub 2008 Aug 19).

Sharma, H.S., Olsson, Y., 1990. Edema formation and cellular alterations following spinal cord injury in the rat and their modification with p-chlorophenylalanine. Acta Neuropathol. 79 (6), 604–610. https://doi.org/10.1007/BF00294237.

Sharma, H.S., Sharma, A., 2010. Breakdown of the blood-brain barrier in stress alters cognitive dysfunction and induces brain pathology: new perspectives for neuroprotective strategies. In: Ritsner, M.S. (Ed.), Brain Protection in Schizophrenia, Mood and Cognitive Disorders. Springer Science+Business Media B.V, New York, USA, pp. 243–304, https://doi.org/10.1007/978-90-481-8553-5.

Sharma, A., Sharma, H.S., 2012a. Monoclonal antibodies as novel neurotherapeutic agents in CNS injury and repair. Int. Rev. Neurobiol. 102, 23–45. https://doi.org/10.1016/B978-0-12-386986-9.00002-8.

Sharma, H.S., Sharma, A., 2012b. Nanowired drug delivery for neuroprotection in central nervous system injuries: modulation by environmental temperature, intoxication of nanoparticles, and comorbidity factors. Wiley Interdisip. Rev. Nanomed. Nanobiotechnol. 4 (2), 184–203. https://doi.org/10.1002/wnan.172 (Epub 2011 Dec 8) 22162425.

Sharma, H.S., Sharma, A., 2013. New perspectives of nanoneuroprotection, nanoneuropharmacology and nanoneurotoxicity: modulatory role of amino acid neurotransmitters, stress, trauma, and co-morbidity factors in nanomedicine. Amino Acids 45 (5), 1055–1071. https://doi.org/10.1007/s00726-013-1584-z.

Sharma, H.S., Sharma, A., 2019. New therapeutic strategies for brain edema and cell injury. Int. Rev. Neurobiol. 146, 2–295. https://doi.org/10.1016/S0074-7742(19)30061-3.

Sharma, H.S., Sjöquist, P.O., 2002. A new antioxidant compound H-290/51 modulates glutamate and GABA immunoreactivity in the rat spinal cord following trauma. Amino Acids 23 (1–3), 261–272. https://doi.org/10.1007/s00726-001-0137-z.

Sharma, H.S., Westman, J., 2004. The Blood-Spinal Cord and Brain Barriers in Health and Disease. Academic Press, San Diego, pp. 1–617 (Release date: Nov. 9, 2003).

Sharma, H.S., Olsson, Y., Dey, P.K., 1990a. Changes in blood-brain barrier and cerebral blood flow following elevation of circulating serotonin level in anesthetized rats. Brain Res. 517 (1–2), 215–223. https://doi.org/10.1016/0006-8993(90)91029-g.

Sharma, H.S., Olsson, Y., Dey, P.K., 1990b. Early accumulation of serotonin in rat spinal cord subjected to traumatic injury. Relation to edema and blood flow changes. Neuroscience 36 (3), 725–730. https://doi.org/10.1016/0306-4522(90)90014-u.

Sharma, H.S., Cervós-Navarro, J., Dey, P.K., 1991. Acute heat exposure causes cellular alteration in cerebral cortex of young rats. Neuroreport 2 (3), 155–158. https://doi.org/10.1097/00001756-199103000-00012.

Sharma, H.S., Kretzschmar, R., Cervós-Navarro, J., Ermisch, A., Rühle, H.J., Dey, P.K., 1992a. Age-related pathophysiology of the blood-brain barrier in heat stress. Prog. Brain Res. 91, 189–196. https://doi.org/10.1016/s0079-6123(08)62334-1.

Sharma, H.S., Zimmer, C., Westman, J., Cervós-Navarro, J., 1992b. Acute systemic heat stress increases glial fibrillary acidic protein immunoreactivity in brain: experimental observations in conscious normotensive young rats. Neuroscience 48 (4), 889–901. https://doi.org/10.1016/0306-4522(92)90277-9.

Sharma, H.S., Olsson, Y., Cervós-Navarro, J., 1993. Early perifocal cell changes and edema in traumatic injury of the spinal cord are reduced by indomethacin, an inhibitor of prostaglandin synthesis. Experimental study in the rat. Acta Neuropathol. 85 (2), 145–153. https://doi.org/10.1007/BF00227761.

Sharma, H.S., Olsson, Y., Persson, S., Nyberg, F., 1995. Trauma-induced opening of the the blood-spinal cord barrier is reduced by indomethacin, an inhibitor of prostaglandin biosynthesis. Experimental observations in the rat using [131I]-sodium, Evans blue and lanthanum as tracers. Restor. Neurol. Neurosci. 7 (4), 207–215. https://doi.org/10.3233/RNN-1994-7403.

Sharma, H.S., Westman, J., Nyberg, F., 1998a. Pathophysiology of brain edema and cell changes following hyperthermic brain injury. Prog. Brain Res. 115, 351–412. https://doi.org/10.1016/s0079-6123(08)62043-9.

Sharma, H.S., Nyberg, F., Gordh, T., Alm, P., Westman, J., 1998b. Neurotrophic factors attenuate neuronal nitric oxide synthase upregulation, microvascular permeability disturbances, edema formation and cell injury in the spinal cord following trauma. In: Stålberg, E., Sharma, H.S., Olsson, Y. (Eds.), Spinal Cord Monitoring. Basic Principles, Regeneration, Pathophysiology and Clinical Aspects. Springer, Wien New York, pp. 118–148.

Sharma, H.S., Alm, P., Westman, J., 1998c. Nitric oxide and carbon monoxide in the brain pathology of heat stress. Prog. Brain Res. 115, 297–333. https://doi.org/10.1016/s0079-6123(08)62041-5.

Sharma, H.S., Patnaik, R., Patnaik, S., Mohanty, S., Sharma, A., Vannemreddy, P., 2007. Antibodies to serotonin attenuate closed head injury induced blood brain barrier disruption and brain pathology. Ann. N. Y. Acad. Sci. 1122, 295–312. https://doi.org/10.1196/annals.1403.022.

Sharma, H.S., Muresanu, D.F., Patnaik, R., Stan, A.D., Vacaras, V., Perju-Dumbrav, L., Alexandru, B., Buzoianu, A., Opincariu, I., Menon, P.K., Sharma, A., 2011a. Superior neuroprotective effects of cerebrolysin in heat stroke following chronic intoxication of Cu or Ag engineered nanoparticles. A comparative study with other neuroprotective agents using biochemical and morphological approaches in the rat. J. Nanosci. Nanotechnol. 11 (9), 7549–7569. https://doi.org/10.1166/jnn.2011.5114.

Sharma, H.S., Miclescu, A., Wiklund, L., 2011b. Cardiac arrest-induced regional blood-brain barrier breakdown, edema formation and brain pathology: a light and electron microscopic study on a new model for neurodegeneration and neuroprotection in porcine brain. J. Neural. Transm. (Vienna) 118 (1), 87–114. https://doi.org/10.1007/s00702-010-0486-4.

Sharma, H.S., Castellani, R.J., Smith, M.A., Sharma, A., 2012. The blood-brain barrier in Alzheimer's disease: novel therapeutic targets and nanodrug delivery. Int. Rev. Neurobiol. 102, 47–90. https://doi.org/10.1016/B978-0-12-386986-9.00003-X.

Sharma, H.S., Muresanu, D.F., Lafuente, J.V., Sjöquist, P.O., Patnaik, R., Sharma, A., 2015a. Nanoparticles exacerbate both ubiquitin and heat shock protein expressions in spinal cord injury: neuroprotective effects of the proteasome inhibitor carfilzomib and the antioxidant compound H-290/51. Mol. Neurobiol. 52 (2), 882–898. https://doi.org/10.1007/s12035-015-9297-9.

Sharma, H.S., Feng, L., Lafuente, J.V., Muresanu, D.F., Tian, Z.R., Patnaik, R., Sharma, A., 2015b. TiO$_2$-nanowired delivery of mesenchymal stem cells thwarts diabetes-induced exacerbation of brain pathology in heat stroke: an experimental study in the rat using morphological and biochemical approaches. CNS Neurol. Disord. Drug Targets 14 (3), 386–399. https://doi.org/10.2174/1871527314666150318114335.

Sharma, H.S., Muresanu, D.F., Lafuente, J.V., Nozari, A., Patnaik, R., Skaper, S.D., Sharma, A., 2016a. Pathophysiology of blood-brain barrier in brain injury in cold and hot environments: novel drug targets for neuroprotection. CNS Neurol. Disord. Drug Targets 15 (9), 1045–1071. https://doi.org/10.2174/1871527315666160902145145.

Sharma, H.S., Muresanu, D.F., Sharma, A., 2016b. Alzheimer's disease: cerebrolysin and nanotechnology as a therapeutic strategy. Neurodegener. Dis. Manag. 6 (6), 453–456. https://doi.org/10.2217/nmt-2016-0037 (Epub 2016 Nov 9).

Sharma, A., Menon, P.K., Patnaik, R., Muresanu, D.F., Lafuente, J.V., Tian, Z.R., Ozkizilcik, A., Castellani, R.J., Mössler, H., Sharma, H.S., 2017. Novel treatment strategies using TiO(2)-nanowired delivery of histaminergic drugs and antibodies to tau with cerebrolysin for superior neuroprotection in the pathophysiology of Alzheimer's disease. Int. Rev. Neurobiol. 137, 123–165. https://doi.org/10.1016/bs.irn.2017.09.002 (Epub 2017 Nov 6).

Sharma, H.S., Muresanu, D.F., Lafuente, J.V., Patnaik, R., Tian, Z.R., Ozkizilcik, A., Castellani, R.J., Mössler, H., Sharma, A., 2018. Co-administration of TiO2 nanowired mesenchymal stem cells with cerebrolysin potentiates neprilysin level and reduces brain pathology in Alzheimer's disease. Mol. Neurobiol. 55 (1), 300–311. https://doi.org/10.1007/s12035-017-0742-9.

Sharma, H.S., Muresanu, D.F., Castellani, R.J., Nozari, A., Lafuente, J.V., Tian, Z.R., Ozkizilcik, A., Manzhulo, I., Mössler, H., Sharma, A., 2019a. Nanowired delivery of cerebrolysin with neprilysin and p-Tau antibodies induces superior neuroprotection in Alzheimer's disease. Prog. Brain Res. 245, 145–200. https://doi.org/10.1016/bs.pbr.2019.03.009 (Epub 2019 Apr 2).

Sharma, A., Muresanu, D.F., Ozkizilcik, A., Tian, Z.R., Lafuente, J.V., Manzhulo, I., Mössler, H., Sharma, H.S., 2019b. Sleep deprivation exacerbates concussive head injury induced brain pathology: neuroprotective effects of nanowired delivery of cerebrolysin with alpha-melanocyte-stimulating hormone. Prog. Brain Res. 245, 1–55. https://doi.org/10.1016/bs.pbr.2019.03.002 (Epub 2019 Apr 2).

Sharma, H.S., Muresanu, D.F., Nozari, A., Castellani, R.J., Dey, P.K., Wiklund, L., Sharma, A., 2019c. Anesthetics influence concussive head injury induced blood-brain barrier breakdown, brain edema formation, cerebral blood flow, serotonin levels, brain pathology and functional outcome. Int. Rev. Neurobiol. 146, 45–81. https://doi.org/10.1016/bs.irn.2019.06.006 (Epub 2019 Jul 8).

Sharma, A., Muresanu, D.F., Castellani, R.J., Nozari, A., Lafuente, J.V., Sahib, S., Tian, Z.R., Buzoianu, A.D., Patnaik, R., Wiklund, L., Sharma, H.S., 2020a. Mild traumatic brain injury exacerbates Parkinson's disease induced hemeoxygenase-2 expression and brain pathology: neuroprotective effects of co-administration of TiO(2) nanowired mesenchymal stem cells and cerebrolysin. Prog. Brain Res. 258, 157–231. https://doi.org/10.1016/bs.pbr.2020.09.010 (Epub 2020 Nov 9).

Sharma, A., Muresanu, D.F., Sahib, S., Tian, Z.R., Castellani, R.J., Nozari, A., Lafuente, J.V., Buzoianu, A.D., Bryukhovetskiy, I., Manzhulo, I., Patnaik, R., Wiklund, L., Sharma, H.S., 2020b. Concussive head injury exacerbates neuropathology of sleep deprivation: superior neuroprotection by co-administration of TiO(2)-nanowired cerebrolysin, alpha-melanocyte-stimulating hormone, and mesenchymal stem cells. Prog. Brain Res. 258, 1–77. https://doi.org/10.1016/bs.pbr.2020.09.003 (Epub 2020 Nov 2).

Sharma, H.S., Sahib, S., Tian, Z.R., Muresanu, D.F., Nozari, A., Castellani, R.J., Lafuente, J.-V., Wiklund, L., Sharma, A., 2020c. Protein kinase inhibitors in traumatic brain injury and repair: new roles of nanomedicine. Prog. Brain Res. 258, 233–283. https://doi.org/10.1016/bs.pbr.2020.09.009 (Epub 2020 Oct 24).

Shaw, M.A., Gao, Z., McElhinney, K.E., Thornton, S., Flick, M.J., Lane, A., Degen, J.L., Ryu, J.K., Akassoglou, K., Mullins, E.S., 2017. Plasminogen deficiency delays the onset and protects from demyelination and paralysis in autoimmune neuroinflammatory disease. J. Neurosci. 37 (14), 3776–3788. https://doi.org/10.1523/JNEUROSCI.2932-15.2017 (Epub 2017 Mar 8).

Shibata, M., Yamada, S., Kumar, S.R., Calero, M., Bading, J., Frangione, B., Holtzman, D.M., Miller, C.A., Strickland, D.K., Ghiso, J., Zlokovic, B.V., 2000. Clearance of Alzheimer's amyloid-ss(1-40) peptide from brain by LDL receptor-related protein-1 at the blood-brain barrier. J. Clin. Invest. 106 (12), 1489–1499. https://doi.org/10.1172/JCI10498.

Shimada, M., Meguro, K., Inagaki, H., Ishizaki, J., Yamadori, A., 2001. Global intellectual deterioration in Alzheimer's disease and a reverse model of intellectual development: an applicability of the Binet scale. Psychiatry Clin. Neurosci. 55 (6), 559–563. https://doi.org/10.1046/j.1440-1819.2001.00906.x.

Shishido, H., Ueno, M., Sato, K., Matsumura, M., Toyota, Y., Kirino, Y., Tamiya, T., Kawai, N., Kishimoto, Y., 2019. Traumatic brain injury by weight-drop method causes transient amyloid-beta deposition and acute cognitive deficits in mice. Behav. Neurol. 2019, 3248519. https://doi.org/10.1155/2019/3248519 (eCollection 2019).

Shokri-Kojori, E., Wang, G.J., Wiers, C.E., Demiral, S.B., Guo, M., Kim, S.W., Lindgren, E., Ramirez, V., Zehra, A., Freeman, C., Miller, G., Manza, P., Srivastava, T., De Santi, S., Tomasi, D., Benveniste, H., Volkow, N.D., 2018. Beta-amyloid accumulation in the human brain after one night of sleep deprivation. Proc. Natl. Acad. Sci. U. S. A. 115 (17), 4483–4488. https://doi.org/10.1073/pnas.1721694115 (Epub 2018 Apr 9).

Silverberg, N.D., Iaccarino, M.A., Panenka, W.J., Iverson, G.L., McCulloch, K.L., Dams-O'Connor, K., Reed, N., McCrea, M., American Congress of Rehabilitation Medicine Brain Injury Interdisciplinary Special Interest Group Mild TBI Task Force, 2020. Management of concussion and mild traumatic brain injury: a synthesis of practice guidelines. Arch. Phys. Med. Rehabil. 101 (2), 382–393. https://doi.org/10.1016/j.apmr.2019.10.179 (Epub 2019 Oct 23).

Simão, F., Ustunkaya, T., Clermont, A.C., Feener, E.P., 2017. Plasma kallikrein mediates brain hemorrhage and edema caused by tissue plasminogen activator therapy in mice after stroke. Blood 129 (16), 2280–2290. https://doi.org/10.1182/blood-2016-09-740670 (Epub 2017 Jan 27).

Šimić, G., Babić Leko, M., Wray, S., Harrington, C., Delalle, I., Jovanov-Milošević, N., Bažadona, D., Buée, L., de Silva, R., Di Giovanni, G., Wischik, C., Hof, P.R., 2016. Tau protein hyperphosphorylation and aggregation in Alzheimer's disease and other tauopathies, and possible neuroprotective strategies. Biomolecules 6 (1), 6. https://doi.org/10.3390/biom6010006.

Šimić, G., Babić Leko, M., Wray, S., Harrington, C.R., Delalle, I., Jovanov-Milošević, N., Bažadona, D., Buée, L., de Silva, R., Di Giovanni, G., Wischik, C.M., Hof, P.R., 2017. Monoaminergic neuropathology in Alzheimer's disease. Prog. Neurobiol. 151, 101–138. https://doi.org/10.1016/j.pneurobio.2016.04.001 (Epub 2016 Apr 12).

Simon, D.W., McGeachy, M.J., Bayır, H., Clark, R.S., Loane, D.J., Kochanek, P.M., 2017. The far-reaching scope of neuroinflammation after traumatic brain injury. Nat. Rev. Neurol. 13 (3), 171–191. https://doi.org/10.1038/nrneurol.2017.13 (Epub 2017 Feb 10).

Skaper, S.D., 2018. Neurotrophic factors: an overview. Methods Mol. Biol. 1727, 1–17. https://doi.org/10.1007/978-1-4939-7571-6_1.

Skaper, S.D., Facci, L., Zusso, M., Giusti, P., 2017. Synaptic plasticity, dementia and Alzheimer disease. CNS Neurol. Disord. Drug Targets 16 (3), 220–233. https://doi.org/10.2174/1871527316666170113120853.

Sopova, K., Gatsiou, K., Stellos, K., Laske, C., 2014. Dysregulation of neurotrophic and haematopoietic growth factors in Alzheimer's disease: from pathophysiology to novel treatment strategies. Curr. Alzheimer Res. 11 (1), 27–39. https://doi.org/10.2174/1567205010666131120100743.

Soria Lopez, J.A., González, H.M., Léger, G.C., 2019. Alzheimer's disease. Handb. Clin. Neurol. 167, 231–255. https://doi.org/10.1016/B978-0-12-804766-8.00013-3.

Sporns, O., Tononi, G., Kötter, R., 2005. The human connectome: a structural description of the human brain. PLoS Comput. Biol. 1 (4), e42. https://doi.org/10.1371/journal.pcbi.0010042.

Stampanoni Bassi, M., Iezzi, E., Gilio, L., Centonze, D., Buttari, F., 2019. Synaptic plasticity shapes brain connectivity: implications for network topology. Int. J. Mol. Sci. 20 (24), 6193. https://doi.org/10.3390/ijms20246193.

Stein, T.D., Montenigro, P.H., Alvarez, V.E., Xia, W., Crary, J.F., Tripodis, Y., Daneshvar, D.H., Mez, J., Solomon, T., Meng, G., Kubilus, C.A., Cormier, K.A., Meng, S., Babcock, K., Kiernan, P., Murphy, L., Nowinski, C.J., Martin, B., Dixon, D., Stern, R.A., Cantu, R.C., Kowall, N.W., McKee, A.C., 2015. Beta-amyloid deposition in chronic traumatic encephalopathy. Acta Neuropathol. 130 (1), 21–34. https://doi.org/10.1007/s00401-015-1435-y (Epub 2015 May 6).

Stern, R.A., Otvos Jr., L., Trojanowski, J.Q., Lee, V.M., 1989. Monoclonal antibodies to a synthetic peptide homologous with the first 28 amino acids of Alzheimer's disease beta-protein recognize amyloid and diverse glial and neuronal cell types in the central nervous system. Am. J. Pathol. 134 (5), 973–978.

Stern, R.A., Adler, C.H., Chen, K., Navitsky, M., Luo, J., Dodick, D.W., Alosco, M.L., Tripodis, Y., Goradia, D.D., Martin, B., Mastroeni, D., Fritts, N.G., Jarnagin, J., Devous Sr., M.D., Mintun, M.A., Pontecorvo, M.J., Shenton, M.E., Reiman, E.M., 2019. Tau positron-emission tomography in former national football league players. N. Engl. J. Med. 380 (18), 1716–1725. https://doi.org/10.1056/NEJMoa1900757 (Epub 2019 Apr 10).

Stokum, J.A., Gerzanich, V., Simard, J.M., 2016. Molecular pathophysiology of cerebral edema. J. Cereb. Blood Flow Metab. 36 (3), 513–538. https://doi.org/10.1177/0271678X15617172. Epub 2015.

Straten, G., Eschweiler, G.W., Maetzler, W., Laske, C., Leyhe, T., 2009. Glial cell-line derived neurotrophic factor (GDNF) concentrations in cerebrospinal fluid and serum of patients with early Alzheimer's disease and normal controls. J. Alzheimers Dis. 18 (2), 331–337. https://doi.org/10.3233/JAD-2009-1146.

Straten, G., Saur, R., Laske, C., Gasser, T., Annas, P., Basun, H., Leyhe, T., 2011. Influence of lithium treatment on GDNF serum and CSF concentrations in patients with early Alzheimer's disease. Curr. Alzheimer Res. 8 (8), 853–859. https://doi.org/10.2174/156720511798192754.

Sturchler-Pierrat, C., Abramowski, D., Duke, M., Wiederhold, K.H., Mistl, C., Rothacher, S., Ledermann, B., Bürki, K., Frey, P., Paganetti, P.A., Waridel, C., Calhoun, M.E., Jucker, M., Probst, A., Staufenbiel, M., Sommer, B., 1997. Two amyloid precursor protein transgenic mouse models with Alzheimer disease-like pathology. Proc. Natl. Acad. Sci. U. S. A. 94 (24), 13287–13292. https://doi.org/10.1073/pnas.94.24.13287.

Su, B., Wang, X., Nunomura, A., Moreira, P.I., Lee, H.G., Perry, G., Smith, M.A., Zhu, X., 2008. Oxidative stress signaling in Alzheimer's disease. Curr. Alzheimer Res. 5 (6), 525–532. https://doi.org/10.2174/156720508786898451.

Sulhan, S., Lyon, K.A., Shapiro, L.A., Huang, J.H., 2020. Neuroinflammation and blood-brain barrier disruption following traumatic brain injury: pathophysiology and potential therapeutic targets. J. Neurosci. Res. 98 (1), 19–28. https://doi.org/10.1002/jnr.24331 (Epub 2018 Sep 27).

Sulimai, N., Lominadze, D., 2020. Fibrinogen and neuroinflammation during traumatic brain injury. Mol. Neurobiol. 57 (11), 4692–4703. https://doi.org/10.1007/s12035-020-02012-2 (Epub 2020 Aug 10).

Sung, P.S., Lin, P.Y., Liu, C.H., Su, H.C., Tsai, K.J., 2020. Neuroinflammation and neurogenesis in Alzheimer's disease and potential therapeutic approaches. Int. J. Mol. Sci. 21 (3), 701. https://doi.org/10.3390/ijms21030701.

Süß, P., Schlachetzki, J.C.M., 2020. Microglia in Alzheimer's disease. Curr. Alzheimer Res. 17 (1), 29–43. https://doi.org/10.2174/1567205017666200212155234.

Sutton, E.T., Thomas, T., Bryant, M.W., Landon, C.S., Newton, C.A., Rhodin, J.A., 1999. Amyloid-beta peptide induced inflammatory reaction is mediated by the cytokines tumor necrosis factor and interleukin-1. J. Submicrosc. Cytol. Pathol. 31 (3), 313–323.

Suzuki, Y., Nakamura, Y., Yamada, K., Igarashi, H., Kasuga, K., Yokoyama, Y., Ikeuchi, T., Nishizawa, M., Kwee, I.L., Nakada, T., 2015. Reduced CSF water influx in Alzheimer's disease supporting the beta-amyloid clearance hypothesis. PLoS One 10 (5). https://doi.org/10.1371/journal.pone.0123708, e0123708 (eCollection 2015).

Sweeney, M.D., Sagare, A.P., Zlokovic, B.V., 2018. Blood-brain barrier breakdown in Alzheimer disease and other neurodegenerative disorders. Nat. Rev. Neurol. 14 (3), 133–150. https://doi.org/10.1038/nrneurol.2017.188 (Epub 2018 Jan 29).

Sweeney, M.D., Zhao, Z., Montagne, A., Nelson, A.R., Zlokovic, B.V., 2019. Blood-brain barrier: from physiology to disease and Back. Physiol. Rev. 99 (1), 21–78. https://doi.org/10.1152/physrev.00050.2017.

Syrovets, T., Lunov, O., Simmet, T., 2012. Plasmin as a proinflammatory cell activator. J. Leukoc. Biol. 92 (3), 509–519. https://doi.org/10.1189/jlb.0212056 (Epub 2012 May 4).

Tabatabaei-Jafari, H., Walsh, E., Shaw, M.E., Cherbuin, N., Alzheimer's Disease Neuroimaging Initiative (ADNI), 2017. The cerebellum shrinks faster than normal ageing in Alzheimer's disease but not in mild cognitive impairment. Hum. Brain Mapp. 38 (6), 3141–3150. https://doi.org/10.1002/hbm.23580 (Epub 2017 Mar 21).

Tanila, H., 2017. The role of BDNF in Alzheimer's disease. Neurobiol. Dis. 97 (Pt. B), 114–118. https://doi.org/10.1016/j.nbd.2016.05.008 (Epub 2016 May 13).

Teng, Y., Zhao, J., Ding, L., Ding, Y., Zhou, P., 2019. Complex of EGCG with Cu(II) suppresses amyloid aggregation and Cu(II)-induced cytotoxicity of alpha-synuclein. Molecules 24 (16), 2940. https://doi.org/10.3390/molecules24162940.

Thal, D.R., Fändrich, M., 2015. Protein aggregation in Alzheimer's disease: Abeta and tau and their potential roles in the pathogenesis of AD. Acta Neuropathol. 129 (2), 163–165. https://doi.org/10.1007/s00401-015-1387-2.

Thoenen, H., Barde, Y.A., Davies, A.M., Johnson, J.E., 1987. Neurotrophic factors and neuronal death. CIBA Found. Symp. 126, 82–95. https://doi.org/10.1002/9780470513422.ch6.

Timaru-Kast, R., Luh, C., Gotthardt, P., Huang, C., Schäfer, M.K., Engelhard, K., Thal, S.C., 2012. Influence of age on brain edema formation, secondary brain damage and inflammatory response after brain trauma in mice. PLoS One 7 (8). https://doi.org/10.1371/journal.pone.0043829, e43829 (Epub 2012 Aug 30).

Tolar, M., Abushakra, S., Hey, J.A., Porsteinsson, A., Sabbagh, M., 2020. Aducanumab, gantenerumab, BAN2401, and ALZ-801-the first wave of amyloid-targeting drugs for Alzheimer's disease with potential for near term approval. Alzheimers Res. Ther. 12 (1), 95. https://doi.org/10.1186/s13195-020-00663-w.

Tong, B.C., Wu, A.J., Li, M., Cheung, K.H., 2018. Calcium signaling in Alzheimer's disease & therapies. Biochim. Biophys. Acta, Mol. Cell Res. 1865 (11 Pt. B), 1745–1760. https://doi.org/10.1016/j.bbamcr.2018.07.018 (Epub 2018 Jul 29).

Tönnies, E., Trushina, E., 2017. Oxidative stress, synaptic dysfunction, and Alzheimer's disease. J. Alzheimers Dis. 57 (4), 1105–1121. https://doi.org/10.3233/JAD-161088.

Tõugu, V., Karafin, A., Palumaa, P., 2008. Binding of zinc(II) and copper(II) to the full-length Alzheimer's amyloid-beta peptide. J. Neurochem. 104 (5), 1249–1259. https://doi.org/10.1111/j.1471-4159.2007.05061.x.

Tromp, D., Dufour, A., Lithfous, S., Pebayle, T., Després, O., 2015. Episodic memory in normal aging and Alzheimer disease: insights from imaging and behavioral studies. Ageing Res. Rev. 24 (Pt. B), 232–262. https://doi.org/10.1016/j.arr.2015.08.006 (Epub 2015 Aug 28).

Tucker, H.M., Kihiko, M., Caldwell, J.N., Wright, S., Kawarabayashi, T., Price, D., Walker, D., Scheff, S., McGillis, J.P., Rydel, R.E., Estus, S., 2000. The plasmin system is induced by and degrades amyloid-beta aggregates. J. Neurosci. 20 (11), 3937–3946. https://doi.org/10.1523/JNEUROSCI.20-11-03937.2000.

Ubukata, S., Oishi, N., Higashi, T., Kagawa, S., Yamauchi, H., Okuyama, C., Watanabe, H., Ono, M., Saji, H., Aso, T., Murai, T., Ueda, K., 2020. Spatial patterns of amyloid deposition in patients with chronic focal or diffuse traumatic brain injury using (18)F-FPYBF-2 PET. Neuropsychiatr. Dis. Treat. 16, 2719–2732. https://doi.org/10.2147/NDT.S268504 (eCollection 2020).

Uemura, M.T., Maki, T., Ihara, M., Lee, V.M.Y., Trojanowski, J.Q., 2020. Brain microvascular pericytes in vascular cognitive impairment and dementia. Front. Aging Neurosci. 12, 80. https://doi.org/10.3389/fnagi.2020.00080 (eCollection 2020).

Vakalopoulos, C., 2017. Alzheimer's disease: the alternative serotonergic hypothesis of cognitive decline. J. Alzheimers Dis. 60 (3), 859–866. https://doi.org/10.3233/JAD-170364.

van Dyck, C.H., 2018. Anti-amyloid-beta monoclonal antibodies for Alzheimer's disease: pitfalls and promise. Biol. Psychiatry 83 (4), 311–319. https://doi.org/10.1016/j.biopsych.2017.08.010 (Epub 2017 Aug 24).

van de Haar, H.J., Burgmans, S., Jansen, J.F., van Osch, M.J., van Buchem, M.A., Muller, M., Hofman, P.A., Verhey, F.R., Backes, W.H., 2016a. Blood-brain barrier leakage in patients with early Alzheimer disease. Radiology 281 (2), 527–535. https://doi.org/10.1148/radiol.2016152244 (Epub 2016 May 31).

van de Haar, H.J., Jansen, J.F.A., van Osch, M.J.P., van Buchem, M.A., Muller, M., Wong, S.M., Hofman, P.A.M., Burgmans, S., Verhey, F.R.J., Backes, W.H., 2016b. Neurovascular unit impairment in early Alzheimer's disease measured with magnetic resonance imaging. Neurobiol. Aging 45, 190–196. https://doi.org/10.1016/j.neurobiolaging.2016.06.006 (Epub 2016 Jun 17).

van de Haar, H.J., Jansen, J.F.A., Jeukens, C.R.L.P.N., Burgmans, S., van Buchem, M.A., Muller, M., Hofman, P.A.M., Verhey, F.R.J., van Osch, M.J.P., Backes, W.H., 2017. Subtle blood-brain barrier leakage rate and spatial extent: considerations for dynamic contrast-enhanced MRI. Med. Phys. 44 (8), 4112–4125. https://doi.org/10.1002/mp.12328 (Epub 2017 Jul 10).

van der Kant, R., Goldstein, L.S.B., Ossenkoppele, R., 2020. Amyloid-beta-independent regulators of tau pathology in Alzheimer disease. Nat. Rev. Neurosci. 21 (1), 21–35. https://doi.org/10.1038/s41583-019-0240-3 (Epub 2019 Nov 28).

van Engelen, M.E., Gossink, F.T., de Vijlder, L.S., Meursing, J.R.A., Scheltens, P., Dols, A., Pijnenburg, Y.A.L., 2020. End stage clinical features and cause of death of behavioral variant frontotemporal dementia and young-onset Alzheimer's disease. J. Alzheimers Dis. 77 (3), 1169–1180. https://doi.org/10.3233/JAD-200337.

Vanmali, B.H., Romanova, E.V., Messner, M.C., Singh, M., Maruniak, J., Sweedler, J., Kirk, M.D., 2003. Endogenous neurotrophic factors enhance neurite growth by bag cell neurons of Aplysia. J. Neurobiol. 56 (1), 78–93. https://doi.org/10.1002/neu.10221.

Veitch, D.P., Friedl, K.E., Weiner, M.W., 2013. Military risk factors for cognitive decline, dementia and Alzheimer's disease. Curr. Alzheimer Res. 10 (9), 907–930. https://doi.org/10.2174/15672050113109990142.

Veitch, D.P., Weiner, M.W., Aisen, P.S., Beckett, L.A., Cairns, N.J., Green, R.C., Harvey, D., Jack Jr., C.R., Jagust, W., Morris, J.C., Petersen, R.C., Saykin, A.J., Shaw, L.M., Toga, A.W., Trojanowski, J.Q., Alzheimer's Disease Neuroimaging Initiative, 2019. Understanding disease progression and improving Alzheimer's disease clinical trials: recent highlights from the Alzheimer's disease neuroimaging initiative. Alzheimers Dement. 15 (1), 106–152. https://doi.org/10.1016/j.jalz.2018.08.005 (Epub 2018 Oct 13).

Villemagne, V.L., Doré, V., Burnham, S.C., Masters, C.L., Rowe, C.C., 2018. Imaging tau and amyloid-beta proteinopathies in Alzheimer disease and other conditions. Nat. Rev. Neurol. 14 (4), 225–236. https://doi.org/10.1038/nrneurol.2018.9 (Epub 2018 Feb 16).

Walsh, D.M., Klyubin, I., Fadeeva, J.V., Cullen, W.K., Anwyl, R., Wolfe, M.S., Rowan, M.J., Selkoe, D.J., 2002. Naturally secreted oligomers of amyloid beta protein potently inhibit hippocampal long-term potentiation in vivo. Nature 416 (6880), 535–539. https://doi.org/10.1038/416535a.

Wang, C., Holtzman, D.M., 2020. Bidirectional relationship between sleep and Alzheimer's disease: role of amyloid, tau, and other factors. Neuropsychopharmacology 45 (1), 104–120. https://doi.org/10.1038/s41386-019-0478-5 (Epub 2019).

Wang, Z., Xiao, Z.P., 2010. Magnetic resonance imaging study of hippocampus structural alterations in post-traumatic stress disorder: a brief review (translated version). East Asian Arch. Psychiatr. 20 (3), 138–144.

Wang, Y.H., Zhang, Y.G., 2018. Amyloid and immune homeostasis. Immunobiology 223 (3), 288–293. https://doi.org/10.1016/j.imbio.2017.10.038 (Epub 2017 Oct 16).

Wang, X., Wang, W., Li, L., Perry, G., Lee, H.G., Zhu, X., 2014. Oxidative stress and mitochondrial dysfunction in Alzheimer's disease. Biochim. Biophys. Acta 1842 (8), 1240–1247. https://doi.org/10.1016/j.bbadis.2013.10.015 (Epub 2013 Nov 1).

Wang, W.Y., Tan, M.S., Yu, J.T., Tan, L., 2015. Role of pro-inflammatory cytokines released from microglia in Alzheimer's disease. Ann. Transl. Med. 3 (10), 136. https://doi.org/10.3978/j.issn.2305-5839.2015.03.49.

Wang, A.Y.L., Loh, C.Y.Y., Shen, H.H., Hsieh, S.Y., Wang, I.K., Chuang, S.H., Wei, F.C., 2019. Topical application of human Wharton's jelly mesenchymal stem cells accelerates mouse sciatic nerve recovery and is associated with upregulated neurotrophic factor expression. Cell Transplant. 28 (12), 1560–1572. https://doi.org/10.1177/0963689719880543 (Epub 2019 Sep 30).

Wardlaw, J.M., Benveniste, H., Nedergaard, M., Zlokovic, B.V., Mestre, H., Lee, H., Doubal, F.N., Brown, R., Ramirez, J., MacIntosh, B.J., Tannenbaum, A., Ballerini, L., Rungta, R.L., Boido, D., Sweeney, M., Montagne, A., Charpak, S., Joutel, A., Smith, K.J., Black, S.E., Colleagues from the Fondation Leducq Transatlantic Network of Excellence on the Role of the Perivascular Space in Cerebral Small Vessel Disease, 2020. Perivascular spaces in the brain: anatomy, physiology and pathology. Nat. Rev. Neurol. 16 (3), 137–153. https://doi.org/10.1038/s41582-020-0312-z (Epub 2020 Feb 24).

Webers, A., Heneka, M.T., Gleeson, P.A., 2020. The role of innate immune responses and neuroinflammation in amyloid accumulation and progression of Alzheimer's disease. Immunol. Cell Biol. 98 (1), 28–41. https://doi.org/10.1111/imcb.12301 (Epub 2019 Nov 20).

Webster, S.J., Van Eldik, L.J., Watterson, D.M., Bachstetter, A.D., 2015. Closed head injury in an age-related Alzheimer mouse model leads to an altered neuroinflammatory response and persistent cognitive impairment. J. Neurosci. 35 (16), 6554–6569. https://doi.org/10.1523/JNEUROSCI.0291-15.2015.

Wei, M., Zhao, B., Huo, K., Deng, Y., Shang, S., Liu, J., Li, Y., Ma, L., Jiang, Y., Dang, L., Chen, C., Wei, S., Zhang, J., Yang, H., Gao, F., Qu, Q., 2017. Sleep deprivation induced plasma amyloid-beta transport disturbance in healthy young adults. J. Alzheimers Dis. 57 (3), 899–906. https://doi.org/10.3233/JAD-161213.

Weiner, M.W., Friedl, K.E., Pacifico, A., Chapman, J.C., Jaffee, M.S., Little, D.M., Manley, G.T., McKee, A., Petersen, R.C., Pitman, R.K., Yaffe, K., Zetterberg, H., Obana, R., Bain, L.J., Carrillo, M.C., 2013. Military risk factors for Alzheimer's disease. Alzheimers Dement. 9 (4), 445–451. https://doi.org/10.1016/j.jalz.2013.03.005.

Weiner, M.W., Veitch, D.P., Aisen, P.S., Beckett, L.A., Cairns, N.J., Cedarbaum, J., Green, R.-C., Harvey, D., Jack, C.R., Jagust, W., Luthman, J., Morris, J.C., Petersen, R.C., Saykin, A.J., Shaw, L., Shen, L., Schwarz, A., Toga, A.W., Trojanowski, J.Q., Alzheimer's Disease Neuroimaging Initiative, 2015. 2014 Update of the Alzheimer's disease neuroimaging initiative: a review of papers published since its inception. Alzheimers Dement. 11 (6), e1–120. https://doi.org/10.1016/j.jalz.2014.11.001.

Welikovitch, L.A., Do Carmo, S., Maglóczky, Z., Malcolm, J.C., Lőke, J., Klein, W.L., Freund, T., Cuello, A.C., 2020. Early intraneuronal amyloid triggers neuron-derived inflammatory signaling in APP transgenic rats and human brain. Proc. Natl. Acad. Sci. U. S. A. 117 (12), 6844–6854. https://doi.org/10.1073/pnas.1914593117 (Epub 2020 Mar 6).

Wiśniewski, H.M., Maślińska, D., 1996. Beta-protein immunoreactivity in the human brain after cardiac arrest. Folia Neuropathol. 34 (2), 65–71.

Wilkins, H.M., Swerdlow, R.H., 2016. Relationships between mitochondria and neuroinflammation: implications for Alzheimer's disease. Curr. Top. Med. Chem. 16 (8), 849–857. https://doi.org/10.2174/1568026615666150827095102.

Windisch, M., Gschanes, A., Hutter-Paier, B., 1998. Neurotrophic activities and therapeutic experience with a brain derived peptide preparation. J. Neural Transm. Suppl. 53, 289–298. https://doi.org/10.1007/978-3-7091-6467-9_25.

Wu, Z., Wang, Z.H., Liu, X., Zhang, Z., Gu, X., Yu, S.P., Keene, C.D., Cheng, L., Ye, K., 2020. Traumatic brain injury triggers APP and Tau cleavage by delta-secretase, mediating Alzheimer's disease pathology. Prog. Neurobiol. 185, 101730. https://doi.org/10.1016/j.pneurobio.2019.101730 (Epub 2019 Nov 25) 31778772.

Xie, Z., Xu, Z., 2013. General anesthetics and beta-amyloid protein. Prog. Neuro-Psychopharmacol. Biol. Psychiatry 47, 140–146. https://doi.org/10.1016/j.pnpbp.2012.08.002 (Epub 2012 Aug 14).

Xin, S.H., Tan, L., Cao, X., Yu, J.T., Tan, L., 2018. Clearance of amyloid beta and tau in Alzheimer's disease: from mechanisms to therapy. Neurotox. Res. 34 (3), 733–748. https://doi.org/10.1007/s12640-018-9895-1 (Epub 2018 Apr 7).

Yang, S.T., Hsiao, I.T., Hsieh, C.J., Chiang, Y.H., Yen, T.C., Chiu, W.T., Lin, K.J., Hu, C.J., 2015. Accumulation of amyloid in cognitive impairment after mild traumatic brain injury. J. Neurol. Sci. 349 (1–2), 99–104. https://doi.org/10.1016/j.jns.2014.12.032 (Epub 2014 Dec 30).

Yang, Y.J., Li, X.L., Xue, Y., Zhang, C.X., Wang, Y., Hu, X., Dai, Q., 2016. Bone marrow cells differentiation into organ cells using stem cell therapy. Eur. Rev. Med. Pharmacol. Sci. 20 (13), 2899–2907.

Yang, T., Dang, Y., Ostaszewski, B., Mengel, D., Steffen, V., Rabe, C., Bittner, T., Walsh, D.-M., Selkoe, D.J., 2019. Target engagement in an alzheimer trial: crenezumab lowers amyloid beta oligomers in cerebrospinal fluid. Ann. Neurol. 86 (2), 215–224. https://doi.org/10.1002/ana.25513 (Epub 2019 Jun 22).

Yang, D., Zhu, W., Wang, Y., Tan, F., Ma, Z., Gao, J., Lin, X., 2020. Selection of mutant plasmin for amyloid-beta cleavage in vivo. Sci. Rep. 10 (1), 12117. https://doi.org/10.1038/s41598-020-69079-8.

Yasuno, F., Kajimoto, K., Ihara, M., Taguchi, A., Yamamoto, A., Fukuda, T., Kazui, H., Iida, H., Nagatsuka, K., 2019. Amyloid beta deposition in subcortical stroke patients and effects of educational achievement: a pilot study. Int. J. Geriatr. Psychiatry 34 (11), 1651–1657. https://doi.org/10.1002/gps.5178 (Epub 2019 Jul 31).

Ye, B.S., Lee, Y., Kwak, K., Park, Y.H., Ham, J.H., Lee, J.J., Shin, N.Y., Lee, J.M., Sohn, Y.H., Lee, P.H., 2016. Posterior ventricular enlargement to differentiate dementia with lewy bodies from Alzheimer's disease. J. Alzheimers Dis. 52 (4), 1237–1243. https://doi.org/10.3233/JAD-160062.

Yeh, F.L., Hansen, D.V., Sheng, M., 2017. TREM2, microglia, and neurodegenerative diseases. Trends Mol. Med. 23 (6), 512–533. https://doi.org/10.1016/j.molmed.2017.03.008 (Epub 2017 Apr 22).

Yepes, M., 2021. The plasminogen activating system in the pathogenesis of Alzheimer's disease. Neural Regen. Res. 16 (10), 1973–1977. https://doi.org/10.4103/1673-5374.308076.

Yokokawa, K., Iwahara, N., Hisahara, S., Emoto, M.C., Saito, T., Suzuki, H., Manabe, T., Matsumura, A., Matsushita, T., Suzuki, S., Kawamata, J., Sato-Akaba, H., Fujii, H.G., Shimohama, S., 2019. Transplantation of mesenchymal stem cells improves amyloid-beta pathology by modifying microglial function and suppressing oxidative stress. J. Alzheimers Dis. 72 (3), 867–884. https://doi.org/10.3233/JAD-190817.

Yuan, C., Aierken, A., Xie, Z., Li, N., Zhao, J., Qing, H., 2020. The age-related microglial transformation in Alzheimer's disease pathogenesis. Neurobiol. Aging 92, 82–91. https://doi.org/10.1016/j.neurobiolaging.2020.03.024 (Epub 2020 Apr 15).

Yun, H.M., Park, K.R., Kim, E.C., Kim, S., Hong, J.T., 2015. Serotonin 6 receptor controls Alzheimer's disease and depression. Oncotarget 6 (29), 26716–26728. https://doi.org/10.18632/oncotarget.5777.

Yun, S.M., Cho, S.J., Jo, C., Park, M.H., Han, C., Koh, Y.H., 2020. Elevation of plasma soluble amyloid precursor protein beta in Alzheimer's disease. Arch. Gerontol. Geriatr. 87, 103995. https://doi.org/10.1016/j.archger.2019.103995 (Epub 2019 Dec 9).

Zayas-Santiago, A., Díaz-García, A., Nuñez-Rodríguez, R., Inyushin, M., 2020. Accumulation of amyloid beta in human glioblastomas. Clin. Exp. Immunol. 202 (3), 325–334. https://doi.org/10.1111/cei.13493 (Epub 2020 Aug 11).

Zempel, H., Mandelkow, E.M., 2012. Linking amyloid-beta and tau: amyloid-beta induced synaptic dysfunction via local wreckage of the neuronal cytoskeleton. Neurodegener. Dis. 10 (1–4), 64–72. https://doi.org/10.1159/000332816 (Epub 2011 Dec 7).

Zenaro, E., Piacentino, G., Constantin, G., 2017. The blood-brain barrier in Alzheimer's disease. Neurobiol. Dis. 107, 41–56. https://doi.org/10.1016/j.nbd.2016.07.007 (Epub 2016 Jul 15).

Zhang, T., Chen, D., Lee, T.H., 2019. Phosphorylation signaling in APP processing in Alzheimer's disease. Int. J. Mol. Sci. 21 (1), 209. https://doi.org/10.3390/ijms21010209.

Zhang, L., Dong, Z.F., Zhang, J.Y., 2020a. Immunomodulatory role of mesenchymal stem cells in Alzheimer's disease. Life Sci. 246, 117405. https://doi.org/10.1016/j.lfs.2020.117405 (Epub 2020 Feb 5).

Zhang, Y.Q., Wang, C.F., Xu, G., Zhao, Q.H., Xie, X.Y., Cui, H.L., Wang, Y., Ren, R.J., Guo, Q.H., Wang, G., 2020b. Mortality of Alzheimer's disease patients: a 10-year follow-up pilot study in Shanghai. Can. J. Neurol. Sci. 47 (2), 226–230. https://doi.org/10.1017/cjn.2019.333.

Zhao, Y., Bhattacharjee, S., Jones, B.M., Hill, J., Dua, P., Lukiw, W.J., 2014. Regulation of neurotropic signaling by the inducible, NF-kB-sensitive miRNA-125b in Alzheimer's disease (AD) and in primary human neuronal-glial (HNG) cells. Mol. Neurobiol. 50 (1), 97–106. https://doi.org/10.1007/s12035-013-8595-3 (Epub 2013 Nov 29).

Zhou, H., Sun, G., Kong, L., Du, Y., Shen, F., Wang, S., Chen, B., Zeng, X., 2014. Acupuncture and moxibustion reduces neuronal edema in Alzheimer's disease rats. Neural Regen. Res. 9 (9), 968–972. https://doi.org/10.4103/1673-5374.133148.

Zhou, W., Hu, X., Tam, K.Y., 2017. Systemic clearance and brain distribution of carbazole-based cyanine compounds as Alzheimer's disease drug candidates. Sci. Rep. 7 (1), 16368. https://doi.org/10.1038/s41598-017-16635-4.

Zhu, X., Lee, H.G., Casadesus, G., Avila, J., Drew, K., Perry, G., Smith, M.A., 2005. Oxidative imbalance in Alzheimer's disease. Mol. Neurobiol. 31 (1–3), 205–217. https://doi.org/10.1385/MN:31:1-3:205.

Zhu, X., Smith, M.A., Honda, K., Aliev, G., Moreira, P.I., Nunomura, A., Casadesus, G., Harris, P.L., Siedlak, S.L., Perry, G., 2007. Vascular oxidative stress in Alzheimer disease. J. Neurol. Sci. 257 (1–2), 240–246. https://doi.org/10.1016/j.jns.2007.01.039 (Epub 2007 Mar 6).

Zimmer, C., 2011. 100 trillion connections. Sci. Am. 304 (1), 58–63. https://doi.org/10.1038/scientificamerican0111-58.

Zlokovic, B.V., 2008. The blood-brain barrier in health and chronic neurodegenerative disorders. Neuron 57 (2), 178–201. https://doi.org/10.1016/j.neuron.2008.01.003.

Zlokovic, B.V., 2011. Neurovascular pathways to neurodegeneration in Alzheimer's disease and other disorders. Nat. Rev. Neurosci. 12 (12), 723–738. https://doi.org/10.1038/nrn3114.

Zou, K., Abdullah, M., Michikawa, M., 2020. Current biomarkers for Alzheimer's disease: from CSF to blood. J. Pers. Med. 10 (3), 85. https://doi.org/10.3390/jpm10030085.

Glioblastoma: What can we do for these patients today and what will we be able to do in the future?

Igor Bryukhovetskiy[a],*, Aleksandra Kosianova[a], Sergeis Zaitsev[a], Oleg Pak[a], Aruna Sharma[b], and Hari Shanker Sharma[b],*

[a]*School of Life Science & Biomedicine, Medical Center, Far Eastern Federal University (FEFU), Vladivostok, Russia*
[b]*International Experimental Central Nervous System Injury & Repair (IECNSIR), Department of Surgical Sciences, Anesthesiology & Intensive Care Medicine, Uppsala University Hospital, Uppsala University, Uppsala, Sweden*
Corresponding authors: Tel.: +7-914-7230503 (IB)/+46-70-20111801 (HSS), e-mail address: igbryukhovetskiy@gmail.com; sharma@surgsci.uu.se

Abstract

Glioblastoma multiforme (GBM) is an extremely aggressive primary human brain tumor. The median survival of GBM patients is 15 months in case of completing the modern complex treatment protocol. Chemotherapy can help to extend the life expectancy of patients. GBM treatment resistance is associated with cancer stem cells (CSCs). The present paper analyses the main reasons for ineffectiveness of the existing GBM treatment methods and suggests treating CSCs as a complex phenomenon, resulting from the coordinated interaction of normal stem cells and cancer cells (CCs) in immunosuppressive microsurroundings. The GBM treatment strategy is suggested not for only suppressing strategically important signaling pathways in CCs, but also for regulating interaction between normal stem cells and cancer cells. The paper considers the issue of controlling penetrability of the blood-brain barrier that is one of the main challenges in neuro-oncology. Also, the paper suggests the ways of extending life expectancy of GBM patients today and prospects for the near future.

Keywords

Glioblastoma multiforme, Chemotherapy, Treatment resistance, Cancer cells, Stem cells, Cancer stem cells, Blood-brain barrier, Brain edema, Immunotherapy, Nanotechnologies

1 Introduction

Glioblastoma multiforme (GBM) is one of the most unfavorable diagnoses in oncology. Its occurrence in the overall percentage of cancer incidences is relatively low. In European and North American countries, it is registered in 3–5 cases per 100,000 people, and it is more typical for men in their 50s and 60s. The special status of GBM among other types of cancer is determined by its high resistance to treatment. Despite all efforts of doctors, the median survival of GBM patients is 15 months in case of completing the modern complex treatment protocol (Lukas et al., 2019; Stupp et al., 2019). Chemotherapy might extend the life expectancy (Gimple et al., 2019; Le Rhun et al., 2019; Sharifzad et al., 2019), but its effects are limited. This paper attempts to answer the following questions: how can we extend a GBM patient life and what shall we expect in the future?

2 Modern approach to GBM treatment

Surgery is usually a preferred method of treatment for GBM patients (Weller et al., 2019). However, GBM is a rapidly growing and invasive tumor without clear borders, that is why radical removal of the tumor without causing severe neurological damage to the patient is deemed impossible (Bräutigam et al., 2019). Therefore, irradiation remains the main treatment method. A patient's life expectancy correlates with the received radiation dose (Wegner et al., 2019), reaching 60 Gy. There are data (Cammarata et al., 2019; Prezado et al., 2019) on extending GBM patients' life expectancy up to 2–3 years in case of increasing the radiation dose, but that remains doubtful. Higher radiation dose causes radiation necrosis and brain edema (Kanemitsu et al., 2019) that create life-threatening conditions for a patient.

Chemotherapy may extend a patient's life. The treatment involves active dehydration, and temozolomide (TMZ) is a traditional chemotherapeutic agent. Patents are recommended (Mathen et al., 2020) a combination of TMZ ($75\,mg/m^2/d$) and radiation therapy, and at least six cycles of TMZ after completing irradiation treatment. The first cycle involves administering $150\,mg/m^2/d$ of TMZ on days 1–5 of the 28-day cycle with increasing the dosage of TMZ up to $200\,mg/m^2/d$ in the subsequent cycles. The treatment effectiveness is relatively low, and despite all efforts of the medical personnel, tumor relapse usually occurs 3–6 months after the surgery.

A case of a 60-year old patient is a good illustration of the above mentioned practice. He had a large supratentorial IDH-wild-type glioblastoma with multiple metastases into the frontal cortex and partial invasion into the parietal lobe and compression of lateral ventricles (Fig. 1A and B). On the 13th of May 2020, he had a surgery in our neurosurgical clinic where contrast and microsurgical techniques were used along with neuronavigation, intraoperative nerve monitoring with a wake-up test when working in the functionally important brain areas, involvement of a

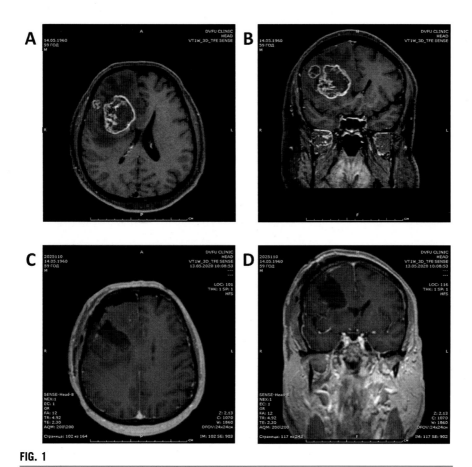

FIG. 1

The MRI results of the patient with GBM (male, 60 y.o.), May 13, 2020. (A and B) Before the surgery on May 13, 2020. A massive tumor with edema and brain compression is visible. (C and D) Condition after the first surgery. Radical removal of the tumor. T1-weighted images, Omniscan® as contrast agent.

neuropsychologist and a series of neuropsychological tests. The results were very impressive, the tumor was almost completely removed (Fig. 1C and D) with a subsequent improvement of his functional status, neurological and mental functions. First signs of a relapse were identified 3 months after the chemoradiotherapy, and 6 months later a massive recurrent tumor was discovered that almost completely filled the postcraniotomy area (Fig. 2A and B) and was surrounded by an extensive swelling. Subsequent chemotherapy detained the progression of the tumor, and the patient had a reoperation on the 19th of February 2021. Despite rather positive MRI

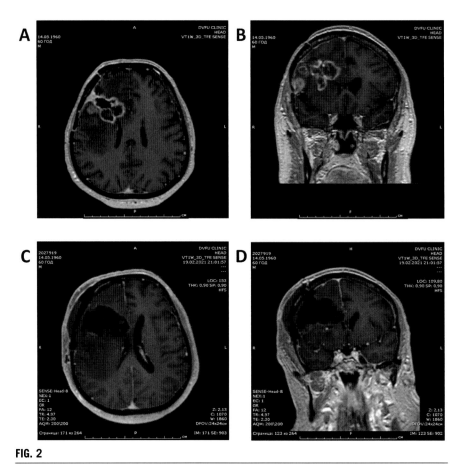

FIG. 2

(A) The MRI results of the patient with a recurrent GBM (male, 60 y.o.), February 19, 2021. First surgery on May 13, 2020. (B) Before the reoperation. A massive recurrent tumor, filling the postcraniotomy area surrounded by swelling, is visible. (C and D) Condition after the reoperation. Radical removal of the tumor. T1-weighted images, Omniscan® as contrast agent.

results, indicating a radical removal of the recurrent tumor (Fig. 2C and D), at the moment some signs of the invasive process are detected. These statistics are typical for the majority of neurosurgical hospitals (Lu et al., 2019; Woodroffe et al., 2020). Choosing a treatment method in case of a relapse is a challenging task. Reoperation is possible not for every patient, but this option should be considered nonetheless, since even partial tumor removal will alleviate the gravity of the liquor-dynamic disturbances and improve the patient's quality of life.

Repeating irradiation treatment usually produces no effect (Morris et al., 2019). Patients are recommended to take 40–100 mg/m^2/d of TMZ with a 7 days on/7 days off schedule. Increasing the number of cycles, frequency of administration or dosage of TMZ does not extend the life expectancy. A combination of TMZ with CCNU (Umphlett et al., 2020) and angiogenic targeted drug bevacizumab is used as the second-line chemotherapy. Derivatives of platinum, procarbazine and vincristine could be used as drugs for the third-line chemotherapy. But numbers do not lie (Lesueur et al., 2019), and combinations of TMZ with other cytostatics, that inhibit DNA repair, increases toxicity levels and do not help to prolong patients' life. The data on combining chemotherapy with alternating electric fields therapy (Tan et al., 2020) and its ability to extend a patient's life up to 1–3 months appear doubtful as well.

The main cause of GBM patients' death is edema, compression and dislocation of the brain that happen due to the tumor growth in the enclosed cranial space, the contents of which are incompressible and do not change their volume. Cerebral edema is one of the most serious issues in GBM treatment. N.N. Burdenko, the Honorary member of the Royal Society of Surgeons in London and the Paris Society of Surgery, internationally renowned neurosurgeon and academician, said that "those who master the art of treating and preventing cerebral edema hold the key to a patient's life and death." But in reality this key is yet to be discovered. The median survival of patients with IDH-mutant GBM is 23 months, and 13 months for those with IDH-wild GBM (Cammarata et al., 2019; Stupp et al., 2017). The crucial factor of treatment resistance is the presence of cancer stem cells in the heterogeneous cell population of GBM, and these cells are traditionally assumed to have unique signaling and morphogenetic properties.

3 Cancer stem cells

Cancer stem cells (CSCs) have been discovered more than 20 years ago (Bonnet and Dick, 1997). Even today there is no complete understanding of the CSC nature, but in order to provide this understanding it is necessary to deal with two issues that the whole history of neuro-oncology is based on. First of all, the main criterion of brain tumor malignancy is the proliferation speed that is inversely proportional to the differentiation degree of the cancer cells of the tumor. Based on this (Scheithauer, 2009), the WHO classification of 2007 assigned Grade IV to the heterogeneous group of primary tumors with such fast proliferation speed and high level of CC anaplasia, so it was almost impossible to define their origin. GBM is the prime example of a tumor, pertaining to that group.

Secondly, the CC differentiation is determined not by genetic mutations (Mata et al., 2020), but their epigenetic consequences. For that reason IDH-mutation was chosen as the main criterion for the WHO classification of brain tumors in 2016 since due to it an excess of 2-hydroxyglutarate appears in CCs that leads to the hypermethylation of the whole genome, promoting longer differentiation of the CCs, that is why the IDH-mutant type of GBM is somewhat similar to less malignant gliomas.

IDH-wild type of GBM occurs in the majority of GBM cases. It is characterized by an extremely high level of DNA demethylation, if compared to the normal brain tissue (Waitkus et al., 2018), and this level is even higher in recurrent tumors (Tan et al., 2020). In fact, these CCs have nothing to do with both nervous tissue and glial one. DNA demethylation triggers a plethora of genome repair mechanisms and activates the processes of living matter self-organization, typical for embryonic cells (Brown et al., 2019; Gimple et al., 2019; Linkous et al., 2019; Sharifzad et al., 2019). This is accompanied by the creation of a primitive hierarchical structure in the cell population, higher level of heterogeneity and development of differentiation potential, similar to that of normal stem cells, with corresponding mimicry of immunohistochemical markers of the cell surface. Cancer stem cells could be deemed as cancer cells that have been dedifferentiated to the level of primitive stem cells (SC), but this is just one side of the coin.

On the other hand, the part of normal stem cells in GBM pathogenesis is obvious, and here we enter one of the most fundamental areas of cellular biology. Firstly, most data (Altmann et al., 2019; Álvarez-Satta et al., 2019; Tuazon et al., 2019; Wang et al., 2019) point to the fact that normal neural stem cells in the germinal zone of the brain are the "source" of GBM. Primary mutations, causing differentiation avoidance and creation of the primary stem line of CCs, are likely to derive from stem cells of that type. Secondly, intensified proliferation of CCs creates competition for oxygen and hypoxia that induces migration and SCs homing to the tumor lesion, especially, red bone marrow SCs, that allows GBM to recruit them from the blood flow and involve them into the neoplastic process. These two arguments are enough to entertain the idea of CSCs being a result of coordinated interaction of CCs and normal SCs.

4 Interaction of stem cells and cancer cells

Mechanisms of interaction between SCs and CSs are relatively unknown. There has been described exchange of proteins and microRNA via microvesicles, fusion of cells (Zhang et al., 2019) and a set of microtubes (Ahn, 2020), connecting the groups of cells into one system where it is possible to exchange not only informational molecules, but also organelles and cell nuclei that could be critical in the process of adapting to treatment. However, all these mechanisms are secondary, since originally migration of normal SCs to the tumor lesion is responsible for much more important outcomes.

Fulfilling the repair potential of the normal SCs is possible only at the final stage of the inflammatory process that starts when their amount in the lesion is big enough. Tissue damage during the GBM growth, products of tumor metabolism and hypoxia induce active migration and homing of normal SCs into the tumor where they interact with other cells and trigger inflammation due to the modification of cytokine inflammatory area. The transforming growth factor (TGF) β—an anti-inflammatory cytokine—is the most important regulator of this process. SCs induce TGFβ synthesis while CCs with receptors to this cytokine start self-activation process and increase its synthesis, based on autocrine induction principle. TGFβ inactivates immunocytes, astroglia and microglia, triggers the epithelial-mesenchymal transition in CCs, increases production of extracellular matrix (ECM) in CCs and interaction with it. But, most importantly, TGFβ increases the stemness of CCs (Shevchenko et al., 2020) and activates CCs invasion into the surrounding tissues.

Besides, hypoxia differentiates some SCs into endothelial cells and induces the production of vascular endothelial growth factor (VEGF) that triggers angiogenesis. Hyperstimulation of VEGF leads to hyperplasia of newly developed microvessels that together with cerebral edema damages close contacts between endotheliocytes and creates pathological fenestrations among them that allows GBM to recruit other bone marrow cells—lymphocytes and monocytes (Prionisti et al., 2019; Sankowski et al., 2019) that produce VEGF under the influence of TGFβ.

The majority of recruited monocytes are transformed into macrophages (Schiffer et al., 2018), M2-activated by TGF-β, accompanied by an increased production of *Wnt*-proteins (Matias et al., 2017), triggering the key stemness mechanism in CCs—*Wnt*-signaling pathway. In response to this stimulation CCs start production of Wnt-ligands that closes the "secretion loop" and activates proliferation and angiogenesis that now have an avalanche-like nature. In the light of the above mentioned, suppression of TGFβ synthesis in the tumor is the most important way to control interaction of CCs and SCs. Of course, VEGF, TGFβ and WNT-signaling pathways inhibitors we should have to use in the complex treatment of GBM. There are many drugs available to target these mechanisms (Fig. 3). However, it should be noted that all these signaling pathways are activated only in the final, reparative phase of the inflammation and they are inactive in the acute phase of this process. For this reason, the approach to treatment of GBM can only be complex, and it is necessary to start with immunotherapy.

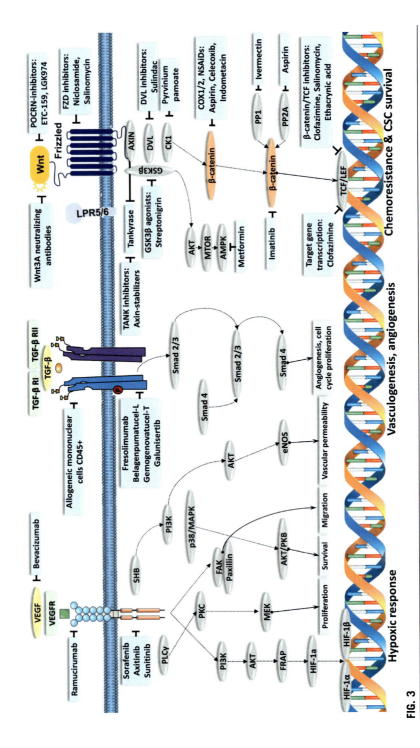

FIG. 3

Targeted drugs that can affect the key components of TRK (VEGF), TGFβ and *wnt*-signaling pathways in the cancer cells of GBM.

5 Immunotherapy and TGFβ

Anti-tumor potential of the immune system should be used in GBM treatment. Cytotoxic T-lymphocytes, dendrite vaccine, macrophage-based therapy, as well as their combinations are actively used in oncology. Such treatment is not very effective (Laval et al., 2020): 10 cytotoxic lymphocytes can destroy only one GBM cell, therefore, when developing a treatment strategy, it should be taken into consideration that elimination of CCs is an important, but not a primary objective of immunotherapy.

Primary objectives of GBM immunotherapy include controlling intercellular interaction, mostly between CCs and SCs. This thesis is based on our own research where immunocompetent animals with a transplanted C6 glioma were injected with autologous mononuclear CD45 + cells, increasing the number of cells with microglia markers (Zaitsev et al., 2020). Subsequent injection of rats with γ-interferon and LPS from *Escherichia Coli* lead to inflammatory activation of microglia with higher expression of CD86 antigen by microglia, lower level of TGFβ in the tumor and longer life expectancy of rats.

Even more pronounced effect of locally suppressing TGFβ synthesis in the tumor tissue could be achieved if animals with C6 glioma are injected with allogeneic and xenogeneic CD45 + mononuclear cells. In our first experiments (Bryukhovetskiy et al., 2016) introduction of mononuclear CD45 + human cells into the blood flow of rats with transplanted C6 glioma led to concentration of these cells in the areas of the tumor invasive growth into the brain tissue with their subsequent penetration into the necrotic area of neoplastic tissue. Tumor infiltration with mononuclear cells continued for 3–5 days while the amount of transplanted cells, stained with fluorescent tracer, was increasing, and on days 7–14 of the experiment we observed their destruction, indicated by pericellular location of the fluorescent material. It should be specifically pointed out that these processes were accompanied by the progressive decrease of TGFβ1 amount in the tumor that was minimal by the end of the experiment. This could be associated with developing reciprocal reactions of graft-vs-tumor and host-vs-graft.

Without doubt, xenotransplantation of CD45 + cells is possible only in experimental conditions, but patients with leucosis, who received transplanted allogeneic bone marrow, had longer and better remission period than those patients who received allogeneic material. This phenomenon is based on the ability of allogeneic T-lymphocytes to attack remaining tumor tissue of the patient that is accompanied by synthesis inhibition of all immunosuppressive cytokines. One of the most important conditions for using allogeneic cell material is HLA-compatibility of the donor and the recipient, but even if these conditions are met, the procedure of introducing allogeneic bone marrow cells into the blood stream is dangerous for the patient. Therefore, allogeneic mononuclear cells could be administered intrathecally, as an addition to other immunotherapy methods, for instance, together with personalized autologous dendritic cell-based vaccines.

Transplanting allogeneic bone marrow cells into the post-operative area is a promising method of GBM biotherapy that provides good prospects for targeting the remaining GBM tissue. But, despite the promising nature of this method, it should be noted that there are no allogeneic immunocytes that are aggressive only toward CCs, and even if they are transduced by suicidal genes, such therapy causes cerebral edema that could turn fatal for the patient. However, proinflammatory modification of the remaining tumor tissue destabilizes interaction between CCs and SCs, providing options for chemotherapy.

Allogeneic CD45+ mononuclear cells could be used for GBM treatment even today, but their extensive application for biomedical cell-based products is a task for the future. CD34+ hematopoietic SCs are a crucial component of CD45+ mononuclear cells. In natural conditions their amount in the blood flow does not exceed 0.1%. However, it can be increased up to 1% after stimulation with granulocyte colony-stimulating factor that, on the one hand, makes these cells invaluable for reconstruction of the hematopoietic system, and, on the other hand, increases anticancer properties of allogeneic mononuclear cells.

Antitumor effect of these cells is yet unclear, but a reasonable assumption is that the main factor of their anticancer activity is graft-vs-tumor reaction, and the accompanying mechanisms might involve: (a) the phenomenon, described by *Yair Reisne* (Reisner and Or-Geva, 2019), when the donor-derived stem cells exercise veto against patient's SCs, e.g., suppressing the ability to stop inflammation; (b) transformation of allogeneic SCs into stromal cells that produce ECM of a different type, preventing adhesion of CCs; (c) epigenetic reprogramming of CCs during intercellular interaction between SCs (Laval et al., 2020) that have no epigenetic memory about the immune response. However, these are no more than assumptions while future development requires more specific and definite answers.

Therefore, immunotherapy might destabilize interaction between CCs and SCs, but properly planned chemotherapy should play an important part in it.

---

6 TGFβ and targeted therapy

Trabedersen—an antisense compound that is one of leading members of the TGFβ synthesis inhibitor class—is more effective than the standard therapy when injected into the tumor even in small doses. However, its effect was less pronounced for patients with GBM. Other TGFβ inhibitors—galunisertib (Wick et al., 2020), fresolimumab, belagenpumatucel-L and gemogenovatucel-T—are yet to prove their worth, that could be attributed to some tactical missteps in developing treatment scenarios. Firstly, GBM pathogenesis suggests that TGFβ inhibitors should be treated as basic medication for targeted therapy protocols and be used together with other targeted drugs, most importantly—VEGF antagonists and other suppressors of tyrosine-kinase signaling. Secondly, systemic immunosuppressive effect of glucocorticoids should be eliminated, and that is not always possible. Thirdly, such therapy must be personalized and targeted at key proteins that are really responsible for resistance to radiation and chemotherapy. For instance, experimental irradiation of GBM cells with 60 Gy significantly decreases the sensitivity of CCs to TMZ, strongly induces the synthesis of proteasome proteins that break down oxidized proteins into short peptides. Suppressing this target with bortezomib increases the sensitivity of CCs to TMZ that should be taken into account in planning the subsequent chemotherapy.

7 Wnt-inhibitors in GBM treatment

Wnt-signaling cascade—one of the basic mechanisms of controlling CCs differentiation potential—is a promising target in GBM cells. The principles of activating Wnt-cascade in GBM cells are relatively unknown. There have been studies (Chen et al., 2020; Manoranjan et al., 2020) showing dramatic increase of some Wnt-proteins expression in GBM cells, especially of β-catenin (Shevchenko et al., 2020), making it a very attractive target for therapy. But we know very little about effective Wnt-inhibitors—unfortunately, the majority of medical agents are still in pre-trial stage. That is why our focus should be directed at drug repositioning that could be used in treatment of patients even today.

Lower incidence of rectal cancer (Katona and Weiss, 2020), breast cancer (Cai et al., 2020) and pancreatic cancer (Qorri et al., 2020) among people, who took nonsteroidal anti-inflammatory drugs (NSAIDs) has been described. The majority of such drugs are non-selective inhibitors of both isoforms of cyclooxygenase, causing lower production of prostaglandin E2 and destabilization of β-catenin (Maniewska and Jeżewska, 2021), but antitumor effects of NSAIDs do not end here.

Indometacin (Beyer et al., 2019) decreases the amount of β-catenin in CCs regardless of the cyclooxygenase suppression level and disrupts the connection between β-catenin-TCF4 and DNA. Celecoxib (Egashira et al., 2017) suppresses cyclin D1 expression in CCs, induces proteome degradation of transcription factors TCF 1 and 4, blocks the PGE 2-GSK3β-β-catenin signaling pathway, reduces the level of stemness and immobilizes CCs (Deng et al., 2013). Sulindac (Maniewska and Jeżewska, 2021) prevents nuclear translocation of β-catenin, inhibits DVL/FZD interaction, preventing the destruction of β-catenin, decreases the expression of TCF target genes. Therefore, NSAIDs could significantly decrease the level of intracellular β-catenin in CCs that does not rule out the possibility of using other drugs.

Anti-leprosy drug clofazimine (Koval and Katanaev, 2018) can inhibit Wnt/β-catenin pathway. Salinomycin, a polyether antibiotic, activates transcription factor FOXO3, disrupting the interaction between β-catenin and TCF, inhibits the expression of Wnt target genes: LEF1, CycD1 and fibronectin (Wang et al., 2020). Tigecycline, a glycylcycline antibiotic (Dong et al., 2019), and ethacrynic acid, a loop diuretic (Yu et al., 2021), are also able to inhibit Wnt target genes. Niclosamide, an antihelminthic drug (Cheng et al., 2017), prevents proliferation of GBM cells that could be attributed to both destabilization of β-catenin and blockade of LEF/TCF reporting. Apart from drug repositioning, special interest should be paid to plant-based Wnt-inhibitors.

Curcumin—a diarylheptanoid, extracted from *Curcuma longa*—impedes CCs growth, causing β-catenin degradation (Hu et al., 2019), decreasing the amount of nuclear β-catenin and the levels of β-catenin transcription cofactors—TCF4, CBP and p300 (Hesari et al., 2019). Resveratrol induces proteasome degradation of TCF4 (Xie et al., 2017), disrupting proliferation and CCs mobility (Yang et al., 2019). Quercetin—a plant-based flavonol (Srivastava and Srivastava, 2019)—inhibits TCF, suppresses CCs proliferation, decreases viability of CCs with epithelial-mesenchymal transition phenotype and impedes *cycD1* and *c-Myc* genes expression.

Genistein—soy isoflavone—inhibits epithelial-mesenchymal transition in CCs (Tuli et al., 2019), promotes β-catenin degradation and suppresses *c-Myc* and *cycD1* genes expression. Apigenin—a natural flavone—prevents β-catenin penetration of the nucleus and Wnt target genes expression in CCs.

However, the ground flora is not the only source of Wnt-inhibitors. In 2017 the Russian-Swiss research team, headed by Vladimir Katanaev (Blagodatski et al., 2017), discovered Wnt-inhibiting properties of *Ophiura Irrorata*—a species of benthic marine animals, belonging to the phylum Echinodermata, that live in the Pacific Ocean. This discovery is not uncommon, since there are a lot of data on Wnt-inhibiting properties of marine organisms extracts, however, active usage of Wnt-signaling inhibitors in complex GBM treatment is a task to be accomplished in the future. But the overwhelming majority of the drugs will be completely useless without solving the issue of their effective transportation through the blood-brain barrier (BBB).

8 Nanotechnologies, cerebral edema and blood-brain barrier

There is a misconception that the issue of the blood-brain barrier penetrability in case of GBM is almost non-existent. It is based on the results of a half-a-century old experiment (Vick et al., 1977) that proved the concentration of intravenously injected contrast material in the tumor. Undoubtedly, hyperplasia of endothelium cells disrupts the blood-brain barrier function due to VEGF hyperstimulation. A surgery contributes to an extensive damage of the blood-brain barrier structure, allowing to make the remaining CCs more chemotherapy-sensitive. But the blood-brain barrier importance in the complex GBM treatment must not be underestimated.

Even if a surgery removes the biggest part of the tumor, CCs that migrate to the brain tissue are completely protected by the blood-brain barrier, and their eradication is almost impossible. Surely, some results might be achieved via intrathecal injection of TMZ or CCNU. There is a method of administering cytostatics after the blood-brain barrier disruption with mannitol, but this technique significantly increases interstitial osmotic pressure and fluid flow, stimulates swelling and intracranial hypertension that threaten a patient's life. Therefore, effective delivery of drugs to CCs is no less important than their development.

Creation of nano-drugs is one of the most promising methods of chemotherapeutic agents delivery through the blood-brain barrier, Nanoparticles of less than 200 nm are able to penetrate the brain on their own (Ramalho et al., 2020), and here we should mention silicon (Zilony-Hanin et al., 2019), silicon-graphene and silicon-calcium nanoparticles with porous structure, allowing to load them with TMZ and other cytostatics (Ozkizilcik et al., 2019; Sharma et al., 2020). Nanoparticles, surrounding the drug, can be injected directly into the tumor tissue or placed into the post-operative area of the tumor (Fig. 4). Using biocompatible polymers, such as dextran, could reduce interaction of injected nanoparticles with ECM and guarantee their wider spread in the lesion. A negative effect of such interaction with a tumor is dissemination of a medical agent due to the internal pressure in the tumor that requires controlled placement and concentration of nanoparticles with a magnetic field.

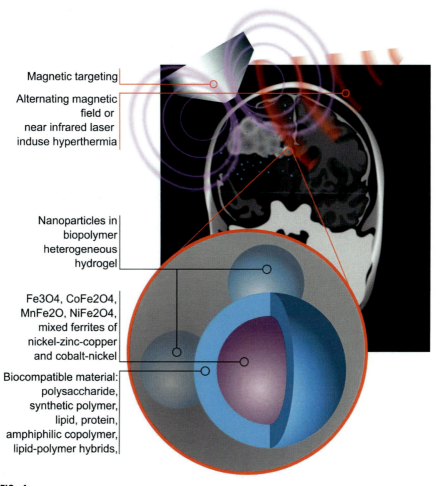

Magnetic targeting

Alternating magnetic field or near infrared laser induse hyperthermia

Nanoparticles in biopolymer heterogeneous hydrogel

Fe3O4, CoFe2O4, MnFe2O, NiFe2O4, mixed ferrites of nickel-zinc-copper and cobalt-nickel

Biocompatible material: polysaccharide, synthetic polymer, lipid, protein, amphiphilic copolymer, lipid-polymer hybrids,

FIG. 4

Schematic diagram of a nanotechnology-based anticancer therapy method. At the first stage, a biodegradable biopolymer hydrogel is loaded into the postoperative cavity which containing nanoparticles based on ferrites and/or magnetite enclosed in a shell of biocompatible materials (polysaccharides, synthetic polymers, lipids, lipid-polymer hybrids). At the second stage, nanoparticles are directed using a magnetic field to the zones of tumor invasion, where they are activated by an alternating magnetic field or an infrared laser, generating hyperthermia and amplification of the affected tissue.

The magnetic field increases the blood-brain barrier permeability (Qiao et al., 2020), so the possibilities of targeted delivery of drugs to CCs could be significantly extended due to nanoparticles, based on such magnetic materials as iron oxide, manganese, cobalt, nickel, zinc, magnesium and their oxides. There have been developed some methods of using nanoparticles with iron oxide (Fe_3O_4) or cobalt ferrite ($CoFe_2O_4$), manganese ferrite ($MnFe_2O_4$), nickel ferrite ($NiFe_2O_4$), or other magnetic materials as a base, surrounded by the polymeric layer with conjugated antibodies to specific CC proteins, e.g., EGFRvIII. When reaching their target, such nanoparticles create immune complexes, land on CCs and release the drug. Unfortunately, heterogenic nature of GBM cell population and polymorphism of immunohistochemical markers of the CC surface significantly limit the possibilities of nanoparticles targeting, and the issue of ultra-precise concentration of nanoparticles in the tumor tissue, using a magnetic field, remains unresolved even in experimental conditions.

Nevertheless, this approach seems promising. Accumulation of nanoparticles in the GBM tissue allows to increase their temperature via using a magnetic field with a certain amplitude and frequency, that induces the release of the transported drug, causes thermal damage and destruction of CCs, improves the effect of irradiation therapy. It is likely that in future these particles will function as nanorobots, able to selectively destruct CCs; that is why nanotherapy is the future of dealing with this disease, but nanotechnologies provide only a partial solution.

Speaking of cerebral edema and the blood-brain barrier permeability, there is a series of contributing factors, the main one being the blood pressure. Fast growth of GBM in the enclosed cranial space is accompanied by the increasing intracranial pressure and hypoxia of brain structures. Intracranial contents are almost incompressible, and arterial hypertension may partially compensate for this process. Undoubtedly, high arterial pressure increases permeability of the blood-brain barrier for medical agents (Sharma et al., 2020), but GBM is an angiogenic tumor where dysplasia of the blood circulation causes swelling that increases due to arterial hypertension and, thus, diminishes chemotherapy effect. Lack of oxygen and ATP molecules disrupts the function of ionic pump, allowing the excess of $Na+$ ions to enter the cells, increasing interacellular osmotic pressure, fluid access and developing cytotoxic swelling. Constant monitoring of arterial pressure and moderate hyperventilation are strategically important for GBM therapy, and these factors might contribute even more than properly selected chemotherapeutic agents.

A factor, no less important for the blood-brain barrier permeability, is glucose intolerance and diabetes (Sharma et al., 2020), disintegrating interaction between pericytes, astrocytes, perivascular macrophages and endotheliocytes that create a neurovascular unit due to their close interaction with the basal membrane. A crucial reason for glucose intolerance is a long-term systemic therapy with glucocorticoids, and in this case diabetes damages the blood-brain barrier functioning and diminishes therapy effect on GBM. Arterial pressure and glucose intolerance are defining factors in the second life period when GBM occurs more frequently, and their importance is usually underrated. There are some studies (Seliger et al., 2020), describing metformin effect on GBM cells, but there are almost no research papers on the effectiveness of antitumor drugs on gliomas, even animal ones, in cases, complicated by hypertension and diabetes.

Therefore, the blood-brain barrier is not only a hystohematic barrier, mechanically preventing drugs from accessing their target, but also one of the most important components for GBM pathogenesis. Increasing the blood-brain barrier penetrability leads to cerebral edema, while decreasing it makes treatment ineffective. For that reason, the effective delivery of drugs through the blood-brain barrier is possible with introduction of automatic smart systems with biological feedback that could coordinate the parameters of administering drugs based on the severity of cerebral edema, concentration of the drug in tissues, blood pressure, glucose level and other key health factors.

9 Conclusion

GBM remains one of the most dangerous brain tumors, almost completely treatment-resistant, and even the most optimistic prognosis does not suggest any cardinal changes in the situation in the upcoming years. GBM patients must have a maximally radical surgery with intraoperative contrast, microsurgery tools, neuronavigation and neurophysiological monitoring. It is crucial to try to avoid causing neurological damage to the patient during the surgery, since it deteriorates their quality of life and survival chances (Fig. 5).

Radiation with 60 Gy is recommended for all GBM patients, and even though exceptions are possible, they are strategically unjustified. Radiation therapy must be complemented with chemotherapy that should be personalized to the maximum possible extent. Tumor aggression can be countered only with equally aggressive

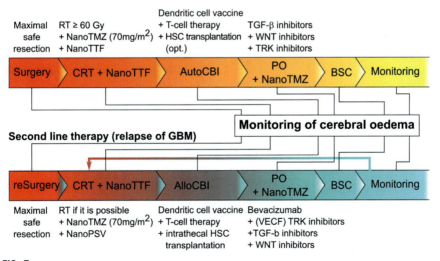

FIG. 5

Biomedicine and nanotechnology-based algorithm of the treatment of GBM. AlloCBI, allogenic cell-based immunotherapy; AutoCBI, autologous cell-based immunotherapy; BSC, best supportive care; CRT, chemoradiotherapy; NanoPCV, nanoform of procarbazine, lomustine, and vincristine; NanoTMZ, nanoform of temozolomide; NTTF, nano-based tumor-treating fields; RT, radiotherapy.

treatment. Personalized immunotherapy is recommended for all GBM patients without any counterindications. The prognosis for GBM patients is not favorable but extending their life expectancy will be possible only after developing new approaches to treatment that aims not only to directly eliminate CCs or suppress the crucial intracellular signaling pathways, but also to disrupt interaction between stem cells and cancer cells, using TGFβ antagonists and Wnt-signaling. The issues of cerebral edema and penetrability of the blood-brain barrier remain the most challenging ones. However, solving these issues will be possible only after creating of special smart devices, adjusting chemotherapy effect according to the severity of cerebral edema and general condition of the patient.

References

Ahn, S.Y., 2020. The role of MSCs in the tumor microenvironment and tumor progression. Anticancer Res. 40 (6), 3039–3047. https://doi.org/10.21873/anticanres.14284.

Altmann, C., Keller, S., Schmidt, M.H.H., 2019. The role of SVZ stem cells in glioblastoma. Cancers (Basel) 11 (4), 448. https://doi.org/10.3390/cancers11040448.

Álvarez-Satta, M., Moreno-Cugnon, L., Matheu, A., 2019. Primary cilium and brain aging: role in neural stem cells, neurodegenerative diseases and glioblastoma. Ageing Res. Rev. 52, 53–63. https://doi.org/10.1016/j.arr.2019.04.004.

Beyer, M., Romanski, A., Mustafa, A.M., Pons, M., Büchler, I., Vogel, A., Pautz, A., Sellmer, A., Schneider, G., Bug, G., Krämer, O.H., 2019. HDAC3 activity is essential for human leukemic cell growth and the expression of β-catenin, MYC, and WT1. Cancers (Basel) 11 (10), 1436. https://doi.org/10.3390/cancers11101436.

Blagodatski, A., Cherepanov, V., Koval, A., Kharlamenko, V.I., Khotimchenko, Y.S., Katanaev, V.L., 2017. High-throughput targeted screening in triple-negative breast cancer cells identifies Wnt-inhibiting activities in Pacific brittle stars. Sci. Rep. 7 (1), 11964. https://doi.org/10.1038/s41598-017-12232-7.

Bonnet, D., Dick, J.E., 1997. Human acute myeloid leukemia is organized as a hierarchy that originates from a primitive hematopoietic cell. Nat. Med. 3 (7), 730–737. https://doi.org/10.1038/nm0797-730.

Bräutigam, E., Lampl, C., Track, C., Nieder, C., Pichler, J., Hammer, J., Geinitz, H., 2019. Re-irradiation of recurrent glioblastoma as part of a sequential multimodality treatment concept. Clin. Transl. Oncol. 21 (5), 582–587. https://doi.org/10.1007/s12094-018-1957-6.

Brown, D.V., Stylli, S.S., Kaye, A.H., Mantamadiotis, T., 2019. Multilayered heterogeneity of glioblastoma stem cells: biological and clinical significance. Adv. Exp. Med. Biol. 1139, 1–21. https://doi.org/10.1007/978-3-030-14366-4_1.

Bryukhovetskiy, I.S., Dyuizen, I.V., Shevchenko, V.E., Bryukhovetskiy, A.S., Mischenko, P.V., Milkina, E.V., Khotimchenko, Y.S., 2016. Hematopoietic stem cells as a tool for the treatment of glioblastoma multiforme. Mol. Med. Rep. 14 (5), 4511–4520. https://doi.org/10.3892/mmr.2016.5852.

Cai, Y., Yousef, A., Grandis, J.R., Johnson, D.E., 2020. NSAID therapy for PIK3CA-altered colorectal, breast, and head and neck cancer. Adv. Biol. Regul. 75, 100653. https://doi.org/10.1016/j.jbior.2019.100653.

Cammarata, F.P., Torrisi, F., Forte, G.I., Minafra, L., Bravatà, V., Pisciotta, P., Savoca, G., Calvaruso, M., Petringa, G., Cirrone, G.A.P., Fallacara, A.L., Maccari, L., Botta, M., Schenone, S., Parenti, R., Cuttone, G., Russo, G., 2019. Proton therapy and Src family kinase inhibitor combined treatments on U87 human glioblastoma multiforme cell line. Int. J. Mol. Sci. 20 (19), 4745. https://doi.org/10.3390/ijms20194745.

Chen, Y., Fang, R., Yue, C., Chang, G., Li, P., Guo, Q., Wang, J., Zhou, A., Zhang, S., Fuller, G.N., Shi, X., Huang, S., 2020. Wnt-induced stabilization of KDM4C is required for Wnt/β-catenin target gene expression and glioblastoma tumorigenesis. Cancer Res. 80 (5), 1049–1063. https://doi.org/10.1158/0008-5472.CAN-19-1229.

Cheng, B., Morales, L.D., Zhang, Y., Mito, S., Tsin, A., 2017. Niclosamide induces protein ubiquitination and inhibits multiple pro-survival signaling pathways in the human glioblastoma U-87 MG cell line. PLoS One 12 (9), e0184324. https://doi.org/10.1371/journal.pone.0184324.

Deng, Y., Su, Q., Mo, J., Fu, X., Zhang, Y., Lin, E., 2013. Celecoxib downregulates CD133 expression through inhibition of the Wnt signaling pathway in colon cancer cells. Cancer Investig. 31 (2), 97–102. https://doi.org/10.3109/07357907.2012.754458.

Dong, Z., Abbas, M.N., Kausar, S., Yang, J., Li, L., Tan, L., Cui, H., 2019. Biological functions and molecular mechanisms of antibiotic tigecycline in the treatment of cancers. Int. J. Mol. Sci. 20 (14), 3577. https://doi.org/10.3390/ijms20143577.

Egashira, I., Takahashi-Yanaga, F., Nishida, R., Arioka, M., Igawa, K., Tomooka, K., Nakatsu, Y., Tsuzuki, T., Nakabeppu, Y., Kitazono, T., Sasaguri, T., 2017. Celecoxib and 2,5-dimethylcelecoxib inhibit intestinal cancer growth by suppressing the Wnt/β-catenin signaling pathway. Cancer Sci. 108 (1), 108–115. https://doi.org/10.1111/cas.13106.

Gimple, R.C., Bhargava, S., Dixit, D., Rich, J.N., 2019. Glioblastoma stem cells: lessons from the tumor hierarchy in a lethal cancer. Genes Dev. 33 (11 – 12), 591–609. https://doi.org/10.1101/gad.324301.119.

Hesari, A., Rezaei, M., Rezaei, M., Dashtiahangar, M., Fathi, M., Rad, J.G., Momeni, F., Avan, A., Ghasemi, F., 2019. Effect of curcumin on glioblastoma cells. J. Cell. Physiol. 234, 10281–10288.

Hu, P., Ke, C., Guo, X., Ren, P., Tong, Y., Luo, S., He, Y., Wei, Z., Cheng, B., Li, R., Luo, J., Meng, Z., 2019. Both glypican-3/Wnt/beta-catenin signaling pathway and autophagy contributed to the inhibitory effect of curcumin on hepatocellular carcinoma. Dig. Liver Dis. 51, 120–126. https://doi.org/10.1016/j.dld.2018.06.012.

Kanemitsu, T., Kawabata, S., Fukumura, M., et al., 2019. Folate receptor-targeted novel boron compound for boron neutron capture therapy on F98 glioma-bearing rats. Radiat. Environ. Biophys. 58 (1), 59–67.

Katona, B.W., Weiss, J.M., 2020. Chemoprevention of colorectal cancer. Gastroenterology 158 (2), 368–388.

Koval, A., Katanaev, V.L., 2018. Dramatic dysbalancing of the Wnt pathway in breast cancers. Sci. Rep. 8 (1), 7329. https://doi.org/10.1038/s41598-018-25672-6.

Laval, B., Maurizio, J., Kandalla, P.K., et al., 2020. C/EBPβ-dependent epigenetic memory induces trained immunity in hematopoietic stem cells. Cell Stem Cell 26 (5), 657–674.e8.

Le Rhun, E., Preusser, M., Roth, P., Reardon, D.A., van den Bent, M., Wen, P., Reifenberger, G., Weller, M., 2019. Molecular targeted therapy of glioblastoma. Cancer Treat. Rev. 80, 101896. https://doi.org/10.1016/j.ctrv.2019.101896.

Lesueur, P., Lequesne, J., Grellard, J.M., et al., 2019. Phase I/IIa study of concomitant radiotherapy with olaparib and temozolomide in unresectable or partially resectable glioblastoma: OLA-TMZ-RTE-01 trial protocol. BMC Cancer 19 (1), 198.

Linkous, A., Balamatsias, D., Snuderl, M., et al., 2019. Modeling patient-derived glioblastoma with cerebral organoids. Cell Rep. 26 (12), 3203–3211.e5.

Lu, V.M., Goyal, A., Graffeo, C.S., et al., 2019. Survival benefit of maximal resection for glioblastoma reoperation in the temozolomide era: a meta-analysis. World Neurosurg. 127, 31–37.

Lukas, R.V., Wainwright, D.A., Ladomersky, E., et al., 2019. Newly diagnosed glioblastoma: a review on clinical management. Oncology (Williston Park) 33 (3), 91–100.

Maniewska, J., Jeżewska, D., 2021. Non-steroidal anti-inflammatory drugs in colorectal cancer chemoprevention. Cancers (Basel) 13 (4), 594.

Manoranjan, B., Chokshi, C., Venugopal, C., et al., 2020. A CD133-AKT-Wnt signaling axis drives glioblastoma brain tumor-initiating cells. Oncogene 39 (7), 1590–1599.

Mata, D.A., Benhamida, J.K., Lin, A.L., et al., 2020. Genetic and epigenetic landscape of IDH-wildtype glioblastomas with FGFR3-TACC3 fusions. Acta Neuropathol. Commun. 8, 186.

Mathen, P., Rowe, L., Mackey, M., et al., 2020. Radiosensitizers in the temozolomide era for newly diagnosed glioblastoma. Neurooncol. Pract. 7 (3), 268–276.

Matias, D., Predes, D., Niemeyer Filho, P., Lopes, M.C., Abreu, J.G., Lima, F.R.S., Moura, N.V., 2017. Microglia-glioblastoma interactions: new role for Wnt signaling. Biochim. Biophys. Acta, Rev. Cancer 1868 (1), 333–340. https://doi.org/10.1016/j.bbcan.2017.05.007.

Morris, S.L., Zhu, P., Rao, M., et al., 2019. Gamma knife stereotactic radiosurgery in combination with bevacizumab for recurrent glioblastoma. World Neurosurg. 127, e523–e533.

Ozkizilcik, A., Sharma, A., Lafuente, J.V., et al., 2019. Nanodelivery of cerebrolysin reduces pathophysiology of Parkinson's disease. Prog. Brain Res. 245, 201–246.

Prezado, Y., Gregory, J., Consuelo, G., et al., 2019. Tumor control in RG2 glioma-bearing rats: a comparison between proton minibeam therapy and standard proton therapy. Int. J. Radiat. Oncol. Biol. Phys. 104 (2), 266–271.

Prionisti, I., Bühler, L.H., Walker, P.R., Jolivet, R.B., 2019. Harnessing microglia and macrophages for the treatment of glioblastoma. Front. Pharmacol. 10, 506. https://doi.org/10.3389/fphar.2019.00506.

Qiao, L., Yu, Q., Wang, Y., et al., 2020. A brain glioma gene delivery strategy by angiopep-2 and TAT-modified magnetic lipid-polymer hybrid nanoparticles. RSC Adv. 10, 41471–41481.

Qorri, B., Harless, W., Szewczuk, M.R., 2020. Novel molecular mechanism of aspirin and celecoxib targeting mammalian neuraminidase-1 impedes epidermal growth factor receptor signaling axis and induces apoptosis in pancreatic cancer cells. Drug Des. Devel. Ther. 14, 4149–4167.

Ramalho, M.J., Andrade, S., Loureiro, J.A., do Carmo Pereira, M., 2020. Nanotechnology to improve the Alzheimer's disease therapy with natural compounds. Drug Deliv. Transl. Res. 10 (2), 380–402.

Reisner, Y., Or-Geva, N., 2019. Veto cells for safer nonmyeloablative haploidentical HSCT and CAR T cell therapy. Semin. Hematol. 56 (3), 173–182. https://doi.org/10.1053/j.seminhematol.2019.03.003.

Sankowski, R., Böttcher, C., Masuda, T., Geirsdottir, L., Sagar, Sindram, E., Seredenina, T., Muhs, A., Scheiwe, C., Shah, M.J., Heiland, D.H., Schnell, O., Grün, D., Priller, J.,

Prinz, M., 2019. Mapping microglia states in the human brain through the integration of high-dimensional techniques. Nat. Neurosci. 22 (12), 2098–2110. https://doi.org/10.1038/s41593-019-0532-y.

Scheithauer, B.W., 2009. Development of the WHO classification of tumors of the central nervous system: a historical perspective. Brain Pathol. 19 (4), 551–564.

Schiffer, D., Mellai, M., Bovio, E., Bisogno, I., Casalone, C., Annovazzi, L., 2018. Glioblastoma: microenvironment and niche concept. Neurol. Sci. 39 (7), 1161–1168. https://doi.org/10.1007/s10072-018-3408-0.

Seliger, C., Genbrugge, E., Gorlia, T., et al., 2020. Use of metformin and outcome of patients with newly diagnosed glioblastoma: pooled analysis. Int. J. Cancer 146 (3), 803–809.

Sharifzad, F., Ghavami, S., Verdi, J., et al., 2019. Glioblastoma cancer stem cell biology: potential theranostic targets. Drug Resist. Updat. 42, 35–45.

Sharma, H.S., Muresanu, D.F., Castellani, R.J., et al., 2020. Pathophysiology of blood-brain barrier in brain tumor. Novel therapeutic advances using nanomedicine. Int. Rev. Neurobiol. 151, 1–66.

Shevchenko, V., Arnotskaya, N., Pak, O., et al., 2020. Molecular determinants of the interaction between glioblastoma CD133+ cancer stem cells and the extracellular matrix. Int. Rev. Neurobiol. 151, 155–169.

Stupp, R., Taillibert, S., Kanner, A., et al., 2017. Effect of tumor-treating fields plus maintenance temozolomide vs maintenance temozolomide alone on survival in patients with glioblastoma: a randomized clinical trial. JAMA 318 (23), 2306–2316.

Srivastava, N., Srivastava, R.A.K., 2019. Curcumin and quercetin synergistically inhibit cancer cell proliferation in multiple cancer cells and modulate Wnt/beta-catenin signaling and apoptotic pathways in A375 cells. Phytomedicine 52, 117–128. https://doi.org/10.1016/j.phymed.2018.09.224.

Stupp, R., Lukas, R.V., Hegi, M.E., 2019. Improving survival in molecularly selected glioblastoma. Lancet 393 (10172), 615–617.

Tan, A.C., Ashley, D.M., López, G.Y., Malinzak, M., Friedman, H.S., Khasraw, M., 2020. Management of glioblastoma: state of the art and future directions. CA Cancer J. Clin. 70 (4), 299–312. https://doi.org/10.3322/caac.21613.

Tuazon, J.P., Castelli, V., Lee, J.Y., et al., 2019. Neural stem cells. Adv. Exp. Med. Biol. 1201, 79–91.

Tuli, H.S., Tuorkey, M.J., Thakral, F., Sak, K., Kumar, M., Sharma, A.K., Sharma, U., Jain, A., Aggarwal, V., Bishayee, A., 2019. Molecular mechanisms of action of genistein in cancer: recent advances. Front. Pharmacol. 10, 1336. https://doi.org/10.3389/fphar.2019.01336.

Umphlett, M., Shea, S., Tome-Garcia, J., et al., 2020. Widely metastatic glioblastoma with BRCA1 and ARID1A mutations: a case report. BMC Cancer 20 (1), 47.

Vick, N.A., Khandekar, J.D., Bigner, D.D., 1977. Chemotherapy of brain tumors. Arch. Neurol. 34 (9), 523–526.

Waitkus, M.S., Diplas, B.H., Yan, H., 2018. Biological role and therapeutic potential of IDH mutations in cancer. Cancer Cell 34 (2), 186–195.

Wang, J., Liu, J., Meng, H., et al., 2019. Neural stem cells promote glioblastoma formation in nude mice. Clin. Transl. Oncol. 21 (11), 1551–1560.

Wang, Z., Feng, T., Zhou, L., et al., 2020. Salinomycin nanocrystals for colorectal cancer treatment through inhibition of Wnt/β-catenin signaling. Nanoscale 12 (38), 19931–19938.

Wegner, R.E., Abel, S., Horne, Z.D., et al., 2019. National trends in radiation dose escalation for glioblastoma. Radiat. Oncol. J. 37 (1), 13–21.

Weller, M., Le Rhun, E., Preusser, M., et al., 2019. How we treat glioblastoma. ESMO Open 4 (Suppl. 2), e000520.

Wick, A., Desjardins, A., Suarez, C., et al., 2020. Phase 1b/2a study of galunisertib, a small molecule inhibitor of transforming growth factor-beta receptor I, in combination with standard temozolomide-based radiochemotherapy in patients with newly diagnosed malignant glioma. Investig. New Drugs 38 (5), 1570–1579.

Woodroffe, R.W., Zanaty, M., Soni, N., et al., 2020. Survival after reoperation for recurrent glioblastoma. J. Clin. Neurosci. 73, 118–124.

Xie, D., Zheng, G.Z., Xie, P., Zhang, Q.H., Lin, F.X., Chang, B., Hu, Q.X., Du, S.X., Li, X.D., 2017. Antitumor activity of resveratrol against human osteosarcoma cells: a key role of Cx43 and Wnt/beta-catenin signaling pathway. Oncotarget 8, 111419–111432.

Yang, H.C., Wang, J.Y., Bu, X.Y., Yang, B., Wang, B.Q., Hu, S., Yan, Z.Y., Gao, Y.S., Han, S.Y., Qu, M.Q., 2019. Resveratrol restores sensitivity of glioma cells to temozolomide through inhibiting the activation of Wnt signaling pathway. J. Cell. Physiol. 234, 6783–6800.

Yu, L., Kim, H.J., Park, M.K., et al., 2021. Ethacrynic acid, a loop diuretic, suppresses epithelial-mesenchymal transition of A549 lung cancer cells via blocking of NDP-induced WNT signaling. Biochem. Pharmacol. 183, 114339.

Zaitsev, S., Sharma, H.S., Sharma, A., et al., 2020. Pro-inflammatory modification of cancer cells microsurroundings increases the survival rates for rats with low differentiated malignant glioma of brain. Int. Rev. Neurobiol. 151, 253–279.

Zhang, L.N., Kong, C.F., Zhao, D., et al., 2019. Fusion with mesenchymal stem cells differentially affects tumorigenic and metastatic abilities of lung cancer cells. J. Cell. Physiol. 234 (4), 3570–3582.

Zilony-Hanin, N., Rosenberg, M., Richman, M., Yehuda, R., Schori, H., Motiei, M., Rahimipour, S., Groisman, A., Segal, E., Shefi, O., 2019. Neuroprotective effect of nerve growth factor loaded in porous silicon nanostructures in an Alzheimer's disease model and potential delivery to the brain. Small 15 (45), e1904203. https://doi.org/10.1002/smll.201904203.

Network pharmacological mechanism of Cinobufotalin against glioma

3

Cong Li[a,†], Hanyu Guo[a,†], Chao Wang[a,†], Wengang Zhan[a], Qijia Tan[a], Caijun Xie[a], Aruna Sharma[b], Hari Shanker Sharma[b,*], Lin Chen[c,*], and Zhiqiang Zhang[a,*]

[a]*The Second Affiliated Hospital of Guangzhou University of Chinese Medicine, Guangdong Province Hospital of Chinese Medical, Guangzhou, China*
[b]*International Experimental Central Nervous System Injury & Repair (IECNSIR), Department of Surgical Sciences, Anesthesiology & Intensive Care Medicine, Uppsala University Hospital, Uppsala University, Uppsala, Sweden*
[c]*Department of Neurosurgery, Dongzhimen Hospital of Beijing University of Traditional Chinese Medicine, Beijing, China*
Corresponding authors: Tel.: +86-13826296218; Fax: +86-02081887233 (Zhiqiang Zhang); Tel.: +46702011801; Fax: +4618243899 (Hari Shanker Sharma), e-mail address: Sharma@surgsci.uu.se; doctorzzq@163.com

Abstract

Objective: Cinobufotalin was extracted from the skin of Chinese giant salamander or black sable with good clinical effect against tumor. This study aims to explore the mechanism of Cinobufotalin components and predict the target of action of Cinobufotalin on glioma.

Methods: The active components of Cinobufotalin were screened by the Chinese medicine pharmacology database and analysis platform (TCMSP), PubChem database, etc. The potential molecular components and targets were identified and enrichment analysis was conducted through the construction of related networks and analysis of their characteristics. Relevant targets of glioma were searched through TTD, DRUGBANK, and other databases, and the intersection was found and the key targets were found too.

Results: A total of 21 active components and 184 target genes of Cinobufotalin were found. According to the enrichment analysis results, the pharmacological mechanism of Cinobufotalin mainly includes inhibition of the cell cycle, promotion of cell apoptosis, and regulation of immunity. On this basis, RAC1, FOS, and NOS3 can be preliminarily predicted as potential targets of Cinobufotalin in the treatment of glioma.

[†]Contributed equally to this article.

Conclusions: The screening of active ingredients and target prediction based on network pharmacology can provide a new research idea for the multi-target treatment of glioma with Cinobufotalin.

Keywords

Network pharmacology, Cinobufotalin, Glioma, Target prediction

1 Introduction

Traditional Chinese medicine (TCM) has a long history in China and East Asian countries. It has the characteristics of a wide range of medicinal effects and less toxic and side effects (Zhang et al., 2019). In recent years, more and more TCM plays an important role in the treatment of cancer. However, it is difficult to study TCM because of its complex chemical composition and numerous pharmacological mechanisms. Recent studies have shown that traditional Chinese medicine can fight against tumors by regulating tumor cell apoptosis and controlling autophagy, with many advantages such as fewer adverse reactions and good tolerance (So et al., 2019; Su and Tai, 2020). Cinobufagin is a water decocting extract from *Bufobufo gargarizans* cantor or *Bufo melanostictus* Schneider, which has the functions of clearing away heat and detoxification, promoting blood circulation and removing blood stasis, improving water and reducing swelling, softening and dispersing conformation (Qi et al., 2010).

Brain glioma is formed by the deterioration and evolution of neuroepithelial cells, accounting for 40–50% of intracranial tumors, and is the most common malignant tumor in the adult central nervous system (Ostrom et al., 2020). The 5-year survival rate of malignant glioma is only 5%, which is characterized by rapid infiltration and growth, rapid proliferation and easy recurrence, and is also the main reason for the extremely high mortality (Hanna et al., 2020; Stupp et al., 2005). According to the molecular and morphological classification standard of the World Health Organization (WHO), brain gliomas can be divided into four grades in several types, with different degrees of malignancy and great differences in prognosis after systematic treatment (van den Bent et al., 2017). At this moment, selected treatments for glioma includes surgical excision, radiation and chemotherapy. At the same time, there are many methods including tumor-treating fields (TTF), immunotherapy, targeted drug therapy, methods of Neurorestoratology, and traditional Chinese medicine can be used as adjuvant therapy (Tan et al., 2020; Xinyu et al., 2020; Zhong et al., 2019). The dismal prognosis and inevitable recurrence of glioma as well as resistance to chemoradiotherapy may be related to its highly cellular heterogeneity and multiple subclonal populations. In particular, as a method with a long history and positive efficacy in China, TCM has its unique advantages in the treatment of it and has attracted much attention in the treatment of brain glioma in recent years (Wang et al., 2019). Wang et al. reviewed the same active ingredient of TCM can act on different signaling pathways, such as ginsenoside Rg3 inhibited proliferation and induced apoptosis via the AKT, MEK signal pathway (Wang et al., 2019). Hence, the multi-target and multi-level pathway effects of TCM may be performed in the treatment of glioma.

Cinobufotalin is a traditional Chinese medicine extracted from the skin of Chinese giant salamander or black sable, and there are lots of studies have confirmed that it showed antitumor effects in lung cancer, liver cancer, breast cancer, ovarian cancer, and gastric cancer through inhibit tumor cell proliferation, induce tumor cell differentiation, promote tumor cell apoptosis and other functions (Afroze et al., 2018; Kai et al., 2014; Li et al., 2019, 2021; Mao et al., 2020; Sun et al., 2019; Wang et al., 2020). However, whether Cinobufotalin can treat glioma remains unclear and there is few studies. Therefore, given Cinobufotalin's effective active ingredient diverse and complex mechanism, we decided to explore the molecular targets of the Cinobufotalin element against glioma forecast and provide data support for the treatment of glioma through network pharmacology.

2 Materials and methods

2.1 Compounds of Cinobufotalin and screening of related targets of Cinobufotalin

We searched the literature on compounds of Cinobufotalin using PubMed (https://pubmed.ncbi.nlm.nih.gov/), CNKI (https://www.cnki.net/), VIP (https://pubmed.ncbi.nlm.nih.gov/), wanfang (http://www.wanfangdata.com.cn/index.html). The keywords used were "dry scalp," "Cinobufotalin" and "Huachansu." The literature search screened 76 compounds. The compounds were input into the TCMSP database (https://tcmspw.com/), TCMID database (http://www.megabionet.org/tcmid/), Pubchem database (https://pubchem.ncbi.nlm.nih.gov/), and we obtained the molecular linear notation (SMILES) and related information for 20 compounds after duplicate data and no structural information compounds have been removed. At the same time, the 20 compounds screened out will be connected to the DRUGBANK database (https://go.drugbank.com/) and using TCMSP database to find relevant targets through the DRUGBANK database. For those compounds that were not included in the DRUGBANK database, we entered their active ingredients into the TCMID and PubChem databases, respectively, and then entered the relevant targets that we found into the universal protein database Uniprot (https://www.uniprot.org/). The conditions for selection were "Reviewed" and "Human" to correct the names of the genetic targets. Finally, the retrieval results were combined to establish a chemical component-target database of Cinobufotalin.

2.2 Target network map of Cinobufotalin

Cytoscape 3.8.0 software was used to construct "Cinobufotalin—Related Chemical Constituents—Target." In the network, the nodes represent the active chemical constituents and target proteins, while the edges represent the relationships among them.

2.3 GO biological process and KEGG metabolic pathway of Cinobufotalin

The joint nodes screened were sorted according to the P-value in the GO analysis by David database, and the top 10 data were selected for analysis in the Biological process, Cell Component, and Molecular Function. The analysis of the KEGG metabolic pathway was still based on P-value <0.05 as the standard.

2.4 Prediction target genes of glioma

Using "brain glioma" or "glioma" as the keywords to search in the providing database (http://db.idrblab.net/ttd/), GeneCards database (https://www.genecards.org/), Drugbank database, NCBI database (https://www.ncbi.nlm.nih.gov/), OMIM database (https://omim.org/) and Disgenet database (https://www.disgenet.org/). 1731 human-related gene targets were obtained after decontamination.

3 Results

3.1 Main active components and prediction targets of Cinobufotalin

A total of 184 potential targets of 20 compounds were identified by TCMSP, TCMID, PubChem, and Uniprot. The 21 chemical components of Cinobufotalin with 184 potential targets are shown in Table 1. The active components are replaced by simple letters to construct the network analysis diagram. When the interaction score was greater than 0.4, a complex network of 184 potential targets was constructed using Cytoscape 3.8.0 software after deleting the isolated and partially loosely connected gene nodes. According to the Degree, a Cinobufotalin-chemical compositional target plot was generated, including 205 nodes and 242 edges (Fig. 1).

3.2 Analysis results of GO biological process and KEGG metabolic pathway

GO function analysis was conducted on Cinobufotalin targets in the David database. There were 329 items related to Biological Process, 23 items related to Cell Component, and 88 items related to Molecular Function, according to the $P < 0.05$, and only the top 10 were selected according to P-value. The potential therapeutic targets of Cinobufotalin were significantly enriched in the "serotonin receptor signaling pathway," "drug response" and "protein phosphorylation" from the results showed in the Biological Process. In terms of Cell Composition, they are mainly "plasma membrane," "component of plasma membrane" and "cytoplasm." In terms of Molecular Function, it mainly includes "the binding of 5-HT," "the binding of drugs" and "the activity of neurotransmitter receptors." The details are shown in Table 2.

Table 1 Summary of active components of Cinobufacini.

Components name	Code
Proline	HCS1
Hypoxanthine	HCS2
Xanthine	HCS3
Uridine	HCS4
Adenine	HCS5
Adenosine	HCS6
Thymidine	HCS7
Guanine	HCS8
Thymine	HCS9
Adipyl-arginine	HCS10
5-Methoxy-tryptamine	HCS11
Bufotenine	HCS12
N-acetyl-5-hydroxy-tryptamine	HCS13
Riboflavin	HCS14
Indole-3-acetic acid	HCS15
Cinobufagin	HCS16
Linoleamide	HCS17
7α-Hydroxycholesterol	HCS18
Oleic-acid	HCS19
Cholesterol	HCS20

394 genes were enriched in 42 pathways by KEGG analysis. Fifteen entries were found as potential therapeutic targets though the condition of statistical analysis as the false discovery rate (FDR) corrected $P < 0.05$ for enrichment. Fig. 2 shows the relationship between drug targets and the KEGG pathway. It can be seen that the drugs are mainly involved in neuroactive ligand-receptor interaction (36 genes), serotonergic synapse (24 genes), calcium signaling pathway (24 genes), nitrogen metabolism (10 genes), cAMP signaling pathway (23 genes), the pathways of viral myocarditis (9 genes), the pathway of cocaine addiction pathway (8 genes), regulation of lipolysis in adipocytes (8 genes), the pathways of the dopaminergic synapse (11 genes), the pathway of the amphetamine addiction (8 genes), the pathways of the autoimmune thyroid diseases (7 genes), the pathway of the allograft rejection (6 genes), the pathway of the cholinergic synapse (9 genes), PPAR signaling pathway (7 genes), the pathway of arginine and proline metabolism (6 genes). Specific target genes in each pathway are shown in Table 3. The KEGG metabolism analysis diagram (Fig. 3) and GO biological process analysis diagram (Fig. 4) of the Cinobufotalin target gene can clear show the target function process of Cinobufotalin.

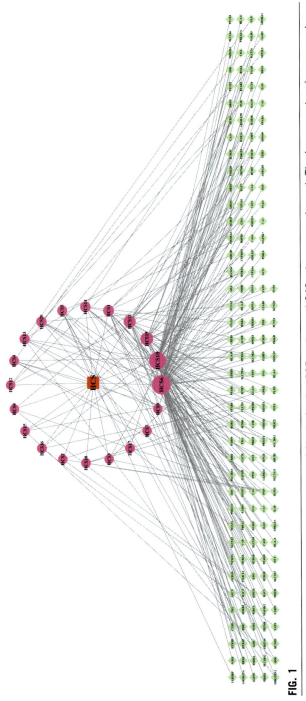

FIG. 1

Cinobufotalin-chemical composition-target plot of Cinobufotalin. 205 nodes and 242 edges were showed. The brown rectangle represents the drug, and the pink circle represents the number of the compounds. The larger the circle is, the more genes are enriched. The light green diamond represents the target gene for each compound.

Table 2 The biological function of Cinobufacini.

GOterm	Subgroup	Enrichment gene count	PValue
GO:0007210 ~ serotonin receptor signaling pathway	Biological process	12	1.0627E-20
GO:0042493 ~ response to drug	Biological process	27	1.56796E-16
GO:0007200 ~ phospholipase C-activating G-protein coupled receptor signaling pathway	Biological process	14	1.72041E-13
GO:0006730 ~ one-carbon metabolic process	Biological process	10	1.62959E-11
GO:0043401 ~ steroid hormone mediated signaling pathway	Biological process	11	3.9025E-10
GO:0015701 ~ bicarbonate transport	Biological process	10	7.10945E-10
GO:0007187 ~ G-protein coupled receptor signaling pathway, coupled to cyclic nucleotide second messenger	Biological process	10	1.08494E-09
GO:0007271 ~ synaptic transmission, cholinergic	Biological process	9	4.04426E-09
GO:0042310 ~ vasoconstriction	Biological process	7	1.4648E-08
GO:0001973 ~ adenosine receptor signaling pathway	Biological process	6	1.54039E-08
GO:0005886 ~ plasma membrane	Cellular component	95	2.66501E-18
GO:0005887 ~ integral component of plasma membrane	Cellular component	54	4.57943E-18
GO:0030425 ~ dendrite	Cellular component	20	9.44904E-10
GO:0042612 ~ MHC class I protein complex	Cellular component	5	2.87764E-06
GO:0032279 ~ asymmetric synapse	Cellular component	4	5.11693E-05
GO:0005829 ~ cytosol	Cellular component	55	7.43458E-05
GO:0005892 ~ acetylcholine-gated channel complex	Cellular component	5	8.38021E-05
GO:0009986 ~ cell surface	Cellular component	17	9.81754E-05
GO:0031901 ~ early endosome membrane	Cellular component	8	0.000153899
GO:0071556 ~ integral component of lumenal side of endoplasmic reticulum membrane	Cellular component	5	0.00018023
GO:0004993 ~ G-protein coupled serotonin receptor activity	Molecular function	15	4.37199E-21

Continued

Table 2 The biological function of Cinobufacini.—cont'd

GOterm	Subgroup	Enrichment gene count	PValue
GO:0051378 ~ serotonin binding	Molecular function	10	1.30162E-17
GO:0008144 ~ drug binding	Molecular function	17	7.11938E-17
GO:0030594 ~ neurotransmitter receptor activity	Molecular function	12	8.72714E-16
GO:0004089 ~ carbonate dehydratase activity	Molecular function	10	2.50298E-15
GO:0004879 ~ RNA polymerase II transcription factor activity, ligand-activated sequence-specific DNA binding	Molecular function	10	9.69419E-11
GO:0003707 ~ steroid hormone receptor activity	Molecular function	11	3.08699E-10
GO:0019899 ~ enzyme binding	Molecular function	18	9.00453E-08
GO:0005232 ~ serotonin-activated cation-selective channel activity	Molecular function	5	1.76362E-07
GO:0042166 ~ acetylcholine binding	Molecular function	7	2.44106E-07

3.3 Protein-protein interaction (PPI) network construction of Cinobufotalin

The 184 effective genes were searched by the Bisogenet plugin, and 5262 nodes and 194,410 edges were obtained. And then screened by $2 \times$ degree value, 1183 nodes and 52,347 edges were found; Then 302 nodes and 10,551 edges were obtained according to $1 \times$ degree value, $1 \times$ Betweenness value and $1 \times$ Closeness value. Finally, screened by $1 \times$ Betweenness value and $2 \times$ LAC value, 124 nodes and 3261 edges were obtained. Then the final PPI network construction is shown in Fig. 5.

3.4 Prediction of targets between Cinobufotalin and glioma

The 184 effective target genes of Cinobufotalin and the 1731 effective target genes of glioma were used for plotting the Venny diagram with a total of 65 interactive genes (Fig. 6) and 14 interactive active compounds (Table 4). Then, the 65 interactive genes were analyzed using the Network Analyzer tool in the Cytoscape 3.8.0 database.

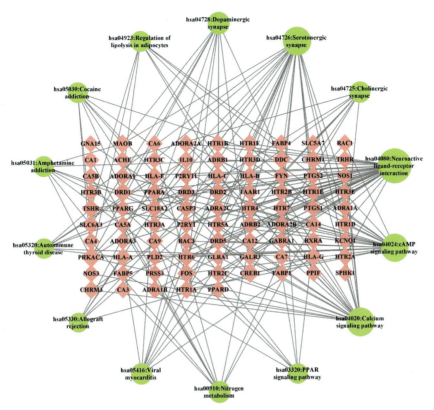

FIG. 2

Target diagram of Cinobufotalin on KEGG pathway. Peripheral green circle represents the KEGG metabolic pathway (the larger the circle is, the more genes were involved). The orange diamonds inside represent the gene that is the target of the Cinobufotalin. KEGG, Kyoto Encyclopedia of Genes and Genomes.

The top 12 cross target gene of cinobufagin were identified as CREB1, CXCL8, CXCL8, DRD2, EPO, ESR1, ESR1, HLA-A, HLA-A, HLA-A, HMOX1, and Rac1. The top 12 cross target gene of glioma were Fos, IL13, IL10, SLC6A3, HIF1A, NOS3, FOS, HLA-C, HLA-G, HLA-B, NFE2L2, and Rac3. Their connection score were 0.99, 0.983, 0.996, 0.99, 0.984, 0.985, 0.982, 0.984, 0.982, 0.986 and 0.982, respectively (Fig. 7).

Table 3 Distribution of Cinobufacini target gene in KEGG pathway.

KEGG_PATHWAY	PValue	Genes
Neuroactive ligand-receptor interaction	9.85E-18	CHRM3, CHRM1, HTR2B, HTR2C, ADRB1, ADRB2, HTR2A, ADRA1B, HTR4, ADRA1A, GLRA1, HTR6, HTR7, GALR3, ADORA3, P2RY1, ADORA1, DRD1, PRSS3, DRD2, DRD3, DRD5, GABRA1, P2RY11, HTR1E, HTR1F, HTR1D, HTR1A, HTR1B, ADRA2C, HTR5A, TAAR1, TSHR, TRHR, ADORA2A, ADORA2B
Serotonergic synapse	1.18E-16	DDC, MAOB, HTR1E, HTR1F, HTR3E, HTR1D, HTR3C, HTR1A, HTR2B, HTR3D, HTR1B, HTR2C, HTR3A, HTR3B, HTR2A, HTR4, HTR5A, PTGS2, PTGS1, HTR6, HTR7, CASP3, PRKACA, SLC18A2
Calcium signaling pathway	5.67E-12	CHRM3, CHRM1, NOS3, SPHK1, HTR2B, HTR2C, ADRB1, ADRB2, HTR2A, HTR4, HTR5A, ADRA1B, ADRA1A, TRHR, HTR6, GNA15, HTR7, ADORA2A, ADORA2B, PPIF, NOS1, DRD1, PRKACA, DRD5
Nitrogen metabolism	2.20E-11	CA12, CA1, CA5B, CA3, CA5A, CA4, CA7, CA6, CA9, CA14
cAMP signaling pathway	3.10E-10	CHRM1, HTR1E, HTR1F, HTR1D, HTR1A, HTR1B, ADRB1, FOS, ADRB2, HTR4, TSHR, PLD2, HTR6, CREB1, ADORA2A, ADORA1, RAC3, DRD1, RAC1, DRD2, PPARA, PRKACA, DRD5
Viral myocarditis	3.27E-05	CASP3, HLA-B, HLA-C, RAC3, FYN, HLA-A, RAC1, HLA-F, HLA-G
Cocaine addiction	9.21E-05	DDC, CREB1, MAOB, DRD1, DRD2, PRKACA, SLC6A3, SLC18A2
Regulation of lipolysis in adipocytes	2.18E-04	FABP4, ADORA1, ADRB1, ADRB2, PRKACA, PTGS2, TSHR, PTGS1
Dopaminergic synapse	5.29E-04	DDC, CREB1, MAOB, FOS, DRD1, DRD2, PRKACA, DRD3, SLC6A3, SLC18A2, DRD5
Amphetamine addiction	6.09E-04	DDC, CREB1, MAOB, FOS, DRD1, PRKACA, SLC6A3, SLC18A2
Autoimmune thyroid disease	9.60E-04	IL10, HLA-B, HLA-C, HLA-A, HLA-F, HLA-G, TSHR
Allograft rejection	0.001250896	IL10, HLA-B, HLA-C, HLA-A, HLA-F, HLA-G
Cholinergic synapse	0.003179032	SLC5A7, ACHE, CHRM3, CHRM1, CREB1, KCNQ1, FYN, FOS, PRKACA
PPAR signaling pathway	0.003582296	FABP1, RXRA, FABP4, FABP5, PPARG, PPARA, PPARD
Arginine and proline metabolism	0.00482425	GATM, DAO, MAOB, P4HA1, NOS3, NOS1

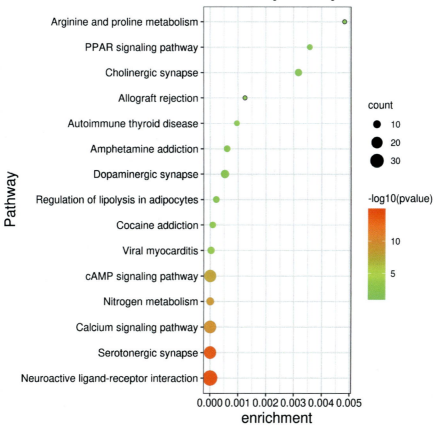

FIG. 3

Enrichment of KEGG metabolic pathway with Cinobufotalin. KEGG, Kyoto Encyclopedia of Genes and Genomes.

3.5 KEGG pathway enrichment analysis between Cinobufotalin and glioma

Using David and KEGG pathway analysis of 65 interactive genes between Cinobufotalin and glioma, 43 biological pathways were obtained. According to the $P < 0.05$ and the relevant literature for screening, we got the top 15 pathways were shown in Table 5. The broad pathways were excluded, and the pathways with high fit were VEGF signaling pathway, cAMP signaling pathway, cAMP signaling pathway, and HIF-1 signaling pathway.

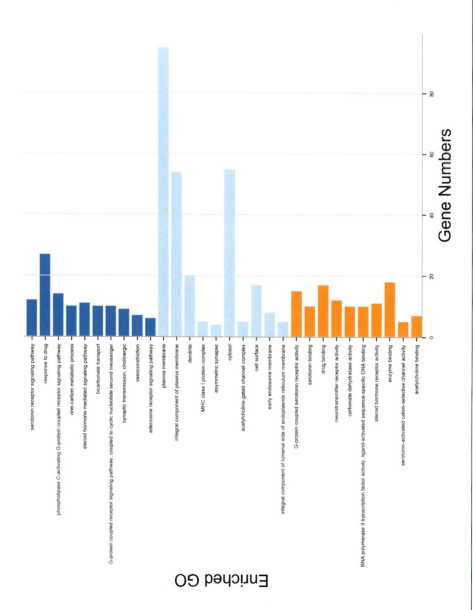

FIG. 4

GO bioprocess analysis of potential target genes of cinobufotalin. The 10 remarkably enriched items in the biological processes, cell component, and molecular function were showed. GO, gene ontology.

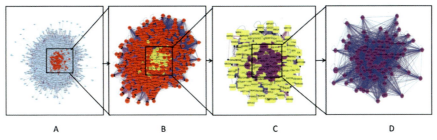

FIG. 5

Construction and screening of a Bisogenet-based PPI Network for Cinobufotalin. (A) to (B) were screened by $2 \times$ degree value. (B) to (C) was screened by $1 \times$ degree value, $1 \times$ Betweenness value and $1 \times$ Closeness value. (C) to (D) was screened by $1 \times$ Betweenness value and $2 \times$ LAC value. (A: 5262 nodes, 194,410 edges; B: 1183 nodes, 52,347 edges; C: 302 nodes, 10,551 edges; D: 124 nodes, 3261 edges.)

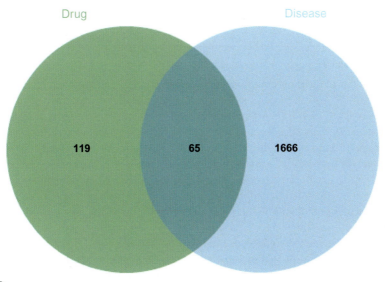

FIG. 6

Common gene target between Cinobufotalin and glioma. The green represents the Cinobufotalin targets, the blue represents the glioma targets, and the cyan in the middle represents the common targets of Cinobufotalin and the glioma.

Table 4 Basic information of 14 interactive active compounds between Cinobufacini and glioma.

Order	Compounds name	Count
1	Proline	5
2	Xanthine	1
3	Uridine	1
4	Bufogargarizanine-D	3
5	Adenine	27
6	Guanosine	17
7	Thymine	7
8	4，9-Anhydro-tetrodotoxin	3
9	Cyclo-Pro-Gly-dipeptide	1
10	Bufo-tenine nitrogen oxide	1
11	Adipyl-arginine	3
12	5-Methoxy-tryptamine	1
13	Bufotenine	5
14	Ethenzamide	3

4 Discussion

Glioma is one of the most common and invasive primary brain tumors. It can occur in any part of the central nervous system and at any age. The prevalence of malignant gliomas is about 29.5/100000 (Stupp et al., 2015), and the incidence rate is increasing in the past 20 years (Taylor and Ramaswamy, 2017). Because glioma, especially high-grade glioma, is highly invasive and vascular, it is difficult to remove completely, the recurrence rate is high, and the survival rate of patients is still not ideal. TCM as one of the effective means of supplementary treatment plays an effective role in the treatment. Cinobufacin is a water-soluble preparation made from the extraction and processing of *Bufo gargarizans* skin. Its main components are Bufo toxin, Cinobufacin, amino acids, and indole alkaloids. Among them, Bufo toxin and Cinobufacin are closely related to their anti-tumor activities (Hao et al., 2019). Its anti-tumor effect is reflected in many aspects, inducing tumor cell apoptosis, blocking tumor cell growth cycle, inhibiting tumor cell metastasis, and inhibiting angiogenesis (Wang et al., 2018). Therefore, cinobufagin has the potential to be used in glioma treatment.

In this study, we found the potential therapeutic targets of Cinobufagin through network pharmacology and molecular docking and explored the mechanism of Cinobufagin anti-glioma. By constructing a complex multi-center gene network and gene docking, we identified 14 main compounds of Cinobufagin in the treatment of glioma. In this study, we extracted 184 gene targets of Cinobufagin.

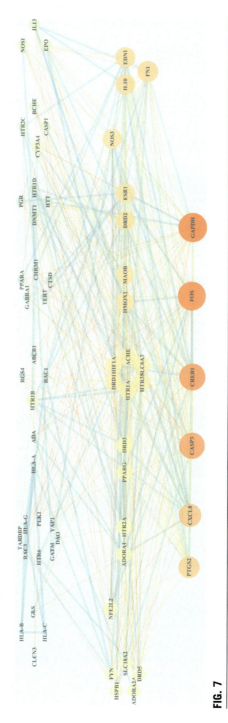

FIG. 7

Interaction target gene network between Cinobufotalin and disease. The darker the color, the larger the circle represents the greater the association between genes.

Table 5 Enrichment analysis of interactive pathways between Cinobufacini and glioma.

Pathways	P	Gene
cAMP signaling pathway	9.17E-10	CHRM1, HTR1D, HTR1A, HTR1B, FOS, HTR6, CREB1, ADORA2A, ADORA1, RAC3, DRD1, RAC1, DRD2, PPARA, DRD5
Serotonergic synapse	3.05E-08	HTR6, MAOB, CASP3, HTR1D, HTR1A, HTR1B, HTR2C, HTR3A, HTR2A, PTGS2, SLC18A2
Viral myocarditis	5.05E-07	CASP3, HLA-B, HLA-C, RAC3, FYN, HLA-A, RAC1, HLA-G
Neuroactive ligand-receptor interaction	5.22E-07	GABRA1, CHRM1, HTR1D, HTR1A, HTR1B, HTR2C, HTR2A, HTR6, ADORA2A, ADORA1, DRD1, DRD2, DRD3, DRD5
Dopaminergic synapse	1.31E-05	CREB1, MAOB, FOS, DRD1, DRD2, DRD3, SLC6A3, SLC18A2, DRD5
Cocaine addiction	6.06E-05	CREB1, MAOB, DRD1, DRD2, SLC6A3, SLC18A2
Calcium signaling pathway	1.45E-04	HTR6, CHRM1, ADORA2A, NOS3, HTR2C, DRD1, NOS1, HTR2A, DRD5
Amphetamine addiction	2.53E-04	CREB1, MAOB, FOS, DRD1, SLC6A3, SLC18A2
Allograft rejection	2.79E-04	IL10, HLA-B, HLA-C, HLA-A, HLA-G
Arginine and proline metabolism	8.93E-04	GATM, DAO, MAOB, NOS3, NOS1
Autoimmune thyroid disease	0.001035958	IL10, HLA-B, HLA-C, HLA-A, HLA-G
HIF-1 signaling pathway	0.001419877	EDN1, EPO, NOS3, HMOX1, HIF1A, GAPDH
VEGF signaling pathway	0.001883785	NOS3, HSPB1, RAC3, RAC1, PTGS2
Graft-vs-host disease	0.002857767	HLA-B, HLA-C, HLA-A, HLA-G
Sphingolipid signaling pathway	0.003775725	NOS3, ADORA1, RAC3, FYN, RAC1, CTSD

Sixty five therapeutic targets of Cinobufacini with glioma were screened by gene interaction. According to the results of metabolic pathway enrichment analysis, Cinobufacini anti-glioma involves a variety of signaling pathways, including VEGF signaling pathway, cAMP signaling pathway, cAMP signaling pathway, and HIF-1 signaling pathway. It is worth noting that Rac1, Fos, and NOS3 were highly expressed in the above pathways. Rac1 is a Ras-related C3 botulinum toxin substrate 1, which is a small GTPase associated with the plasma membrane and circulates between active GTP binding state and ineffective GDP binding state. In its active state, it binds with a variety of effector proteins to regulate cellular responses, such as secretion process and epithelial cell polarization. Rac1 is highly expressed in glioblastoma and is closely related to the proliferation activity and invasion ability of glioma cells (Qin et al., 2017; Yukinaga et al., 2014). The results suggest that Cinobufacini can significantly reduce the proliferation and invasion of glioblastoma cells by inhibiting the activity of Rac1 protein. It has been reported that Fos is highly expressed in gliomas (the higher the malignant degree of gliomas, the higher the expression of the Fos gene) and it plays an important role in regulating cell growth, proliferation, and differentiation (Debinski and Gibo, 2005; Liu et al., 2016; Tao et al., 2013). Therefore, Cinobufacini may induce glioma cancer cell death by inhibiting high expression of Fos to improve patients' survival. NOS3 was highly expressed in the VEGF signaling pathway, cAMP signaling pathway, and HIF-1 signaling pathway. NOS3 can affect glioma growth, invasion, angiogenesis, immunosuppression, differentiation state, and therapeutic resistance (Tran et al., 2017). These multifaceted effects NOS3 on gliomas both in vitro and in vivo suggest the potential of modulating the pathway for anti-glioma patient therapies.

However, the mechanism of Cinobufacini and glioma both are very complex. The network pharmacology study is helps to better explore the application of cinobufagin in glioma. The anti-glioma effects of Cinobufacini need to be verified by cell experiments, animal experiments, and clinical methods.

Funding

Scientific research project of Guangdong Bureau of traditional Chinese Medicine (No.20211159).

References

Afroze, S.H., Peddaboina, C., McDowell, A.J., Ashraf, A., McCormick, T.C., Newell-Rogers, M.K., Zawieja, D.C., Kuehl, T.J., Uddin, M.N., 2018. Differential effects of in vitro treatment with cinobufotalin on three types of ovarian cancer cells. Anticancer Res. 38, 5717–5724.

Debinski, W., Gibo, D.M., 2005. Fos-related antigen 1 modulates malignant features of glioma cells. Mol. Cancer Res. 3, 237–249.

Hanna, C., Lawrie, T.A., Rogozinska, E., Kernohan, A., Jefferies, S., Bulbeck, H., Ali, U.M., Robinson, T., Grant, R., 2020. Treatment of newly diagnosed glioblastoma in the elderly: a network meta-analysis. Cochrane Database Syst. Rev. 3, D13261.

Hao, Y., Luo, X., Ba, X., et al., 2019. Huachansu suppresses TRPV1 up-regulation and spinal astrocyte activation to prevent oxaliplatin-induced peripheral neuropathic pain in rats. Gene 680, 43–50.

Kai, S., Lu, J.H., Hui, P.P., Zhao, H., 2014. Pre-clinical evaluation of cinobufotalin as a potential anti-lung cancer agent. Biochem. Biophys. Res. Commun. 452, 768–774.

Li, J., Rong, M.H., Dang, Y.W., He, R.Q., Lin, P., Yang, H., Li, X.J., Xiong, D.D., Zhang, L.J., Qin, H., Feng, C.X., Chen, X.Y., Zhong, J.C., Ma, J., Chen, G., 2019. Differentially expressed gene profile and relevant pathways of the traditional Chinese medicine cinobufotalin on MCF7 breast cancer cells. Mol. Med. Rep. 19, 4256–4270.

Li, L.L., Su, Y.X., Mao, Y., Jiang, P.Y., Chu, X.L., Xue, P., Jia, B.H., Zhu, S.J., 2021. The effectiveness and safety of Cinobufotalin injection as an adjunctive treatment for lung cancer: a meta-analysis of randomized controlled trials. Evid. Based Complement. Alternat. Med. 2021, 8852261.

Liu, Z.G., Jiang, G., Tang, J., Wang, H., Feng, G., Chen, F., Tu, Z., Liu, G., Zhao, Y., Peng, M.-J., He, Z.W., Chen, X.Y., Lindsay, H., Xia, Y.F., Li, X.N., 2016. C-Fos over-expression promotes radioresistance and predicts poor prognosis in malignant glioma. Oncotarget 7, 65946–65956.

Mao, Y., Peng, X., Xue, P., Lu, D., Li, L., Zhu, S., 2020. Network pharmacology study on the pharmacological mechanism of cinobufotalin injection against lung cancer. Evid. Based Complement. Alternat. Med. 2020, 1246742.

Ostrom, Q.T., Patil, N., Cioffi, G., Waite, K., Kruchko, C., Barnholtz-Sloan, J.S., 2020. CBTRUS statistical report: primary brain and other central nervous system tumors diagnosed in the United States in 2013–2017. Neuro-Oncology 22, v1–v96.

Qi, F., Li, A., Zhao, L., Xu, H., Inagaki, Y., Wang, D., Cui, X., Gao, B., Kokudo, N., Nakata, M., Tang, W., 2010. Cinobufacini, an aqueous extract from Bufo bufo gargarizans Cantor, induces apoptosis through a mitochondria-mediated pathway in human hepatocellular carcinoma cells. J. Ethnopharmacol. 128, 654–661.

Qin, W., Rong, X., Dong, J., Yu, C., Yang, J., 2017. miR-142 inhibits the migration and invasion of glioma by targeting Rac1. Oncol. Rep. 38, 1543–1550.

So, T.H., Chan, S.K., Lee, V.H., Chen, B.Z., Kong, F.M., Lao, L.X., 2019. Chinese medicine in cancer treatment—how is it practised in the east and the west? Clin. Oncol. (R. Coll. Radiol.) 31, 578–588.

Stupp, R., Mason, W.P., van den Bent, M.J., Weller, M., Fisher, B., Taphoorn, M.J., Belanger, K., Brandes, A.A., Marosi, C., Bogdahn, U., Curschmann, J., Janzer, R.C., Ludwin, S.K., Gorlia, T., Allgeier, A., Lacombe, D., Cairncross, J.G., Eisenhauer, E., Mirimanoff, R.O., 2005. Radiotherapy plus concomitant and adjuvant temozolomide for glioblastoma. N. Engl. J. Med. 352, 987–996.

Stupp, R., Taillibert, S., Kanner, A.A., et al., 2015. Maintenance therapy with tumor-treating fields plus temozolomide vs temozolomide alone for glioblastoma: A randomized clinical trial. JAMA 314 (23), 2535–2543.

Su, P.H., Tai, C.J., 2020. Current development in integrative therapy of traditional Chinese medicine for cancer treatment: a mini-review. J. Tradit. Complement. Med. 10, 429–433.

Sun, H., Wang, W., Bai, M., Liu, D., 2019. Cinobufotalin as an effective adjuvant therapy for advanced gastric cancer: a meta-analysis of randomized controlled trials. Onco. Targets. Ther. 12, 3139–3160.

Tan, A.C., Ashley, D.M., Lopez, G.Y., Malinzak, M., Friedman, H.S., Khasraw, M., 2020. Management of glioblastoma: state of the art and future directions. CA Cancer J. Clin. 70, 299–312.

Tao, T., Lu, X., Yao, L., Wang, J., Shi, Y., Luo, H., Liu, N., You, Y., 2013. Expression of FOS protein in glioma and its effect on the growth of human glioma cells. Zhonghua Yi Xue Yi Chuan Xue Za Zhi 30, 293–296.

Taylor, M.D., Ramaswamy, V., 2017. Cancer: drain the swamp to beat glioma. Nature 549 (7673), 460–461.

Tran, A.N., Boyd, N.H., Walker, K., Hjelmeland, A.B., 2017. NOS expression and NO function in glioma and implications for patient therapies. Antioxid. Redox Signal. 26, 986–999.

van den Bent, M.J., Weller, M., Wen, P.Y., Kros, J.M., Aldape, K., Chang, S., 2017. A clinical perspective on the 2016 WHO brain tumor classification and routine molecular diagnostics. Neuro-Oncology 19, 614–624.

Wang, Z., Qi, F., Cui, Y., et al., 2018. An update on Chinese herbal medicines as adjuvant treatment of anticancer therapeutics. Biosci. Trends 12 (3), 220–239.

Wang, J., Qi, F., Wang, Z., Zhang, Z., Pan, N., Huai, L., Qu, S., Zhao, L., 2019. A review of traditional Chinese medicine for treatment of glioblastoma. Biosci. Trends 13, 476–487.

Wang, P.P., Wang, Y.H., Wang, L.S., Wu, T., 2020. Anti-tumor effect and its related mechanisms of cinobufotalin combined with cisplatin on H22 liver cancer mice. Zhongguo Zhong Yao Za Zhi 45, 3945–3951.

Xinyu, W., Nan, S., Xiangqi, M., Meng, C., Chuanlu, J., Jinquan, C., 2020. Review of clinical nerve repair strategies for neurorestoration of central nervous system tumor damage. J. Neurorestoratol. 08, 172–181.

Yukinaga, H., Shionyu, C., Hirata, E., Ui-Tei, K., Nagashima, T., Kondo, S., Okada-Hatakeyama, M., Naoki, H., Matsuda, M., 2014. Fluctuation of Rac1 activity is associated with the phenotypic and transcriptional heterogeneity of glioma cells. J. Cell Sci. 127, 1805–1815.

Zhang, Z., Zhang, S., Lui, C.N., Zhu, P., Zhang, Z., Lin, K., Dai, Y., Yung, K.K., 2019. Traditional Chinese medicine-based neurorestorative therapy for Alzheimer's and Parkinson's disease. J. Neurorestoratol. 07, 207–222.

Zhong, D., Hai, Y., Ning, W., Wahap, A., Jia, W., Tuo, W., Changwang, D., Maode, W., 2019. Impact of preoperative Karnofsky performance scale (KPS) and American Society of Anesthesiologists (ASA) scores on perioperative complications in patients with recurrent glioma undergoing repeated operation. J. Neurorestoratol. 07, 143–152.

Nanodelivery of oxiracetam enhances memory, functional recovery and induces neuroprotection following concussive head injury

Feng Niu[a], Aruna Sharma[b],*, Zhenguo Wang[a], Lianyuan Feng[c],
Dafin F. Muresanu[d,e], Seaab Sahib[f], Z. Ryan Tian[f], José Vicente Lafuente[g],
Anca D. Buzoianu[h], Rudy J. Castellani[i], Ala Nozari[j], Preeti K. Menon[k],
Ranjana Patnaik[l], Lars Wiklund[b], and Hari Shanker Sharma[b],*

[a]CSPC NBP Pharmaceutical Medicine, Shijiazhuang, China
[b]International Experimental Central Nervous System Injury & Repair (IECNSIR),
Department of Surgical Sciences, Anesthesiology & Intensive Care Medicine,
Uppsala University Hospital, Uppsala University, Uppsala, Sweden
[c]Department of Neurology, Bethune International Peace Hospital, Shijiazhuang, China
[d]Department of Clinical Neurosciences, University of Medicine & Pharmacy, Cluj-Napoca,
Romania
[e]"RoNeuro" Institute for Neurological Research and Diagnostic, Cluj-Napoca, Romania
[f]Department of Chemistry & Biochemistry, University of Arkansas, Fayetteville, AR, United States
[g]LaNCE, Department of Neuroscience, University of the Basque Country (UPV/EHU),
Leioa, Bizkaia, Spain
[h]Department of Clinical Pharmacology and Toxicology, "Iuliu Hatieganu" University of Medicine
and Pharmacy, Cluj-Napoca, Romania
[i]Department of Pathology, University of Maryland, Baltimore, MD, United States
[j]Anesthesiology & Intensive Care, Massachusetts General Hospital, Boston, MA, United States
[k]Department of Biochemistry and Biophysics, Stockholm University, Stockholm, Sweden
[l]Department of Biomaterials, School of Biomedical Engineering, Indian Institute of Technology,
Banaras Hindu University, Varanasi, India
*Corresponding authors: Tel.: +46-70-21-95-963; Fax: +46-18-24-38-99 (Aruna Sharma);
Tel.: +46-70-2011-801 (Hari Shanker Sharma),
e-mail address: aruna.sharma@surgsci.uu.se; sharma@surgsci.uu.se;
harishanker_sharma55@icloud.com; hssharma@aol.com

Progress in Brain Research, Volume 265, ISSN 0079-6123, https://doi.org/10.1016/bs.pbr.2021.06.004

Abstract

Military personnel are the most susceptible to concussive head injury (CHI) caused by explosion, blast or missile or blunt head trauma. Mild to moderate CHI could induce lifetime functional and cognitive disturbances causing significant decrease in quality of life. Severe CHI leads to instant death and lifetime paralysis. Thus, further exploration of novel therapeutic agents or new features of known pharmacological agents are needed to enhance quality of life of CHI victims.

Previous reports from our laboratory showed that mild CHI induced by weight drop technique causing an impact of 0.224 N results in profound progressive functional deficit, memory impairment and brain pathology from 5 h after trauma that continued over several weeks of injury.

In this investigation we report that TiO2 nanowired delivery of oxiracetam (50 mg/kg, i.p.) daily for 5 days after CHI resulted in significant improvement of functional deficit on the 8th day. This was observed using Rota Rod treadmill, memory improvement assessed by the time spent in finding hidden platform under water. The motor function improvement is seen in oxiracetam treated CHI group by placing forepaw on an inclined mesh walking and foot print analysis for stride length and distance between hind feet. TiO2-nanowired oxiracetam also induced marked improvements in the cerebral blood flow, reduction in the BBB breakdown and edema formation as well as neuroprotection of neuronal, glial and myelin damages caused by CHI at light and electron microscopy on the 7th day after 5 days TiO2 oxiracetam treatment. Adverse biochemical events such as upregulation of CSF nitrite and nitrate, IL-6, TNF-a and p-Tau are also reduced significantly in oxiracetam treated CHI group. On the other hand post treatment of 100 mg/kg dose of normal oxiracetam in identical conditions after CHI is needed to show slight but significant neuroprotection together with mild recovery of memory function and functional deficits on the 8th day. These observations are the first to point out that nanowired delivery of oxiracetam has superior neuroprotective ability in CHI. These results indicate a promising clinical future of TiO2 oxiracetam in treating CHI patients for better quality of life and neurorehabilitation, not reported earlier.

Keywords

Concussive head injury, Oxiracetam, TiO2 nanowired delivery, Blood-brain barrier, Brain edema, Brain pathology, Neuroprotection, Functional deficit

1 Introduction

Military personnel are prone to various kinds of central nervous system (CNS) insults during combat operation often resulting in lifetime disabilities (Chapman and Diaz-Arrastia, 2014; Clausen et al., 2020; Janak et al., 2019). Traumatic brain injuries (TBI) are one of the prominent insults to military personnel leading to slowly developing Alzheimer' disease (AD) and Parkinson's disease (PD) in their later periods of life (Al-Dahhak et al., 2018; Elder et al., 2019; Gardner and Yaffe, 2015; Gardner et al., 2018). TBI induced physical disability and rehabilitation of veterans requires prolonged therapy and care to maintain their quality of life (Kempuraj et al., 2020;

Sibener et al., 2014). Thus, effort should be made to explore novel therapy and new medical strategies to treat the victims of TBI to enhance their quality of life with minimum disturbances.

TBI basically could be classified into two brad categories. These include concussive head injury (CHI) or perforating brain injury (PBI) (Sharma et al., 2007, 2016). In CHI vital brain organs are damaged within the skull due to counter coup injury forces resulting in severe brain swelling in the closed cranial compartment (Kerr, 2013; Schmidt et al., 2005). Compression of brain vital centers within the closed cranium lead to influence cardiac and respiratory centers located within the medullary-brain stem regions often resulting in instant death (Fehily and Fitzgerald, 2017; Giza and Hovda, 2014; Sharma et al., 2007). However, those who survive showed severe neurological consequences leading to paraparesis, paralysis, hemiplegia, quadriplegia, coma or brain death symptoms (see Sharma et al., 2016). This suggests that CHI appears to be very complicated and so far no suitable therapeutic strategies are available to effectively treat these patients or to save their lives.

Perforating brain injury (PBY) represents piercing of cerebral tissue, damage to vital neuronal networks, microhemorrhages, rupture of cerebral microvessels and surrounding tissues leading to disruption of neuronal tracts in ascending and descending fibers (Pandey et al., 2018; Sharma et al., 2016; Weston et al., 2021). These changes lead to severe functional disabilities including paraparesis, partial or complete paralysis and dysfunction of autonomic control system of the vital organs (Meyfroidt et al., 2017; Solstrand Dahlberg et al., 2018).

The brain pathology following CHI and PBI appears to be of same patters such as break down of the blood-brain barrier (BBB) permeability, brain edema formation, alteration in local cerebral blood flow (CBF) and imbalances of neurochemical or immunological factors disrupting brain cellular homeostasis (Guerriero et al., 2015; Kozlov et al., 2017; Portbury and Adlard, 2015; Sharma et al., 2016). However, regional variability and perforating lesion in PBI and impact forces on skull determine the magnitude and intensity of local brain pathologies (Sharma et al., 2016). Thus, drugs able to reduce BBB function and brain edema could be able to alter the pathomechanisms of TBI induced brain pathology (Sharma and Sharma, 2019, 2020).

We have shown previously that DL-NBP when delivered through TiO2-nanowired in CHI of 48 h the drug was able to induce profound neuroprotection and improved functional parameters (Feng et al., 2018; Niu et al., 2019). TiO2 nanowired delivery of DL-NBP also induced significant neuroprotection in Parkinson's disease (PD) (Niu et al., 2020). When nanowired DL-NBP was administered with nanowired mesenchymal stem cells the combination of drugs induced superior neuroprotection in PD (Niu et al., 2020). This suggests that DL-NBP is equally neuroprotective in CHI and in PD cases (Niu et al., 2019, 2020).

Recently, oxiracetam a nootropic drug is used in clinical practice to enhance mood elevation, attention, and improved memory and cognition (Ponzio et al., 1989; Li et al., 2017). Oxiracetam is also shown beneficial effects in dementia (Malykh and Sadaie, 2010). This suggests that oxiracetam could be neuroprotective in neurodegenerative diseases as well as in CHI (Wang et al., 2019).

The present investigation was under taken to examine the potential neuroprotective effects of oxiracetam in mild CHI. We have used both conventional oxiracetam administration and also using TiO2 nanowired delivery in CHI to evaluate its neuroprotective effects.

We used two different types of TiO2 nanowired labeled with oxiracetam to understand whether the composition of nanowires may have similar or different neuroprotective effects in CHI. Accordingly Ti-Na and Ti-K nanowired were labeled separately with oxiracetam and delivered under identical condition in CHI and evaluated neuroprotective effects in CHI on brain pathology and biobehavioral functions. Our observations suggest that nanodelivery of oxiracetam has superior neuroprotective ability and CHI and there are subtle positive differences between Ti-Na and Ti-K nanowired delivery of the drug on CHI induced pathophysiological parameters. The functional significance of our findings and significances are discussed below.

1.1 Dementia and neurodegeneration

Dementia is a clinical condition described by at least two psychological domains, including but not limited to memory, visuospatial capacity, and altered language capabilities that leads to a significant decline in carrying out day to day normal activities (Ferencz and Gerritsen, 2015; Gale et al., 2018; Hachinski, 2019). Alzheimer's disease (AD) is the most well-known precipitation of dementia around the world that could be responsible for up to almost 80% of demented cases (Crous-Bou et al., 2017). Although, the mortality rate in the US from stroke and cardiovascular diseases is well controlled now with active health care facilities and proactive approaches, the death rate from AD continue to enhance among the elderly populations comprising more than 89% of cases during the last two decades (Alzheimer's, 2015, 2016). This causes a heavy burden on our society on health care and rehabilitation together with necessary medical services amounting about 320 billion US dollar per year for AD patients (Alzheimer's Association Report, 2020; https://www.alz.org/alzheimer_s_dementia,; Takizawa et al., 2015).

Our knowledge on AD for the improving the quality of lie of patients is largely based on extensive diagnostic techniques and protocols such as evaluation of biomarkers within the cerebrospinal liquid (CSF) or using positron emission tomography (PET) as well as histopathological diagnosis of cases at postmortem (Budson and Solomon, 2012; Draper et al., 2016; Galvin et al., 2012; McKhann et al., 2011; Rossor et al., 2010). Although some specific and nonspecific drugs are available to treat AD at different stages of the disease, exploration of new therapeutic methods and pharmacotherapeutic agents either alone or in combination are still the need of the hour in reducing progression and persistence of AD brain pathology (Chu, 2012; Fessel, 2018; Gupta et al., 2017; Sharma et al., 2016, 2018b, 2019b; Weller and Budson, 2018). Prevalence of AD, however, varies among different populations across the Globe (Bredesen et al., 2016; Camacho-Mercado et al., 2016; Chen et al., 2016; Chhetri et al., 2019; Miki, 1998; Riva et al., 2014). For instance, in

USA, the highest incidence rate is reported among Hispanics followed by African Americans as compare to the white Americans (Blue et al., 2019; Farrer et al., 1997; Kamara et al., 2018; Stepler and Robinson, 2019; Vega et al., 2017; Zissimopoulos et al., 2017). This suggests that the hereditary variables significantly affect the AD pathology and/or its intensity and the onset (Cacace et al., 2016; Karch et al., 2014; Stoccoro and Coppede, 2018). Apart from the genetic variations among AD patients, co-morbidity factors such as hypertension, diabetes, and environmental conditions could also influence the disease progression and persistence (Baglietto-Vargas et al., 2016; Javanshiri et al., 2018; Jiang et al., 2013; Lennon et al., 2019; Mayeux and Stern, 2012; Pugazhenthi et al., 2017; Sery et al., 2014).

Several reports have stated that Alzheimer's disease like symptoms could be precipitated following a mild or moderate traumatic brain injury (TBI) (Al-Dahhak et al., 2018; Kokiko-Cochran and Godbout, 2018; Ramos-Cejudo et al., 2018; Ruozi et al., 2015). This is especially prevalent in the military on combat operation that are prone to head injury resulting the development of AD a few years later (Menon et al., 2012; Patnaik et al., 2018; Sharma et al., 2016, 2019b). Treatment with neurotrophic factors or commercially available cerebrolysin, a balanced composition of several neurotrophic factors improving the symptoms and induce neuroprotection following TBI or concussive head injury (CHI) (Menon et al., 2012; Sharma et al., 2012a,c; Sharma and Sharma, 2011). However, due to the complex biological mechanisms involved in AD, finding a suitable therapy is still a challenge for clinicians and the researchers alike (Esquerda-Canals et al., 2017; Moreira, 2018; Tian et al., 2013). It appears that the efficacy of suitable drugs may be further enhanced by nanodelivery in AD cases to improve therapeutic strategy for neuroprotection, recovery time, and neuronal regeneration (Sharma et al., 2012a,b,c, 2019a,b).

2 Alzheimer's disease

Alzheimer's disease (AD) is a chronic neurodegenerative pathology that affects neuronal system in progressive manner and responsible for early dementia (Jiang et al., 2013; Sharma et al., 2016, 2018b, 2019b; Soria Lopez et al., 2019; Tomiyama, 2010). About an average of 65% of dementia cases are triggered by Alzheimer's disease (Kim and Factora, 2018; McKhann et al., 2011; Mollers et al., 2019; Oboudiyat et al., 2013; Yilmaz, 2015). Alzheimer's disease could be categorized into two forms namely the sporadic and the familial forms (Dorszewska et al., 2016; Iqbal and Grundke-Iqbal, 1997). The sporadic form of AD is attributed to the majority of the cases (Cruchaga et al., 2018; Dorszewska et al., 2016; Iqbal and Grundke-Iqbal, 1997). Sporadic form of AD might be triggered due to some environmental or metabolic causes (Iqbal and Grundke-Iqbal, 1997; Roher et al., 2016). On the other hand, the familial form that has been reported in minority of cases are developed due to genetic dysfunctions (Chen et al., 2013; Iqbal and Grundke-Iqbal, 1997; Nikolac Perkovic and Pivac, 2019).

Common symptoms include but not limited to cognitive dysfunction, disorientation, motivation decrease, mood disruption, and behavioral related problems (Lee et al., 2017; Lyketsos et al., 2011; Shinagawa, 2016; Wolinsky et al., 2018). However, intensity of symptoms is progressive that lead to death within less than 10 years (Atri, 2019; Hoyert and Rosenberg, 1997; Lyketsos et al., 2011). The prevalence of AD has increased during the recent years (Niu et al., 2017; Reitz et al., 2011). In the USA, the number of the cases diagnosed with Alzheimer's disease is elevated from seven millions in 2015 to thirteen millions in 2020 (Sharma et al., 2016). Also, it seems likely that the incidence rate of AD in females is higher than in males (Scheyer et al., 2018; Sharma et al., 2016).

This may be due to some hormonal and genetic related variations between the different genders (Scheyer et al., 2018; Tao et al., 2018). For several decades, the explanation of the mechanism of AD pathology is mostly focused on the effects of amyloid beta and tau proteins and their depositions during the course of the illness (Bloom, 2014; Gallardo and Holtzman, 2019; Hardy and Selkoe, 2002). Recently, several other factors found to be significantly contributing to the AD pathology (Silva et al., 2019b). These factors include the biological molecules in relation to the genetic and other co-morbidity factors that indirectly influence the development of the disease (Blades, 2015; Reitz, 2015; Reitz et al., 2011; Reitz and Mayeux, 2014). Therefore, understanding the mechanisms of AD is needed to attenuate the pathology and/or to limit its progression.

3 Molecular mechanisms associated with neurodegenerative diseases

3.1 Amyloid precursor protein

Amyloid precursor protein (APP) is a transmembrane protein that synthesized in the endoplasmic reticulum and translocated to Golgi apparatus that further modified and stored there (O'Brien and Wong, 2011; Wilkins and Swerdlow, 2017). The expression of APP occurs in several tissues and organs, but the high concentration is observed in the neuronal tissues where APP plays a major role in synaptic development, mineral transfer, neuronal plasticity, and regulation of cell signaling (Nixon, 2017; O'Brien and Wong, 2011; Tcw and Goate, 2017). Under normal physiological conditions, the C-terminus of APP is critical key factor for neuronal development and regulation of genes expression (Galvan et al., 2006, 2008; Hsiao, 1998; Nguyen et al., 2008).

Several reports suggest that the susceptibility rate of mutation is very high during the development stages of the APP (Bagyinszky et al., 2019; Roher et al., 2017; Tcw and Goate, 2017). Mutated APP is responsible of 15% of the early onset of AD (Jonsson et al., 2012; Tcw and Goate, 2017; Wu et al., 2012). The development of AD in individuals at the mid-age is highly attributed to the amyloid deposits (Reitz, 2015; Rossor et al., 1993). These deposits are categorized based on their morphological variations into three subtypes namely the diffuse deposits, primitive deposits, and classic deposits (Gouras et al., 2015; Oakley et al., 2006; Paroni et al., 2019). The deposition of the beta amyloid is found to be the key factor for the induction of AD pathology and also other neurological dysfunctions such as development of senile plaques, neurofibrillary tangles, axonal destruction, and dementia (Braak and Braak, 1991; Duyckaerts et al., 1994; Manzano-Leon and Mas-Oliva, 2006; Tomiyama, 2010). Accumulation of the amyloid into insoluble beta-sheet fibrils immediately triggers the neuronal dysfunction (Fedele et al., 2015; Karran et al., 2011; Selkoe and Hardy, 2016; Tomiyama, 2010).

3.2 α-Secretase

α-secretase is a proteolytic enzyme under the family of metalloprotease that plays a major role in the cleavage process of APP into small subunits called sAPPα (Tan and Gleeson, 2019; Wilkins and Swerdlow, 2017; Yuksel and Tacal, 2019). α-secretase targets APP within the Aβ domain that results in diminishing the formation of Aβ aggregates that is the key factor in the development of AD (Peron et al., 2018; Zhang et al., 2011). Therefore, it seems likely that upregulation of the activity of α-secretase boosts neuroprotection and inhibits the induction and progression of AD (Ahmad et al., 2019; Lichtenthaler, 2011; Zhang et al., 2011). Previous study suggests that potentiation of metalloprotease expression in APP transgenic mice significantly attenuates neuronal tissue Aβ level and enhances cognition and other central nervous system (CNS) functions (Postina et al., 2004). Also, it appears that the sAPPα, the final product of APP cleavage remarkably ameliorates neuronal plasticity (Fragkouli et al., 2012), neuroprotection (Cimdins et al., 2019), neurorestoration (Llufriu-Daben et al., 2018), and cell signaling activities (Hesse et al., 2018). Moreover, sAPPα is found to be a key factor in the development and regeneration process of the neuronal tissues through the potentiation of neural stem cell propagation and differentiation during the early stages of the development (Ohsawa et al., 1999).

3.3 **Tau protein**

Tau protein is the microtubules-stabilizing protein essential for the development of neuronal cytoskeleton under normal physiological conditions (Johnson and Stoothoff, 2004). Tau protein is mostly phosphorylated by the action of regulatory proteins such as protein kinases and phosphatases (Avila et al., 2004). Phosphorylation, in some abnormal cases, results in modifying the structure of Tau protein, which in turn leads to enhance the aggregation of Tau protein (Andorfer et al., 2003; Buee et al., 2010). The resulted aggregates are extremely toxic for axons and deteriorates the normal functionality of neuronal tissues. Dysfunction of Tau protein results in the development of neurological disorders including Alzheimer's disease, Parkinson's disease, and frontotemporal dementias (Arendt et al., 1998; Irwin et al., 2013; Naseri et al., 2019; Silva et al., 2019a). Hyperphosphorylation state of Tau protein is dominant during the developmental stages of AD pathology (Andorfer et al.,

2003). The hyperphosphorylated Tau is a malfunctioned form that lacks the ability to attach to microtubules and regulate their construction (Qiang et al., 2018; Tan et al., 2019). Moreover, this form of Tau protein disrupts the assembly and the development of the normal Tau and other microtubules associated proteins (MAPs) such as MAP1A, MAP1B, and MAP2 (Bonnet et al., 2001; Cravchik et al., 1994; Iqbal et al., 2008). This results in interruption of the assembly of microtubules, diminish the axonal functions, and perturb the neuronal circuitry and network (Johnson and Stoothoff, 2004; Scott et al., 2016; Venkatramani and Panda, 2019). Another destructive mechanism is induced by the accumulation of the Tau oligomers into the extracellular fluid (Cortes et al., 2018). Tau protein potentiates the neuronal damage resulting in neuronal death and leaking of Tau affects microglial cells and affects the components of the CNS (Cortes et al., 2018; Illes et al., 2019; Maccioni et al., 2010). Other factors including high protein kinases activity and attenuated protein phosphatase activity, either alone or in combination are involved in the development of tauopathies (Ma et al., 2017b; Martin et al., 2013; Sontag and Sontag, 2014). Tau phosphorylation reaches increased levels with the help of the MAP-kinase, GSK-3β, and/or Cdk5 and targets the amino acid residues for the action of theses kinases (Arendt et al., 1998). The amino acid residues include the serine and the threonine as main targets for the phosphorylation by kinases in animal models of AD (Arendt et al., 1998). However, several factors including oxidative stress (Haque et al., 2019), iron overload (Becerril-Ortega et al., 2014), Aβ aggregation (Arendt et al., 1998), cholesterol imbalance (Xue-Shan et al., 2016), and immune responses (Heneka et al., 2015) are other causative factors for tauopathies. This complexity on the mechanisms of tauopathies causes serious problems in diagnosis and treatment of AD.

4 Traumatic brain injury induces Alzheimer's disease

Traumatic brain injury (TBI) is one of the most common causative factors for several neurological disorders (Azouvi et al., 2017; Capizzi et al., 2020; Sharma et al., 2006). Annually, about 10 million cases of traumatic brain injury is reported around the world (Masel and DeWitt, 2010). Severity of the traumatic injury is ranged from mild to severe based on the accident type and velocity of the impact (Dixon, 2017; Pavlovic et al., 2019; Rissanen et al., 2017). Damage resulted from the TBI occurs either immediately at time of the injury or delayed due to secondary damage that develops later as a complication of the primary injury (Blennow et al., 2016, 2012). Most of the secondary complications are serious and results in death (Blennow et al., 2016, 2012; Katzenberger et al., 2015; Krishnamoorthy et al., 2015). Secondary injury involves intracranial hemorrhage, formation of edema, disruption of cellular hemostasis and network, cardiac arrest, and induction of cellular and molecular mechanisms apoptosis or neuronal death (Al-Mufti et al., 2017; Jha and Kochanek, 2018; Krishnamoorthy et al., 2015; Martin et al., 2019; Pijet et al., 2019; Salehi et al., 2017; Wolf et al., 2017).

Initiation of tauopathies is quite common following traumatic brain injury (Edwards 3rd et al., 2017, 2020; Kulbe and Hall, 2017). Several studies show an increase in the levels of p-Tau and neurofibrillary tangles (NFTs) immediately following TBI (Albayram et al., 2016; Gomes et al., 2019; Smith et al., 2003). Around 33% of TBI induced death showed elevated level of NFTs and accumulation of tau in the brain tissues (Edwards 3rd et al., 2020; Johnson et al., 2012). Development of dementia in TBI survived cases are related to tauopathies (Graham and Sharp, 2019; Mendez, 2017). In addition, the development of AD is highly related to the TBI severity (Becker et al., 2018; Zysk et al., 2019). The magnitude and severity of the primary injury following TBI induces chronic traumatic encephalopathy and functional disability (McKee et al., 2009, 2013; VanItallie, 2019).

Tau protein aggregation shares the same characteristics of the prion in terms of seeding and proliferating (DeVos et al., 2017; Holmes et al., 2014; Nizynski et al., 2018). These findings suggest that in brain tau aggregation deteriorate other essential proteins activity via interrupting protein-folding processes (Dujardin et al., 2018; Morales et al., 2013). Neuronal injury after TBI perturbs neuronal tissue hemostasis and trigger tau aggregation that disturb molecular pathways resulting in exacerbation of further neuronal damage and precipitating neurodegenerative diseases (Irwin, 2016; Irwin et al., 2013; Kovacs, 2017; Wang and Mandelkow, 2016).

5 Inflammation and oxidative stress following neurodegenerative diseases

Several reports suggests an interrelationship between Aβ accumulation and other cellular disturbances including free radical generation, release of inflammatory mediators, and axonal and microglial aggregations (Chen and Zhong, 2014; Lue et al., 2001; Pohanka, 2018; Wang et al., 2014b). Altered performance and functions of microglia and astrocytes derived from the white and gray matters of AD patients result in enhanced inflammatory processes such as cytokines release, activation of immune scavenger cells, nitric oxide and stimulation of macrophages (Choubey, 2019; Katsuse et al., 2003; Liu et al., 2019; Lue et al., 2001; Regen et al., 2017; Stamouli and Politis, 2016). Neuroinflammation triggered by AβP during AD or TBI contributes to the development of further neuronal damage (Calsolaro and Edison, 2016; Nazem et al., 2015; Wilkins and Swerdlow, 2016). Following the accumulation of Aβ in the brain tissue, microglia is activated and induces uncontrolled release of the inflammatory mediators such as interleukin-1b (IL-1b), IL-6 tumor necrosis factor-α (TNF-α), and nitric oxide (NO) (Asiimwe et al., 2016; Babić Leko et al., 2020; Bellucci et al., 2011; Bhaskar et al., 2014; Hayes et al., 2004).

These high levels of inflammatory mediators exacerbate neuronal degeneration and disrupt neuronal tissue environment (Asiimwe et al., 2016; Babić Leko et al., 2020; Bellucci et al., 2011; Bhaskar et al., 2014; Choubey, 2019). Microglia appears to be a key factor in the Aβ-mediated neuroinflammation and induction of oxidative stress (Chen and Zhong, 2014; Haque et al., 2019; Kamat et al., 2016a; Pohanka,

2018). Neuronal tissue is highly sensitive to free radicals induced cell damage as compared to other tissues (Asiimwe et al., 2016; Sahib et al., 2019; Salim, 2017; Sharma et al., 2019a, 2018a, 2020d, 2010, 2007). In fact, several factors such as Aβ accumulation, mitochondrial dysfunction, inflammation, and increased metal contents trigger the oxidative stress in AD or TBI patients (Chen and Zhong, 2014). The mechanism underlying the exacerbation of the neuronal degeneration in AD or TBI appears due to tau hyperphosphorylation in association with the oxidative stress (Alavi Naini and Soussi-Yanicostas, 2015; Kang et al., 2017; Tönnies and Trushina, 2017).

Oxidative stress changes cellular hemostasis via disrupting different kinds of kinases and intracellular pathways such as calmodulin-dependent protein kinase (CaMKII) (Zhang et al., 2016), extracellular receptor kinase (ERK) (Moniruzzaman et al., 2018), mitogen-activated protein kinase (MAPK) (Runchel et al., 2011), glycogen synthase-3β (GSK-3β) (Kamat et al., 2016a,b), cAMP response element-binding protein (CREB) (Mo et al., 2019; Taheri et al., 2018), and calcineurin pathways (Kamat et al., 2016a). As protein phosphatase (PP) inhibition and tau hyperphosphorylation are the main mechanisms of the pathogenesis of AD, several intracellular alterations occur including the N-methyl-D-aspartate receptor (NMDAR) activity and increase in Aβ aggregation that together exacerbate cognitive dysfunction via synaptic signaling (Kamat et al., 2016a; Kumar et al., 2019). NMDA receptors (NMDARs) are glutamate receptors for excitatory neurons that play important role in maintaining neuronal plasticity and regulating memory functions (Lau et al., 2009).

High level of post synaptic Ca^{2+} due to the activation of the NMDAR receptor and glutamate release is one of the key factors contributing synaptic dysfunction in neurodegenerative diseases (Wang and Reddy, 2017; Wang and Mandelkow, 2016). Aggregation of AβP induces NMDA-mediated Ca^{2+} influx, excitotoxicity, and disturbances in neuronal signaling, generation of free radicals, reduction in energy utilization, and intracellular ionic hemostasis (Bezprozvanny and Mattson, 2008). NMDARs activation induce the stimulation of extracellular signal regulated kinases (ERK) that catalyze phosphorylation of several downstream regulating proteins (Chen et al., 2008; Kravchick et al., 2016; Sun et al., 2016). Numerous critical cellular processes such as cell proliferation, development, and neuronal survival are highly regulated via the action of ERK (Rai et al., 2019; Zhong, 2016). High rate of NMDARs activation is associates with an excessive release of ERK that in turn cause disturbances in mitochondrial membrane potential, dendritic and synaptic dysfunctions (Ivanov et al., 2006; Leveille et al., 2008).

Free radical affects neuronal damages during neurodegenerative diseases the mechanisms of which are still not fully understood. Most of the neuronal dysfunction observed during the neurodegenerative diseases are driven by direct or indirect deleterious effects of free radicals that triggers apoptosis cell death (Cadenas and Davies, 2000; Radi et al., 2014). Oxidative stress enhances excitotoxicity and disturbances in synaptic networks (Forder and Tymianski, 2009; Khatri et al., 2018). These events also contribute to activation of microglia as seen following moderate or severe TBI (Fesharaki-Zadeh, 2019). Immunological reactions such as formation of complement system occurs following induction of microgliosis (Alawieh et al., 2018). Neuronal damage results after damage to the synapses and neuronal conductivity. In addition, activated microglia interrupt tau protein folding processes and enhances its aggregation by the phosphorylation process (Huang et al., 2016).

6 Oxiracetam and brain function

Oxiracetam is a potent nootropic agent of the racetam family and exerts superior ability to minimize symptoms associated with neurological pathologies such as stress and anxiety, epilepsy, seizure, cognition impairment, memory dysfunction associated with AD and dementia and neuronal tissue hypoxia (Malykh and Sadaie, 2010; Mondadori et al., 1986, 1996). Recently, nootropic drug is used for the treatment of the other brain pathologies, such as traumatic head injury and cerebral stroke. Studies show that oxiracetam, a novel agent of the family ameliorates regional glucose metabolism, regulates amino acid and nucleic acids production, and induces neuroprotective effects (Hokonohara et al., 1992a; Kometani et al., 1991). Wang et al. reported that oxiracetam administration thwarts programed cell death in neuronal tissues and enhances neurorecovery (Wang et al., 2014a). Moreover, oxiracetam improves learning and memory function in rodents with preconditioned vascular dementia and restores the behavioral activities in rats with induced cerebral hypoperfusion (Chen et al., 2015; Yao et al., 2016).

This memory enhancement effects of oxiracetam is further reported by several other investigators. Oxiracetam significantly improves neuropsychological activities in human and diminishes intellectual dysfunction in elderly (Mondadori et al., 1994, 1996; Rozzini et al., 1993). Interestingly, the efficient penetrating ability of oxiracetam across the blood brain barrier makes easy to reach hippocampus and cerebral cortex quickly is idea as a promising drug candidate for other neurological disorders (Gouliaev and Senning, 1994). Rodents preconditioned to acute or chronic cerebral impairments receiving short-term oxiracetam treatment exhibited significant improvement in learning and memory functions (Li et al., 2017; Xu et al., 2019). The possible key mechanism for oxiracetam in enhancing memory function is likely due to its ability in inducing turnover of phospholipids in the neuronal tissues (Li et al., 2013; Ma and Wang, 2019; Pugliese et al., 1990). Another mechanism of oxiracetam could in minimizing cognitive dysfunction is due to an increased acetylcholine utilization (Spignoli and Pepeu, 1986, 1987). This effect is triggered by oxiracetam either in attenuating acetylcholine production or inhibiting muscarinic receptors (Raiteri et al., 1992; Spignoli and Pepeu, 1986, 1987).

A clinical study shows that one dose of (1 g) oxiracetam remarkably elevates the alpha-EEG activity during the course of diazepam treatment (Giaquinto et al., 1986). This suggested that oxiracetam ameliorates cognitive behaviors of patients under sedation (Giaquinto et al., 1986). On the other hand, long-term administration of oxiracetam for 3 months significantly improves the cognitive behaviors in patients with neurocognitive disorder (Mondadori et al., 1986; Spignoli et al., 1986).

Previous investigations suggest that oxiracetam regulates the expression of several essential cellular proteins. Oxiracetam ameliorate the production of phosphatidylcholine and phosphatidylethanolamine, amino acids, DNA, and RNA. (Gabryel et al., 1999). This suggests that oxiracetam enhances energy production of the neuronal cells via modifying cells morphology and intracellular membranes surface area that in turn influences the rate of cell metabolism (Gabryel et al., 1999). Enhancement of neuronal tissue metabolism following oxiracetam administration may also be related to the improvement in neurotransmitter activity observed during treatment (Gouliaev and Senning, 1994).

In addition, oxiracetam helps recovery in cognitive dysfunction through elevating cellular levels of high-energy phosphates, membrane-bound protein kinase C (PKC), and the hippocampal PKC functionality (Fordyce et al., 1995; Gabryel et al., 1999).

It is likely that the neuroprotective and cognitive restoration by oxiracetam is due to improvement in neural plasticity, synaptic microstructure, and white matter architecture (Bayat et al., 2015; Hai et al., 2010; Jing et al., 2015; Ueno et al., 2002; Yao et al., 2016). This hypothesis gets support by Yao et al. after administering oxiracetam in rats with preconditioned cognitive dysfunction that results in amelioration of cognitive dysfunction following chronic cerebral hypoperfusion (Yao et al., 2016). This effect of oxiracetam may also be due to substantial potentiation of neuronal networks, ATP utilization, glutamine-glutamate ratio, anti-oxidants activity, and/or synaptic plasticity (Li et al., 2017; Yao et al., 2016).

Oxiracetam under hypoxic conditions, is able to boost neuronal tissue metabolism and enhances neuroprotection. This effects may be related to oxiracetam induced upregulation of neurotrophic factors such as Brain-derived neurotrophic factor (BDNF) (Hokonohara et al., 1992a,b; Kida, 2012). Oxiracetam could affect level of BDNF via regulating the expression of cAMP response element-binding protein (CREB) that is an essential signaling agent for BDNF production (Kida, 2012). Also, CREB exerts superior ameliorating effect on cognitive functional ions and memory regulation (Kida, 2012). This indicates that oxiracetam is a potent neurotrophic factors regulator.

Oxiracetam maintains and protects neuronal structures and dendritic spine that plays a major role in cognitive function and memory formation. Dendritic spines are small membranous projections from the neurons that act as a neuronal signal receiver at the site of synapses (Herms and Dorostkar, 2016; Hlushchenko et al., 2016) and maintain neuronal plasticity and functional networks (Herms and Dorostkar, 2016; Hlushchenko et al., 2016). Oxiracetam indirectly improves the development and functions of dendrites structures by production of neurotrophic factors that in turn enhances the neuronal communication, boosts memory, and reduces cognitive impairment (Luine and Frankfurt, 2013). Attenuation of cognitive impairment by oxiracetam is largely due to its enhancement of the BDNF production and regulation of energy utilization by neurons (Yao et al., 2016).

Oxiracetam regulates the ATP/ADP ratio in brain and neuronal tissue to maintain the metabolic pathways and energy production that is severely disturbed under pathological conditions (Moglia et al., 1986; Yao et al., 2016). Oxiracetam boost the synthesis and catalysis of energy producing molecules by changing the metabolic pathways for sufficient energy production and minimize cognitive dysfunction following to ischemia (Moglia et al., 1986; Yao et al., 2016).

The density of glutamatergic synapses in CNS is another key factor influencing cognitive performance (Benarroch, 2018; Moretto et al., 2018; Sando et al., 2017). These glutamatergic synapses form an electron-dense network beneath the postsynaptic layer (Kellermayer et al., 2018; Takahashi-Nakazato et al., 2019). Subjecting preconditioned rats to chronic cerebral hypoperfusion with oxiracetam improves the density and activity of the glutamatergic synapses and restores cognition and memory functions (Carvalho et al., 2008; Yao et al., 2016). Since oxiracetam increases production of neurotrophic factors it is likely that improvement in the post synaptic functions could be due to the enhanced BDNF effect (Carvalho et al., 2008).

7 Nanodelivery of drug for neurological dysfunctions

Recent decade witnessed a remarkable improvement in drug delivery techniques to precise organs targeting (De Jong and Borm, 2008; Pandit et al., 2017). Several clinical trials suggests that nanodevices could be a promising method for the treatment of AD (Sahib et al., 2019; Sharma et al., 2020a, 2012a, 2019b). Our research group with fabricated titanate nanowires tagged drugs exhibited remarkable neuroprotection and enhance the delivery of the Neprilysin and p-Tau antibodies to the CNS across the BBB in AD (Sharma et al., 2019b). Also, a combination of titanate nanowires and cerebrolysin administered in rats in AD resulted in a significant improvement in cognitive function with reduced brain pathology (Sharma et al., 2018b). In spinal cord injury, nanowired delivery of DL-3-n-butylphthalide (DL-NBP) induces superior neuroprotection and enhanced spinal cord conduction (Sahib et al., 2019). In this study blood spinal cord barrier function after spinal cord injury was also improved following administration of the nanowires loaded DL-NBP (Sahib et al., 2019).

These observations suggest that potential neuroprotective drugs when delivered through biocompatible nanostructures such as titanate nanowires induce superior neuroprotection and improve neuronal tissue restoration and recovery (Sharma et al., 2012a,b, 2013, 2016). Bonda et al. showed that usage of nanoparticles regardless of their dimensions or solubility as nanocarriers for metal chelating agents in AD improves BBB penetration and enhance antioxidant capability of the loaded drug (Bonda et al., 2011, 2012).

Several types of nanostructures fabricated from inorganic or organic materials with a combination of drugs for therapeutic delivery in neuronal disease models (Menon et al., 2012, 2017; Sharma et al., 2009a, 2013; Sharma and Sharma, 2011). These different techniques used to modify nanocarriers in enhancing their loading capability, penetration ability, or tissue targeting specificity (Martin-Rapun et al., 2017; Sharma, 2011a,b; Sharma et al., 2009b). Modification of nanocarriers for AD diagnosis and revealed a remarkable increase in the affinity of theses nanodevices toward the brain capillary endothelial cells (Brambilla et al., 2011).

Furthermore, modifying nanoparticles affinity toward amyloid-β for the treatment in AD facilitates the recovery through the "sink effect" that eliminate the deposition of amyloid aggregations (Bertuccio and Tilton, 2017; Carradori et al., 2018; Mancini et al., 2016).

Cerebrovascular amyloid deposits (CVA) are major inducers for neurological dysfunction including AD and dementia (Al-Shawi et al., 2016; Liu et al., 2015). Amyloid β protein antibody-coated iron nanoparticles are very effective material for the detection and therapy in rodents AD model (Poduslo et al., 2011). These inorganic based nanoparticles could be detected via high-field strength magnetic resonance imaging (MRI) that in turn provides diagnostic image for the histological changes induced before the development of AD pathology (Poduslo et al., 2011). Using nanostructure in combination with neurological drugs thus appears to be a promising strategy in enhancing neuronal tissue recovery and restoration, facilitate early diagnosis, and potentially enhance drug efficacy (Niu et al., 2020; Ozkizilcik et al., 2019; Sahib et al., 2019, 2020; Sharma et al., 2018a, 2020a,b,c,d).

Different types of nanostructures such as nanowires, nanoparticles, nanotubes, and hollow nanospheres are developed for the treatment of neurological diseases (Hu et al., 2019; Khosa et al., 2018; Pachioni-Vasconcelos Jde et al., 2016). Morphological variation among the developed nanostructure also impact the therapeutic effects (Banerjee et al., 2016; Sahib et al., 2019; Truong et al., 2015). Shape of nanoparticles greatly affects cellular uptake behavior and the amount along with the time required for delivering the drugs through the biological membranes (Niu et al., 2020; Sharma et al., 2020a,b,c,d). For instant, spherical gold nanoparticles are faster in term of cancer cell membrane penetration as compared to same sized gold nanospikes and nanorods (Ma et al., 2017a). Thus, manipulating shape, surface area, and the fabrication material of nanostructure greatly influence the drug delivery capacity, drug potency, and membrane penetration ability across the BBB.

7.1 Materials and methods

7.1.1 Animals

Experiments were carried out on Male Sprague-Dawley Rats (weighing 320–350 g body weight; Age 10–12 weeks) were housed at controlled Room Temperature ($21 \pm 1\,^\circ C$) with 12h light and 12h dark schedule. The Rat feed and tap water was supplied ad libitum.

All experiments were conducted according to National Institute of Health (NIH) Guide for the Care and Use of Laboratory Animal (2011) and approved by Local Ethics Committee.

7.1.2 Concussive head injury

For concussive head injury (CHI) rats were anesthetized with Equithesin (3 mL/kg, i.p.) and their head was fixed in a Stereotaxic Apparatus fixed with ear bars. The underneath of head cotton wool was placed so that impact injury may prevent and head movements during trauma. The skin over the head is incised and the skull bon was exposed. The skull bone was cleaned and freed from muscles if any and wiped with 70% alcohol. The exposed parietal bone was used to induce trauma (Sharma et al., 2006, 2010).

A tapered carbon steel rod weighing 114.6 g was dropped over the right parietal bone from 20 cm height through an aluminum guide tube to induce impact head injury. These weight and height combinations inflict an impact of 0.224 N as described earlier (Dey and Sharma, 1984; Sharma et al., 2007, 2010, 2016; Vannemreddy et al., 2006).

Control rats were treated identically regarding parietal skull opening and head fixing in the Stereotaxic Apparatus but no injury was inflicted.

7.1.3 Maintenance of CHI animals and survival period

The control and CHI rats were allowed to survive 2, 4 and 8 day after the primary insult.

All injured animals were kept individually in the cage with food and water and taken care daily at 8:00 AM and 8:00 PM for their maintenance and help if any. These CHI animals were moderately injured thus, checking daily weight, rectal temperature and their movement inside the cage twice daily and flushing the cannula with heparinized saline once daily keep them fit during the survival periods. Occasionally, tail-flick test or hot-plate test was carried out to check their pain sensitivity as compared to control or intact animals (Sharma and Westman, 2004; Sharma et al., 2004).

7.1.4 Body weight and rectal temperature

The control and CHI animals were handled about 1 week before for body temperature measurement and daily body weight recording. For rectal temperature measurement thermistor probes (Yellow Springfield, USA) were used that was connected to a digital Telethermometer (Harvard Apparatus, Holliston, USA) as described earlier (Sharma, 2005, 2006, 2007). These animals were handled daily for 1 week before actual experiment so that stress induced increase in rectal temperature will not affect the data recording. For this purpose, animals were partially restrained in a loose Perspex cage where tail is placed outside. About 4–6 cm rat thermistor probe were inserted into the rectum and secured in position with the tail using adhesive tape to minimize probe movement during temperature recording. Rats were accustomed to this procedure for 1 week and thus, the rectal temperature recording were quite stable (Sharma, 1987; Sharma and Dey, 1986a,b). The core body temperature recorded in this manner represent deep visceral temperature or liver temperature that is quite stable in nature (Sharma and Dey, 1986b).

The body weight was taken using a digital Rodent weighing machine (Mettler-Toledo, USA) and for this the animals were also accustomed daily for 1 week handling. This minimizes and stress on the animals and their movement on the weighing machine (Sharma and Dey, 1984; Sharma et al., 1991).

7.1.5 Pain sensitivity test for CHI

To test the pain sensitivity of CHI animals thermal nociception and mechanical nociception test were used. For thermal nociception a radiant heat was focus on the tail of the rats that was cleaned and applied paraffin oil for better sensitivity using an Analgesiometer (Harvard Apparatus, Holliston, MA, USA). The cut-off time was set 30 s. The tail flick latency was recoded manually. The control rats tail-flick latency is about 7–10 s. Hyperalgesic rats may flick their tail earlier (Sharma et al., 2004).

For mechanoreceptors stimulation a tail pressure analgesiometer was used where a pressure was gradually applied and slowly increased until the animal flicks it tail (Harvard Apparatus, Holliston, MA, USA) usually within the 3–5 s as normal response. These tests will determine the hypersensitivity to pain, if any in CHI rats over 1 week (Sharma, 2004; Sharma et al., 2004).

Animals that do not show increased rectal temperature beyond 37.5 °C or any kind of hypersensitivity to pain measurement or do not loose weight during the period of survival after CHI was used for further experimental procedures to record behavioral and pathophysiological data.

7.1.6 Oxiracetam treatment

Oxiracetam powder (99.9% pure) was obtained as a gift from CSPC Pharmaceutical Group Ltd. (Shijiazhuang, Hubei, China). Oxiracetam was dissolved in double distilled sterile water (100–200 mg/mL) and given intraperitoneally 0.5 mL per animal according to dosage as desired. The nanowired dose of oxiracetam was also adjusted in a manner that maximum volume should not exceed >0.5 mL per animal.

Oxiracetam solution in sterile doubled distilled water was administered in control and experimental group of rats in a dose of 100 mg/kg, i.p. once daily after 24 h CHI for 5 days. On the 8th day functional deficits, brain pathology and physiological parameters were analyzed.

In a separate group of animals Oxiracetam (50 mg/kg, i.p.) was delivered using TiO2 nanowired technology under identical conditions as describe earlier. Thus the TiO2 nanowired oxiracetam was administered intraperitoneally starting from 24 h after CHI once daily for 5 days. On the 8th day pathophysiological data and functional parameters were evaluated using standard procedures.

8 Nanoformulation of oxiracetam

8.1 Materials

Titanium dioxide TiO_2 powder (Degussa, P25); Sodium hydroxide pellet (reagent grade, Carolina); Chloramphenicol (AB Bergman Labora AB, Danderyd, Sweden); Lysogeny broth medium (Lennox, Thermo Fisher Scientific, US); Deionized water prepared by ion exchange in the lab.

8.2 Methods

8.2.1 Nanowires synthesis

A 0.375 g portion of Titanium Dioxide nanopowder (Degussa P25) was mixed with 50 mL of 10 M NaOH solution and left on the magnetic stirrer for 24 h. Then the obtained solution was sonicated for 15 min before the thermal treatment. After sonication, the resulting solution was transferred to a 100 mL stainless steel Teflon linear container and sealed well before cooking. The sealed Teflon linear container was then cooked in the oven at 240 °C for 72 h. After the cooking was done, the obtained suspension was washed with DI water several times alternating between sonication and vertexing for 5 min between the washes. The washing process is repeated approximately 20 times until the resulting material becomes neutralized (pH 7.2). The same procedure was performed to obtain the potassium version of the titanium dioxide nanowires with the exception that a 10-M concentration of potassium hydroxide solution was used instead of sodium hydroxide mentioned previously.

8.2.2 Drug loading, release, cell assay

Equal amounts of Oxiracetam were mixed with each nanomaterial (Na-Ti nw and K-Ti nw) with a w/w ratio of 1:10. The mixtures were rotated gently for 48 h at room temperature. Amount of oxiracetam or nanomaterials alone were also rotated at the same temperature, and with the same amount of time to serve as controls.

8.3 Characterization

For material characterization purposes, XRD and SEM imaging have been considered in this study. X-ray Diffractometer (XRD, Rigaku MiniFlex II) was used to analyze the phase composition of the as-prepared titanate nanowires (Fig. 1). Morphologies of titanate nanowires were obtained using a scanning electron microscope (SEM, Tescan VEGA II SBH) (Fig. 2).

8.3.1 Nanowired delivery of oxiracetam

TiO2 nanowired oxiracetam either produced using Na-titanate or K-titanate were administered in into the control or CHI animals in a dose of 50 mg/kg, intraperitoneally once for 5 days starting from 24 h after trauma in identical conditions according to standard protocol as described earlier (Sharma et al., 1997a).

8.3.2 Parameters measured

The following behavioral and pathophysiological parameters are examined in Oxiracetam treatment with control and CHI rats.

8.3.3 Physiological variables

The mean arterial blood pressure (MABP) arterial pH and arterial PaO2 and PaCO2 were measured in control, CHI and drug treated rats using standard protocol (Sharma, 1987). In brief, for MABP measurement, a polythene cannula (PE10) was implanted into the left common carotid artery retrogradely toward heart under

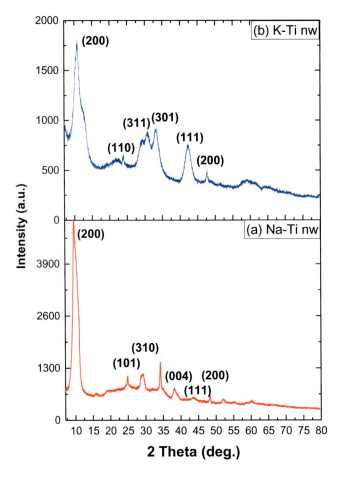

FIG. 1

XRD pattern for (A) Sodium titanate nanowires (B) Potassium titanate nanowires.

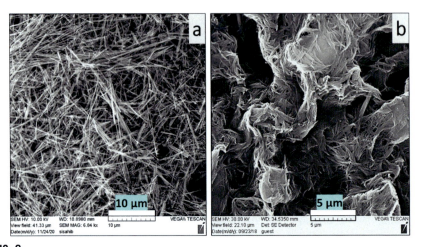

FIG. 2

SEM image of Na-Ti nw (A) and K-Ti nw (B).

aseptic conditions 1 week before the experiments. At the time of the experiment the arterial cannula was connected to a Pressure transducer (Strain Gauge P23, Statham, USA) and the blood pressure was recorded to a chart recorder (Electromed, UK). At the time of connecting the arterial cannula to pressure transducer, about 1 mL of arterial blood was withdrawn for later determination of arterial pH and blood gases (Sharma, 1987).

The hear rate and respiration cycle was also recorded simultaneously using standard protocol on the chart recorder (Dey et al., 1980).

8.3.4 Blood-brain barrier permeability

The blood brain barrier (BBB) permeability was examined using Evans blue albumin (EBA) and radioiodine ([131]-Iodine-Na) tracers (Sharma, 1987; Sharma and Dey, 1986a,b). For this purpose 2% of EBA solution in 0.9% physiological saline (pH 7.4) was administered (3 mL/kg) into the right femoral vein through a polythene cannula (PE10) implanted under aseptical condition 1 week before the experiment. For radioiodine tracer 100 μCi/kg was administered either separately through the femoral vein cannula or mixed into the EBA solution and injected together. Both tracers were allowed to circulate 5–10 min. After the end of the experiments, the animas were deeply anesthetized and the chest was rapidly opened. The intravascular tracer was washed out by intracardiac perfusion of cold (4°C) 0.9% physiological saline (about 50 mL) at the pressure of 90 Torr as described earlier (Sharma and Dey, 1986a,b).

After perfusion the brain were removed and examined manually the extravasation of EBA on the brain surfaces and within the ventricles and other organs. After that the desired areas of brain from left and the right hemispheres were cut and the radioactivity counted in a 3-in well type Gamma Counter (Packard, USA). Immediately before saline perfusion, about 1 mL of whole blood was withdrawn from the left ventricle after cardiac puncture for later determination of whole blood radioactivity. The BBB leakage to radioiodine was expressed as percentage of radioactivity in the brain as compared from the whole blood radioactivity (Sharma, 1987; Sharma et al., 1990a).

After counting the radioactivity, the samples were homogenized in 0.5% Sodium sulfate pH 7.0 and Pure Grade Acetone and centrifuged at $900 \times g$ to separate the supernatant containing EBA dye. The samples were then measured in a Spectrophotometer (Thermo Fisher Scientific, Waltham, MA, USA) at 620 nm and the dye concentration was calculated according to EBA standard curve as described earlier (Sharma and Dey, 1986a,b).

8.3.5 Cerebral blood flow

Cerebral blood flow was measured in control, CHI and drug treated animals using radioiodine labeled carbonized microspheres (15 ± 0.6 μm o.d.) as described earlier (Sharma and Dey, 1986a, 1987, 1988). In brief about 1×106 [125]-Iodine labeled microspheres were administered as a bolus into the left common carotid artery implanted retrogradely toward heart 1 week before the experiment (Sharma, 1987).

Microspheres were administered within 45 s and peripheral artery reference blood flow was collected at every 30 s intervals beginning from 45 s before microsphere injection and up to 90 s after completion of the microsphere administration (Sharma and Dey, 1986a,b). For this purpose, right femoral artery was cannulated with a PE10 catheter 1 week before the experiment. The reference blood sample (RBS) was collected at a rate of 0.8 mL/min and all the timed arterial samples were collected on a filter paper and the radioactivity counted at the end of the experiment (Sharma and Dey, 1986a,b, 1987).

Immediately after termination of microsphere injection, the rats were decapitated and the brains were removed and placed on a cold saline wetted filter paper. The large superficial blood vessels and blood clots, if any were removed immediately. The desired portions of the brain were dissected and weighed immediately. The radioactivity of brain samples was counted in a 3-in Gamma Counter (Packard, USA).

The cerebral blood flow (CBF) was calculated from the radioactivity trapped within the brain samples in relation to the reference blood flow radioactivity (Sharma and Dey, 1986a,b). The CBF is expressed as mL/g/min (see Sharma, 1987).

8.3.6 Brain edema formation

Brain edema was determined from measuring increase in the brain water content and volume swelling as described earlier (Sharma and Olsson, 1990; Sharma et al., 1990b). For this purpose, after the experiments rays were decapitated and the brain were removed and placed on a filter paper soaked with 0.9% cold saline (4 °C). The major blood vessels and blood clots, if any were removed. The desired brain areas were dissected rapidly and weighed using a preweighed Filter paper (Whatman No. 1, USA). After taking wet weight of the tissue samples the specimens were placed in an oven maintained at 90 °C for 72 h to obtain dry weight. When the dry weights of the samples become constant in three subsequent determinations, the brain waster was calculated from the differences between wet and dry weights of the sample (Sharma et al., 1991).

Volume swelling was calculated from the differences between control and experimental brain water using formula of Elliott and Jasper (1949) as described earlier (Sharma et al., 1991). In general about 1% change in the brain water denotes approximately 4% increases in volume swelling (Sharma and Olsson, 1990).

8.3.7 Brain pathology

Brain pathology was examined using standard Light Microcopy and Transmission Electron Microscopy (TEM) as described earlier (Sharma and Olsson, 1990; Sharma et al., 1998a,b). For this purpose, after the experiments, animals were deeply anesthetized with Equithesin (3 mL/kg, i.p.) and chest was rapidly opened and the heart was exposed. The right auricle was cut and a butterfly cannula (21G) was inserted into the left ventricle. About 100 mL ice cold (4 °C) of 0.1 M Phosphate buffer saline (PBS) was perfused at 90 Torr to washout intravascular blood. After that either 4% buffered paraformaldehyde (about 150 mL at 4 °C) of Somogyi fixative containing picric (4 °C) acid was used under identical conditions. After the perfusion, the animals were wrapped in an aluminum foil and kept in the refrigerator (4 °C) overnight. On the next day, the brains were dissected out for morphological examinations (Sharma and Cervós-Navarro, 1990).

Several coronal sections were cut from the brain passing through caudate nucleus; hippocampus, colliculi, cerebellum and brainstem were embedded in paraffin (Sakura Tissue-Tek, Fine-Tek, Japan). About 3-μm thick sections were cut and stained with Hematoxylin & Eosin (H&E) or Nissl Stain and examined in a Zeiss Inverted Bright field Microscope connected with a digital camera (Olympus, Germany).

For glial fibrillary acidic protein (GFAP) immunoreactivity for reactive astrocytes, the sections were deparaffinized and endogenous peroxidase activity was blocked with 0.3% hydrogen peroxide with 1% non-immune horse serum in phosphate buffer saline (PBS, pH 7.4) for 20 min. After that the sections were incubated with monoclonal anti-GFAP antibody (DAKO, Hamburg, Germany) in dilution of 1:500 in PBS (Sharma et al., 1992a,b). The sections were washed in PBS three times and incubated with biotinylated horse anti-mouse immunoglobulin (IgG) 1:50 dilution and avidin-biotin complex (ABC, Vector, Burlingame, CA, USA) for 45 min. The brown reaction product was developed with 3,3'-tetra-amino-benzidine and hydrogen peroxide in 0.05: Tris-HCl buffer (pH 7.4) for 4 min (Sharma et al., 1992a,b).

The light microscopic images were taken at $20 \times$ or $40 \times$ original magnifications and stored in an Apple Macintosh Power Book System Mc Os El Capitan version 10.11.6. The images from control, experimental and drug treated group were simultaneously processed using commercial software Adobe Photoshop 12.4×64 using identical color balance and filters (Sharma and Sjöquist, 2002).

For TEM, small tissue pieces from the desired area were post fixed with Osmium tetraoxide (OsO4) and embedded in Epon 812. Semithin sections (1 μm) were cur and stained with Toluidine blue and examined under a Bench Microscope for trimming the Epon blocks for ultrathin sections. Ultrathin sections (50 nm) were cut on an ultramicrotome (LKB, Sweden) using diamond knife (Sharma and Olsson, 1990). Serial ultrathin sections were collected on one hole copper Grid and counterstained with uranyl acetate and lead citrate. Some sections were left unstained. These Grids were examined under Phillips 400 TEM connected with a digital camera (Gatan Technology, USA). The original images were captured at $4000 \times$ $8000 \times$ or $12,000 \times$ magnifications and stored in an Apple Mac Book Pro computer system Mc Os El Capitan version 10.11.6. Images from the controls, experimental and drug treated groups were processed simultaneously using commercial software Adobe Photoshop version 12.04×64 using identical filters and brightness control (Sharma and Sjöquist, 2002).

8.3.8 Behavioral parameters
The following behavioral parameters were examined in CHI rats with or without drug treatment.

8.3.9 Memory function

Learning and memory function was examined using Morris Water Maze (MWM) test. For this purpose about 1 m circular Perspex water tank was used that is filled with 50 cm deep tap water and a hidden platform was situated in the center of the pond that was 24 cm tall and with the circular platform was 10 cm × 10 cm on which rats could rest. The water temperature was maintained at 29 °C that is thermoneutral zone for the rats.

Rats were trained in MWM to search platform daily 10 trials within 30 min intervals from different locations to find the platform and stay there. Rats were allowed maximum 120 s to search the platform and stay there for 30 s before they were rescued. In some cases the water was made opaque using dry milk powder. The test is repeated daily for 10 trials each rat for 1 week so that the animals memorize quickly to reach hidden platform as soon as possible (Smith et al., 1991). Trained rats took only 4–5 s to reach the hidden platform (Morris, 1984; H.S. Sharma, unpublished observations).

Untreated or drug treated CHI rats were subjected to MWM test at 3 and 7 days after trauma. Normal controls saline or drug treated were also examined for MWM test under identical conditions for comparison (D'Hooge and De Deyn, 2001).

8.3.10 Rota rod treadmill

A Rota Rod treadmill (Harvard Apparatus, Holliston, MA, USA) was used at 16 rotations per minute (rpm) speed for rats and allowed to stay 120 s as cut of time (Sharma, 2006). Experimental CHI rats either treated or untreated were subjected to Rota Rod Treadmill and their stay time at 16 rpm was monitored manually (Sharma, 2006).

8.3.11 Inclined plane angle test

Inclined plane angle test was measured using a platform on which rats could stay comfortably without falling for 5 s. The angle at witch rats will stay for 5 s was measured. Control rats could stay for 5 s easily at steep angles whereas the angle of inclined plane needs to raise for CHI animals. The angle at which the experimental animals could stay over the platform was manually recorded. For this test animals were trained daily in three sessions separated by 1 h for 1 week before the experiment (Sharma, 2006).

8.3.12 Walking on mesh grid and placement error

A stainless steel mesh grid (1 m × 1 m) inclined at 45° was used with mesh size 1 cm × 1 cm on which control and experimental rats were allowed to walk for 1 min. The numbers of steps were counted manually. In CHI rats forepaw placement error normally occurs while walking and also the number of steps taken are reduced. The numbers of placement error was recorded in addition nr of steps taken during 1 min manually (Sharma, 2006).

For this purpose rats were trained daily for three sessions with 1 h interval on grid walking for 1 week before the experiment. Injured rats were allowed to walk after 48 h of surgery and were free of pain and any kind of infection as tested for body temperature and pain tests.

8.3.13 Foot print analysis

The stride length and distances between hind feet are markers of normal gait (Sharma, 2006). For this purpose foot print analysis was done on an acetone paper stained with bromophenol blue. Rats were allowed to walk on this paper to obtain their footprint on two subsequent steps. The distances between hind feet and stride length were measured manually (Johnston et al., 1991; Sharma et al., 2016).

8.3.14 Biochemical measurements in the cerebrospinal fluid

The following biochemical parameters were measured in the cerebrospinal fluid (CSF) of control, CHI and drug treated rats using standard protocol as described earlier (Sharma and Sharma, 2019, 2020). To obtain CSF from rats, cisterna magna route was used to collect 1 mL CSF after the end of the experiment. For this purpose, under Equithesin anesthesia (3 mL/kg, i.p.) the rat head was fixed in a Stereotaxic Apparatus (Harvard Apparatus, Holliston, MA, USA) and the atlanto-occipital membrane was exposed using surgical operation carefully. After exposing the membrane a sterile needle (26G) attached to a syringe with polythene catheter the membrane was punctured carefully and about 1 mL CSF was collected slowly. Care was taken to hold syringe and needle perfectly so that underlying brain tissue is not injured. Also the CSF samples are not contaminated with blood. The CSF samples collected were stored at $-80\,^{\circ}$C for further analysis of biochemical as described below.

8.3.15 Nitrate and nitrite in the CSF

Previous studies from our laboratory show that nitric oxide is one of the key agent causing cell and tissue injuries following brain or spinal cord injury (Sharma, 2004; Sharma et al., 1996, 1998b). Thus, the nitric oxide metabolism is measured using nitrate and nitrite concentration in the CSF following CHI with and without oxiracetam treatment.

The CSF samples were processed according to standard protocol and the nitrate and nitrite were assayed using HPLC with electrochemical detection as described in detail elsewhere (Sharma et al., 1998b; Troutman et al., 2018).

8.3.16 Interleukin-6 and tumor necrosis factor-alpha

Traumatic brain injury induces marked alteration in the harmful cytokines within the brain and CSF (Sharma and Sharma, 2019, 2020; Woodcock and Morganti-Kossmann, 2013). We measured interleukin-6 (IL-6) and tumor necrosis factor-alpha (TNF-a) using commercial ELISA kits, respectively using commercial protocol (Thermo-Fisher Scientific, Waltham, MA USA) (Sharma et al., 2020a,b,c,d).

8.3.17 Phosphorylated tau protein (p-Tau)

Brain injury causes accumulation of p-Tau in the CSF (Caprelli et al., 2018). We measured p-tau using commercial p-Tau ELISA kit (Thermo-Fischer Scientific, Waltham, MA, USA) according to standard protocol as directed (Sharma et al., 2019a,b, 2020a,b,c,d).

8.3.18 Statistical analysis

SPSS 13.0 software for IBM and/or StatView 5 (Abacus concepts Ltd.) for Mac Classic environment were used for data analysis to calculate and compare the standard deviations and average mean values of the obtained data. Levels of significance were determined using one-way ANOVA and post hoc test. For multiple comparison, ANOVA followed Dunnett's test using one control was used for statistical analysis of the significance of the data. A p-value <0.05 was considered significant.

9 Results

9.1 Nanowired oxiracetam characterization

9.1.1 X-ray diffraction (XRD)

The XRD patterns of titanate nanowires are shown in Fig. 1. From the obtained diffraction patterns of the two different nanowires, it seems like that the crystal structure of the titanate nanofibers mainly depends on the $K^+_{(aq.)}$ or $Na^+_{(aq.)}$ in the interlayer space of each crystal lattice. The strongest diffraction peak of the Na-Titanate nanowires (NWs) appeared at $2\theta=9.788°$ and 27.858 from the (200) and (310) planes, respectively (Fig. 1A), corresponding to the phase of $Na_2Ti_2O_5 \cdot nH_2O$ (JCPDS No. 47-0124) (Miao et al., 2012; Xu et al., 2012), which is supported by the rest of diffraction peaks (Miao et al., 2012). On the other hand, the K-Ti nw in (Fig.1B) shows diffraction peaks at $2\theta=24.118$, 29.948, and 30.138 which correspond to (110) plane and (311) plane, respectively, of $K_2Ti_6O_{13} \cdot nH_2O$ (JCPDS no 40-0403) (Xu et al., 2012; Zhang et al., 2019). Given fact that the $K^+_{(aq.)}$ is smaller than the $Na^+_{(aq.)}$ (Cotton et al., 1988), the K-Titanate NW's d-space is smaller than the Na-titanate's, resulting in the Na-titanate's strongest diffraction peak (200) with a slightly higher 2θ, i.e., lower d-space (Fig.1B), which matches the data in literature (Xu et al., 2012; Zhang et al., 2019).

9.1.2 Scanning electron microscopy (SEM) imaging

The morphologies of the two prepared samples obtained via the SEM imaging were quite different. The sodium-based titanate nanowire (Na-Ti nw) revealed a diameter of about 0.2–0.5 µm and a length of about 15–20 µm (Fig. 2A), while the potassium version (K-Ti nw) recorded a diameter ranging from 0.1 to 0.3 µm and a length ranging from 5 to 10 µm (Fig. 2B). Additionally, the Na-Ti nw showed a very straight form of wires that crossed to form a mesh-like structure (Fig. 2A). While K-TiO$_2$ nanowires had a fluffy shape due to the smaller dimensions that weaved together to form a well-compacted scaffold (Fig. 2B).

9.2 Oxiracetam and physiological variables in concussive head injury

The following physiological variables were measured in control, CHI and oxiracetam treated groups as follows.

(a) Body weight

Body weight taken by sensitive rodent balance with precision (± 1 Gram) did not vary significantly among control or oxiracetam treated either given 50 and 100 mg or TiO2 nanowired oxiracetam (Fig. 3).

In CHI group from 48 h to 1 week survival animals body weight were slightly increased from preinjury level indicating that CHI did not decrease body weight after trauma over 1 week period (Fig. 3).

In oxiracetam treated group with CHI showed significant increase in the body weight over 1 week injury. However, there is no much difference between oxiracetam low, high or nanowire delivery of oxiracetam in CHI (Fig. 3).

(b) Rectal temperature

Rectal temperature using thermistor probe (sensitivity $\pm 0.01\,°C$) recorded in control, CHI or oxiracetam treated group over 1 week period did not show hyperthermia (Fig. 3). Thus, in control group the average rectal temperature was $36.81 \pm 0.23\,°C$. Treatment with oxiracetam with 50 mg dose daily for 5 days resulted in a small rise of the rectal temperature to $37.05 \pm 0.04\,°C$ (n.s.) whereas treatment with 100 mg dose under identical condition caused a slight drop of the rectal temperature to $36.93 \pm 0.04\,°C$. Daily administration of nanowired oxiracetam 50 mg also lowered the body temperature by small amount to $36.76 \pm 0.04\,°C$ (Fig. 3).

After CHI in animals who survived for 48 h to 1 week after injury, the rectal temperature did not vary much indicating no pyrogen has affected them throughout their survival period of injury (Fig. 3). The body temperature after 48 h CHI was $37.04 \pm 0.02\,°C$ and after 96 h the rectal temperature shows $37.12 \pm 0.21\,°C$. At the end of 1 week CHI the rectal temperature recorded was $37.06 \pm 0.11\,°C$. These data indicate that CHI animals from 48 h to 1 week were healthy without any temperature changes (see Fig. 3).

When oxiracetam was administered in CHI animals for days during 1 week post injury survival their body temperature was $37.06 \pm 0.12\,°C$ after 50 mg dose, $37.01 \pm 0.08\,°C$ for 100 mg dose or $37.13 \pm 0.14\,°C$ after nanowired delivery of oxiracetam under identical condition (Fig. 3).

(c) Pain sensitivity

The pain sensitivity of control, CHI and oxiracetam traded animals were determined using thermal and noxious nociceptive test using standard protocol (Fig. 3). As evident from the table that control, CHI or oxiracetam treated group including nanowired delivery of oxiracetam did not alter thermal or noxious pain sensation from 48 h after injury to 1 week after CHI survival periods. The thermal nociception was within the range of 5–7 s very similar to that of control group. Likewise noxious stimulus was within the range of 3–5 s that is very similar to normal control animals. This suggests that CHI untreated or treated animals were devoid of pain perception throughout their survival (Fig. 3).

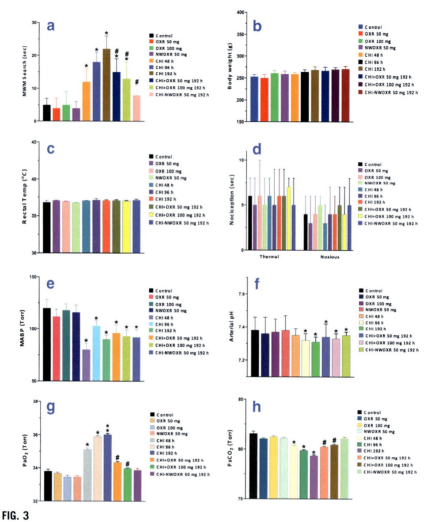

FIG. 3

Effect of TiO2 nanowired oxiracetam (NWOXR) on Morris Water Maze (MWM) search (A), Body weight (B), rectal temperature (C), pain perception (D) and the Physiological variables mean arterial blood pressure (MABP), (E), Arterial pH (F), PaO2 (G) and PaCO2 (H) following concussive head injury (CHI) in the rat on the 8th day after trauma and their modification with conventional or TiO2 nanowired delivery of oxiracetam. CHI was inflicted using a 0.224 N impact injury over Right parietal skull and the animals were allowed to survive for 1 week after the primary insult. Oxiracetam (OXR) was given in a dose of 50 or 100 mg/kg, i.p. once daily after 1 day of CHI for 5 days and the parameters were evaluated on the 8th day. For TiO2 nanowired delivery of oxiracetam (NWOXR) 50 mg/kg (i.p.) was given under identical conditions. Values are Mean ± SD of 6–8 rats at each point. *MABP*, mean arterial blood pressure; *OXR*, oxiracetam; *NWOXR*, nanowired oxiracetam. * $P < 0.05$ from Control; # $P < 0.05$ from CHI at respective group. ANOVA followed by Dunnett's test for multiple group comparison using one control. For details see text.

(d) Mean arterial blood pressure

Mean arterial blood pressure (MABP) in control animals was 120 ± 8 Torr that was slightly reduced by oxiracetam 50 or 100 mg doses given over 5 days to 112 ± 7 and 116 ± 6 Torr, respectively. Nanodelivery of oxiracetam also reduced the MABP slightly by 116 ± 7 Torr.

Subjection of animals to CHI significantly reduced the MABP by mean 40 Torr ($P < 0.05$ from control) that was evident at 48 h after trauma (Fig. 3). At the end of 96 h after CHI the MABP was only mean 17 Torr ($P < 0.05$ from control) declined from the control group. On the other hand at 192 h after CHI the MABP was further down by mean 30 Torr ($P < 0.05$ from control) from the control values significantly.

In oxiracetam treated group in CHI the MABP was significantly improved at 48 h by mean 16 Torr ($P < 0.05$ from control) from the untreated CHI group (Fig. 3). Whereas after 96 h CHI in oxiracetam 100 mg treated CHI group the MABP was 93 ± 6 Torr ($P < 0.05$ from control) that is mean 10 Torr ($P < 0.05$ from control) lower than untreated CHI group at this point. On the other hand, after 192 h of CHI the MABP values did not differ between untreated (90 ± 4 Torr) or nanowired delivered 50 mg oxiracetam group (92 ± 7 Torr).

(e) Arterial pH

The arterial pH in control group was 7.38 ± 0.08 that was slightly lower with oxiracetam 50 (7.36 ± 0.10) and 100 mg (7.37 ± 0.08) doses given for 5 days. On the other hand no differences in arterial pH was noted between control group and animals treated with TiO2-oxiracetam 50 mg for 5 days (7.38 ± 0.09) (Fig. 3).

However, CHI progressively reduced the arterial pH as compared to the control group. Thus, after 48 h CHI the arterial pH was 7.35 ± 0.04 ($P < 0.05$, from control) that was reduced to 7.32 ± 0.04 ($P < 0.05$ from control) at 48 h that further decreased to 7.31 ± 0.03 ($P < 0.05$ from control) after 192 h injury (Fig. 3).

Treatment with oxiracetam 50 and 100 mg doses did not affect arterial pH from the untreated CHI group after 192 h of CHI. Thus, the arterial pH in 50 mg was 7.34 ± 0.08 ($P < 0.05$ from control) whereas 100 mg was 7.33 ± 0.03 ($P < 0.05$ from control) after 182 h of CHI. Interestingly, TiO2-oxiracetam treatment in CHI for 5 days also resulted in lower arterial pH (7.35 ± 0.092, $P < 0.05$ from control) at 192 h after CHI (Fig. 3).

(f) Arterial PaCO2

The arterial PaCO2 was not affected by oxiracetam in control group. Thus, the control group PaCO2 was 33.81 ± 0.08 Torr that was slightly altered by oxiracetam treatment for 5 days using 50 (33.67 ± 0.07 Torr) and 100 mg (33.45 ± 0.10 Torr) doses. Nanodelivery of oxiracetam in 50 mg doses the PaCO2 value in control group was 33.46 ± 0.08 (Fig. 3).

On the other hand, CHI significantly elevated PaCO2 levels significantly from 48–192 h after insult. Thus, the PaCO2 value after 48 h CHI was increased to 35.13 ± 0.08 Torr ($P < 0.05$ from control) and further elevated by 96 h to 35.89 ± 0.08 Torr ($P < 0.05$ from control). The PaCO2 reached to 36.01 ± 0.07 Torr ($P < 0.05$ from control) after 192 h of CHI (Fig. 3).

Treatment with oxiracetam in CHI significantly reduced the PaCO2 elevation at 192h after CHI following oxiracetam 50 (34.34 ± 0.06 Torr, $P < 0.05$ from CHI) and 100mg (33.98 ± 0.03 Torr, $P < 0.05$ from CHI) treatment. Nanodelivery of oxiracetam in 50mg dose has superior reduction in PaCO2 (33.87 ± 0.10 Torr, $P < 0.05$ from CHI) after 192h of CHI (Fig. 3).

(g) Arterial PaO2:

The arterial PaO2 in control group was not affected by oxiracetam conventional or nanodelivery 5 days treatment. Thus, the PaO2 level in control group was 81.56 ± 0.23 Torr. Oxiracetam 50mg resulted in PaO2 value to 81.04 ± 0.08 Torr and 100mg to 81.22 ± 0.10 Torr in arterial PaO2. Nanodelivery of oxiracetam in 50mg dose resulted in a slight increase inPaO2 values to 81.99 ± 0.06 Torr (Fig. 3).

However, CHI resulted in a significant decrease in PaO2 values following 48–192h after insult. Thus, After 48h of CHI the PaO2 value was 80.34 ± 0.06 Torr ($P < 0.05$ from control) and this was further decreased at 96h CHI by 79.87 ± 0.08 Torr ($P < 005$ from Control) followed by additional decrease at 192h CHI (79.34 ± 0.10 Torr, $P < 0.05$ from control) (Fig. 3).

Treatment with oxiracetam either conventional or nanodrug delivery significantly elevated the PaO2 values in CHI. Thus, the PaO2 values in conventional oxiracetam treated group after 192h with 50mg doses was 80.23 ± 0.07 Torr ($P < 0.05$ from control) and 100mg dose was 80.45 ± 0.08 Torr ($P < 0.05$ from CHI). Using nanodrug delivery of oxiracetam 50mg doses resulted in elevation in PaO2 values in CHI at 192h to 81.08 ± 0.12 Torr ($P < 0.05$ from CHI) (Fig. 3).

9.3 Oxiracetam and the blood-brain barrier in CHI

The blood-brain barrier (BBB) in control, CHI and oxiracetam treated injured group was examined suing Evans blue albumin (EBA) and radioiodine ([131]Iodine-Na) extravasation. The BBB breakdown to EBA or radioiodine in control or oxiracetam either conventional or nanowired treatment did not alter the course of normal permeability of these tracers in the brain. Thus, the BBB in control group to EBA was 0.22 ± 0.02 mg % in Left hemisphere (LH) and 0.20 ± 0.05 mg % in Right hemisphere (RH). The radioiodine concentration in RH was $0.30 \pm 0.04\%$ and in LH was $0.32 \pm 0.03\%$ of whole blood value (Fig. 4).

In general CHI results in extravasation of BBB tracers in the whole brain but the most pronounced leakage was observed in the contralateral uninjured left half as compared to the injured right half (Fig. 4). Also, radioiodine leakage was seen in the brain irrespective of RH or LH a little higher extravasation that EBA in the brain.

Thus, in the right half EBA leakage was 2.10 ± 0.10 mg % whereas radioiodine showed $2.38 \pm 0.12\%$ extravasation after 48h CHI. On the other hand LH showed 2.56 ± 0.08 mg % leakages of EBA and radioiodine exhibited $2.70 \pm 0.12\%$ penetration after 48h CHI ($P < 0.01$ from Control). At 96h CHI the EBA showed 3.08 ± 0.11 mg % in LH and 2.34 ± 0.08 mg % in RH as compared to radioiodine leakage $3.45 \pm 0.14\%$ in LH and $2.64 \pm 0.07\%$ in RH ($P < 0.01$ from Control)

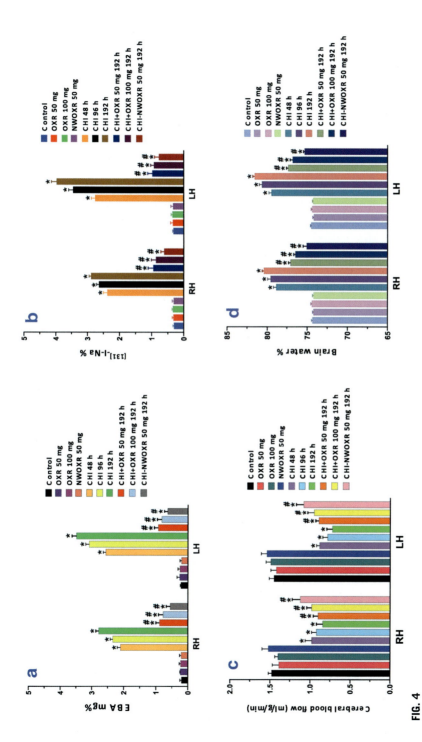

FIG. 4

Effect of TiO2 nanowired oxiracetam (NWOXR) on blood brain barrier permeability to Evans blue albumin (EBA, A) and Radioiodine (B), cerebral blood flow (C) and brain edema (D) following concussive head injury (CHI) in the rat on 8th day after trauma. CHI was inflicted using a 0.224 N impact injury over Right parietal skull and the animals were allowed to survive for 1 week after the primary insult. Oxiracetam (OXR) was given in a dose of 50 or 100 mg/kg, i.p. once daily after 1 day of CHI for 5 days and the parameters were evaluated on the 8th day. For TiO2 nanowired delivery of oxiracetam (NWOXR) 50 mg/kg (i.p.) was given under identical conditions. Values are Mean ± SD of 6–8 rats at each point. *MABP*, mean arterial blood pressure; *OXR*, oxiracetam; *NWOXR*, nanowired oxiracetam; *nd*, not done. * $P < 0.05$ from Control; # $P < 0.05$ from CHI at respective group. ANOVA followed by Dunnett's test for multiple group comparison using one control. For details see text.

(Fig. 4). These tracers extravasation were further augmented in CHI at 192 h in both RH and LH. Thus, EBA leakage in LH was 3.48 ± 0.14 mg % and in RH 2.78 ± 0.09 mg % as compared to radioiodine permeation 2.89 ± 0.06% in RH and 3.98 ± 0.14% in LH ($P < 0.01$ from Control) (Fig. 4).

Treatment with nanowired delivery of oxiracetam 50 mg dose for 5 days during 1 week after CHI significantly reduced the leakage of EBA and radioiodine across the BBB in RH and LH. Thus, nanowired oxiracetam the leakage of EBA was 0.56 ± 0.12 mg % in RH and 0.64 ± 0.10 in LH as compared to radioiodine extravasation 0.62 ± 0.10% in RH and 0.78 ± 0.10 in LH ($P < 0.05$ from CHI) as compared to conventional oxiracetam treatment irrespective of low or high doses (Fig. 4). Oxiracetam 50 mg in CHI was able to reduce BBB breakdown to EBA by 0.89 ± 0.11 mg % in RH and 0.92 ± 0.12% in LH as compared to radioiodine leakage by 0.95 ± 0.10% in RH and 0.98 ± 0.14% in LH ($P < 0.05$ from CHI). Oxiracetam in 100 mg doses further reduced this BBB leakage to EBA in RH by 0.78 ± 0.14 mg % and 0.82 ± 0.14 mg % in LH as compared to radioiodine extravasation by 0.87 ± 0.14% in RH and 0.93 ± 0.10% in LH ($P < 0.05$ from CHI) (Fig. 4).

9.4 Oxiracetam and cerebral blood flow in CHI

The cerebral blood flow (CBF) was measured using radiolabeled carbonized microspheres in control, CHI and following oxiracetam treated. In control naïve or oxiracetam treated animals the CBF is almost similar in both the RH and LH without much fluctuations. Thus, the control CBF in RH was 1.48 ± 0.04 mL/g and in LH this was 1.48 ± 0.06 mL/g (Fig. 4). Oxiracetam treatment for 5 days with 50 mg dose slightly reduced the CBF to 1.39 ± 0.08 mL/g in RH and 1.42 ± 0.07 mL/g in LH whereas with 100 mg dose the CBF in RH was 1.40 ± 0.04 mL/g and in LH the values are 1.49 ± 0.06 mL/g. Interestingly nanodelivery of oxiracetam in 50 mg doses slightly elevated the CBF to 1.52 ± 0.08 mL/g in RH and 1.54 ± 0.07 mL/g in LH (Fig. 4).

Infliction of CHI significantly decreased the CBF progressively in both halves of the brain that was more marked in the LH as compared to the RH. Thus, 48 h CHI resulted in reduction of CBF in RH to 0.98 ± 0.04 mL/g ($P < 0.05$ from control) and in LH 0.88 ± 0.04 mL/g ($P < 0.05$ from control and RH) (Fig. 4). At 96 h CHI the CBF was further declined in both halves that was the most prominent in the LH. Thus, at this point RH showed decline to CBF to 0.92 ± 0.06 mL/g and in LH values are 0.78 ± 0.06 mL/g ($P < 0.05$ from control and RH). The CBF values in RH further declined to 0.84 ± 0.08 mL/g and in LH to 0.72 ± 0.07 mL/g ($P < 0.05$ from control and RH) at 192 h of CHI (Fig. 4).

Treatment with conventional oxiracetam in CHI following 192 h enhanced the CBF in both halves depending on the doses used. Interestingly, nanowired delivery of oxiracetam in 192 h CHI almost restored the CBF values near normal levels. Thus, the CBF in oxiracetam using 50 mg was 0.90 ± 0.06 mL/g in RH and 0.89 ± 0.04 mL/g ($P < 0.05$ from CHI) in LH whereas with 100 mg dose the CBF values significantly increased further in RH to 0.98 ± 0.06 and in LH 0.95 ± 0.08 ($P < 0.05$ from CHI). On the other hand when nanowired delivery of oxiracetam was done in 50 mg doses the CBF values reached almost near normal level in RH to 1.12 ± 0.10 mL/g and in LH to 1.08 ± 0.09 mL/g ($P < 0.05$ from CHI) (Fig. 4).

9.5 Oxiracetam and brain edema formation in CHI

The brain edema formation was evaluated using measurement of brain water content and volume swelling as described earlier (Sharma and Cervós-Navarro, 1990). The brain water content in control and oxiracetam treated group was almost similar in both halves of the brain. Thus, control animals exhibited brain water content of $74.34 \pm 0.12\%$ in RH and $74.51 \pm 0.08\%$ in LH. Conventional oxiracetam treatment or nanowired delivery did not influence the brain water content in normal animals. Thus, oxiracetam 50 mg treatment for 5 days the brain water content in RH was $74.23 \pm 0.11\%$ and in LH was $74.19 \pm 0.12\%$ whereas 100 mg dose showed RH water content of $74.43 \pm 0.13\%$ and LH exhibiting $74.38 \pm 0.14\%$. Nanowired delivery of oxiracetam in 50 mg doses the brain water content showed $74.21 \pm 0.13\%$ in RH and $74.26 \pm 0.09\%$ in LH (Fig. 4).

However, the brain water content following CHI was increased progressively with time in both halves of the brain but the LH showed higher increase in brain water than the RH. Thus, 48 h after CHI the brain water in RH was $78.87 \pm 0.35\%$ and in LH this was $79.48 \pm 0.32\%$ ($P < 0.05$ from control) corresponding to volume swelling ($\%f$) of 18 and 20%, respectively (Fig. 5). At 96 h after CHI the brain water content increased further by $79.59 \pm 0.36\%$ in RH and $80.67 \pm 0.31\%$ ($P < 0.05$ from control) in LH corresponding to 21 and 24% increase in volume swelling, respectively (Fig. 5). After 192 h of CHI the RH showed $80.43 \pm 0.21\%$ brain water and in LH this was $81.56 \pm 0.12\%$ ($P < 0.05$ from control) indicating a volume swelling increase by 24% and 28%, respectively from the control values (Fig. 5). This indicates a progressive increase in brain edema formation in CHI from 2 days to 1 week in our study.

Treatment with conventional oxiracetam treatment for 5 days after the onset of CHI significantly reduced the brain water content in both halves and volume swelling significantly at 192 h after trauma that is dose dependent. Thus, 50 ng oxiracetam treatments in CHI resulted in reduced brain water content of $77.06 \pm 0.25\%$ in RH and $77.34 \pm 0.21\%$ ($P < 0.05$ from CHI) in LH corresponding to about 11% volume swelling in both halves at 192 h (Figs. 4 and 5). When the oxiracetam dose was increased to 100 mg the brain water content after 192 h of CHI was $76.46 \pm 0.21\%$ in RH and in $76.78 \pm 0.12\%$ ($P < 0.05$ from CHI) in LH that is equal to 6% and 7% increase in volume swelling, respectively (Figs. 4 and 5).

On the other hand, when nanowired delivery of oxiracetam 50 mg dose was administered in CHI group of 192 h significantly higher reduction in the brain water content was seen. Thus, in nanowired delivery of oxiracetam the brain water content in RH was $75.08 \pm 0.43\%$ and in LH was $75.32 \pm 0.12\%$ ($P < 0.05$ from CHI) after 192 of CHI. This corresponds to about 3% increase in volume swelling in both halves as compared from control group. This indicates that nanowired delivery of oxiracetam has superior antiedematous effects (Figs. 4 and 5).

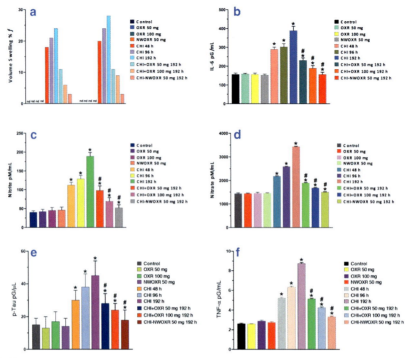

FIG. 5

Effect of TiO2 nanowired oxiracetam (NWOXR) on volume swelling in the brain (% f) (A) and biochemical change in the cerebrospinal fluid (CSF) Interleukin-6 (IL-6, B), Nitrite (C), Nitrate (D), phosphorylated Tau (p-Tau, E) and tumor necrosis factor-alpha (TNF-α, F) concussive head injury (CHI) in the rat on 8th day after trauma. CHI was inflicted using a 0.224 N impact injury over Right parietal skull and the animals were allowed to survive for 1 week after the primary insult. Oxiracetam (OXR) was given in a dose of 50 or 100 mg/kg, i.p. once daily after 1 day of CHI for 5 days and the parameters were evaluated on the 8th day. For TiO2 nanowired delivery of oxiracetam (NWOXR) 50 mg/kg (i.p.) was given under identical conditions. For details see text. Volume swelling (% f) was calculated from brain water changes between control and CHI with or without OXR or NWOXR treatment. %f=about 1% increase in brain water is comparable to approximately 4% increase in brain swelling. Values are Mean±SD of 6–8 rats at each point. *MABP*, mean arterial blood pressure; *OXR*, oxiracetam; *NWOXR*, nanowired oxiracetam; *nd*, not done. * $P < 0.05$ from Control; # $P < 0.05$ from CHI at respective group. ANOVA followed by Dunnett's test for multiple group comparison using one control. For details see text. §=Impact injury 0.224 N; %f=about 1% increase in brain water is comparable to approximately 4% increase in brain swelling. Volume swelling were calculated from brain water changes between control and CHI with or without OXR or NWOXR treatment. For details see text.

9.6 Oxiracetam influences regional brain functions

In CHI almost all brain areas are affected leading to gradual neurodegeneration and functional disability. Thus, we examined regional changes in the brain faction in eight areas namely cerebral cortex, hippocampus, cerebellum, thalamus, hypothalamus, caudate nucleus, medulla and brain stem following 192 h of CHI on right parietal bone in control and following TiO2-nanowired delivery of oxiracetam for 5 days. Since composition of TiO2 nanowired with Na and K have different properties as mentioned above, it appears that TiO2-Na and TiO2-K nanowires may also have varied ability on neuroprotection. This effect was also examined in CHI of 192 h using Ti-Na and Ti-K nanowired delivery of oxiracetam under identical conditions in various brain regions (see below).

a. *Regional blood-brain barrier permeability*

The BBB was examined in eight brain regions (as above) in control, CHI 192 h and its modification with Ti-Na nanowired or Ti-K-nanowired oxiracetam. In normal animals the BBB permeability to EBA or radioiodine is minimal in almost all brain areas. Cerebellum shows medulla and brain stem has very low level of permeability to EBA (0.6–0.08 mg %) and radioiodine (0.08–0.15%) followed by cerebellum (EBA 0.12 mg %, radioiodine 0.14%), caudate nucleus (EBA 0.12 mg %, radioiodine 0.15%), cerebral cortex (EBA 0.24 mg %, radioiodine 0.28%), hippocampus (EBA 0.32 mg %, radioiodine 0.43%), thalamus (EBA 0.43 mg %, radioiodine 0.54%), hypothalamus (0.54 mg %, radioiodine 0.59%). In control group these brain regions exhibit very similar distribution of tracers between RH and LH (Figs. 6 and 7).

After CHI the regional BBB breakdown to EBA or radioiodine increased significantly in almost all brain regions at 192 h. However, the LH brain regions always exhibit greater extravasation of these tracers as compared to identical RH brain areas. Thus, the cerebral cortex showed the highest leakage of EBA (RH 1.89 mg %, LH 2.14 mg %), or radioiodine (RH 2.17 &, LH 2.45%) followed by caudate nucleus (EBA RH 1.10 mg %, LH 1.23 mg %; radioiodine RH 1.28%, LH 1.34%), cerebellum (EBA RH 0.86 mg %, LH 0.90 mg %, radioiodine RH 1.12%, LH 1.23%), thalamus (EBA RH 0.88 mg %, LH 0.94 mg %; radioiodine RH 0.92 &, LH 1.09%), hypothalamus (EBA RH 0.78 mg %, LH0.85 mg %; radioiodine 1.03%, LH 1.12%), medulla (EBA RH 0.19 mg %, LH 0.23 mg %; radioiodine RH 0.24%, LH 0.35%) and brain stem (EBA RH 0.13 mg %, LH 0.21 mg %; radioiodine RH 0.22%, LH 0.28%) (Figs. 6 and 7).

Treatment with TiO2 nanowired delivery of oxiracetam in 192 h CHI significantly reduced the regional BBB breakdown in almost all brain areas. However, we observed that Ti-K nanowired oxiracetam has some superior neuroprotective activity in reducing the BBB breakdown to EBA and radioiodine in several brain areas following CHI at 192 h after trauma (see below).

Treatment with Ti-Na nanowired oxiracetam significantly attenuated increased BBB breakdown to EBA and radioiodine in almost all the brain regions (Figs. 6 and 7). Thus, the cerebral cortex showed EBA extravasation in RH to 0.70 mg %

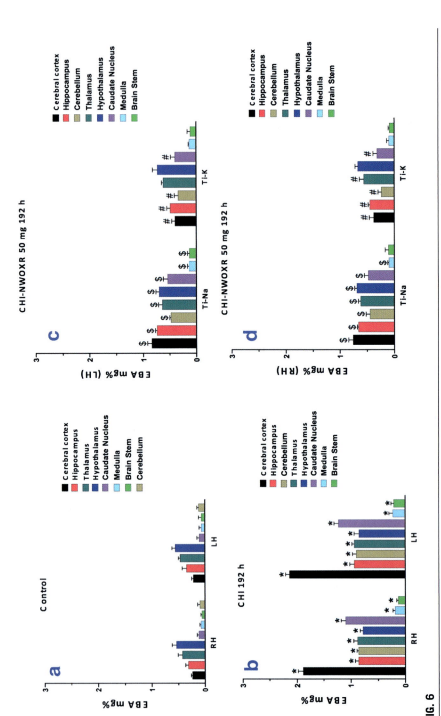

FIG. 6

Effect of Na-TiO2 vs K-TiO2 nanowired oxiracetam (NWOXR) (C, D) on Regional blood brain barrier permeability to Evans blue albumin (EBA) in control (A) and following concussive head injury (CHI, B) in the rat on 8th day after trauma. CHI was inflicted using a 0.224 N impact injury over Right parietal skull and the animals were allowed to survive for 1 week after the primary insult. Oxiracetam (OXR) was given in a dose of 50 or 100 mg/kg, i.p. once daily after 1 day of CHI for 5 days and the parameters were evaluated on the 8th day. For Na-TiO2 or K-TiO2 nanowired delivery of oxiracetam (NWOXR) 50 mg/kg (i.p.) was given under identical conditions. For details see text. Values are Mean ± SD of 6–8 rats at each point. *MWM*, Morris Water Maze platform search; *OXR*, oxiracetam; *NWOXR*, nanowired oxiracetam; *TiO2-Na*, sodium titanate nanowires; *TiO2-K*, potassium titanate nanowires. Volume swelling (%f) was calculated according to Elliott and Jasper (1949). * $P < 0.05$ from Control; # $P < 0.05$ from untreated CHI; § $P < 0.05$ from CHI + NWOXR (TiO2·Na) at respective group. ANOVA followed by Dunnett's test for multiple group comparison using one control. For details see text.

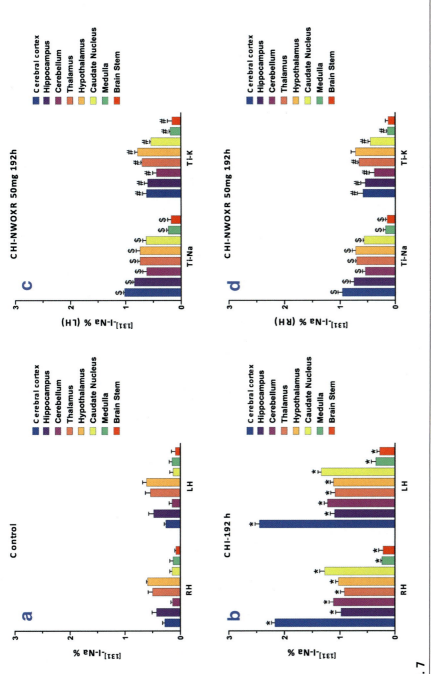

FIG. 7

Effect of Na-TiO2 vs K-TiO2 nanowired oxiracetam (NWOXR) (C, D) on Regional blood brain barrier permeability to radioiodine in control (A) and following concussive head injury (CHI, B) in the rat on 8th day after trauma. CHI was inflicted using a 0.224 N impact injury over Right parietal skull and the animals were allowed to survive for 1 week after the primary insult. Oxiracetam (OXR) was given in a dose of 50 or 100 mg/kg, i.p. once daily after 1 day of CHI for 5 days and the parameters were evaluated on the 8th day. For Na-TiO2 or K-TiO2 nanowired delivery of oxiracetam (NWOXR) 50 mg/kg (i.p.) was given under identical conditions. For details see text. Values are Mean ± SD of 6–8 rats at each point. *MWM,* Morris Water Maze platform search; *OXR,* oxiracetam; *NWOXR,* nanowired oxiracetam; *TiO2-Na,* sodium titanate nanowires; *TiO2-K,* potassium titanate nanowires. Volume swelling (%*f*) was calculated according to Elliott and Jasper (1949). * $P < 0.05$ from Control; # $P < 0.05$ from untreated CHI; § $P < 0.05$ from CHI+NWOXR (TiO2·Na) at respective group. ANOVA followed by Dunnett's test for multiple group comparison using one control. For details see text.

and in LH this was 0.84 mg % whereas the radioiodine leakage was limited to 0.96% in RH and 1.02% in LH ($P<0.05$ from CHI). Likewise caudate nucleus exhibited EBA extravasation only 0.48 mg % in RH and 0.54 mg % in LH and the radioiodine leakage was limited to 0.56% in RH and 0.63% in LH after 192 h of CHI. The cerebellum showed EBA extravasation of 0.45 mg % in RH and 0.48 mg % in LH and radioiodine extravasation in RH was 0.54% in RH and 0.62% in LH ($P<0.05$ from CHI). Hippocampus EBA leakage was limited for EBA to 0.66 mg % in RH and 0.74 mg % in LH whereas radioiodine showed 0.75% in RH and 0.75% in LH ($P<0.05$ from CHI) permeability after trauma. The EBA permeability in thalamus for EBA was 0.62 mg % in RH and 0.64 mg % in LH and the radioiodine leakage was 0.70% in RH and 0.74% in LH ($P<0.05$ from CHI). The EBA extravasation in hypothalamus was 0.69 mg % in RH and 0.70 mg % in LH and radioiodine permeability is limited to 0.72% in RH and 0.74% in LH ($P<0.05$ from CHI) after trauma. In medulla and brain stem the EBA permeability was 0.10 and 0.11 mg % in RH and 0.14 mg and 0.13 mg % in LH whereas the radioactivity leak was 0.18 and 0.15% in RH and 0.23 and 0.18% in LH ($P<0.05$ from CHI), respectively (Figs. 6 and 7).

When Ti-K nanowired oxiracetam was administered for 5 days under identical conditions in 192 h CHI group, significant lower BBB breakdown to EBA and radioiodine was observed in several brain areas as compared to the Ti-Na nanowired oxiracetam. Thus, the EBA permeability was lower in Ti-K nanowired oxiracetam treated CHI group in cerebral cortex, hippocampus, cerebellum, thalamus and caudate nucleus in RH and in the cerebral cortex, hippocampus, cerebellum and caudate nucleus in the LH ($P<0.05$ from Ti-Na nanowired oxiracetam). The radioiodine extravasation was lowered in this group in cerebral cortex, hippocampus, cerebellum, thalamus, caudate nucleus and medulla in RH and all the eight brain regions in the LH ($P<0.05$ from Ti-Na nanowired oxiracetam). These observation supports the idea that Ti-K nanowired delivery of drug has somehow superior neuroprotective effects than Ti-Na nanowired drug under identical conditions (Figs. 6 and 7).

b. *Regional cerebral blood flow changes*

The regional cerebral blood flow (rCBF) was measured using carbonized microspheres labeled with radioiodine (Sharma, 1987). In control of oxiracetam treated normal animals showed minor variations in the rCBF among different brain regions depending upon the gray matter and capillary densities (Sharma and Dey, 1986a,b; Sharma and Dey, 1987). Thus, the cerebellum exhibited the highest CBF of 1.63 mL/g/min followed by cerebral cortex (1.54 mL/g/min), caudate nucleus (1.23 mL/g/min), hippocampus (0.85 mL/g/min), medulla (0.84 mL/g/min), brain stem (0.80 mL/g/min), thalamus (0.76 mL/g/min) and hypothalamus (0.70 mL/g/min). This pattern of rCBF was very similar between RH and LH of control group. Treatment with conventional oxiracetam (50 or 100 mg) or Ti-Na or Ti-K nanowired delivery of oxiracetam did not influence this pattern of rCBF in any regions examined (H S Sharma unpublished results).

The rCBF reduced significantly in several regions after 192 h of CHI that was the most pronounced in LH as compared to RH (Fig. 8). Thus, cerebellum showed highest decrease in the rCBF followed by cerebral cortex, caudate nucleus, medulla, brain

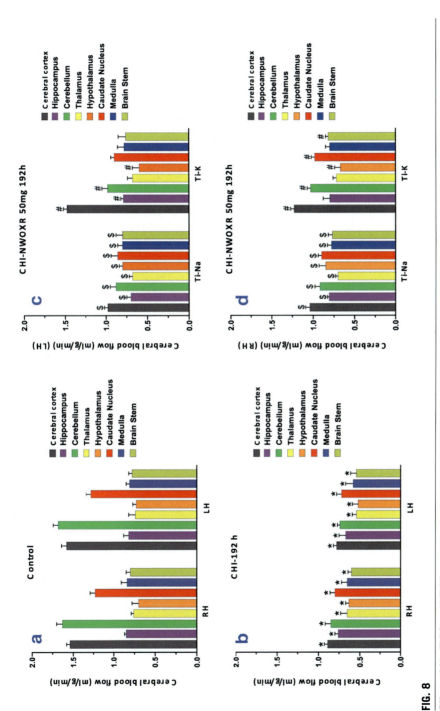

FIG. 8

Effect of Na-TiO2 vs K-TiO2 nanowired oxiracetam (NWOXR) (C, D) on Regional cerebral blood flow in control (A), following concussive head injury (CHI, B) in the rat on 8th day after trauma. CHI was inflicted using a 0.224N impact injury over Right parietal skull and the animals were allowed to survive for 1 week after the primary insult. Oxiracetam (OXR) was given in a dose of 50 or 100 mg/kg, i.p. once daily after 1 day of CHI for 5 days and the parameters were evaluated on the 8th day. For Na-TiO2 or K-TiO2 nanowired delivery of oxiracetam (NWOXR) 50 mg/kg (i.p.) was given under identical conditions. For details see text. Values are Mean ± SD of 6–8 rats at each point. *MWM*, Morris Water Maze platform search; *OXR*, oxiracetam; *NWOXR*, nanowired oxiracetam; *TiO2-Na*, sodium titanate nanowires; *TiO2-K*, potassium titanate nanowires. Volume swelling (%f) was calculated according to Elliott and Jasper (1949). * $P < 0.05$ from Control; § $P < 0.05$ from untreated CHI; # $P < 0.05$ from CHI + NWOXR (TiO2:Na) at respective group. ANOVA followed by Dunnett's test for multiple group comparison using one control. For details see text.

stem, hippocampus, thalamus and hypothalamus in RH that was also seen in LH with greater intensity. Thus, the cerebral cortex showed rCBF reduction leading to 0.98 mL/g/min in RH and 0.78 mL/g/min in LH ($P < 0.05$ from control), hippocampus (0.76 mL/g/min RH, 0.67 mL/g/min LH, $P < 0.05$ from control), cerebellum (0.85 mL/g/min RH, 0.74 mL/g/min LH, $P < 0.05$ from control), thalamus (0.65 mL/g/min RH, 0.54 mL/g/min LH, $P < 0.05$ from control), hypothalamus (0.63 mL/g/min RH, 0.54 mL/g/min, $P < 0.05$ from control), caudate nucleus (0.80 mL/g/min RH, 0.72 mL/g/min LH, $P < 0.05$ from control), medulla (0.65 mL/g/min RH, 0.58 mL/g/min LH, $P < 0.05$ from control) and brain stem (0.60 mL/g/min RH, 0.54 mL/g/min LH, $P < 0.05$ from control) (Fig. 8).

Nanodelivery of oxiracetam either using either Ti-Na or Ti-K nanowires significantly enhanced the rCBF in almost all brain regions examined after CHI. However, Ti-K nanowired delivery appears to have significant higher increase in the rCBF in some of the brain regions in both RH and LH after CHI (Fig. 8).

Thus, Ti-Na nanowired delivery of oxiracetam resulted in increased rCBF after 192 h CHI in cerebral cortex (1.04 mL/g/min RH, 0.98 mL/g/min LH, $P < 0.05$ from CHI), hippocampus (0.81 mL/g/min RH, 0.70 mL/g/min LH, $P < 0.05$ from CHI), cerebellum (0.92 mL/g/min RH, 0.88 mL/g/min LH, $P < 0.05$ from CHI), thalamus (0.70 mL/g/min RH, 0.68 mL/g/min LH, $P < 0.05$ from CHI), hypothalamus (0.85 mL/g/min RH, 0.80 mL/g/min LH, $P < 0.05$ from CHI), caudate nucleus (0.90 mL/g/min RH, 0.86 mL/g/min LH, $P < 0.05$ from CHI), medulla (0.78 mL/g/min RH, 0.80 mL/g/min LH, $P < 0.05$ from CHI) and brain stem (0.77 mL/g/min RH, 0.89 mL/g/min LH, $P < 0.05$ from CHI) (Fig. 8).

Interestingly, Ti-K nanowired delivery of oxiracetam maintained significantly higher rCBF after 192 h CHI in cerebral cortex (1.23 mL/g/min RH, 1.48 mL/g/min LH, $P < 0.05$ from Ti-Na), cerebellum (1.03 mL/g/min RH, 0.98 mL/g/min LH, $P < 0.05$ from Ti-Na), hippocampus in LH (0.79 mL/g/min, $P < 0.05$ from Ti-Na), caudate nucleus in RH (0.98 mL/g/min, $P < 0.05$ from Ti-Na), and in medulla RH (0.82 mL/g/min, $P < 0.05$ from Ti-Na). On the other hand, the rCBF was significantly declined in Ti-K nanowired treated CHI group in hypothalamus (0.67 mL/g/min RH, 0.60 mL/g/min LH, $P < 0.05$ from Ti-Na) (Fig. 8).

c. *Regional brain edema formation*

The regional brain water content varies widely depending on their gray vs white matter ration. Thus, the brain water content of cerebral cortex was 75.14% in RH and 75.19% in RH, where as other brain regions showed higher brain water content in normal control group. Thus, brain water content recorded in hippocampus was 80.14% RH, and 80.09% LH followed by hypothalamus (82.33% RH, 82.30% LH), brain stem (81.23% RH, 81.32% in LH), medulla (80.34% RH, 80.53% LH), caudate nucleus (79.34% RH, 79.36% LH), thalamus (77.31% RH, 77.20% LH), and cerebellum (78.34% RH, 78.45% LH) (Fig. 9).

The regional brain water content (rBWC) increased significantly after 192 h CHI that shows profound increase in volume swelling (% *f*) of 4–16% in the RH and 6–21% increase in the LH as compared between changes in the brain water content from control group (Figs. 9 and 10). The cerebral cortex showed the highest amount

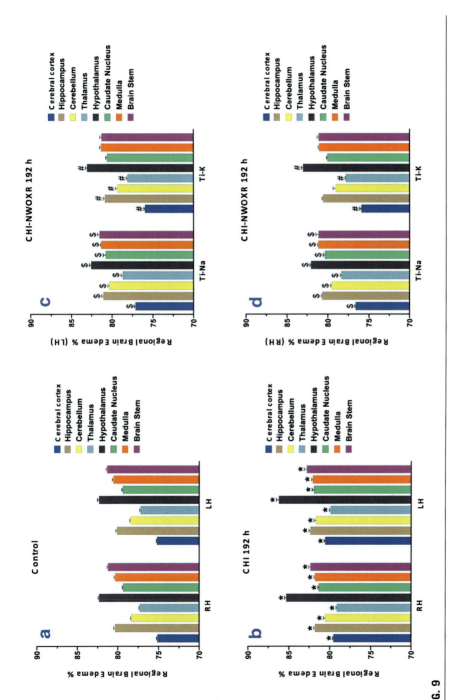

FIG. 9

Effect of Na-TiO2 vs K-TiO2 nanowired oxiracetam (NWOXR) (C, D) on Regional brain edema in control (A), following concussive head injury (CHI, B) in the rat on 8th day after trauma. CHI was inflicted using a 0.224N impact injury over Right parietal skull and the animals were allowed to survive for 1 week after the primary insult. Oxiracetam (OXR) was given in a dose of 50 or 100mg/kg, i.p. once daily after 1 day of CHI for 5 days and the parameters were evaluated on the 8th day. For Na-TiO2 or K-TiO2 nanowired delivery of oxiracetam (NWOXR) 50mg/kg (i.p.) was given under identical conditions. For details see text. Values are Mean ± SD of 6–8 rats at each point. *MWM*, Morris Water Maze platform search; *OXR*, oxiracetam; *NWOXR*, nanowired oxiracetam; *TiO2-Na*, sodium titanate nanowires; *TiO2-K*, potassium titanate nanowires. Volume swelling (%*f*) was calculated according to Elliott and Jasper (1949). * $P < 0.05$ from Control; § $P < 0.05$ from untreated CHI; # $P < 0.05$ from CHI + NWOXR (TiO2-Na) at respective group. ANOVA followed by Dunnett's test for multiple group comparison using one control. For details see text.

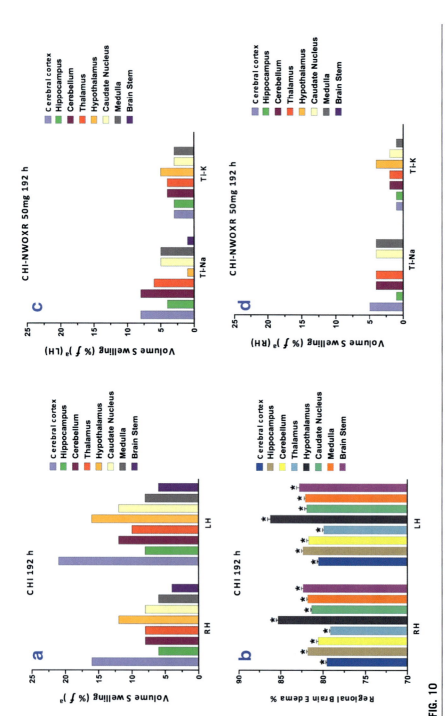

FIG. 10

Effect of Na-TiO2 vs K-TiO2 nanowired oxiracetam (NWOXR) (C, D) on Regional volume swelling (% f) in control (A), following concussive head injury (CHI, B) in the rat on 8th day after trauma. CHI was inflicted using a 0.224 N impact injury over Right parietal skull and the animals were allowed to survive for 1 week after the primary insult. Oxiracetam (OXR) was given in a dose of 50 or 100 mg/kg, i.p. once daily after 1 day of CHI for 5 days and the parameters were evaluated on the 8th day. For Na-TiO2 or K-TiO2 nanowired delivery of oxiracetam (NWOXR) 50 mg/kg (i.p.) was given under identical conditions. For details see text. Values are Mean ± SD of 6–8 rats at each point. *MWM*, Morris Water Maze platform search; *OXR*, oxiracetam; *NWOXR*, nanowired oxiracetam; *TiO2-Na*, sodium titanate nanowires; *TiO2-K*, potassium titanate nanowires. Volume swelling (% f) was calculated according to Elliott and Jasper (1949). * $P < 0.05$ from Control; § $P < 0.05$ from untreated CHI; # $P < 0.05$ from CHI+NWOXR (TiO2·Na) at respective group. ANOVA followed by Dunnett's test for multiple group comparison using one control. For details see text.

of volume swelling of 16% in RH (79.54%) and 21% (80.54%) in LH followed by hypothalamus 12% RH (85.32%) and 16% LH (86.21%), cerebellum and caudate nucleus both by 8% in RH and 12% in LH. The volume swelling in thalamus was 8% in RH and 10% in LH whereas hippocampus showed a volume swelling of 6% in RH and 8% in LH. The volume swelling in medulla and brain stem showed 6% and 4% in RH and 8% and 6% in LH, respectively after 192 h of CHI (Figs. 9 and 10).

Treatment with nanodelivery of oxiracetam in CHI significantly recued the rBWC in several regions. Ti-K nanowired oxiracetam appears to be superior in reducing rBWC in CHI in several regions as compared to Ti-Na nanowired drug delivery. Treatment with Ti-Na nanowired oxiracetam reduced volume swelling in the cerebral cortex to 5% in RH and 8% in LH, hippocampus 1% in RH and 4% in LH, cerebellum 4% in RH and 8% in LH, thalamus 4% in RH and 6% in LH, hypothalamus completely in RH and 1% in LH, caudate nucleus and medulla 4% in RH and 5% in LH, and brain stem complete in RH and 1% in LH at 192 h of CHI as compared to untreated injured group (Figs. 9 and 10).

Treatment with Ti-K nanowired further significantly reduced volume swelling in several brain regions after CHI. Thus, the volume swelling is reduced in cerebral cortex and hippocampus to 1% in RH and 3% in LH, cerebellum and thalamus to 2% in RH and 4% in LH, hypothalamus 4% in RH and 5% in LH, caudate nucleus 2 & in RH and 3 & in LH, medulla 1% in RH and 3% in LH and in brain stem is completely prevented (Fig. 10). This suggests that the composition of nanowires for dug delivery could influence the neuroprotective ability of the compound.

9.7 Oxiracetam influence biobehavioral changes in CHI

The biobehavioral functions are evaluated in CHI group of animals using memory retention, sensory, motor and cognitive functions as well as locomotor functions and their modification with conventional or nanowired delivery of oxiracetam.

a. *Memory function*

Morris water maze (MWM) trained rats were allowed to search hidden platform under a pool of water from normal, CHI and conventional or nanowired oxiracetam treated group. Control group of rats either naïve or oxiracetam treated group find hidden platform within 4–5 s whereas, CHI rats took more time to find platform in the MWM as compared to controls. Thus, CHI progressively delays in finding the hidden platform depending on the deration of survival after trauma. Accordingly, 12, 18 and 22 s are needed to find CHI rats the hidden platform in MWM following 48, 96 and 192 h period, respectively ($P < 0.05$ from control). This suggests that CHI progressively impairs memory function (Fig. 3).

Treatment with conventional oxiracetam either 50 or 100 mg doses for 5 days results in significant attenuation of time in finding the platform in MWM depending on the dose. Thus, after 192 h of CHI rats treated with 50 mg oxiracetam find the platform in 15 s ($P < 0.05$ from CHI 192 h) whereas under identical condition 100 mg oxiracetam treated injured rats could search the platform within 13 s ($P < 0.05$ from CHI 192 h). On the other hand TiO2-naowired oxiracetam 50 mg treated 192 h CHI rats find the platform in MWM within 8 s ($P < 0.05$ from CHI 192 h) (Fig. 3). This suggests that oxiracetam treatment in CHI improve memory function and this effect is the most pronounced with nanodelivery of the drug.

b. *Cognitive function on Rota-Rod treadmill*

Rota-Rod test evaluates the cognitive and motor functions of animals, as they have to balance on the moving platform from falling. This requires balance skill and motor function to stay on moving platform (Sharma, 2006; Shiotsuki et al., 2010). Control rats when placed on the Rota-Rod treadmill for 120 s at 16 rpm thy could easily stay for 118 s. Treatment with conventional oxiracetam 50 and 100 mg or TiO2-nanowired oxiracetam 50 mg dose in normal rats did not affect their staying time over the Rota-Rod treadmill. These treated rats easily stayed over the treadmill for 116–118 s (Fig. 11).

Placement of CHI rats on the Rota-Rod treadmill significantly reduced their staying time depending on the injury duration. Thus, 48 h CHI rats could not stay beyond 80 s on Rota-Rod treadmill whereas 96 and 192 h CHI animals maintained their stay for 72 and 60 s ($P < 0.05$ from control), respectively (Fig. 11). This suggests that CHI over time reduce the ability of cognitive and motor skill in animals.

However, treatment with oxiracetam 50 or 100 mg doses significantly enhanced staying time over the Rota-Rod treadmill after 192 h CHI in a dose dependent manner. Interestingly nanodelivery of oxiracetam 50 mg dose further enhanced the timing spent on the Rota-Rod treadmill in 192 h CHI rats (Fig. 11). Thus, oxiracetam 50 or 100 mg doses resulted in 192 h CHI rats to stay on the platform for 70 and 84 s, respectively ($P < 0.005$ from control). Whereas TiO2-nanowired oxiracetam 50 mg resulted in 192 h CHI rats to stay over the Rota-Rod treadmill for 92 s (Fig. 11).

c. *Inclined plane angle test*

Another cognitive test was used to evaluate in animals on an incline plane angle platform ($60°$) that allows normal rats to stay for 5 s (Sharma, 2006). Traumatized animals could not stay on this steep angle platform and need to adjust the inclination for them to stay for 5 s. Control animals could stay at $60°$ for 5 s and also oxiracetam treated control group were fine at $58–60°$ on this test.

Subjection of CHI significantly reduced the angle of the platform that was progressive in nature. Thus, 48 h CHI animals require $48°$ to stay where as 96 and 120 h CHI need the angle to be lowered down to $40°$ and $30°$, respectively ($P < 0.05$ from control) (Fig. 11).

Treatment with oxiracetam resulted in significant increase in plane angle following 192 h CHI that was dose dependent. Thus, 50 mg oxiracetam resulted in elevated platform of $38°$ and with 100 mg dose animals could stay on inclined plane angle of $44°$ ($P < 0.05$ from CHI). However, TiO2-nanowired delivery of 50 mg oxiracetam treated animals could easily stayed at the incline plane of $52°$ after 192 h CHI ($P < 0.05$ from CHI) (Fig. 11).

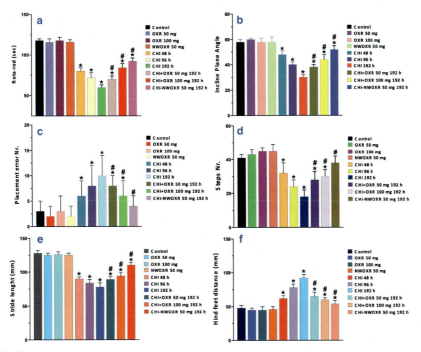

FIG. 11

Effect of Na-TiO2 nanowired oxiracetam (NWOXR) (C, D) on behavioral functions on Rota-Rod treadmill (A), inclined plane angle (B), placement error (C), steps taken (D) on a mesh grid, stride length (E) and hind feet distances (F) while walking and gait analysis following concussive head injury (CHI) in the rat on 8th day after trauma. CHI was inflicted using a 0.224 N impact injury over Right parietal skull and the animals were allowed to survive for 1 week after the primary insult. Oxiracetam (OXR) was given in a dose of 50 or 100 mg/kg, i.p. once daily after 1 day of CHI for 5 days and the parameters were evaluated on the 8th day. For Na-TiO2 nanowired delivery of oxiracetam (NWOXR) 50 mg/kg (i.p.) was given under identical conditions. For details see text. Values are Mean ± SD of 6–8 rats at each point. *MWM*, Morris Water Maze platform search; *OXR*, oxiracetam; *NWOXR*, nanowired oxiracetam; *TiO2-Na*, sodium titanate nanowires. Volume swelling (%*f*) was calculated according to Elliott and Jasper (1949). * $P < 0.05$ from Control; § $P < 0.05$ from untreated CHI; # $P < 0.05$ from CHI+NWOXR (TiO2·Na) at respective group. ANOVA followed by Dunnett's test for multiple group comparison using one control. For details see text.

d. *Motor function and placement error*

The motor skill was evaluated in trained animals by allowing them to walk on an inclined (45°) mesh platform where placement of forepaw was also investigated for placement error (Sharma, 2006). These tests also suggest motor and cognitive skills of animals in walking on an inclined mesh platform with minimized placement error (Sharma, 2006). In these situations number of steps taken and occurrence of placement error was evaluated.

Control group with or without oxiracetam treatment showed average 41–45 steps taken on the inclined mesh within 1 min. In these groups placement error was confined to 2–3 times in 60 s walking (Fig. 11).

On the other hand animals with CHI the number of steps taken in a minute were significantly reduced and the occurrence of placement error increased significantly with time. Thus, 48 h CHI showed 32 steps/min that was reduced to 24 steps/min after 96 h of CHI. At 192 h of CHI only 18 steps/min were counted ($P < 0.05$ from control). Whereas, the number of placement error occurred 6/min after 48 h CHI that increased further to 8 steps/min after 96 h and reduced to 4 steps/min after 192 h (Fig. 11). It may be that less number of steps taken at this time is accountable of decreased placement error.

e. *Gait and locomotion on stride length and hind feet distance*

Brain damage caused by trauma or hyperthermia alters gait and locomotor behavior (Sharma, 2006). We analyzed stride length and distances between hind feet to evaluate gait and locomotion disturbances in control, CHI with and without oxiracetam treatment.

In control group with or without treatment the stride length was very similar ranging from 125–128 mm and the distances between hind feet in two consecutive movements was within 45–48 mm. Rats subjected to CHI showed a progressive decrease in stride length and increase in distances between hind feet. Thus, 48 h after CHI the stride length was 90 mm and the distances between hind feet were 62 mm ($P < 0.05$ from control). These values are further reduced at 96 h CHI (stride length 84 mm, distance between hind feet 78 mm) followed by 192 h CHI (stride length 78 mm, distance between hind feet 92 mm, $P < 0.05$ from control) (Fig. 11).

Treatment with oxiracetam with 50 or 100 mg doses in CHI showed significant increase in stride length by 89 and 94 mm, respectively at 192 h ($P < 0.05$ from CHI). The distances between hid feet were also reduced 65 and 60 mm after oxiracetam treatment with 50 and 100 mg doses, respectively after 192 h of CHI ($P < 0.05$ from CHI). However, treatment with TiO2-nanowired oxiracetam 50 mg resulted in superior protective effect on the gait and locomotion at 192 h of CHI. Thus, in this group the stride length increased to 110 mm and the distance between hind feet was reduced to 54 mm from untreated CHI at this point ($P < 0.05$ from CHI) (Fig. 11).

9.8 Oxiracetam influences biochemical changes in CSF following CHI

Trauma to the brain influences several biochemical parameters in the brain affecting neurotoxicity (Sharma et al., 2016). Alterations in the brain fluid environment are reflected very well in the cerebrospinal fluid (CSF) analysis (Sharma et al., 2019a,b, 2020a,b,c,d). The prominent alterations related to neurotoxicity involves nitric oxide and their metabolites (Sharma et al., 1996), neurotoxic cytokines, e.g., interleukin-6 (IL-6) and tumor necrosis factor-alpha (TNF-a), a phosphorylated tau protein (p-tau) (Sharma et al., 2019a,b, 2020a,b,c,d). Thus, we evaluated changes in the CSF nitrate and nitrite, IL-6, TNF-a and p-tau in control, CHI with or without treatment with oxiracetam.

a. *Nitrate and nitrite*

The concentration of nitrite and nitrate in the CSF is significantly increased progressively after CHI survival duration. Treatment with oxiracetam significantly reduced the CSF nitrate and nitrite levels (Fig. 5). Thus, the control group of animals with or without oxiracetam treatment did not fluctuate significantly. The nitrate level in control and oxiracetam treated normal rats were in the range of 40–46 pM/mL and nitrite levels show 1434–1452 pM/mL. Subjection to animals to CHI has increased these levels significantly in a progressive manner with survival duration. Thus, nitrite levels increased after 48 h CHI to 112 pM/mL and further enhanced to 129 pM/mL at 96 h whereas at 192 h CHI the nitrite level was 189 pM/mL detected ($P < 0.05$ from control). The nitrate levels after CHI were 2178 pM/ml at 48 h, 2578 pM/mL at 96 h and 3452 pM/mL after 192 h CHI ($P < 0.05$ from control) (Fig. 5).

Treatment with oxiracetam in 50 or 100 mg doses significantly reduced the elevation of nitrite and nitrate in the CSF following 192 h CHI. Thus, 192 h after CHI 50 mg oxiracetam resulted in 98 and 69 pM/mL nitrite and 1890 and 1678 pM/mL nitrate, respectively ($P < 0.05$ from CHI). With 100 mg treatment of oxiracetam further reduced the nitrite level to 69 pM/mL and nitrate level to 1678 pM/mL in CHI 192 h CSF ($P < 0.05$ from CHI) (Fig. 5).

b. *Interleukin-6*

Interleukin-6 (IL-6) is a neurotoxic cytokine (Sharma et al., 2020a,b,c,d). Thus, IL-6 levels were significantly elevated after CHI as compared to controls and oxiracetam treatment reduced this increase following CHI.

The control values of IL-6 (156 pG/mL) did not alter significantly after 50 (158 pG/mL) or 100 mg (156 pG/mL) oxiracetam treatment of normal animals. TiO2-nanowired administration in normal animals also showed IL-6 values similar to control group (152 pG/mL) (Fig. 5).

These IL-6 values significantly increased in the CSF in a progressive manner with increase in survival period. Thus, IL-6 values in the CSF rode to 289 pG/mL at CHI 48 h, 302 pG/mL at 96 h and 389 pG/mL after 192 h trauma ($P < 0.05$ from control) (Fig. 5).

Treatment with oxiracetam dose dependently decreased CSF IL-6 after 192 h CHI. Thus, the IL-6 values in CSF after oxiracetam 50 mg is 231 pG/mL and after 100 mg IL-6 level was 189 pG/mL at 192 h CHI ($P < 0.05$ from CHI). Interestingly, TiO2-nanowired oxiracetam treatment resulted in severe reduction in IL-6 values (150 pG/mL) at 192 h CHI ($P < 0.05$ from CHI). This indicates that nanodelivery of oxiracetam has superior effects in lowering IL-6 in the CSF of CHI group (Fig. 5).

c. *Tumor necrosis factor-alpha*

Tumor necrosis factor-alpha (TNF-a) is another neurotoxic cytokine (Sharma et al., 2003). TNF-a levels in control CSF is about 2.61 pG/mL and treatment with either conventional oxiracetam 50 or 100 mg or TiO2-nanowired delivery of 50 mg did not affect normal TNF-a values. Thus, they are within 2.59–2.89 pG/mL levels in the uninjured rat CSF (Fig. 5).

Subjection to CHI significantly increased the TNF-a level progressively with time. Thus, the TNF-a level after CHI 48 h was 5.23 pG/mL, at 96 h was 6.34 and 8.76 pG/mL at 192 h after trauma ($P < 0.05$ from control) (Fig. 5).

Treatment with oxiracetam significantly attenuated TNF-a level in CHI at 192 h survival. Thus, the CSF TNF-a level was 5.14 pG/mL with 50 mg dose and 4.23 pG/mL with 100 mg oxiracetam treatment ($P < 0.05$ from CHI). Interestingly, TiO2-nanowired oxiracetam 50 mg was able to induce superior reduction in the CSF TNF-a to 3.32 pG/mL ($P < 0.05$ from CHI) (Fig. 5).

d. *Phosphorylated-tau*

Phosphorylated-tau (p-tau) is a toxic protein and induces neurodegeneration in Alzheimer's and Parkinson's disease (Sharma et al., 2019a,b, 2020a,b,c,d). Brain injury also increases p-tau concentration in the brain (Sharma et al., 2016). We measured p-tau in the CSF of control, CHI an oxiracetam treated group in this investigation.

This control p-tau level in the CSF was 15 pG/μL and this was not altered significantly following treatment with either conventional 50 or 100 mg oxiracetam or TiO2-nanowired 50 mg delivery in normal rats. The values of p-tau in the CSF were 13–17 pG/μL with oxiracetam treatment (Fig. 5).

A significant increase in CSF p-tau was seen progressively after CHI with increased survival period. Thus, 48 h CHI resulted in elevation of p-tau in the CSF to 30 pG/μL and 96 h after the value was 38 pG/μL. After 192 h of CHI p-tau was elevated in the CSF to 45 pG/μL ($P < 0.05$ from control) (Fig. 5).

Treatment with oxiracetam 50 and 100 mg significantly reduced the p-tau levels in the CSF to 28 and 24 pG/μL, respectively at 192 h CHI ($P < 0.05$ from CHI) (Fig. 5).

Interestingly TiO2-nanowired delivery of oxiracetam 50 mg dose markedly reduced the p-tau level in the CSF to 18 pG/μL at 192 h of CHI ($P < 0.05$ from CHI) (Fig. 5).

9.9 Oxiracetam and brain pathology

Brain pathology in CHI and oxiracetam treatment for neuronal and glial reactions was evaluated using light microscopy and myelin vesiculation and endothelial cell reactions were investigated by transmission electron microscopy.

a. *Light microscopy*

Using light microscopy neuronal and astrocytic changes were examined in eight different brain regions in CHI 192 h and its modification with Ti-Na or Ti-K nanowired delivery of oxiracetam under identical conditions.

Neuronal changes

The neuronal changes as seen using HE staining exhibited profound neuronal damages in CHI 192 h and nanodelivery of oxiracetam significantly attenuated these neuronal reactions effectively (Fig. 12). In general, the normal control animals exhibited only few altered neurons (2–4) in eight different brain areas examined. However, significant neuronal abnormalities were seen in animals after CHI 192 h. Thus, the cerebral cortex showed about 198 altered neurons in the RH and 234 abnormal neurons in the LH at this time of injury ($P < 0.05$ from control). These neuronal abnormalities include swollen or shrunken neurons with or without a visible nucleus with distinct nucleolus. The eccentric nucleolus was a common finding in CHI 192 h animals (Fig. 13). Perineuronal edema is quite frequent in both RH and LH at 192 h of CHI. Appearance of several dark and distorted neurons in the neuropil is clearly seen. The neuropil exhibits sponginess and edema and general expansion was also quite evident. Loss of neurons and density of nerve cells is quite reduced within the neuropil. These neuronal changes are quite frequent in almost all brain regions examined (Fig. 14). The LH appears to have more neuronal damage than the RH in most of the brain areas seen.

Hippocampus exhibited about 84 altered neurons in the RH and 104 nerve cells appeared abnormal in the LH. These hippocampal changes in nerve cells ware present in CA1, CA4 and dentate gyrus after CHI (results not shown). Cerebellum appears to be quite sensitive of CHI. This is evident from about 134 abnormal Purkinje or granule cells in the RH and 189 reactive cells in the LH (H.S. Sharma, unpublished observations). There were 138 abnormal neurons in thalamus RH and 189 nerve cells in LH were distorted. Hypothalamus showed 34 abnormal cells in RH and 47 distorted neurons in the LH ($P < 0.05$ from control). About 38 neurons were distorted in the caudate nucleus in RH and 48 nerve cells were looking abnormal in LH. The abnormal nerve cells in the medulla were 12 in RH and 18 in LH. The brain stem showed 8 abnormal nerve cells in RH and 14 in LH ($P < 0.05$ from control) (Figs. 12 and 14). This indicates that several brain regions are showing neuronal alterations following CHI over 192 h survival.

Treatment with TiO2-Na nanowired oxiracetam effectively reduced neuronal changes in several brain regions at 192 h of CHI (Figs. 12 and 14). Thus, the cerebral cortex showed only 46 abnormal neurons in RH and 67 altered nerve cells in LH ($P < 0.05$ from CHI). Several nerve cells were round in appearance and exhibited clear nucleus with a distinct nucleolus. The sponginess, expansion and edematous swelling or shrinkage of the neurons is much less evident. Appearances of dark and distorted nerve cells are much less frequent in nanowired oxiracetam treated group. The hippocampus showed only 23 distorted nerve cells in RH and 32 in LH that are sporadically appear in CA1, CA4 and dentate gyrus regions (Results not shown). The number of distorted cells were also much reduced in the cerebellum. Thus, only 28 abnormal Purkinje or granule cells are evident in RH and 32 distorted cells were seen in LH ($P < 0.05$ from CHI). Thalamus showed only 24 and 32 abnormal cells in RH and LH, respectively. In hypothalamus RH exhibited 12 distorted nerve cells and 16 abnormal cells were found in LH. About 8 cells were altered in RH caudate nucleus and LH showed 14 abnormal cells. Medulla exhibited 4

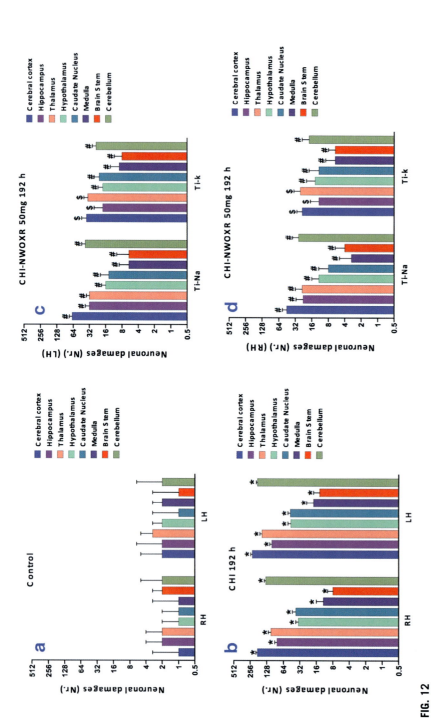

FIG. 12

Effect of Na-TiO2 vs K-TiO2 nanowired oxiracetam (NWOXR) (C, D) on regional neuronal damages in control (A) and following concussive head injury (CHI, B) in the rat on 8th day after trauma. CHI was inflicted using a 0.224 N impact injury over Right parietal skull and the animals were allowed to survive for 1 week after the primary insult. Oxiracetam (OXR) was given in a dose of 50 or 100mg/kg, i.p. once daily after 1 day of CHI for 5 days and the parameters were evaluated on the 8th day. For Na-TiO2 or K-TiO2 nanowired delivery of oxiracetam (NWOXR) 50mg/kg (i.p.) was given under identical conditions. For details see text. Values are Mean ± SD of 6–8 rats at each point. *MWM*, Morris Water Maze platform search; *OXR*, oxiracetam; *NWOXR*, nanowired oxiracetam; *TiO2-Na*, sodium titanate nanowires; *TiO2-K*, potassium titanate nanowires. Volume swelling (%) was calculated according to Elliott and Jasper (1949). * $P < 0.05$ from Control; § $P < 0.05$ from untreated CHI; # $P < 0.05$ from CHI+NWOXR (TiO2-Na) at respective group. ANOVA followed by Dunnett's test for multiple group comparison using one control. For details see text.

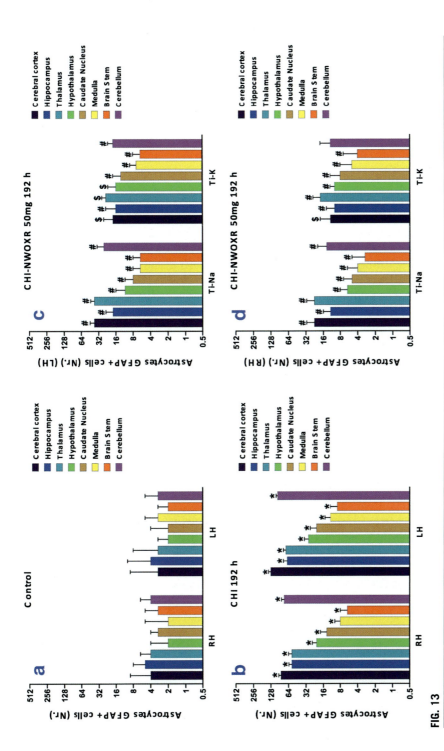

FIG. 13

Effect of Na-TiO2 vs K-TiO2 nanowired oxiracetam (NWOXR) (C, D) on regional astrocytes activation as seen with glial fibrillary acidic protein (GFAP) immunoreactivity in control (A) and following concussive head injury (CHI, B) in the rat on 8th day after trauma. CHI was inflicted using a 0.224N impact injury over Right parietal skull and the animals were allowed to survive for 1 week after the primary insult. Oxiracetam (OXR) was given in a dose of 50 or 100 mg/kg, i.p. once daily after 1 day of CHI for 5 days and the parameters were evaluated on the 8th day. For Na-TiO2 or K-TiO2 nanowired delivery of oxiracetam (NWOXR) 50 mg/kg (i.p.) was given under identical conditions. For details see text. Values are Mean ± SD of 6–8 rats at each point. *MWM*, Morris Water Maze platform search; *OXR*, oxiracetam; *NWOXR*, nanowired oxiracetam; *TiO2-Na*, sodium titanate nanowires; *TiO2-K*, potassium titanate nanowires. Volume swelling (%f) was calculated according to Elliott and Jasper (1949). * $P < 0.05$ from Control; § $P < 0.05$ from untreated CHI; # $P < 0.05$ from CHI+NWOXR (TiO2·Na) at respective group. ANOVA followed by Dunnett's test for multiple group comparison using one control. For details see text.

neuronal distortions in RH and LH showed 6 abnormal appearances of nerve cells. Medulla contains only 4 and 6 altered neurons in RH and LH, respectively (results not shown).

Interestingly, TiO2-K nanowired oxiracetam in identical doses appears to have superior neuroprotective effects in 192 h CHI. This is evident from 24 abnormal neurons in RH and 36 distorted nerve cells in LH in the cerebral cortex ($P < 0.05$ from TiO2-Na) (Fig. 12). The neuronal alterations in hippocampus were also significantly less in Ti-K nanowired oxiracetam as compared to Ti-Na nanowired drug under identical conditions. Thus, only 12 and 16 neurons showed abnormal reaction in hippocampus after CHI in Ti-K nanowired oxiracetam treated group ($P < 0.05$ from Ti-Na nanowired drug). Cerebellum showed 18 and 24 disrobed Purkinje or granule cells in RH and LH, respectively in this groups ($P < 0.05$ from Ti-K nanowired drug). About 26 and 34 damaged nerve cells were seen in thalamus in RH and LH, respectively. Hypothalamus exhibited almost similar number of nerve cell reaction in this group as in Ti-Na nanowired drug delivery. Caudate nucleus showed 12 and 21 abnormal cells in Ti-K nanowired drug group that is significantly higher than Ti-Na nanowired drug delivery. No apparent differences in nerve cell reactions in medulla and brainstem in RH or LH was seen in Ti-K or Ti-Na nanowired drug delivery at 192 h CHI (Fig. 13). This suggests that the composition of nanowires affects neuroprotection of oxiracetam.

Astrocytic changes

The astrocytic changes are seen using immunohistochemistry of GFAP. Our results showed only a few astrocytes are positive to GFAP immunoreactivity in different brain regions ranging from 3 to 6 cells in both RH and LH of normal control group.

The number of GFAP positive astrocytes significantly enhanced after CHI 192 h in almost all brain regions examined (Fig. 13). The LH in general sees greater number of GFAP-positive astrocytes than the RH (Figs. 13 and 15). Thus, the cerebral cortex showed 86 GFAP positive cells in RH while LH displayed 129 immunoreactive cells ($P < 0.05$ from control). In hippocampus about 56 astrocytic reactions in cells was seen whereas LH showed 67 GFAP positive cells in CA1, CA4 and dentate gyrus area. Cerebellum exhibited 76 reactive astrocytes in RH and 98 GFAP positive cells in LH. These astrocytic reactions are seen between both the Purkinje and granule cells. There were about 56 and 71 GFAP immunoreactive cells in the thalamus RH and LH, respectively ($P < 0.05$ from control). In hypothalamus 24 in RH and 29 in LH immunoreactive astrocytes were seen. Caudate nucleus exhibited GFAP positive cells about 14 in RH and 21 in LH. In medulla RH showed 8 GFAP positive astrocytes while 12 positive cells were present in RH. Brain stem RH had 6 GAP positive cells and LH displayed 9 such cells after CHI 192 h survival (Fig. 13).

Treatment with Ti-Na-nanowired oxiracetam significantly reduced the number of GFAP positive astrocytes in different brain regions after CHI 192 h. Thus, cerebral cortex showed only 23 in RH and 38 in LH GFAP immunoreactive astrocytes. Hippocampus positive astrocytes are limited to 12 in RH and 18 in LH. ($P < 0.05$ from CHI). The number of GFAP positive astrocytes were 14 in RH and 26 in LH between

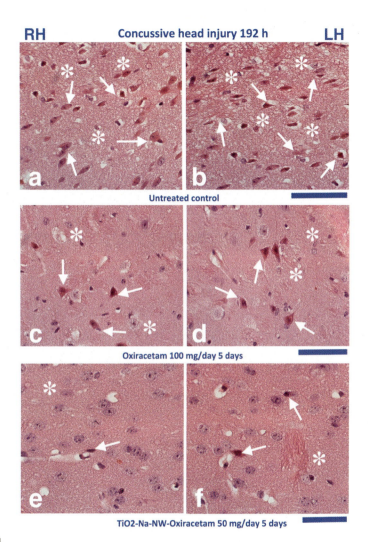

FIG. 14

High power light micrograph showing Hematoxylin and Eosin (H&E) stained 3-μm thick paraffin sections from 192 h concussive head injury in the injured parietal cerebral cortex of right hemisphere (RH, A) and the uninjured left hemisphere (LH, B) neuronal pathology and their modification with conventional oxiracetam (C, D) or titanium nanowired sodium (TiO2-Na-NW) oxiracetam (E, F). CHI exhibited neuronal pathology (arrows), sponginess, expansion (*) that was the most pronounced in the LH as compared to RH. Dark and distorted neurons with perineuronal edema is prominent in CHI. In conventional oxiracetam (100 mg) treatment neuronal pathology (arrows), expansion and sponginess (*) were reduced considerable (C. D). However, TiO2·Na-NW.oxiracetam (50 mg) induced superior neuroprotection in 192 h CHI. Neuronal pathology (arrows) and sponginess or edema (*) is much less seen in NW-oxiracetam treated CHI (e, f). Bars = 40 μm.

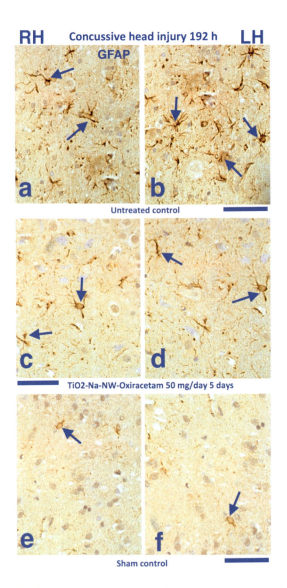

FIG. 15

High power light micrograph of glial fibrillary acidic protein (GFAP) immunoreactivity on 3-μm thick paraffin sections from 192h concussive head injury (CHI) in right hemisphere (RH) injured parietal cerebral cortex (A) and left hemisphere (LH) uninjured parietal cerebral cortex (B) and their modification with titanium nanowired sodium delivery of oxiracetam (TiO2-Na-NW-oxiracetam C, D) as compared to the sham control (E, F). CHI induced pronounced activation of astrocytes as evident with GFAP immunostaining around cells and microvessels (arrows) that is most prominent in the LH as compared to the RH. Treatment with TiO2-Na-NW-oxiracetam significantly attenuated GFAP activation (arrows) in the RH (C) and LH (D). Sham control RH (E) and LH (F) showed only mild GFAP staining of astrocytes in very few cells (arrows). Bars = 40 μm.

Purkinje and Granule cells in cerebellum. In thalamus RH showed 23 GFAP positive astrocytes whereas LH exhibited 38 immunoreactive cells. Hypothalamus displayed 6 positive GFAP cells in RH and 11 cells in LH. In caudate nucleus only 5 cells in RH and 8 cells in LH were GFAP positive. In medulla 4 in RH and 6 in LH whereas in brain stem 3 in RH and 6 in LH GFAP positive cells were seen (Fig. 13).

Interestingly, treatment with Ti-K nanowired oxiracetam appears to have superior neuroprotective effects of astrocytes in some brain regions at 192 h CHI. Thus, only 12 in RH and 18 in LH in cerebral cortex and 10 in RH and 16 in LH of hippocampus showed GFAP immunoreactive cells. Cerebellum showed 12 in RH and 18 in LH, while thalamus exhibited 18 and 24 GFAP-positive cells in TH and LH, respectively. Hypothalamus exhibited 10 in RH and 16 in LH whereas caudate nucleus showed 8 in RH and 13 in LH positive GFAP cells. In medulla 5 in RH and 7 in LH and in brainstem 4 in RH and 6 in LH GFAP immunoreactive cells were present. This observation is inline with the idea that Ti-Na and Ti-K nanowires affect the oxiracetam neuroprotective ability on astrocytes in CHI as well (Figs. 13 and 15).

b. *Electron microscopy*

Transmission electron microscopy (TEM) was used to evaluate myelin reaction and endothelial cell morphology at ultrastructural level in CHI and their modification with oxiracetam treatment.

Myelin alteration and vesiculation

We examined myelin changes showing vesiculation, alteration and splitting in areas of desired brain regions in RH and LH and counted manually. In control group of animals myelinated nerve fibers deformity was occasionally present in some cases. Thus, the myelin changes including vesiculation, or deformity was limited to 2–4 incidents in every brain areas examined. There were no differences between RH and LH regarding myelin damage in control group (Fig. 16).

Animals subjected to CHI exhibited significant structural changes and vesiculation of myelin in several brain regions progressively at 48, 96 and 192 h. These myelin changes include alteration in shape, splitting, vesiculation and degeneration located within the areas of neuropil exhibiting vacuolation and edema. In general these pathological changes in myelin at the ultrastructural level were more frequent in the LH as compared to RH within the same brain regions. Thus, cerebral cortex exhibited 34 instances of myelin damage in RH and 49 incidences of myelin pathology after CHI 192 h survival. At this time hippocampal myelin changes were seen in 8 instances in RH and 14 cases of myelin distortion in LH ($P < 0.05$ from control). Cerebellum showed 12 abnormal myelinated fibers in RH and 18 cases in LH. In thalamus 18 myelinated fibers appear distorted in RH and 29 damaged cases of myelin were seen in LH. Hypothalamus and caudate nucleus exhibited 6 and 8 cases of myelin vesiculation in RH and 10 and 12 cases of myelin damage in LH, respectively. Medulla and brain stem richly innervated by myelinated nerve fibers display large numbers of myelin damage in both RH and LH in CHI. Thus, medulla showed 26 instances of myelin damage in RH and 38 cases in LH regarding nerve fibers

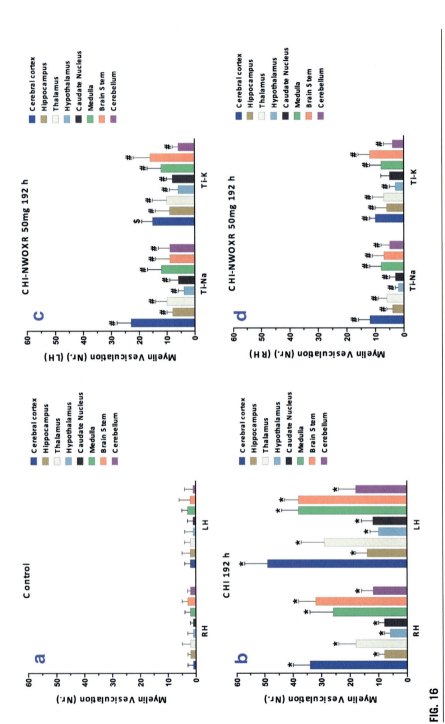

FIG. 16

Effect of Na-TiO2 vs K-TiO2 nanowired oxiracetam (NWOXR) (C, D) on regional myelin vesiculation in control (A) and following concussive head injury (CHI, B) in the rat on 8th day after trauma. CHI was inflicted using a 0.224 N impact injury over Right parietal skull and the animals were allowed to survive for 1 week after the primary insult. Oxiracetam (OXR) was given in a dose of 50 or 100 mg/kg, i.p. once daily after 1 day of CHI for 5 days and the parameters were evaluated on the 8th day. For Na-TiO2 or K-TiO2 nanowired delivery of oxiracetam (NWOXR) 50 mg/kg (i.p.) was given under identical conditions. For details see text. Values are Mean ± SD of 6–8 rats at each point. *MWM*, Morris Water Maze platform search; *OXR*, oxiracetam; *NWOXR*, nanowired oxiracetam; *TiO2-Na*, sodium titanate nanowires; *TiO2-K*, potassium titanate nanowires. Volume swelling (%f) was calculated according to Elliott and Jasper (1949). * $P < 0.05$ from Control; § $P < 0.05$ from untreated CHI; # $P < 0.05$ from CHI + NWOXR (TiO2-Na) at respective group. ANOVA followed by Dunnett's test for multiple group comparison using one control. For details see text.

distortion. In brain stem 32 myelinated fibers are distorted in RH and this was further increased in 38 instances in LH ($P < 0.05$ from control) (Fig. 16).

Treatment with nanowired delivery of oxiracetam significantly reduced CHI induced myelin damages in almost all brain regions examined at the ultrastructural level. Thus, the cerebral cortex showed only 12 in RH and 23 in LH cases of myelin splitting, vesiculation or damage ($P < 0.05$ from CHI). Hippocampus, Cerebellum and thalamus myelin damage was limited in this group by 4 to 6 incidences in RH and 8 to 10 cases in LH, respectively ($P < 0.05$ from CHI). Only 2 to 3 cases of myelin alterations was seen in hypothalamus and caudate nucleus in RH and only 4 to 6 incidences occur in LH in these regions, respectively after CHI 192h following Ti-Na nanowired oxiracetam treatment. Myelin damage was also limited in medulla and brain stem in this group. Thus, only 8 and 7 cases in RH and 12 and 9 cases in LH of myelin alterations were evident in medulla and brain stem, respectively ($P < 0.05$ from CHI) (Figs. 16 and 17).

Interestingly, Ti-K-nanowired delivery of oxiracetam under identical conditions in CHI appears to have superior protective effects on myelin in the cerebral cortex after CHI 192h (Figs. 18 and 19). Thus, only 10 instances of myelin disruption in RH and 15 cases of myelin damage in LH was seen in the cerebral cortex after CHI. However, the effects of Ti-K nanowired oxiracetam on myelin in hippocampus, cerebellum, thalamus, hypothalamus, caudate nucleus and medulla myelin protection after CHI was very similar to that of Ti-Na nanowired drug delivery ($P < 0.05$ from CHI). In brain stem, however, Ti-K nanowired oxiracetam the number of myelin damage although significantly lower than untreated CHI ($P < 0.05$ from CHI) but the numbers were higher than Ti-Na nanowired drug treated group. Thus, 12 cases of myelin damage in RH and 16 instances of myelin disruption in LH were seen in brain stem as compared to Ti-Na nanowired drug delivery ($P < 0.05$). This suggests that composition of nanowires may affect neuroprotection of oxiracetam (Fig. 16).

Cerebral endothelial cell abnormality

The endothelial cells in the brain of several regions are analyzed for collapse, structure anomaly, cellular vacuolation, perivascular edema causing capillary abnormality were examined at the ultrastructural level using TEM. Our results show that in control animals, endothelial cell abnormalities are largely absent. Screening of TEM grids from several brain regions for endothelial cell ultrastructure and capillary morphology, only two to three capillaries with abnormal endothelial cells were seen (Figs. 18 and 19).

However, the number of endothelial cell structural abnormalities is much more evident at 192h of CHI (Fig. 19). These endothelial cell abnormalities are present in both RH and LH of each brain region where LH showed greater capillaries or endothelial cell alterations than the RH of the brain areas. Thus, about 34 endothelial cells in capillaries of small or middle sizes exhibit abnormality in RH and 48 endothelial cells were altered in LH. These alterations include swelling, irregular lumen, vacuolation within the endothelial cell, cavity formation, perivascular edema and activated pericytes (Figs. 18 and 19). However, in many cases only one endothelial cell within the cerebral capillary showed these changes and the most often the

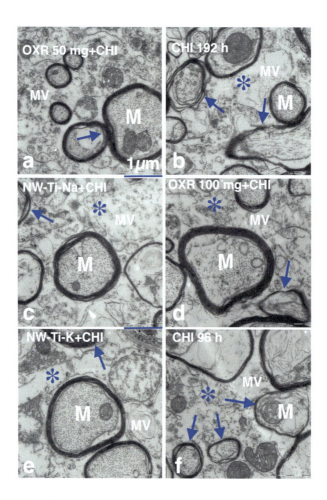

FIG. 17

High power transmission electron micrographs showing myelin (M) and membrane vacuolation (MV) after 192 h concussive head injury (CHI, B, F) and neuroprotection by oxiracetam (OXR) 50 mg (A), 100 mg (D) and titanium nanowired-Na-OXR (NW-Ti-Na) or NW-Ti-K-OXR (E). Membrane vacuolation (MV) and myelin damage (M) are prominent in CHI (B, F) and conventional OXR treatment induced profound neuroprotection in dose dependent manner (A, D). However, superior neuroprotection is seen following NW-Ti delivered OXR that was the most prominent in NW-Ti-K (E) as compared to NW-Ti-Na.OXR in CHI (c). Membrane vacuolation and edema (*) are considerable reduced in OXR treated CHI (A, C, D, E) as compared to untreated 192 h CHI group (B, F). Bars = 1 μm.

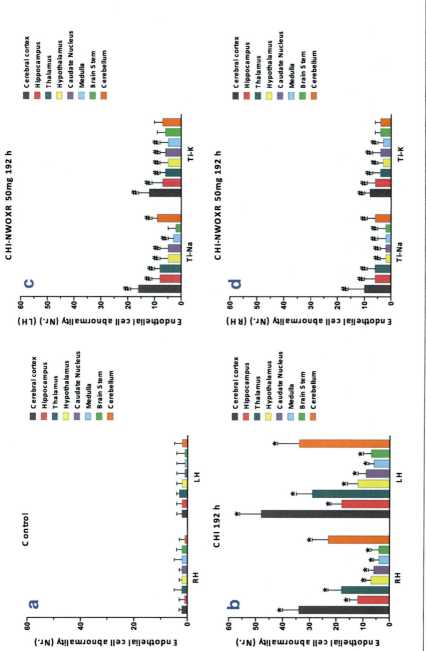

FIG. 18

Effect of Na-TiO2 vs K-TiO2 nanowired oxiracetam (NWOXR) (C, D) on regional endothelial cell deformity in control (A) and following concussive head injury (CHI, B) in the rat on 8th day after trauma. CHI was inflicted using a 0.224N impact injury over Right parietal skull and the animals were allowed to survive for 1 week after the primary insult. Oxiracetam (OXR) was given in a dose of 50 or 100mg/kg, i.p. once daily after 1 day of CHI for 5 days and the parameters were evaluated on the 8th day. For Na-TiO2 or K-TiO2 nanowired delivery of oxiracetam (NWOXR) 50mg/kg (i.p.) was given under identical conditions. For details see text. Values are Mean±SD of 6–8 rats at each point. *MWM*, Morris Water Maze platform search; *OXR*, oxiracetam; *NWOXR*, nanowired oxiracetam; *TiO2-Na*, sodium titanate nanowires; *TiO2-K*, potassium titanate nanowires. Volume swelling (%f) was calculated according to Elliott and Jasper (1949). * $P < 0.05$ from Control; # $P < 0.05$ from untreated CHI; § $P < 0.05$ from CHI+NWOXR (TiO2-Na) at respective group. ANOVA followed by Dunnett's test for multiple group comparison using one control. For details see text.

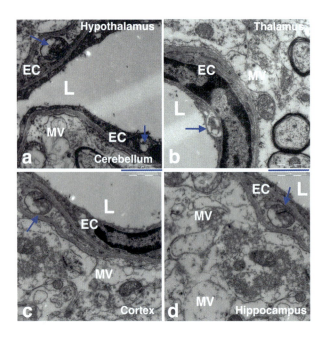

FIG. 19

High power Transmission electron micrographs showing endothelial cell (EC) deformity and membrane vacuolation (MV) in hypothalamus, cerebellum (A), thalamus (B), parietal cerebral cortex (C) and hippocampus (D). The luminal (L) surface of the endothelium show vacuolation (arrow) that is prominent within the endothelial cells (arrows) in concussive head injury (CHI) at 192 h after initial insults in the left hemisphere. Altered pericytes structure is also seen in some cerebral capillaries (B, C). Perivascular edema is also prominent. Bars = 2 μm.

adjacent endothelial cell looks normal in appearance (H.S. Sharma, unpublished observations). The second high number of endothelial cell alteration was seen in the cerebellum (Fig. 18). Thus, about 23 endothelial cells in RH capillaries and 34 cells in LH microvessels were abnormal in appearance ($P < 0.05$ from control). This was followed by thalamus showing 18 abnormal endothelial cells in RH and 29 altered endothelial cells in LH. In hippocampal capillaries 12 endothelial cells look distorted in RH and 18 abnormal cells were seen in LH after 192 h of CHI ($P < 0.05$ from control). Hypothalamus showed 7 altered endothelial cells in RH and 12 endothelial cells exhibited structural anomalies in LH. About 6 endothelial cells were damaged in RH and 9 in LH of caudate nucleus capillaries. Medulla showed 4 abnormal endothelial cells in RH and 6 altered cells in LH. Whereas brain stem exhibited 4 altered endothelial cell in capillaries in RH and 7 distorted endothelial cells in LH after 192 h of CHI (Fig. 18).

Treatment with TiO2 nanowired oxiracetam significantly reduced CHI induced endothelial cells abnormalities in several brain regions after 192 h CHI. Thus, the cerebral cortex displayed only 10 abnormal endothelial cells in RH and 16 altered endothelial cells in LH ($P < 0.05$ from CHI). Hippocampus, cerebellum and thalamus each showed only 6 abnormal endothelial cells in capillaries of RH while in LH 8, 9 and 8 endothelial cells, respectively were seen distorted in appearance. Hypothalamus, caudate nucleus, medulla and brain stem each exhibited 2-endothelial cells distortion in RH where as in LH 5, 5, 3 and 2 endothelial cells showed structural anomalies, respectively after 192 h of CHI (Figs. 18 and 19).

Interestingly, treatment with Ti-K-nanowired oxiracetam further reduced the number of endothelial cells abnormalities in the cerebral cortex in RH (8 cells) and in LH (12 cells), there were not so much differences in endothelial cell abnormalities in hippocampus, cerebellum, thalamus, hypothalamus, caudate nucleus and medulla in CHI as compared to Ti-Na nanowired oxiracetam treatment. There was a slight increase in number of endothelial cell abnormalities in brain stem in RH (4 cells) and in LH (6 cells) in Ti-K nanowired oxiracetam treatment than Ti-Na nanowired delivery of oxiracetam in CHI (Figs. 18 and 19). This indicates that composition of nanowires are important factors in drug delivery.

10 Discussion

The salient feature of this investigation show that oxiracetam when administered 48 h after injury daily for 5 days is capable to induce neuroprotection in CHI seen 1 week after injury. These observations are the first to show that oxiracetam is a neuroprotective agent in CHI not reported earlier. This effect of oxiracetam is dose related. This is evident from the findings that administration of 100 mg dose of oxiracetam induces better neuroprotective effects in CHI at 192 h as compared to 50 mg drug in identical situation. Furthermore when the oxiracetam is delivered through TiO2 nanowired delivery in 50 mg doses for 5 days this treatment induced superior neuroprotection in CHI than 100 mg conventional oxiracetam. This suggests for the first time that TiO2-nanowired delivery of oxiracetam in small doses is quite effective as neuroprotective agent in CHI up to 192 h of survival. This finding indicates that nanowired delivery of oxiracetam enhances its neuroprotective effects in CHI, not described earlier.

From our studies this is unclear whether TiO2 nanowired delivery of 50 mg oxiracetam could further extend its neuroprotective ability beyond 192 h CHI. It would be interesting to see whether this dose of nanowired delivery of oxiracetam can also induce neuroprotection in CHI after 2 weeks (384 h) of trauma. In addition, if we elevate the dose of TiO2 nanowired delivery of oxiracetam to 100 mg whether effective neuroprotection could be seen beyond 2 or 4 weeks of survival (768 h). To examine these issues further investigations are needed in this model of CHI.

Another novel information came out from this study is the finding that Ti-K nanowired delivery of oxiracetam in identical conditions (50 mg) enhanced some aspects of neuroprotection in CHI 192 h as compared to Ti-Na nanowired delivery of the drug. This suggests for the first time that the composition of the nanowires is significantly influencing the drug ability to induce neuroprotection. Based on these data it is interesting to examine whether nanowired prepared from other nanomaterials, e.g., Au, Ag, Cu or Al could also influence the neuroprotective capacity of oxiracetam in CHI. Our laboratory is currently engaged in finding answers to this question in CHI.

Nanowired delivery of oxiracetam has superior neuroprotective effects. This suggests that nanowired delivery could reach brain tissues faster and for longer duration that is needed for superior neuroprotection (Sharma et al., 2006). There are several reasons to believe that nanowires loaded with drug could penetrate endothelial cell membrane by piercing and could reach to the brain extracellular fluid environment rapidly (Sharma et al., 1997b; Tian et al., 2012). From there they could also target neurons, astrocytes or microglia where they can release the drugs either around them or intracellularly be piercing their cell membranes. Thus, nanowired delivery of oxiracetam could work like mini osmotic pump regarding release of drugs slowly for long duration. Since the drug is attached to nanowires while passing through various biological membranes, endogenous catabolic enzymes are unable to metabolize the drug before it reaches to its target tissue (Sharma et al., 1997b, 2009a,b; Sharma et al., 2016; Sharma et al., 2019a,b; Sharma et al., 2020a,b,c,d). These factors seem to play crucial role for exerting superior neuroprotective effects of the nanowired delivery of oxiracetam as compared to its conventional delivery.

We have observed that Ti-K nanowired delivery of oxiracetam appears to have some significant differences in inducing neuroprotection in CHI as compared to Ti-Na nanowired delivery of drugs. Thus, in some brain regions, e.g., cerebral cortex, cerebellum and hippocampus Ti-K nanowired delivery has improved neuroprotection where as in other brain regions such as medulla, brain stem, and this effect was not seen. In rest of the brain regions there was no differences in brain protection between Ti-K and Ti-Na nanowired delivery of oxiracetam in CHI. These observations suggests that the composition of nanowires for drug delivery affect biological process of neuroprotection. The probable mechanisms of such differences in neuroprotective ability of nanowires are not well known. However, it appears that the differences in XRD patterns of Ti-Na and Ti-K as well as their theta potential may affect their ability in neuroprotection by drugs. It would be interesting to see whether these differences in neuroprotection is either due to the characteristics of nanowires itself or the nanowires alter the drug ability in some way when delivered though these nanowires affecting neuroprotection. This is a subject that requires additional investigation using different drug types to these nanowires in CNS injury.

The probable mechanisms of oxiracetam-induced neuroprotection in CHI are still unclear from this investigation. However, the data reported here clearly suggests that oxiracetam reduced BBB breakdown after CHI to protein tracers and also improved regional CBF. Ischemia caused by CHI is due to release of several vasoconstriction agents in the blood leading to ischemic injury (Sharma et al., 2016; Sharma and Westman, 2004; Sharma, 2004, 2009). Regional ischemia is further complicated with breakdown of the BBB to large molecule tracers (Sharma and Dey, 1986a,b, 1987, 1988). Although, the regional increase in BBB breakdown is largely associated with the regional CBF decrease, however, the magnitude of CBF reduction alone is unlikely to increase BBB disruption (Sharma et al., 1990a,b; Sharma, 2004). There are reasons to believe that CHI induced vasoactive substances and neurochemicals are responsible for reduction in the CBF (Sharma et al., 1990a,b). Increased serotonin, prostaglandin, histamine, generation of nitric oxide and carbon monoxide all could affect disruption of the BBB and CBF (Sharma et al., 2007a,b, 1992a,b, 1995, 1996; Sharma, 1999, 2004, 2009).

Leakage of plasma proteins into the brain fluid microenvironment associated with ischemia leads to brain edema formation (Sharma and Cervós-Navarro, 1990; Sharma et al., 1991). Increased brain edema formation, reduction in the CBF and alterations in brain fluid microenvironment all could lead to cellular changes in the neurons, glial cells and nerve fibers leading to brain pathology. Progressive brain pathology over 192 h of CHI suggests that the pathological processes after trauma is continuously advancing with duration of survival. This Ida is supported by progressive increase in the BBB and brain edema formation associated with brain pathology in CHI.

Morphological alterations in neurons following CHI at 192 h shows marked nerve cell damage that is spread all over the brain along with astrocytic activation of GFAP. These neuronal and astrocytic pathological alterations are predominantly found in the cerebral cortex, cerebellum, hippocampus and thalamus indicating that these brain areas are highly affected by CHI and alterations in brain fluid microenvironment. This is supported by higher increase in the BBB breakdown and brain edema formation together with reduction in the CBF. These arrases also exhibited greater brain pathology as compared to other regions in the brain. These observations support the idea that BBB breakdown is the gateway of all pathological changes in the brain (Sharma, 2009).

The probable mechanism behind oxiracetam-induced neuroprotection is related to its ability in reducing the BBB breakdown in CHI. Obviously BBB disruption is attenuated by oxiracetam in CHI leading to neuroprotection. Oxiracetam is a nootropic compound and able to influence cellular and molecular mechanisms by modulating cytokines, oxidative stress parameters and strengthening cellular membrane function (Li et al., 2017; Malykh and Sadaie, 2010; Ponzio et al., 1989; Wang et al., 2019). Thus, there are evidences that oxiracetam works like a potent antioxidant and also function as neurotrophin like molecule at the cellular level. This suggests that oxiracetam has the capacity to induce significant neuroprotection in CHI.

Our study shows that oxiracetam when administered for 5 days after 48 h CHI resulted in effective neuroprotection on the 8th day. This effect of oxiracetam was the most pronounced following nanowired delivery of the drug. In our study, oxiracetam was given 48 h after trauma for 5 days and stopped on 7th day of CHI that resulted in marked neuroprotection on the 8th day. It is still not clear how long after cessation of oxiracetam administration could still induce neuroprotection in CHI. This information is important in clinical settings for neurorecovery of military personnel during combat injuries. This is because of the fact that when soldiers are injured in the battlefield and need to be transported home for super specialty neurotrauma hospital the therapeutic window for oxiracetam treatment and cessation could play critical role in maintenance of their brain function during transit. This is a feature urgently being investigated in our laboratory.

Previous observations from our laboratory suggests that traumatic brain or spinal cord injuries induces pathological changes in the nervous system through nitric oxide medaled mechanism (Sharma et al., 1996, 2000). Nitric oxide is a free radical gas that indices cell injury by causing cell membrane disruption affecting BBB breakdown, edema formation and neuronal damage (Sharma et al., 1998a,b). Release of nitric oxide gas within the CNS is difficult to measure because of its ultrashort half-life of 2–5 s (Sharma et al., 1998a,b). However, its metabolites nitrite and nitrate are stable and could be measured using biochemical approach. Our results are the first to show that CHI of 192 h significantly increased nitrite and nitrate in the CSF indicating release of nitric oxide in the brains that could be responsible for oxidative stress and brain pathology. Treatment of oxiracetam significantly reduced the CSF nitrite and nitrate levels indication that the drug treatment significantly reduced increased release of nitric oxide following CHI. These observations are in line with the idea that oxiracetam antagonized oxidative stress mechanisms in CHI by attenuating nitric oxide production within the brain as reflected by reduced CSF levels of its metabolites, i.e., nitrite and nitrate.

Oxiracetam also influences harmful cytokine levels in cell culture and enhance neuroprotection. This hypothesis is further supported from our findings of reduced neurodestructive cytokines IL-6 and TNF-a in the CSF of CHI animals at 192 h survival. Increased levels of IL-6 and/or TNF-a induce brain pathology in brain injury, stroke or neurodegenerative diseases (Sharma et al., 2020a,b,c,d, 2003). TNF-a and IL-6 elevation is interrelated with nitric oxide function. In spinal cord injury, topical application of antibodies to TNF-a induced neuroprotection and inhibited upregulation of nitric oxide synthase within the spinal cord (Sharma et al., 2003). Elevated IL-6 occurs following blast brain injury that is significantly attenuated by nanowired cerebrolysin-a balanced composition of several neurotrophic factor and active peptide fragments (Muresanu et al., 2020). These observations support the idea that neurotrophins that stabilize cell membranes induce neuroprotection in brain injury. A reduction in IL-6 and TNF-a in the CSF at 192 h of CHI in oxiracetam treated group suggests that the drug is capable to reduce cellular mechanisms leading to production of neurotoxic cytokines in brain injury.

Another evidence of neuroprotection by oxiracetam comes from the findings of reduced p-tau accumulation in the CSF of CHI group at 192 h of survival. Brain injury, Parkinson's disease or Alzheimer's disease are well known to enhance p-tau accumulation in the brain or CSF leading to cell pathology. There are reasons to believe that p-tau is neurotoxic to the cellular microenvironment leading to neurodegeneration. Several reports indicate that brain injury is a predisposing factor for Parkinson's and Alzheimer' disease. Thus, accumulation of p-tau together with amyloid beta peptide (AbP) is quite common in brain injury and Alzheimer's or Parkinson's disease (Ozkizilcik et al., 2018a,b; Ozkizilcik et al., 2019) In present study significant higher increase in p-tau is seen in the CSF of CHI rats at 192 h and oxiracetam is able to decrease the p-tau content within the CSF after CHI. This indicates that biological mechanisms responsible for elevation of p-tau in the CSF after CHI is markedly inhibited by oxiracetam administration. These results support the idea that oxiracetam may be able to induce neuroprotection in Parkinson's or Alzheimer's disease as well. However, this is a new feature and requires additional investigation.

Another possibility of reduced CSF levels of nitrite, nitrate, IL-6, TNF-a and p-tau by oxiracetam could be due to significant reduction in the BBB or blood-CSF-barrier (BCSFB) breakdown in CHI (Sharma and Westman, 2004). Obviously, strengthening of the BCSFB could result in blockade of these agents to enter CSF compartment from the brain compartment. In addition, if these agents are in part coming from vascular compartments to the brain and into the CSF after CHI is thus restricted by the strengthened BCSFB. Further studies are needed to clarify these points.

Oxiracetam is well known to enhance alertness, memory and cognitive functions. These functions are severely compromised in CHI. Thus, we also examined the effects of oxiracetam on sensory-motor, cognitive and memory functions in CHI. Our observations clearly show that oxiracetam effectively improves memory, sensory motor and cognitive functions. Memory, cognitive and sensory motor functions are regulated by the neuronal networks within the brain. These brain structures such as hippocampus, thalamus, caudate nucleus, cerebellum and cerebral cortex are associated with stress, memory, sensory-motor and cognitive functions (Hainmueller and Bartos, 2018; Neves et al., 2008). When these brain regions are damaged by CHI it is obvious that memory, cognitive and sensory-motor functions are compromised. Treatment with nanowired oxiracetam significantly attenuated brain pathology in these areas resulting in improved network connections. Reduction in neuronal, glial, axonal and endothelial cell pathology in these brain areas by oxiracetam in CHI results in improved performances on memory, cognitive and sensory-motor functions. This is further evident from the finding that Ti-Na or Ti-K nanowired delivery of oxiracetam reduces brain pathology in CHI shows close correspondence with behavioral changes seen on memory, cognitive and sensory-motor functions. This indicates that perturbation of neuronal network following brain pathology caused by CHI is the key factor in impairing sensory-motor, memory and cognitive functions. Disruption of the BBB and formation of vasogenic brain edema appears to be the key factors in inducing brain pathology. Oxiracetam significantly reduces the BBB breakdown to protein tracers and reduced brain edema formation leading to neuroprotection in CHI. These effects of oxiracetam appears to be directly related with improvement of behavioral and higher mental functions in CHI.

Apart from brain injury, trauma to the spinal cord also significantly induces sensory-motor and cognitive dysfunctions (Sahib et al., 2019, 2020). It would be interesting to investigate the effects of oxiracetam in spinal cord injury induced cord pathology, spinal edema and motor functions. This is a feature currently being examined in our laboratory.

11 Conclusion

In conclusion, our observation is the first to show that oxiracetam has powerful neuroprotective effects in CHI. This effect is further enhanced by TiO2-nanowired delivery of oxiracetam following CHI. Our results further show that Ti-Na or Ti-K nanowired delivery of the drug has different level of neuroprotection in CHI, not reported before. The probable mechanisms of oxiracetam induced neuroprotection is caused by the ability of the drug to attenuate BBB disruption in CHI. Reduction in the BBB breakdown will attenuate edema formation and cell injury. It appears that oxiracetam could affect oxidative stress in CHI resulting in reduction in nitric oxide mediated mechanism of brain pathology. Oxiracetam is also able to reduce BCSFB breakdown in CHI resulting in attenuation of neurotoxic cytokines like TNF-a and IL-6 and prevented p-tau accumulation in the CSF. Reduction in cellular injury or damage appears to be crucial for enhancing memory, sensory motor and cognitive functions in CHI by oxiracetam. These observations suggest that oxiracetam as a powerful neuroprotective agents depending on its dose and mode of delivery. It appears that nanowired delivery of oxiracetam has superior neuroprotective effects in CHI induced brain pathology and functional disabilities.

12 Future perspectives

It would be interesting to see whether combination of nanowired delivery of oxiracetam and DL-NBP together could further enhance the neuroprotective effects and time window in reducing brain pathology and biobehavioral functions in CHI. In addition, the powerful neuroprotective effects of oxiracetam alone or together with NBP need to be investigated in the pathological processes of AD and PD induced brain pathology and functional disabilities associated with these neurodegenerative diseases. These are the points that are currently being investigated in our laboratory.

Acknowledgments

This investigation is supported by grants from the Ministry of Science & Technology, People Republic of China, the Air Force Office of Scientific Research (EOARD, London, UK), and Air Force Material Command, USAF, under grant number FA8655-05-1-3065; Grants from the Alzheimer's Association (IIRG-09-132087), the National Institutes of Health (R01 AG028679) and the Dr. Robert M. Kohrman Memorial Fund (RJC); Swedish Medical Research Council (Nr 2710-HSS), Göran Gustafsson Foundation, Stockholm, Sweden (HSS), Astra Zeneca, Mölndal, Sweden (HSS/AS), The University Grants Commission, New Delhi, India (HSS/AS), Ministry of Science & Technology, Govt. of India (HSS/AS), Indian Medical Research Council, New Delhi, India (HSS/AS) and India-EU Co-operation Program (RP/AS/HSS) and IT-901/16 (JVL), Government of Basque Country and PPG 17/51 (JVL), JVL thanks to the support of the University of the Basque Country (UPV/EHU) PPG 17/51 and 14/08, the Basque Government (IT-901/16 and CS-2203) Basque

Country, Spain; and Foundation for Nanoneuroscience and Nanoneuroprotection (FSNN), Romania. Technical and human support provided by Dr. Ricardo Andrade from SGIker (UPV/EHU) is gratefully acknowledged. Dr. Seaab Sahib is supported by Research Fellowship at the University of Arkansas Fayetteville AR by Department of Community Health; Middle Technical University; Wassit; Iraq, and The Higher Committee for Education Development in Iraq; Baghdad; Iraq. We thank Suraj Sharma, Blekinge Inst. Technology, Karlskrona, Sweden and Dr. Saja Alshafeay, University of Arkansas Fayetteville, Fayetteville AR, USA for computer and graphic support. The U.S. Government is authorized to reproduce and distribute reprints for Government purpose notwithstanding any copyright notation thereon. The views and conclusions contained herein are those of the authors and should not be interpreted as necessarily representing the official policies or endorsements, either expressed or implied, of the Air Force Office of Scientific Research or the U.S. Government.

Conflict of interest

There is no conflict of interest between any entity and/or organization mentioned here.

References

Ahmad, S.S., Khan, S., Kamal, M.A., Wasi, U., 2019. The structure and function of alpha, beta and gamma-secretase as therapeutic target enzymes in the development of Alzheimer's disease: a review. CNS Neurol. Disord. Drug Targets 18 (9), 657–667. https://doi.org/10.2174/1871527318666191011145941.

Al-Dahhak, R., Khoury, R., Qazi, E., Grossberg, G.T., 2018. Traumatic brain injury, chronic traumatic encephalopathy, and Alzheimer disease. Clin. Geriatr. Med. 34 (4), 617–635. https://doi.org/10.1016/j.cger.2018.06.008.

Al-Mufti, F., Amuluru, K., Changa, A., Lander, M., Patel, N., Wajswol, E., Al-Marsoummi, S., Alzubaidi, B., Singh, I.P., Nuoman, R., Gandhi, C., 2017. Traumatic brain injury and intracranial hemorrhage-induced cerebral vasospasm: a systematic review. Neurosurg. Focus 43 (5), E14. https://doi.org/10.3171/2017.8.FOCUS17431.

Al-Shawi, R., Tennent, G.A., Millar, D.J., Richard-Londt, A., Brandner, S., Werring, D.J., Simons, J.P., Pepys, M.B., 2016. Pharmacological removal of serum amyloid P component from intracerebral plaques and cerebrovascular Abeta amyloid deposits in vivo. Open Biol. 6 (2), 150202. https://doi.org/10.1098/rsob.150202.

Alavi Naini, S.M., Soussi-Yanicostas, N., 2015. Tau hyperphosphorylation and oxidative stress, a critical vicious circle in neurodegenerative tauopathies? Oxid. Med. Cell. Longev. 2015, 151979. https://doi.org/10.1155/2015/151979.

Alawieh, A., Langley, E.F., Weber, S., Adkins, D., Tomlinson, S., 2018. Identifying the role of complement in triggering neuroinflammation after traumatic brain injury. J. Neurosci. 38 (10), 2519–2532. https://doi.org/10.1523/jneurosci.2197-17.2018.

Albayram, O., Herbert, M.K., Kondo, A., Tsai, C.Y., Baxley, S., Lian, X., Hansen, M., Zhou, X.Z., Lu, K.P., 2016. Function and regulation of tau conformations in the development and treatment of traumatic brain injury and neurodegeneration. Cell Biosci. 6, 59. https://doi.org/10.1186/s13578-016-0124-4.

Alzheimer's, A, 2015. 2015 Alzheimer's disease facts and figures. Alzheimers Dement. 11 (3), 332–384. https://doi.org/10.1016/j.jalz.2015.02.003.

Alzheimer's, A, 2016. 2016 Alzheimer's disease facts and figures. Alzheimers Dement. 12 (4), 459–509. https://doi.org/10.1016/j.jalz.2016.03.001.

Alzheimer's Association Report, 2020. Free Access, Alzheimer's disease facts and figures. First published: 10 March 2020. https://doi.org/10.1002/alz.12068.

Andorfer, C., Kress, Y., Espinoza, M., de Silva, R., Tucker, K.L., Barde, Y.A., Duff, K., Davies, P., 2003. Hyperphosphorylation and aggregation of tau in mice expressing normal human tau isoforms. J. Neurochem. 86 (3), 582–590. https://doi.org/10.1046/j.1471-4159.2003.01879.x.

Arendt, T., Holzer, M., Fruth, R., Bruckner, M.K., Gartner, U., 1998. Phosphorylation of tau, abeta-formation, and apoptosis after in vivo inhibition of PP-1 and PP-2A. Neurobiol. Aging 19 (1), 3–13. https://doi.org/10.1016/s0197-4580(98)00003-7.

Asiimwe, N., Yeo, S.G., Kim, M.S., Jung, J., Jeong, N.Y., 2016. Nitric oxide: exploring the contextual link with Alzheimer's disease. Oxid. Med. Cell. Longev. 2016, 7205747. https://doi.org/10.1155/2016/7205747.

Atri, A., 2019. The Alzheimer's disease clinical spectrum: diagnosis and management. Med. Clin. North Am. 103 (2), 263–293. https://doi.org/10.1016/j.mcna.2018.10.009.

Avila, J., Lucas, J.J., Perez, M., Hernandez, F., 2004. Role of tau protein in both physiological and pathological conditions. Physiol. Rev. 84 (2), 361–384. https://doi.org/10.1152/physrev.00024.2003.

Azouvi, P., Arnould, A., Dromer, E., Vallat-Azouvi, C., 2017. Neuropsychology of traumatic brain injury: an expert overview. Rev. Neurol. (Paris) 173 (7–8), 461–472. https://doi.org/10.1016/j.neurol.2017.07.006.

Babić Leko, M., Nikolac Perković, M., Klepac, N., Štrac, D., Borovečki, F., Pivac, N., Hof, P.R., Šimić, G., 2020. IL-1β, IL-6, IL-10, and TNFα single nucleotide polymorphisms in human influence the susceptibility to Alzheimer's disease pathology. J. Alzheimers Dis. 75 (3), 1029–1047. https://doi.org/10.3233/jad-200056.

Baglietto-Vargas, D., Shi, J., Yaeger, D.M., Ager, R., LaFerla, F.M., 2016. Diabetes and Alzheimer's disease crosstalk. Neurosci. Biobehav. Rev. 64, 272–287. https://doi.org/10.1016/j.neubiorev.2016.03.005.

Bagyinszky, E., Kang, M.J., Van Giau, V., Shim, K., Pyun, J.M., Suh, J., An, S.S.A., Kim, S., 2019. Novel amyloid precursor protein mutation, Val669Leu ("Seoul APP"), in a Korean patient with early-onset Alzheimer's disease. Neurobiol. Aging 84, 236.e231–236.e237. https://doi.org/10.1016/j.neurobiolaging.2019.08.026.

Banerjee, A., Qi, J., Gogoi, R., Wong, J., Mitragotri, S., 2016. Role of nanoparticle size, shape and surface chemistry in oral drug delivery. J. Control. Release 238, 176–185. https://doi.org/10.1016/j.jconrel.2016.07.051.

Bayat, M., Sharifi, M.D., Haghani, M., Shabani, M., 2015. Enriched environment improves synaptic plasticity and cognitive deficiency in chronic cerebral hypoperfused rats. Brain Res. Bull. 119 (Pt. A), 34–40. https://doi.org/10.1016/j.brainresbull.2015.10.001.

Becerril-Ortega, J., Bordji, K., Freret, T., Rush, T., Buisson, A., 2014. Iron overload accelerates neuronal amyloid-beta production and cognitive impairment in transgenic mice model of Alzheimer's disease. Neurobiol. Aging 35 (10), 2288–2301. https://doi.org/10.1016/j.neurobiolaging.2014.04.019.

Becker, R.E., Kapogiannis, D., Greig, N.H., 2018. Does traumatic brain injury hold the key to the Alzheimer's disease puzzle? Alzheimers Dement. 14 (4), 431–443. https://doi.org/10.1016/j.jalz.2017.11.007.

Bellucci, A., Bugiani, O., Ghetti, B., Spillantini, M.G., 2011. Presence of reactive microglia and neuroinflammatory mediators in a case of frontotemporal dementia with P301S mutation. Neurodegener Dis 8 (4), 221–229. https://doi.org/10.1159/000322228.

Benarroch, E.E., 2018. Glutamatergic synaptic plasticity and dysfunction in Alzheimer disease: emerging mechanisms. Neurology 91 (3), 125–132. https://doi.org/10.1212/WNL. 0000000000005807.

Bertuccio, A.J., Tilton, R.D., 2017. Silver sink effect of humic acid on bacterial surface colonization in the presence of silver ions and nanoparticles. Environ. Sci. Technol. 51 (3), 1754–1763. https://doi.org/10.1021/acs.est.6b04957.

Bezprozvanny, I., Mattson, M.P., 2008. Neuronal calcium mishandling and the pathogenesis of Alzheimer's disease. Trends Neurosci. 31 (9), 454–463. https://doi.org/10.1016/j.tins. 2008.06.005.

Bhaskar, K., Maphis, N., Xu, G., Varvel, N.H., Kokiko-Cochran, O.N., Weick, J.P., Staugaitis, S.M., Cardona, A., Ransohoff, R.M., Herrup, K., Lamb, B.T., 2014. Microglial derived tumor necrosis factor-α drives Alzheimer's disease-related neuronal cell cycle events. Neurobiol. Dis. 62, 273–285. https://doi.org/10.1016/j.nbd.2013.10.007.

Blades, M., 2015. Diet and Alzheimer's disease. Perspect. Public Health 135 (2), 65. https:// doi.org/10.1177/1757913915569961.

Blennow, K., Hardy, J., Zetterberg, H., 2012. The neuropathology and neurobiology of traumatic brain injury. Neuron 76 (5), 886–899. https://doi.org/10.1016/j.neuron.2012. 11.021.

Blennow, K., Brody, D.L., Kochanek, P.M., Levin, H., McKee, A., Ribbers, G.M., Yaffe, K., Zetterberg, H., 2016. Traumatic brain injuries. Nat. Rev. Dis. Primers. 2, 16084. https:// doi.org/10.1038/nrdp.2016.84.

Bloom, G.S., 2014. Amyloid-beta and tau: the trigger and bullet in Alzheimer disease pathogenesis. JAMA Neurol. 71 (4), 505–508. https://doi.org/10.1001/jamaneurol.2013.5847.

Blue, E.E., Horimoto, A., Mukherjee, S., Wijsman, E.M., Thornton, T.A., 2019. Local ancestry at APOE modifies Alzheimer's disease risk in Caribbean Hispanics. Alzheimers Dement. 15 (12), 1524–1532. https://doi.org/10.1016/j.jalz.2019.07.016.

Bonda, D.J., Lee, H.G., Blair, J.A., Zhu, X., Perry, G., Smith, M.A., 2011. Role of metal dyshomeostasis in Alzheimer's disease. Metallomics 3 (3), 267–270. https://doi.org/10.1039/ c0mt00074d.

Bonda, D.J., Liu, G., Men, P., Perry, G., Smith, M.A., Zhu, X., 2012. Nanoparticle delivery of transition-metal chelators to the brain: oxidative stress will never see it coming. CNS Neurol. Disord. Drug Targets 11 (1), 81–85. https://doi.org/10.2174/187152712799960709.

Bonnet, C., Boucher, D., Lazereg, S., Pedrotti, B., Islam, K., Denoulet, P., Larcher, J.C., 2001. Differential binding regulation of microtubule-associated proteins MAP1A, MAP1B, and MAP2 by tubulin polyglutamylation. J. Biol. Chem. 276 (16), 12839–12848. https://doi. org/10.1074/jbc.M011380200.

Braak, H., Braak, E., 1991. Neuropathological stageing of Alzheimer-related changes. Acta Neuropathol. 82 (4), 239–259. https://doi.org/10.1007/BF00308809.

Brambilla, D., Le Droumaguet, B., Nicolas, J., Hashemi, S.H., Wu, L.P., Moghimi, S.M., Couvreur, P., Andrieux, K., 2011. Nanotechnologies for Alzheimer's disease: diagnosis, therapy, and safety issues. Nanomedicine 7 (5), 521–540. https://doi.org/10.1016/j.nano. 2011.03.008.

Bredesen, D.E., Amos, E.C., Canick, J., Ackerley, M., Raji, C., Fiala, M., Ahdidan, J., 2016. Reversal of cognitive decline in Alzheimer's disease. Aging (Albany NY) 8 (6), 1250–1258. https://doi.org/10.18632/aging.100981.

Budson, A.E., Solomon, P.R., 2012. New criteria for Alzheimer disease and mild cognitive impairment: implications for the practicing clinician. Neurologist 18 (6), 356–363. https://doi.org/10.1097/NRL.0b013e31826a998d.

Buee, L., Troquier, L., Burnouf, S., Belarbi, K., Van der Jeugd, A., Ahmed, T., Fernandez-Gomez, F., Caillierez, R., Grosjean, M.E., Begard, S., Barbot, B., Demeyer, D., Obriot, H., Brion, I., Buee-Scherrer, V., Maurage, C.A., Balschun, D., D'Hooge, R., Hamdane, M., Blum, D., Sergeant, N., 2010. From tau phosphorylation to tau aggregation: what about neuronal death? Biochem. Soc. Trans. 38 (4), 967–972. https://doi.org/10.1042/BST0380967.

Cacace, R., Sleegers, K., Van Broeckhoven, C., 2016. Molecular genetics of early-onset Alzheimer's disease revisited. Alzheimers Dement. 12 (6), 733–748. https://doi.org/10.1016/j.jalz.2016.01.012.

Cadenas, E., Davies, K.J., 2000. Mitochondrial free radical generation, oxidative stress and aging. Free Radic. Biol. Med. 29 (3–4), 222–230. https://doi.org/10.1016/s0891-5849(00)00317-8.

Calsolaro, V., Edison, P., 2016. Neuroinflammation in Alzheimer's disease: current evidence and future directions. Alzheimers Dement. 12 (6), 719–732. https://doi.org/10.1016/j.jalz.2016.02.010.

Camacho-Mercado, C.L., Figueroa, R., Acosta, H., Arnold, S.E., Vega, I.E., 2016. Profiling of Alzheimer's disease patients in Puerto Rico: a comparison of two distinct socioeconomic areas. SAGE Open Med. 4, 2050312115627826. https://doi.org/10.1177/2050312115627826.

Capizzi, A., Woo, J., Verduzco-Gutierrez, M., 2020. Traumatic brain injury: an overview of epidemiology, pathophysiology and medical management. Emerg. Med. Clin. North Am. 104 (2), 213–238. https://doi.org/10.1016/j.mcna.2019.11.001.

Caprelli, M.T., Mothe, A.J., Tator, C.H., 2018. Hyperphosphorylated tau as a novel biomarker for traumatic axonal injury in the spinal cord. J. Neurotrauma. 1535 (16), 1929–1941. https://doi.org/10.1089/neu.2017.5495.

Carradori, D., Balducci, C., Re, F., Brambilla, D., Le Droumaguet, B., Flores, O., Gaudin, A., Mura, S., Forloni, G., Ordonez-Gutierrez, L., Wandosell, F., Masserini, M., Couvreur, P., Nicolas, J., Andrieux, K., 2018. Antibody-functionalized polymer nanoparticle leading to memory recovery in Alzheimer's disease-like transgenic mouse model. Nanomedicine 14 (2), 609–618. https://doi.org/10.1016/j.nano.2017.12.006.

Carvalho, A.L., Caldeira, M.V., Santos, S.D., Duarte, C.B., 2008. Role of the brain-derived neurotrophic factor at glutamatergic synapses. Br. J. Pharmacol. 153 (S1), S310–S324. https://doi.org/10.1038/sj.bjp.0707509.

Chapman JC, Diaz-Arrastia R., 2014. Military traumatic brain injury: a review. Alzheimers Dement, 10(3 Suppl):S97-104. doi: 10.1016/j.jalz.2014.04.012. PMID: 24924680.

Chen, Z., Zhong, C., 2014. Oxidative stress in Alzheimer's disease. Neurosci. Bull. 30 (2), 271–281. https://doi.org/10.1007/s12264-013-1423-y.

Chen, G., Chen, P., Tan, H., Ma, D., Dou, F., Feng, J., Yan, Z., 2008. Regulation of the NMDA receptor-mediated synaptic response by acetylcholinesterase inhibitors and its impairment in an animal model of Alzheimer's disease. Neurobiol. Aging 29 (12), 1795–1804. https://doi.org/10.1016/j.neurobiolaging.2007.04.023.

Chen, Y., Liang, Z., Blanchard, J., Dai, C.L., Sun, S., Lee, M.H., Grundke-Iqbal, I., Iqbal, K., Liu, F., Gong, C.X., 2013. A non-transgenic mouse model (icv-STZ mouse) of Alzheimer's disease: similarities to and differences from the transgenic model (3xTg-AD mouse). Mol. Neurobiol. 47 (2), 711–725. https://doi.org/10.1007/s12035-012-8375-5.

Chen, J., Zhou, S.N., Zhang, Y.M., Feng, Y.L., Wang, S., 2015. Glycosides of cistanche improve learning and memory in the rat model of vascular dementia. Eur. Rev. Med. Pharmacol. Sci. 19 (7), 1234–1240. https://www.ncbi.nlm.nih.gov/pubmed/25912583.

Chen, C., Homma, A., Mok, V.C., Krishnamoorthy, E., Alladi, S., Meguro, K., Abe, K., Dominguez, J., Marasigan, S., Kandiah, N., Kim, S.Y., Lee, D.Y., De Silva, H.A., Yang, Y.H., Pai, M.C., Senanarong, V., Dash, A., 2016. Alzheimer's disease with cerebrovascular disease: current status in the Asia-Pacific region. J. Intern. Med. 280 (4), 359–374. https://doi.org/10.1111/joim.12495.

Chhetri, J.K., Chan, P., Vellas, B., Touchon, J., Gauthier, S., 2019. Report from the first clinical trials on Alzheimer's disease (CTAD) Asia-China 2018: bringing together global leaders. J. Prev Alzheimers Dis. 6 (2), 144–147. https://doi.org/10.14283/jpad.2019.2.

Choubey, D., 2019. Type I interferon (IFN)-inducible absent in melanoma 2 proteins in neuroinflammation: implications for Alzheimer's disease. J. Neuroinflammation 16 (1), 236. https://doi.org/10.1186/s12974-019-1639-5.

Chu, L.W., 2012. Alzheimer's disease: early diagnosis and treatment. Hong Kong Med. J. 18 (3), 228–237. https://www.ncbi.nlm.nih.gov/pubmed/22665688.

Cimdins, K., Waugh, H.S., Chrysostomou, V., Lopez Sanchez, M.I.G., Johannsen, V.A., Cook, M.J., Crowston, J.G., Hill, A.F., Duce, J.A., Bush, A.I., Trounce, I.A., 2019. Amyloid precursor protein mediates neuronal protection from rotenone toxicity. Mol. Neurobiol. 56 (8), 5471–5482. https://doi.org/10.1007/s12035-018-1460-7.

Clausen AN, Clarke E, Phillips RD, Haswell C; VA Mid-Atlantic MIRECC Workgroup, Morey RA. Combat exposure, posttraumatic stress disorder, and head injuries differentially relate to alterations in cortical thickness in military Veterans. Neuropsychopharmacology, 2020, 45 (3), 491-498. doi: 10.1038/s41386-019-0539-9. Epub 2019 Oct 10.

Cortes, N., Andrade, V., Guzman-Martinez, L., Estrella, M., Maccioni, R.B., 2018. Neuroimmune tau mechanisms: their role in the progression of neuronal degeneration. Int. J. Mol. Sci. 19 (4), 956. https://doi.org/10.3390/ijms19040956.

Cotton, F.A., Wilkinson, G., Murillo, C.A., Bochmann, M., Grimes, R., 1988. Advanced Inorganic Chemistry. vol. 5 Wiley, New York.

Cravchik, A., Reddy, D., Matus, A., 1994. Identification of a novel microtubule-binding domain in microtubule-associated protein 1A (MAP1A). J. Cell Sci. 107 (Pt. 3), 661–672. https://www.ncbi.nlm.nih.gov/pubmed/8006079.

Crous-Bou, M., Minguillon, C., Gramunt, N., Molinuevo, J.L., 2017. Alzheimer's disease prevention: from risk factors to early intervention. Alzheimers Res. Ther. 9 (1), 71. https://doi.org/10.1186/s13195-017-0297-z.

Cruchaga, C., Del-Aguila, J.L., Saef, B., Black, K., Fernandez, M.V., Budde, J., Ibanez, L., Deming, Y., Kapoor, M., Tosto, G., Mayeux, R.P., Holtzman, D.M., Fagan, A.M., Morris, J.-C., Bateman, R.J., Goate, A.M., Dominantly Inherited Alzheimer, N, Disease Neuroimaging, I, Study, N.-L. F, Harari, O., 2018. Polygenic risk score of sporadic late-onset Alzheimer's disease reveals a shared architecture with the familial and early-onset forms. Alzheimers Dement. 14 (2), 205–214. https://doi.org/10.1016/j.jalz.2017.08.013.

De Jong, W.H., Borm, P.J., 2008. Drug delivery and nanoparticles: applications and hazards. Int. J. Nanomedicine 3 (2), 133–149. https://doi.org/10.2147/ijn.s596.

DeVos, S.L., Miller, R.L., Schoch, K.M., Holmes, B.B., Kebodeaux, C.S., Wegener, A.J., Chen, G., Shen, T., Tran, H., Nichols, B., Zanardi, T.A., Kordasiewicz, H.B., Swayze, E.E., Bennett, C.F., Diamond, M.I., Miller, T.M., 2017. Tau reduction prevents neuronal loss and reverses pathological tau deposition and seeding in mice with tauopathy. Sci. Transl. Med. 9 (374), eaag0481. https://doi.org/10.1126/scitranslmed.aag0481.

Dey, P.K., Sharma, H.S., 1984. Influence of ambient temperature and drug treatments on brain oedema induced by impact injury on skull in rats. Indian J. Physiol. Pharmacol. 28 (3), 177–186.

Dey, P.K., Sharma, H.S., Rao, K.S., 1980. Effect of indomethacin (a prostaglandin synthetase inhibitor) on the permeability of blood-brain and blood-CSF barriers in rat. Indian J. Physiol. Pharmacol. 24 (1), 25–36.

Dixon, K.J., 2017. Pathophysiology of traumatic brain injury. Phys. Med. Rehabil. Clin. N. Am. 28 (2), 215–225. https://doi.org/10.1016/j.pmr.2016.12.001.

Dorszewska, J., Prendecki, M., Oczkowska, A., Dezor, M., Kozubski, W., 2016. Molecular basis of familial and sporadic Alzheimer's disease. Curr. Alzheimer Res. 13 (9), 952–963. https://doi.org/10.2174/1567205013666160314150501.

Draper, B., Cations, M., White, F., Trollor, J., Loy, C., Brodaty, H., Sachdev, P., Gonski, P., Demirkol, A., Cumming, R.G., Withall, A., 2016. Time to diagnosis in young-onset dementia and its determinants: the INSPIRED study. Int. J. Geriatr. Psychiatry 31 (11), 1217–1224. https://doi.org/10.1002/gps.4430.

D'Hooge, R., De Deyn, P.P., 2001. Applications of the Morris water maze in the study of learning and memory. Brain Res. Brain Res. Rev. 36 (1), 60–90. https://doi.org/10.1016/s0165-0173(01)00067-4.

Dujardin, S., Begard, S., Caillierez, R., Lachaud, C., Carrier, S., Lieger, S., Gonzalez, J.A., Deramecourt, V., Deglon, N., Maurage, C.A., Frosch, M.P., Hyman, B.T., Colin, M., Buee, L., 2018. Different tau species lead to heterogeneous tau pathology propagation and misfolding. Acta Neuropathol. Commun. 6 (1), 132. https://doi.org/10.1186/s40478-018-0637-7.

Duyckaerts, C., Delaere, P., He, Y., Seilhean, D., Hauw, J.J., 1994. Alzheimer disease. Role of beta A4 peptide and cerebral amyloid substance. Presse Med. 23 (18), 848–854. https://www.ncbi.nlm.nih.gov/pubmed/7937606. (Maladie d'Alzheimer. Role du peptide beta A4 et de la substance amyloide cerebrale).

Edwards 3rd, G., Moreno-Gonzalez, I., Soto, C., 2017. Amyloid-beta and tau pathology following repetitive mild traumatic brain injury. Biochem. Biophys. Res. Commun. 483 (4), 1137–1142. https://doi.org/10.1016/j.bbrc.2016.07.123.

Edwards 3rd, G., Zhao, J., Dash, P.K., Soto, C., Moreno-Gonzalez, I., 2020. Traumatic brain injury induces tau aggregation and spreading. J. Neurotrauma 37 (1), 80–92. https://doi.org/10.1089/neu.2018.6348.

Elder GA, Ehrlich ME, Gandy S. 2019. Relationship of traumatic brain injury to chronic mental health problems and dementia in military veterans. Neurosci. Lett. 707:134294. doi: 10.1016/j.neulet.2019.134294. Epub 2019 May 26.

Elliott, K.A., Jasper, H., 1949. Measurement of experimentally induced brain swelling and shrinkage. Am. J. Physiol. 157 (1), 122–129. https://doi.org/10.1152/ajplegacy.1949.157.1.122.

Esquerda-Canals, G., Montoliu-Gaya, L., Guell-Bosch, J., Villegas, S., 2017. Mouse models of Alzheimer's disease. J. Alzheimers Dis. 57 (4), 1171–1183. https://doi.org/10.3233/JAD-170045.

Farrer, L.A., Cupples, L.A., Haines, J.L., Hyman, B., Kukull, W.A., Mayeux, R., Myers, R.H., Pericak-Vance, M.A., Risch, N., van Duijn, C.M., 1997. Effects of age, sex, and ethnicity on the association between apolipoprotein E genotype and Alzheimer disease. A meta-analysis. APOE and Alzheimer disease meta analysis consortium. JAMA 278 (16), 1349–1356. https://www.ncbi.nlm.nih.gov/pubmed/9343467.

Fedele, E., Rivera, D., Marengo, B., Pronzato, M.A., Ricciarelli, R., 2015. Amyloid beta: walking on the dark side of the moon. Mech. Ageing Dev. 152, 1–4. https://doi.org/10.1016/j.mad.2015.09.001.

Fehily, B., Fitzgerald, M., 2017. Repeated mild traumatic brain injury: potential mechanisms of damage. Cell Transplant. 26 (7), 1131–1155. https://doi.org/10.1177/0963689717714092.

Feng, L., Sharma, A., Niu, F., Huang, Y., Lafuente, J.V., Muresanu, D.F., Ozkizilcik, A., Tian, Z.R., Sharma, H.S., 2018. TiO(2)-Nanowired delivery of DL-3-n-butylphthalide (DL-NBP) attenuates blood-brain barrier disruption, brain edema formation, and neuronal damages following concussive head injury. Mol. Neurobiol. 55 (1), 350–358. https://doi.org/10.1007/s12035-017-0746-5.

Ferencz, B., Gerritsen, L., 2015. Genetics and underlying pathology of dementia. Neuropsychol. Rev. 25 (1), 113–124. https://doi.org/10.1007/s11065-014-9276-3.

Fesharaki-Zadeh, A., 2019. Chronic traumatic encephalopathy: a brief overview. Front. Neurol. 10, 713. https://doi.org/10.3389/fneur.2019.00713.

Fessel, J., 2018. Alzheimer's disease combination treatment. Neurobiol. Aging 63, 165. https://doi.org/10.1016/j.neurobiolaging.2017.10.022.

Forder, J.P., Tymianski, M., 2009. Postsynaptic mechanisms of excitotoxicity: involvement of postsynaptic density proteins, radicals, and oxidant molecules. Neuroscience 158 (1), 293–300. https://doi.org/10.1016/j.neuroscience.2008.10.021.

Fordyce, D.E., Clark, V.J., Paylor, R., Wehner, J.M., 1995. Enhancement of hippocampally-mediated learning and protein kinase C activity by oxiracetam in learning-impaired DBA/2 mice. Brain Res. 672 (1–2), 170–176. https://doi.org/10.1016/0006-8993(94)01389-y.

Fragkouli, A., Papatheodoropoulos, C., Georgopoulos, S., Stamatakis, A., Stylianopoulou, F., Tsilibary, E.C., Tzinia, A.K., 2012. Enhanced neuronal plasticity and elevated endogenous sAPPalpha levels in mice over-expressing MMP9. J. Neurochem. 121 (2), 239–251. https://doi.org/10.1111/j.1471-4159.2011.07637.x.

Gabryel, B., Trzeciak, H.I., Pudelko, A., Cieslik, P., 1999. Influence of piracetam and oxiracetam on the content of high-energy phosphates and morphometry of astrocytes in vitro. Pol. J. Pharmacol. 51 (6), 485–495. https://www.ncbi.nlm.nih.gov/pubmed/10817526.

Gale, S.A., Acar, D., Daffner, K.R., 2018. Dementia. Am. J. Med. 131 (10), 1161–1169. https://doi.org/10.1016/j.amjmed.2018.01.022.

Gallardo, G., Holtzman, D.M., 2019. Amyloid-beta and tau at the crossroads of Alzheimer's disease. Adv. Exp. Med. Biol. 1184, 187–203. https://doi.org/10.1007/978-981-32-9358-8_16.

Galvan, V., Gorostiza, O.F., Banwait, S., Ataie, M., Logvinova, A.V., Sitaraman, S., Carlson, E., Sagi, S.A., Chevallier, N., Jin, K., Greenberg, D.A., Bredesen, D.E., 2006. Reversal of Alzheimer's-like pathology and behavior in human APP transgenic mice by mutation of Asp664. Proc. Natl. Acad. Sci. U. S. A. 103 (18), 7130–7135. https://doi.org/10.1073/pnas.0509695103.

Galvan, V., Zhang, J., Gorostiza, O.F., Banwait, S., Huang, W., Ataie, M., Tang, H., Bredesen, D.E., 2008. Long-term prevention of Alzheimer's disease-like behavioral deficits in PDAPP mice carrying a mutation in Asp664. Behav. Brain Res. 191 (2), 246–255. https://doi.org/10.1016/j.bbr.2008.03.035.

Galvin, J.E., Sadowsky, C.H., Nincds, A., 2012. Practical guidelines for the recognition and diagnosis of dementia. J. Am. Board Fam. Med. 25 (3), 367–382. https://doi.org/10.3122/jabfm.2012.03.100181.

Gardner R.C., Yaffe K., 2015. Epidemiology of mild traumatic brain injury and neurodegenerative disease, Mol. Cell Neurosci. 66 (Pt B), 75-80. doi: 10.1016/j.mcn.2015.03.001. Epub 2015 Mar 5.

Gardner RC, Byers AL, Barnes DE, Li Y, Boscardin J, Yaffe K., Mild TBI and risk of Parkinson disease: a chronic effects of neurotrauma consortium study. Neurology. 2018, 90 (20):e1771-e1779. doi: 10.1212/WNL.0000000000005522. Epub 2018 Apr 18.

Giaquinto, S., Nolfe, G., Vitali, S., 1986. EEG changes induced by oxiracetam on diazepam-medicated volunteers. Clin. Neuropharmacol. 9 Suppl 3, S79–S84. https://www.ncbi.nlm.nih.gov/pubmed/3594460.

Giza, C.C., Hovda, D.A., 2014. The new neurometabolic cascade of concussion. Neurosurgery 75 (Suppl 4(0 4)), S24–33. https://doi.org/10.1227/NEU.0000000000000505.

Gomes, L.A., Hipp, S.A., Rijal Upadhaya, A., Balakrishnan, K., Ospitalieri, S., Koper, M.J., Largo-Barrientos, P., Uytterhoeven, V., Reichwald, J., Rabe, S., Vandenberghe, R., von Arnim, C.A.F., Tousseyn, T., Feederle, R., Giudici, C., Willem, M., Staufenbiel, M., Thal, D.R., 2019. Abeta-induced acceleration of Alzheimer-related tau-pathology spreading and its association with prion protein. Acta Neuropathol. 138 (6), 913–941. https://doi.org/10.1007/s00401-019-02053-5.

Gouliaev, A.H., Senning, A., 1994. Piracetam and other structurally related nootropics. Brain Res. Brain Res. Rev. 19 (2), 180–222. https://doi.org/10.1016/0165-0173(94)90011-6.

Gouras, G.K., Olsson, T.T., Hansson, O., 2015. beta-amyloid peptides and amyloid plaques in Alzheimer's disease. Neurotherapeutics 12 (1), 3–11. https://doi.org/10.1007/s13311-014-0313-y.

Graham, N.S., Sharp, D.J., 2019. Understanding neurodegeneration after traumatic brain injury: from mechanisms to clinical trials in dementia. J. Neurol. Neurosurg. Psychiatry 90 (11), 1221–1233. https://doi.org/10.1136/jnnp-2017-317557.

Guerriero, R.M., Giza, C.C., Rotenberg, A., 2015. Glutamate and GABA imbalance following traumatic brain injury. Curr. Neurol. Neurosci. Rep. 15 (5), 27. https://doi.org/10.1007/s11910-015-0545-1.

Gupta, P., Bhattacharjee, S., Sharma, A.R., Sharma, G., Lee, S.S., Chakraborty, C., 2017. miR-NAs in Alzheimer disease—a therapeutic perspective. Curr. Alzheimer Res. 14 (11), 1198–1206. https://doi.org/10.2174/1567205014666170829101016.

Hachinski, V., 2019. Dementia: new vistas and opportunities. Neurol. Sci. 40 (4), 763–767. https://doi.org/10.1007/s10072-019-3714-1.

Hai, J., Su, S.H., Lin, Q., Zhang, L., Wan, J.F., Li, H., Chen, Y.Y., Lu, Y., 2010. Cognitive impairment and changes of neuronal plasticity in rats of chronic cerebral hypoperfusion associated with cerebral arteriovenous malformations. Acta Neurol. Belg. 110 (2), 180–185. https://www.ncbi.nlm.nih.gov/pubmed/20873448.

Hainmueller T, Bartos M. 2018. Parallel emergence of stable and dynamic memory engrams in the hippocampus. Nature, 558 (7709), 292-296. doi: 10.1038/s41586-018-0191-2. Epub 2018 Jun 6.

Haque, M.M., Murale, D.P., Kim, Y.K., Lee, J.S., 2019. Crosstalk between oxidative stress and tauopathy. Int. J. Mol. Sci. 20 (8), 1959. https://doi.org/10.3390/ijms20081959.

Hardy, J., Selkoe, D.J., 2002. The amyloid hypothesis of Alzheimer's disease: progress and problems on the road to therapeutics. Science 297 (5580), 353–356. https://doi.org/10.1126/science.1072994.

Hayes, A., Green, E.K., Pritchard, A., Harris, J.M., Zhang, Y., Lambert, J.C., Chartier-Harlin, M.C., Pickering-Brown, S.M., Lendon, C.L., Mann, D.M., 2004. A polymorphic variation in the interleukin 1A gene increases brain microglial cell activity in Alzheimer's disease. J. Neurol. Neurosurg. Psychiatry 75 (10), 1475–1477. https://doi.org/10.1136/jnnp.2003. 030866.

Heneka, M.T., Carson, M.J., El Khoury, J., Landreth, G.E., Brosseron, F., Feinstein, D.L., Jacobs, A.H., Wyss-Coray, T., Vitorica, J., Ransohoff, R.M., Herrup, K., Frautschy, S.A., Finsen, B., Brown, G.C., Verkhratsky, A., Yamanaka, K., Koistinaho, J., Latz, E., Halle, A., Petzold, G.C., Town, T., Morgan, D., Shinohara, M.L., Perry, V.H., Holmes, C., Bazan, N.G., Brooks, D.J., Hunot, S., Joseph, B., Deigendesch, N., Garaschuk, O., Boddeke, E., Dinarello, C.A., Breitner, J.C., Cole, G.M., Golenbock, D.T., Kummer, M.P., 2015. Neuroinflammation in Alzheimer's disease. Lancet Neurol. 14 (4), 388–405. https://doi.org/10.1016/S1474-4422(15)70016-5.

Herms, J., Dorostkar, M.M., 2016. Dendritic spine pathology in neurodegenerative diseases. Annu. Rev. Pathol. 11 (1), 221–250. https://doi.org/10.1146/annurev-pathol-012615-044216.

Hesse, R., von Einem, B., Wagner, F., Bott, P., Schwanzar, D., Jackson, R.J., Fohr, K.J., Lausser, L., Kroker, K.S., Proepper, C., Walther, P., Kestler, H.A., Spires-Jones, T.L., Boeckers, T., Rosenbrock, H., von Arnim, C.A.F., 2018. sAPPbeta and sAPPalpha increase structural complexity and E/I input ratio in primary hippocampal neurons and alter Ca(2+) homeostasis and CREB1-signaling. Exp. Neurol. 304, 1–13. https://doi.org/10.1016/j.expneurol.2018.02.010.

Hlushchenko, I., Koskinen, M., Hotulainen, P., 2016. Dendritic spine actin dynamics in neuronal maturation and synaptic plasticity. Cytoskeleton 73 (9), 435–441. https://doi.org/10.1002/cm.21280.

Hokonohara, T., Sako, K., Shinoda, Y., Tomabechi, M., Yonemasu, Y., 1992a. The effects of oxiracetam (CT-848) on local cerebral glucose utilization after focal cerebral ischemia in rats. Jpn. J. Pharmacol. 58 (2), 127–135. https://doi.org/10.1254/jjp.58.127.

Hokonohara, T., Shinoda, Y., Hori, N., 1992b. Effects of oxiracetam on the decrease in population spikes in hypoxic and low glucose media. Nihon Yakurigaku Zasshi 99 (3), 123–133. https://doi.org/10.1254/fpj.99.123.

Holmes, B.B., Furman, J.L., Mahan, T.E., Yamasaki, T.R., Mirbaha, H., Eades, W.C., Belaygorod, L., Cairns, N.J., Holtzman, D.M., Diamond, M.I., 2014. Proteopathic tau seeding predicts tauopathy in vivo. Proc. Natl. Acad. Sci. U. S. A. 111 (41), E4376–E4385. https://doi.org/10.1073/pnas.1411649111.

Hoyert, D.L., Rosenberg, H.M., 1997. Alzheimer's disease as a cause of death in the United States. Public Health Rep. 112 (6), 497–505. https://www.ncbi.nlm.nih.gov/pubmed/10822478.

Hsiao, K., 1998. Transgenic mice expressing Alzheimer amyloid precursor proteins. Exp. Gerontol. 33 (7–8), 883–889. https://doi.org/10.1016/s0531-5565(98)00045-x.

Hu, Q., Li, H., Wang, L., Gu, H., Fan, C., 2019. DNA nanotechnology-enabled drug delivery systems. Chem. Rev. 119 (10), 6459–6506. https://doi.org/10.1021/acs.chemrev.7b00663.

Huang, Y., Zhao, Z., Wei, X., Zheng, Y., Yu, J., Zheng, J., Wang, L., 2016. Long-term trihexyphenidyl exposure alters neuroimmune response and inflammation in aging rat: relevance to age and Alzheimer's disease. J. Neuroinflammation 13 (1), 175. https://doi.org/10.1186/s12974-016-0640-5.

Illes, P., Rubini, P., Huang, L., Tang, Y., 2019. The P2X7 receptor: a new therapeutic target in Alzheimer's disease. Expert Opin. Ther. Targets 23 (3), 165–176. https://doi.org/10.1080/14728222.2019.1575811.

Iqbal, K., Grundke-Iqbal, I., 1997. Mechanism of Alzheimer neurofibrillary degeneration and the formation of tangles. Mol. Psychiatry 2 (3), 178–180. https://doi.org/10.1038/sj.mp. 4000269.

Iqbal, K., Alonso Adel, C., Grundke-Iqbal, I., 2008. Cytosolic abnormally hyperphosphorylated tau but not paired helical filaments sequester normal MAPs and inhibit microtubule assembly. J. Alzheimers Dis. 14 (4), 365–370. https://doi.org/10.3233/jad-2008-14402.

Irwin, D.J., 2016. Tauopathies as clinicopathological entities. Parkinsonism Relat. Disord. 22 (S1), S29–S33. https://doi.org/10.1016/j.parkreldis.2015.09.020.

Irwin, D.J., Lee, V.M., Trojanowski, J.Q., 2013. Parkinson's disease dementia: convergence of alpha-synuclein, tau and amyloid-beta pathologies. Nat. Rev. Neurosci. 14 (9), 626–636. https://doi.org/10.1038/nrn3549.

Ivanov, A., Pellegrino, C., Rama, S., Dumalska, I., Salyha, Y., Ben-Ari, Y., Medina, I., 2006. Opposing role of synaptic and extrasynaptic NMDA receptors in regulation of the extracellular signal-regulated kinases (ERK) activity in cultured rat hippocampal neurons. J. Physiol. 572 (Pt. 3), 789–798. https://doi.org/10.1113/jphysiol.2006.105510.

Janak, J.C., Mazuchowski, E.L., Kotwal, R.S., Stockinger, Z.T., Howard, J.T., Butler, F.K., Sosnov, J.A., Gurney, J.M., Shackelford, S.A., 2019. Patterns of anatomic injury in critically injured combat casualties: a network analysis. Sci. Rep. 249 (1), 13767. https://doi.org/10.1038/s41598-019-50272-3.

Javanshiri, K., Waldo, M.L., Friberg, N., Sjovall, F., Wickerstrom, K., Haglund, M., Englund, E., 2018. Atherosclerosis, hypertension, and diabetes in Alzheimer's disease, vascular dementia, and mixed dementia: prevalence and presentation. J. Alzheimers Dis. 65 (4), 1247–1258. https://doi.org/10.3233/JAD-180644.

Jha, R.M., Kochanek, P.M., 2018. A precision medicine approach to cerebral edema and intracranial hypertension after severe traumatic brain injury: quo vadis? Curr. Neurol. Neurosci. Rep. 18 (12), 105. https://doi.org/10.1007/s11910-018-0912-9.

Jiang, T., Yu, J.T., Tian, Y., Tan, L., 2013. Epidemiology and etiology of Alzheimer's disease: from genetic to non-genetic factors. Curr. Alzheimer Res. 10 (8), 852–867. https://doi.org/10.2174/15672050113109990155.

Jing, Z., Shi, C., Zhu, L., Xiang, Y., Chen, P., Xiong, Z., Li, W., Ruan, Y., Huang, L., 2015. Chronic cerebral hypoperfusion induces vascular plasticity and hemodynamics but also neuronal degeneration and cognitive impairment. J. Cereb. Blood Flow Metab. 35 (8), 1249–1259. https://doi.org/10.1038/jcbfm.2015.55.

Johnson, G.V., Stoothoff, W.H., 2004. Tau phosphorylation in neuronal cell function and dysfunction. J. Cell Sci. 117 (Pt. 24), 5721–5729. https://doi.org/10.1242/jcs.01558.

Johnson, V.E., Stewart, W., Smith, D.H., 2012. Widespread tau and amyloid-beta pathology many years after a single traumatic brain injury in humans. Brain Pathol. 22 (2), 142–149. https://doi.org/10.1111/j.1750-3639.2011.00513.x.

Johnston, R.B., Zachary, L., Dellon, A.L., Seiler 4th, W.A., Teplica, D.M., 1991. Improved imaging of rat hindfoot prints for walking track analysis. J. Neurosci. Methods 38 (2–3), 111–114. https://doi.org/10.1016/0165-0270(91)90161-r.

Jonsson, T., Atwal, J.K., Steinberg, S., Snaedal, J., Jonsson, P.V., Bjornsson, S., Stefansson, H., Sulem, P., Gudbjartsson, D., Maloney, J., Hoyte, K., Gustafson, A., Liu, Y., Lu, Y., Bhangale, T., Graham, R.R., Huttenlocher, J., Bjornsdottir, G., Andreassen, O.A., Jonsson, E.G., Palotie, A., Behrens, T.W., Magnusson, O.T., Kong, A., Thorsteinsdottir, U., Watts, R.J., Stefansson, K., 2012. A mutation in APP protects against Alzheimer's disease and age-related cognitive decline. Nature 488 (7409), 96–99. https://doi.org/10.1038/nature11283.

Kamara, D.M., Gangishetti, U., Gearing, M., Willis-Parker, M., Zhao, L., Hu, W.T., Walker, L.C., 2018. Cerebral amyloid angiopathy: similarity in African-Americans and caucasians with Alzheimer's disease. J. Alzheimers Dis. 62 (4), 1815–1826. https://doi.org/10.3233/JAD-170954.

Kamat, P.K., Kalani, A., Rai, S., Swarnkar, S., Tota, S., Nath, C., Tyagi, N., 2016a. Mechanism of oxidative stress and synapse dysfunction in the pathogenesis of Alzheimer's disease: understanding the therapeutics strategies. Mol. Neurobiol. 53 (1), 648–661. https://doi.org/10.1007/s12035-014-9053-6.

Kamat, P.K., Kalani, A., Rai, S., Tota, S.K., Kumar, A., Ahmad, A.S., 2016b. Streptozotocin intracerebroventricular-induced neurotoxicity and brain insulin resistance: a therapeutic intervention for treatment of sporadic Alzheimer's disease (sAD)-like pathology. Mol. Neurobiol. 53 (7), 4548–4562. https://doi.org/10.1007/s12035-015-9384-y.

Kang, S.W., Kim, S.J., Kim, M.S., 2017. Oxidative stress with tau hyperphosphorylation in memory impaired 1,2-diacetylbenzene-treated mice. Toxicol. Lett. 279, 53–59. https://doi.org/10.1016/j.toxlet.2017.07.892.

Karch, C.M., Cruchaga, C., Goate, A.M., 2014. Alzheimer's disease genetics: from the bench to the clinic. Neuron 83 (1), 11–26. https://doi.org/10.1016/j.neuron.2014.05.041.

Karran, E., Mercken, M., De Strooper, B., 2011. The amyloid cascade hypothesis for Alzheimer's disease: an appraisal for the development of therapeutics. Nat. Rev. Drug Discov. 10 (9), 698–712. https://doi.org/10.1038/nrd3505.

Katsuse, O., Iseki, E., Kosaka, K., 2003. Immunohistochemical study of the expression of cytokines and nitric oxide synthases in brains of patients with dementia with Lewy bodies. Neuropathology 23 (1), 9–15. https://doi.org/10.1046/j.1440-1789.2003.00483.x.

Katzenberger, R.J., Chtarbanova, S., Rimkus, S.A., Fischer, J.A., Kaur, G., Seppala, J.M., Swanson, L.C., Zajac, J.E., Ganetzky, B., Wassarman, D.A., 2015. Death following traumatic brain injury in Drosophila is associated with intestinal barrier dysfunction. Elife 4, e04790. https://doi.org/10.7554/eLife.04790.

Kellermayer, B., Ferreira, J.S., Dupuis, J., Levet, F., Grillo-Bosch, D., Bard, L., Linares-Loyez, J., Bouchet, D., Choquet, D., Rusakov, D.A., Bon, P., Sibarita, J.B., Cognet, L., Sainlos, M., Carvalho, A.L., Groc, L., 2018. Differential nanoscale topography and functional role of GluN2-NMDA receptor subtypes at glutamatergic synapses. Neuron 100 (1), 106–119.e107. https://doi.org/10.1016/j.neuron.2018.09.012.

Kempuraj D, Ahmed ME, Selvakumar GP, Thangavel R, Raikwar SP, Zaheer SA, Iyer SS, Burton C, James D, Zaheer A. Psychological stress-induced immune response and risk of Alzheimer's disease in veterans from operation enduring freedom and operation iraqi freedom. Clin. Ther. 2020;42(6):974-982. doi: 10.1016/j.clinthera.2020.02.018. Epub 2020 Mar 14.

Kerr H.A., Closed head injury, Clin. Sports Med. 2013;32(2):273-287. doi: 10.1016/j.csm.2012.12.008. Epub 2013 Feb 8.

Khatri, N., Thakur, M., Pareek, V., Kumar, S., Sharma, S., Datusalia, A.K., 2018. Oxidative stress: major threat in traumatic brain injury. CNS Neurol. Disord. Drug Targets 17 (9), 689–695. https://doi.org/10.2174/1871527317666180627120501.

Khosa, A., Reddi, S., Saha, R.N., 2018. Nanostructured lipid carriers for site-specific drug delivery. Biomed. Pharmacother. 103, 598–613. https://doi.org/10.1016/j.biopha.2018.04.055.

Kida, S., 2012. A functional role for CREB as a positive regulator of memory formation and LTP. Exp. Neurobiol. 21 (4), 136–140. https://doi.org/10.5607/en.2012.21.4.136.

Kim, L.D., Factora, R.M., 2018. Alzheimer dementia: starting, stopping drug therapy. Cleve. Clin. J. Med. 85 (3), 209–214. https://doi.org/10.3949/ccjm.85a.16080.

Kokiko-Cochran, O.N., Godbout, J.P., 2018. The inflammatory continuum of traumatic brain injury and Alzheimer's disease. Front. Immunol. 9, 672. https://doi.org/10.3389/fimmu.2018.00672.

Kometani, M., Okada, M., Takemori, E., Hasegawa, Y., Nakao, N., Inukai, T., 1991. Effect of oxiracetam on cerebrovascular impairment in rats. Arzneimittelforschung 41 (7), 684–689. https://www.ncbi.nlm.nih.gov/pubmed/1772453.

Kovacs, G.G., 2017. Tauopathies. Handb. Clin. Neurol. 145, 355–368. https://doi.org/10.1016/B978-0-12-802395-2.00025-0.

Kozlov AV, Bahrami S, Redl H, Szabo C. Alterations in nitric oxide homeostasis during traumatic brain injury. Biochim. Biophys. Acta Mol. Basis Dis. 2017, 1863(10 Pt B):2627-2632. doi: 10.1016/j.bbadis.2016.12.020. Epub 2017 Jan 5.

Kravchick, D.O., Karpova, A., Hrdinka, M., Lopez-Rojas, J., Iacobas, S., Carbonell, A.U., Iacobas, D.A., Kreutz, M.R., Jordan, B.A., 2016. Synaptonuclear messenger PRR7 inhibits c-Jun ubiquitination and regulates NMDA-mediated excitotoxicity. EMBO J. 35 (17), 1923–1934. https://doi.org/10.15252/embj.201593070.

Krishnamoorthy, V., Prathep, S., Sharma, D., Fujita, Y., Armstead, W., Vavilala, M.S., 2015. Cardiac dysfunction following brain death after severe pediatric traumatic brain injury: a preliminary study of 32 children. Int. J. Crit. Illn. Inj. Sci. 5 (2), 103–107. https://doi.org/10.4103/2229-5151.158409.

Kulbe, J.R., Hall, E.D., 2017. Chronic traumatic encephalopathy-integration of canonical traumatic brain injury secondary injury mechanisms with tau pathology. Prog. Neurobiol. 158, 15–44. https://doi.org/10.1016/j.pneurobio.2017.08.003.

Kumar, A., Thinschmidt, J.S., Foster, T.C., 2019. Subunit contribution to NMDA receptor hypofunction and redox sensitivity of hippocampal synaptic transmission during aging. Aging (Albany NY) 11 (14), 5140–5157. https://doi.org/10.18632/aging.102108.

Lau, C.G., Takeuchi, K., Rodenas-Ruano, A., Takayasu, Y., Murphy, J., Bennett, M.V., Zukin, R.S., 2009. Regulation of NMDA receptor Ca2+ signalling and synaptic plasticity. Biochem. Soc. Trans. 37 (Pt. 6), 1369–1374. https://doi.org/10.1042/BST0371369.

Lee, J., Lee, K.J., Kim, H., 2017. Gender differences in behavioral and psychological symptoms of patients with Alzheimer's disease. Asian J. Psychiatr. 26, 124–128. https://doi.org/10.1016/j.ajp.2017.01.027.

Lennon, M.J., Makkar, S.R., Crawford, J.D., Sachdev, P.S., 2019. Midlife hypertension and Alzheimer's disease: a systematic review and meta-analysis. J. Alzheimers Dis. 71 (1), 307–316. https://doi.org/10.3233/JAD-190474.

Leveille, F., El Gaamouch, F., Gouix, E., Lecocq, M., Lobner, D., Nicole, O., Buisson, A., 2008. Neuronal viability is controlled by a functional relation between synaptic and extrasynaptic NMDA receptors. FASEB J. 22 (12), 4258–4271. https://doi.org/10.1096/fj.08-107268.

Li, J.W., Yang, D.J., Chen, X.Y., Liang, H.Q., 2013. Protective effect of oxiracetam on traumatic brain injury in rats. Zhongguo Ying Yong Sheng Li Xue Za Zhi 29 (4), 298–300. https://www.ncbi.nlm.nih.gov/pubmed/24175546.

Li, W., Liu, H., Jiang, H., Wang, C., Guo, Y., Sun, Y., Zhao, X., Xiong, X., Zhang, X., Zhang, K., Nie, Z., Pu, X., 2017. (S)-oxiracetam is the active ingredient in oxiracetam that alleviates the cognitive impairment induced by chronic cerebral hypoperfusion in rats. Sci. Rep. 7 (1), 10052. https://doi.org/10.1038/s41598-017-10283-4.

Lichtenthaler, S.F., 2011. Alpha-secretase in Alzheimer's disease: molecular identity, regulation and therapeutic potential. J. Neurochem. 116 (1), 10–21. https://doi.org/10.1111/j.1471-4159.2010.07081.x.

Liu, W., Wong, A., Law, A.C., Mok, V.C., 2015. Cerebrovascular disease, amyloid plaques, and dementia. Stroke 46 (5), 1402–1407. https://doi.org/10.1161/STROKEAHA.114. 006571.

Liu, C., Liang, M.C., Soong, T.W., 2019. Nitric oxide, iron and neurodegeneration. Front. Neurosci. 13, 114. https://doi.org/10.3389/fnins.2019.00114.

Llufriu-Daben, G., Carrete, A., Chierto, E., Mailleux, J., Camand, E., Simon, A., Vanmierlo, T., Rose, C., Allinquant, B., Hendriks, J.J.A., Massaad, C., Meffre, D., Jafarian-Tehrani, M., 2018. Targeting demyelination via alpha-secretases promoting sAP-Palpha release to enhance remyelination in central nervous system. Neurobiol. Dis. 109 (Pt. A), 11–24. https://doi.org/10.1016/j.nbd.2017.09.008.

Lue, L.F., Rydel, R., Brigham, E.F., Yang, L.B., Hampel, H., Murphy Jr., G.M., Brachova, L., Yan, S.D., Walker, D.G., Shen, Y., Rogers, J., 2001. Inflammatory repertoire of Alzheimer's disease and nondemented elderly microglia in vitro. Glia 35 (1), 72–79. https://doi. org/10.1002/glia.1072.

Luine, V., Frankfurt, M., 2013. Interactions between estradiol, BDNF and dendritic spines in promoting memory. Neuroscience 239, 34–45. https://doi.org/10.1016/j.neuroscience. 2012.10.019.

Lyketsos, C.G., Carrillo, M.C., Ryan, J.M., Khachaturian, A.S., Trzepacz, P., Amatniek, J., Cedarbaum, J., Brashear, R., Miller, D.S., 2011. Neuropsychiatric symptoms in Alzheimer's disease. Alzheimers Dement. 7 (5), 532–539. https://doi.org/10.1016/j.jalz. 2011.05.2410.

Ma, J.H., Wang, T., 2019. Therapeutic effect of oxiracetam combined with *Ginkgo biloba* extract on acute cerebral hemorrhage and its effect on serum inflammatory factors. Zhong-guo Xian Dai Shen Jing Ji Bing Za Zhi 19 (8), 588–596. https://doi.org/10.3969/ j.issn.1672-6731.2019.08.009.

Ma, N., Wu, F.G., Zhang, X., Jiang, Y.W., Jia, H.R., Wang, H.Y., Li, Y.H., Liu, P., Gu, N., Chen, Z., 2017a. Shape-dependent radiosensitization effect of gold nanostructures in cancer radiotherapy: comparison of gold nanoparticles, nanospikes, and nanorods. ACS Appl. Mater. Interfaces 9 (15), 13037–13048. https://doi.org/10.1021/acsami. 7b01112.

Ma, R.H., Zhang, Y., Hong, X.Y., Zhang, J.F., Wang, J.Z., Liu, G.P., 2017b. Role of microtubule-associated protein tau phosphorylation in Alzheimer's disease. J. Huazhong Univ. Sci. Technolog. Med. Sci. 37 (3), 307–312. https://doi.org/10.1007/s11596-017-1732-x.

Maccioni, R.B., Farias, G., Morales, I., Navarrete, L., 2010. The revitalized tau hypothesis on Alzheimer's disease. Arch. Med. Res. 41 (3), 226–231. https://doi.org/10.1016/j.arcmed. 2010.03.007.

Malykh, A.G., Sadaie, M.R., 2010. Piracetam and piracetam-like drugs: from basic science to novel clinical applications to CNS disorders. Drugs 70 (3), 287–312. https://doi.org/10. 2165/11319230-000000000-00000.

Mancini, S., Minniti, S., Gregori, M., Sancini, G., Cagnotto, A., Couraud, P.O., Ordonez-Gutierrez, L., Wandosell, F., Salmona, M., Re, F., 2016. The hunt for brain Abeta oligomers by peripherally circulating multi-functional nanoparticles: potential therapeutic approach for Alzheimer disease. Nanomedicine 12 (1), 43–52. https://doi.org/10.1016/ j.nano.2015.09.003.

Manzano-Leon, N., Mas-Oliva, J., 2006. Oxidative stress, beta-amyloide peptide and Alzheimer's disease. Gac. Med. Mex. 142 (3), 229–238. https://www.ncbi.nlm.nih.gov/ pubmed/16875352. (Estres oxidativo, peptido beta-amiloide y enfermedad de Alzheimer).

Martin, L., Latypova, X., Wilson, C.M., Magnaudeix, A., Perrin, M.L., Yardin, C., Terro, F., 2013. Tau protein kinases: involvement in Alzheimer's disease. Ageing Res. Rev. 12 (1), 289–309. https://doi.org/10.1016/j.arr.2012.06.003.

Martin, G.E., Pugh, A.M., Moran, R., Veile, R., Friend, L.A., Pritts, T.A., Makley, A.T., Caldwell, C.C., Goodman, M.D., 2019. Microvesicles generated following traumatic brain injury induce platelet dysfunction via adenosine diphosphate receptor. J. Trauma Acute Care Surg. 86 (4), 592–600. https://doi.org/10.1097/TA.0000000000002171.

Martin-Rapun, R., De Matteis, L., Ambrosone, A., Garcia-Embid, S., Gutierrez, L., de la Fuente, J.M., 2017. Targeted nanoparticles for the treatment of Alzheimer's disease. Curr. Pharm. Des. 23 (13), 1927–1952. https://doi.org/10.2174/1381612822666161226151011.

Masel, B.E., DeWitt, D.S., 2010. Traumatic brain injury: a disease process, not an event. J. Neurotrauma 27 (8), 1529–1540. https://doi.org/10.1089/neu.2010.1358.

Mayeux, R., Stern, Y., 2012. Epidemiology of Alzheimer disease. Cold Spring Harb. Perspect. Med. 2 (8), a006239. https://doi.org/10.1101/cshperspect.a006239.

McKee, A.C., Cantu, R.C., Nowinski, C.J., Hedley-Whyte, E.T., Gavett, B.E., Budson, A.E., Santini, V.E., Lee, H.S., Kubilus, C.A., Stern, R.A., 2009. Chronic traumatic encephalopathy in athletes: progressive tauopathy after repetitive head injury. J. Neuropathol. Exp. Neurol. 68 (7), 709–735. https://doi.org/10.1097/NEN.0b013e3181a9d503.

McKee, A.C., Stern, R.A., Nowinski, C.J., Stein, T.D., Alvarez, V.E., Daneshvar, D.H., Lee, H.S., Wojtowicz, S.M., Hall, G., Baugh, C.M., Riley, D.O., Kubilus, C.A., Cormier, K.A., Jacobs, M.A., Martin, B.R., Abraham, C.R., Ikezu, T., Reichard, R.R., Wolozin, B.L., Budson, A.E., Goldstein, L.E., Kowall, N.W., Cantu, R.C., 2013. The spectrum of disease in chronic traumatic encephalopathy. Brain 136 (Pt. 1), 43–64. https://doi.org/10.1093/brain/aws307.

McKhann, G.M., Knopman, D.S., Chertkow, H., Hyman, B.T., Jack Jr., C.R., Kawas, C.H., Klunk, W.E., Koroshetz, W.J., Manly, J.J., Mayeux, R., Mohs, R.C., Morris, J.C., Rossor, M.N., Scheltens, P., Carrillo, M.C., Thies, B., Weintraub, S., Phelps, C.H., 2011. The diagnosis of dementia due to Alzheimer's disease: recommendations from the National Institute on aging-Alzheimer's Association workgroups on diagnostic guidelines for Alzheimer's disease. Alzheimers Dement. 7 (3), 263–269. https://doi.org/10.1016/j.jalz.2011.03.005.

Mendez, M.F., 2017. What is the relationship of traumatic brain injury to dementia? J. Alzheimers Dis. 57 (3), 667–681. https://doi.org/10.3233/JAD-161002.

Menon, P.K., Muresanu, D.F., Sharma, A., Mossler, H., Sharma, H.S., 2012. Cerebrolysin, a mixture of neurotrophic factors induces marked neuroprotection in spinal cord injury following intoxication of engineered nanoparticles from metals. CNS Neurol. Disord. Drug Targets 11 (1), 40–49. https://doi.org/10.2174/187152712799960781.

Menon, P.K., Sharma, A., Lafuente, J.V., Muresanu, D.F., Aguilar, Z.P., Wang, Y.A., Patnaik, R., Mossler, H., Sharma, H.S., 2017. Intravenous administration of functionalized magnetic iron oxide nanoparticles does not induce CNS injury in the rat: influence of spinal cord trauma and cerebrolysin treatment. Int. Rev. Neurobiol. 137, 47–63. https://doi.org/10.1016/bs.irn.2017.08.005.

Meyfroidt, G., Baguley, I.J., Menon, D.K., 2017. Paroxysmal sympathetic hyperactivity: the storm after acute brain injury. Lancet Neurol. 16 (9), 721–729. https://doi.org/10.1016/S1474-4422(17)30259-4.

Miao, H., Hu, X., Shang, Y., Zhang, D., Ji, R., Liu, E., Zhang, Q., Wang, Y., Fan, J., 2012. Preparation of 3D network Na2Ti2O4(OH)2 nanotube film and study on formation mechanism of nanotubes and light absorption properties. J. Nanosci. Nanotechnol. 12 (10), 7927–7931. https://doi.org/10.1166/jnn.2012.6611.

Miki, T., 1998. Progress of the molecular genetic analysis for Alzheimer's disease. Rinsho Byori 46 (10), 1008–1014. https://www.ncbi.nlm.nih.gov/pubmed/9816912.

Mo, J., Enkhjargal, B., Travis, Z.D., Zhou, K., Wu, P., Zhang, G., Zhu, Q., Zhang, T., Peng, J., Xu, W., Ocak, U., Chen, Y., Tang, J., Zhang, J., Zhang, J.H., 2019. AVE 0991 attenuates oxidative stress and neuronal apoptosis via Mas/PKA/CREB/UCP-2 pathway after subarachnoid hemorrhage in rats. Redox Biol. 20, 75–86. https://doi.org/10.1016/j.redox.2018.09.022.

Moglia, A., Sinforiani, E., Zandrini, C., Gualtieri, S., Corsico, R., Arrigo, A., 1986. Activity of oxiracetam in patients with organic brain syndrome: a neuropsychological study. Clin. Neuropharmacol. 9 (Suppl. 3), S73–S78. https://www.ncbi.nlm.nih.gov/pubmed/3594459.

Mollers, T., Perna, L., Stocker, H., Ihle, P., Schubert, I., Schottker, B., Frolich, L., Bauer, J., Brenner, H., 2019. Alzheimer's disease medication and outcomes of hospitalisation among patients with dementia. Epidemiol. Psychiatr. Sci. 29, e73. https://doi.org/10.1017/S2045796019000702.

Mondadori, C., Classen, W., Borkowski, J., Ducret, T., Buerki, H., Schade, A., 1986. Effects of oxiracetam on learning and memory in animals: comparison with piracetam. Clin. Neuropharmacol. 9 (Suppl. 3), S27–S38. https://www.ncbi.nlm.nih.gov/pubmed/3594453.

Mondadori, C., Hengerer, B., Ducret, T., Borkowski, J., 1994. Delayed emergence of effects of memory-enhancing drugs: implications for the dynamics of long-term memory. Proc. Natl. Acad. Sci. U. S. A. 91 (6), 2041–2045. https://doi.org/10.1073/pnas.91.6.2041.

Mondadori, C., Mobius, H.J., Borkowski, J., 1996. The GABAB receptor antagonist CGP 36,742 and the nootropic oxiracetam facilitate the formation of long-term memory. Behav. Brain Res. 77 (1–2), 223–225. https://doi.org/10.1016/0166-4328(95)00222-7.

Moniruzzaman, M., Ghosal, I., Das, D., Chakraborty, S.B., 2018. Melatonin ameliorates H(2)O(2)-induced oxidative stress through modulation of Erk/Akt/NFkB pathway. Biol. Res. 51 (1), 17. https://doi.org/10.1186/s40659-018-0168-5.

Morales, R., Moreno-Gonzalez, I., Soto, C., 2013. Cross-seeding of misfolded proteins: implications for etiology and pathogenesis of protein misfolding diseases. PLoS Pathog. 9 (9), e1003537. https://doi.org/10.1371/journal.ppat.1003537.

Moreira, P.I., 2018. Sweet mitochondria: a shortcut to Alzheimer's disease. J. Alzheimers Dis. 62 (3), 1391–1401. https://doi.org/10.3233/JAD-170931.

Moretto, E., Murru, L., Martano, G., Sassone, J., Passafaro, M., 2018. Glutamatergic synapses in neurodevelopmental disorders. Prog. Neuropsychopharmacol. Biol. Psychiatry 84 (Pt B), 328–342. https://doi.org/10.1016/j.pnpbp.2017.09.014.

Morris, R., 1984. Developments of a water-maze procedure for studying spatial learning in the rat. J. Neurosci. Methods 11 (1), 47–60. https://doi.org/10.1016/0165-0270(84)90007-4.

Muresanu DF, Sharma A, Sahib S, Tian ZR, Feng L, Castellani RJ, Nozari A, Lafuente JV, Buzoianu AD, Sjöquist PO, Patnaik R, Wiklund L, Sharma HS, Diabetes exacerbates brain pathology following a focal blast brain injury: new role of a multimodal drug cerebrolysin and nanomedicine. Prog. Brain Res. 2020;258:285-367. doi: 10.1016/bs.pbr.2020.09.004. Epub 2020 Nov 9.

Naseri, N.N., Wang, H., Guo, J., Sharma, M., Luo, W., 2019. The complexity of tau in Alzheimer's disease. Neurosci. Lett. 705, 183–194. https://doi.org/10.1016/j.neulet.2019.04.022.

Nazem, A., Sankowski, R., Bacher, M., Al-Abed, Y., 2015. Rodent models of neuroinflammation for Alzheimer's disease. J. Neuroinflammation 12, 74. https://doi.org/10.1186/s12974-015-0291-y.

Neves, G., Cooke, S.F., Bliss, T.V., 2008. Synaptic plasticity, memory and the hippocampus: a neural network approach to causality. Nat. Rev. Neurosci. 9 (1), 65–75. https://doi.org/10.1038/nrn2303.

Nguyen, T.V., Galvan, V., Huang, W., Banwait, S., Tang, H., Zhang, J., Bredesen, D.E., 2008. Signal transduction in Alzheimer disease: p21-activated kinase signaling requires C-terminal cleavage of APP at Asp664. J. Neurochem. 104 (4), 1065–1080. https://doi.org/10.1111/j.1471-4159.2007.05031.x.

Nikolac Perkovic, M., Pivac, N., 2019. Genetic markers of Alzheimer's disease. Adv. Exp. Med. Biol. 1192, 27–52. https://doi.org/10.1007/978-981-32-9721-0_3.

Niu, H., Alvarez-Alvarez, I., Guillen-Grima, F., Aguinaga-Ontoso, I., 2017. Prevalence and incidence of Alzheimer's disease in Europe: a meta-analysis. Neurologia 32 (8), 523–532. https://doi.org/10.1016/j.nrl.2016.02.016. (Prevalencia e incidencia de la enfermedad de Alzheimer en Europa: metaanalisis).

Niu F, Sharma A, Feng L, Ozkizilcik A, Muresanu DF, Lafuente JV, Tian ZR, Nozari A, Sharma HS. Nanowired delivery of DL-3-n-butylphthalide induces superior neuroprotection in concussive head injury. Prog. Brain Res. 2019;245:89-118. doi: 10.1016/bs.pbr.2019.03.008. Epub 2019 Apr 2.

Niu, F., Sharma, A., Wang, Z., Feng, L., Muresanu, D.F., Sahib, S., Tian, Z.R., Lafuente, J.V., Buzoianu, A.D., Castellani, R.J., Nozari, A., Patnaik, R., Wiklund, L., Sharma, H.S., 2020. Co-administration of TiO2-nanowired dl-3-n-butylphthalide (dl-NBP) and mesenchymal stem cells enhanced neuroprotection in Parkinson's disease exacerbated by concussive head injury. Prog. Brain Res. 258, 101–155. https://doi.org/10.1016/bs.pbr.2020.09.011.

Nixon, R.A., 2017. Amyloid precursor protein and endosomal-lysosomal dysfunction in Alzheimer's disease: inseparable partners in a multifactorial disease. FASEB J. 31 (7), 2729–2743. https://doi.org/10.1096/fj.201700359.

Nizynski, B., Nieznanska, H., Dec, R., Boyko, S., Dzwolak, W., Nieznanski, K., 2018. Amyloidogenic cross-seeding of tau protein: transient emergence of structural variants of fibrils. PLoS One 13 (7), e0201182. https://doi.org/10.1371/journal.pone.0201182.

O'Brien, R.J., Wong, P.C., 2011. Amyloid precursor protein processing and Alzheimer's disease. Annu. Rev. Neurosci. 34, 185–204. https://doi.org/10.1146/annurev-neuro-061010-113613.

Oakley, H., Cole, S.L., Logan, S., Maus, E., Shao, P., Craft, J., Guillozet-Bongaarts, A., Ohno, M., Disterhoft, J., Van Eldik, L., Berry, R., Vassar, R., 2006. Intraneuronal beta-amyloid aggregates, neurodegeneration, and neuron loss in transgenic mice with five familial Alzheimer's disease mutations: potential factors in amyloid plaque formation. J. Neurosci. 26 (40), 10129–10140. https://doi.org/10.1523/JNEUROSCI.1202-06.2006.

Oboudiyat, C., Glazer, H., Seifan, A., Greer, C., Isaacson, R.S., 2013. Alzheimer's disease. Semin. Neurol. 33 (4), 313–329. https://doi.org/10.1055/s-0033-1359319.

Ohsawa, I., Takamura, C., Morimoto, T., Ishiguro, M., Kohsaka, S., 1999. Amino-terminal region of secreted form of amyloid precursor protein stimulates proliferation of neural stem cells. Eur. J. Neurosci. 11 (6), 1907–1913. https://doi.org/10.1046/j.1460-9568.1999.00601.x.

Ozkizilcik, A., Williams, R., Tian, Z.R., Muresanu, D.F., Sharma, A., Sharma, H.S., 2018a. Synthesis of biocompatible titanate nanofibers for effective delivery of neuroprotective agents. Methods Mol. Biol. 1727, 433–442. https://doi.org/10.1007/978-1-4939-7571-6_35.

Ozkizilcik, A., Sharma, A., Muresanu, D.F., Lafuente, J.V., Tian, Z.R., Patnaik, R., Mössler, H., 2018b. Timed release of cerebrolysin using drug-loaded titanate nanospheres reduces brain pathology and improves behavioral functions in Parkinson's disease. Mol. Neurobiol. 55 (1), 359–369. https://doi.org/10.1007/s12035-017-0747-4. PMID: 28875428.

Ozkizilcik, A., Sharma, A., Lafuente, J.V., Muresanu, D.F., Castellani, R.J., Nozari, A., Tian, Z.R., Mossler, H., Sharma, H.S., 2019. Nanodelivery of cerebrolysin reduces pathophysiology of Parkinson's disease. Prog. Brain Res. 245, 201–246. https://doi.org/10.1016/bs.pbr.2019.03.014.

Pachioni-Vasconcelos Jde, A., Lopes, A.M., Apolinario, A.C., Valenzuela-Oses, J.K., Costa, J.S., Nascimento Lde, O., Pessoa, A., Barbosa, L.R., Rangel-Yagui Cde, O., 2016. Nanostructures for protein drug delivery. Biomater. Sci. 4 (2), 205–218. https://doi.org/10.1039/c5bm00360a.

Pandey, S., Li, L., Cui, D.M., Wang, K., Gao, L., 2018. Perforating brain injury by a rusty steel bar. J. Craniofac. Surg. 29 (4), e372–e375. https://doi.org/10.1097/SCS.0000000000004394.

Pandit, V., Kumar, A., Ashawat, M.S., Verma, C.P., Kumar, P., 2017. Recent advancement and technological aspects of pulsatile drug delivery system—a laconic review. Curr. Drug Targets 18 (10), 1191–1203. https://doi.org/10.2174/1389450117666160208144343.

Paroni, G., Bisceglia, P., Seripa, D., 2019. Understanding the amyloid hypothesis in Alzheimer's disease. J. Alzheimers Dis. 68 (2), 493–510. https://doi.org/10.3233/JAD-180802.

Patnaik, R., Sharma, A., Skaper, S.D., Muresanu, D.F., Lafuente, J.V., Castellani, R.J., Nozari, A., Sharma, H.S., 2018. Histamine H3 inverse agonist BF 2649 or antagonist with partial H4 agonist activity clobenpropit reduces amyloid beta peptide-induced brain pathology in Alzheimer's disease. Mol. Neurobiol. 55 (1), 312–321. https://doi.org/10.1007/s12035-017-0743-8.

Pavlovic, D., Pekic, S., Stojanovic, M., Popovic, V., 2019. Traumatic brain injury: neuropathological, neurocognitive and neurobehavioral sequelae. Pituitary 22 (3), 270–282. https://doi.org/10.1007/s11102-019-00957-9.

Peron, R., Vatanabe, I.P., Manzine, P.R., Camins, A., Cominetti, M.R., 2018. Alpha-secretase ADAM10 regulation: insights into Alzheimer's disease treatment. Pharmaceuticals (Basel) 11 (1), 12. https://doi.org/10.3390/ph11010012.

Pijet, B., Stefaniuk, M., Kaczmarek, L., 2019. MMP-9 contributes to dendritic spine remodeling following traumatic brain injury. Neural Plast. 2019, 3259295. https://doi.org/10.1155/2019/3259295.

Poduslo, J.F., Hultman, K.L., Curran, G.L., Preboske, G.M., Chamberlain, R., Marjanska, M., Garwood, M., Jack Jr., C.R., Wengenack, T.M., 2011. Targeting vascular amyloid in arterioles of Alzheimer disease transgenic mice with amyloid beta protein antibody-coated nanoparticles. J. Neuropathol. Exp. Neurol. 70 (8), 653–661. https://doi.org/10.1097/NEN.0b013e318225038c.

Pohanka, M., 2018. Oxidative stress in Alzheimer disease as a target for therapy. Bratisl. Lek. Listy 119 (9), 535–543. https://doi.org/10.4149/BLL_2018_097.

Ponzio, F., Pozzi, O., Banfi, S., Dorigotti, L., 1989. Brain entry and direct central pharmacological effects of the nootropic drug oxiracetam. Oxiracetam: brain entry and pharmacological effects. Pharmacopsychiatry 22 (Suppl 2), 111–115. https://doi.org/10.1055/s-2007-1014629.

Portbury, S.D., Adlard, P.A., 2015. Traumatic brain injury, chronic traumatic encephalopathy, and Alzheimer's disease: common pathologies potentiated by altered zinc homeostasis. J. Alzheimers Dis. 46 (2), 297–311. https://doi.org/10.3233/JAD-143048.

Postina, R., Schroeder, A., Dewachter, I., Bohl, J., Schmitt, U., Kojro, E., Prinzen, C., Endres, K., Hiemke, C., Blessing, M., Flamez, P., Dequenne, A., Godaux, E., van Leuven, F., Fahrenholz, F., 2004. A disintegrin-metalloproteinase prevents amyloid plaque formation and hippocampal defects in an Alzheimer disease mouse model. J. Clin. Invest. 113 (10), 1456–1464. https://doi.org/10.1172/JCI20864.

Pugazhenthi, S., Qin, L., Reddy, P.H., 2017. Common neurodegenerative pathways in obesity, diabetes, and Alzheimer's disease. Biochim. Biophys. Acta, Mol. Basis Dis. 1863 (5), 1037–1045. https://doi.org/10.1016/j.bbadis.2016.04.017.

Pugliese, A.M., Corradetti, R., Ballerini, L., Pepeu, G., 1990. Effect of the nootropic drug oxiracetam on field potentials of rat hippocampal slices. Br. J. Pharmacol. 99 (1), 189–193. https://doi.org/10.1111/j.1476-5381.1990.tb14676.x.

Qiang, L., Sun, X., Austin, T.O., Muralidharan, H., Jean, D.C., Liu, M., Yu, W., Baas, P.W., 2018. Tau does not stabilize axonal microtubules but rather enables them to have long labile domains. Curr. Biol. 28 (13), 2181–2189.e2184. https://doi.org/10.1016/j.cub.2018.05.045.

Radi, E., Formichi, P., Battisti, C., Federico, A., 2014. Apoptosis and oxidative stress in neurodegenerative diseases. J. Alzheimers Dis. 42 (Suppl. 3), S125–S152. https://doi.org/10.3233/jad-132738.

Rai, S.N., Dilnashin, H., Birla, H., Singh, S.S., Zahra, W., Rathore, A.S., Singh, B.K., Singh, S.P., 2019. The role of PI3K/Akt and ERK in neurodegenerative disorders. Neurotox. Res. 35 (3), 775–795. https://doi.org/10.1007/s12640-019-0003-y.

Raiteri, M., Costa, R., Marchi, M., 1992. Effects of oxiracetam on neurotransmitter release from rat hippocampus slices and synaptosomes. Neurosci. Lett. 145 (1), 109–113. https://doi.org/10.1016/0304-3940(92)90215-s.

Ramos-Cejudo, J., Wisniewski, T., Marmar, C., Zetterberg, H., Blennow, K., de Leon, M.J., Fossati, S., 2018. Traumatic brain injury and Alzheimer's disease: the cerebrovascular link. EBioMedicine 28, 21–30. https://doi.org/10.1016/j.ebiom.2018.01.021.

Regen, F., Hellmann-Regen, J., Costantini, E., Reale, M., 2017. Neuroinflammation and Alzheimer's disease: implications for microglial activation. Curr. Alzheimer Res. 14 (11), 1140–1148. https://doi.org/10.2174/1567205014666170203141717.

Reitz, C., 2015. Genetic diagnosis and prognosis of Alzheimer's disease: challenges and opportunities. Expert Rev. Mol. Diagn. 15 (3), 339–348. https://doi.org/10.1586/14737159.2015.1002469.

Reitz, C., Mayeux, R., 2014. Alzheimer disease: epidemiology, diagnostic criteria, risk factors and biomarkers. Biochem. Pharmacol. 88 (4), 640–651. https://doi.org/10.1016/j.bcp.2013.12.024.

Reitz, C., Brayne, C., Mayeux, R., 2011. Epidemiology of Alzheimer disease. Nat. Rev. Neurol. 7 (3), 137–152. https://doi.org/10.1038/nrneurol.2011.2.

Rissanen, R., Berg, H.Y., Hasselberg, M., 2017. Quality of life following road traffic injury: a systematic literature review. Accid. Anal. Prev. 108, 308–320. https://doi.org/10.1016/j.aap.2017.09.013.

Riva, M., Caratozzolo, S., Cerea, E., Gottardi, F., Zanetti, M., Vicini Chilovi, B., Cristini, C., Padovani, A., Rozzini, L., 2014. Diagnosis disclosure and advance care planning in Alzheimer disease: opinions of a sample of Italian citizens. Aging Clin. Exp. Res. 26 (4), 427–434. https://doi.org/10.1007/s40520-014-0195-1.

Roher, A.E., Maarouf, C.L., Kokjohn, T.A., 2016. Familial presenilin mutations and sporadic Alzheimer's disease pathology: is the assumption of biochemical equivalence justified? J. Alzheimers Dis. 50 (3), 645–658. https://doi.org/10.3233/JAD-150757.

Roher, A.E., Kokjohn, T.A., Clarke, S.G., Sierks, M.R., Maarouf, C.L., Serrano, G.E., Sabbagh, M.S., Beach, T.G., 2017. APP/Abeta structural diversity and Alzheimer's disease pathogenesis. Neurochem. Int. 110, 1–13. https://doi.org/10.1016/j.neuint.2017.08.007.

Rossor, M.N., Newman, S., Frackowiak, R.S., Lantos, P., Kennedy, A.M., 1993. Alzheimer's disease families with amyloid precursor protein mutations. Ann. N. Y. Acad. Sci. 695, 198–202. https://doi.org/10.1111/j.1749-6632.1993.tb23052.x.

Rossor, M.N., Fox, N.C., Mummery, C.J., Schott, J.M., Warren, J.D., 2010. The diagnosis of young-onset dementia. Lancet Neurol. 9 (8), 793–806. https://doi.org/10.1016/S1474-4422(10)70159-9.

Rozzini, R., Zanetti, O., Bianchetti, A., 1993. Treatment of cognitive impairment secondary to degenerative dementia. Effectiveness of oxiracetam therapy. Acta Neurol. 15 (1), 44–52. https://www.ncbi.nlm.nih.gov/pubmed/8456595.

Runchel, C., Matsuzawa, A., Ichijo, H., 2011. Mitogen-activated protein kinases in mammalian oxidative stress responses. Antioxid. Redox Signal. 15 (1), 205–218. https://doi.org/10.1089/ars.2010.3733.

Ruozi, B., Belletti, D., Sharma, H.S., Sharma, A., Muresanu, D.F., Mossler, H., Forni, F., Vandelli, M.A., Tosi, G., 2015. PLGA nanoparticles loaded cerebrolysin: studies on their preparation and investigation of the effect of storage and serum stability with reference to traumatic brain injury. Mol. Neurobiol. 52 (2), 899–912. https://doi.org/10.1007/s12035-015-9235-x.

Sahib, S., Niu, F., Sharma, A., Feng, L., Tian, Z.R., Muresanu, D.F., Nozari, A., Sharma, H.S., 2019. Potentiation of spinal cord conduction and neuroprotection following nanodelivery of DL-3-n-butylphthalide in titanium implanted nanomaterial in a focal spinal cord injury induced functional outcome, blood-spinal cord barrier breakdown and edema formation. Int. Rev. Neurobiol. 146, 153–188. https://doi.org/10.1016/bs.irn.2019.06.009.

Sahib, S., Sharma, A., Menon, P.K., Muresanu, D.F., Castellani, R.J., Nozari, A., Lafuente, J.V., Bryukhovetskiy, I., Tian, Z.R., Patnaik, R., Buzoianu, A.D., Wiklund, L., Sharma, H.S., 2020. Cerebrolysin enhances spinal cord conduction and reduces blood-spinal cord barrier breakdown, edema formation, immediate early gene expression and cord pathology after injury. Prog. Brain Res. 258, 397–438. https://doi.org/10.1016/bs.pbr.2020.09.012.

Salehi, A., Zhang, J.H., Obenaus, A., 2017. Response of the cerebral vasculature following traumatic brain injury. J. Cereb. Blood Flow Metab. 37 (7), 2320–2339. https://doi.org/10.1177/0271678X17701460.

Salim, S., 2017. Oxidative stress and the central nervous system. J. Pharmacol. Exp. Ther. 360 (1), 201–205. https://doi.org/10.1124/jpet.116.237503.

Sando, R., Bushong, E., Zhu, Y., Huang, M., Considine, C., Phan, S., Ju, S., Uytiepo, M., Ellisman, M., Maximov, A., 2017. Assembly of excitatory synapses in the absence of glutamatergic neurotransmission. Neuron 94 (2), 312–321.e313. https://doi.org/10.1016/j.neuron.2017.03.047.

Scheyer, O., Rahman, A., Hristov, H., Berkowitz, C., Isaacson, R.S., Diaz Brinton, R., Mosconi, L., 2018. Female sex and Alzheimer's risk: the menopause connection. J. Prev Alzheimers Dis. 5 (4), 225–230. https://doi.org/10.14283/jpad.2018.34.

Schmidt OI, Heyde CE, Ertel W, Stahel PF. 2005. Closed head injury--an inflammatory disease? Brain Res. Brain Res. Rev. 48(2):388-399. doi: 10.1016/j.brainresrev.2004.12.028. Epub 2005 Jan 28.

Scott, L., Kiss, T., Kawabe, T.T., Hajos, M., 2016. Neuronal network activity in the hippocampus of tau transgenic (Tg4510) mice. Neurobiol. Aging 37, 66–73. https://doi.org/10.1016/j.neurobiolaging.2015.10.002.

Selkoe, D.J., Hardy, J., 2016. The amyloid hypothesis of Alzheimer's disease at 25 years. EMBO Mol. Med. 8 (6), 595–608. https://doi.org/10.15252/emmm.201606210.

Sery, O., Hlinecka, L., Balcar, V.J., Janout, V., Povova, J., 2014. Diabetes, hypertension and stroke—does Alzheimer protect you? Neuro Endocrinol. Lett. 35 (8), 691–696. https://www.ncbi.nlm.nih.gov/pubmed/25702297.

Sharma, H.S., 1987. Effect of captopril (a converting enzyme inhibitor) on blood-brain barrier permeability and cerebral blood flow in normotensive rats. Neuropharmacology 26 (1), 85–92. https://doi.org/10.1016/0028-3908(87)90049-9.

Sharma, H.S., 1999. Pathophysiology of blood-brain barrier, brain edema and cell injury following hyperthermia: new role of heat shock protein, nitric oxide and carbon monoxide. an experimental study in the rat using light and electron microscopy. Acta Universitatis Upsaliensis 830, 1–94.

Sharma, H.S., 2004. Pathophysiology of the blood-spinal cord barrier in traumatic injury. In: Sharma, H.S., Westman, J. (Eds.), The Blood-Spinal Cord and Brain Barriers in Health and Disease. Elsevier Academic Press, San Diego, pp. 437–518.

Sharma HS. Methods to produce brain hyperthermia. Curr. Protoc. Toxicol. 2005;Chapter 11: Unit11.14. doi: 10.1002/0471140856.tx1114s23. PMID: 23045108.

Sharma, H.S., 2006. Hyperthermia influences excitatory and inhibitory amino acid neurotransmitters in the central nervous system. An experimental study in the rat using behavioural, biochemical, pharmacological, and morphological approaches. J. Neural. Transm. (Vienna) 113 (4), 497–519. https://doi.org/10.1007/s00702-005-0406-1.

Sharma, H.S., 2007. Methods to produce hyperthermia-induced brain dysfunction. Prog. Brain Res. 162, 173–199. https://doi.org/10.1016/S0079-6123(06)62010-4.

Sharma, H.S., 2009. Blood–central nervous system barriers: the gateway to neurodegeneration, neuroprotection and neuroregeneration. In: Lajtha, A., Banik, N., Ray, S.K. (Eds.), Handbook of Neurochemistry and Molecular Neurobiology: Brain and Spinal Cord Trauma. Springer Verlag, Berlin, Heidelberg, New York, pp. 363–457.

Sharma, H.S., 2011a. Blood-CNS barrier, neurodegeneration and neuroprotection: recent therapeutic advancements and nano-drug delivery. J. Neural Transm. (Vienna) 118 (1), 3–6. https://doi.org/10.1007/s00702-010-0542-0.

Sharma, H.S., 2011b. Early microvascular reactions and blood-spinal cord barrier disruption are instrumental in pathophysiology of spinal cord injury and repair: novel therapeutic strategies including nanowired drug delivery to enhance neuroprotection. J. Neural Transm. (Vienna) 118 (1), 155–176. https://doi.org/10.1007/s00702-010-0514-4.

Sharma, H.S., Cervós-Navarro, J., 1990. Brain oedema and cellular changes induced by acute heat stress in young rats. Acta Neurochir. Suppl. (Wien) 51, 383–386. https://doi.org/10.1007/978-3-7091-9115-6_129.

Sharma, H.S., Dey, P.K., 1984. Role of 5-HT on increased permeability of blood-brain barrier under heat stress. Indian J. Physiol. Pharmacol. 28 (4), 259–267.

Sharma, H.S., Dey, P.K., 1986a. Influence of long-term immobilization stress on regional blood-brain barrier permeability, cerebral blood flow and 5-HT level in conscious normotensive young rats. J. Neurol. Sci. 72 (1), 61–76. https://doi.org/10.1016/0022-510x(86)90036-5.

Sharma, H.S., Dey, P.K., 1986b. Probable involvement of 5-hydroxytryptamine in increased permeability of blood-brain barrier under heat stress in young rats. Neuropharmacology 25 (2), 161–167. https://doi.org/10.1016/0028-3908(86)90037-7.

Sharma, H.S., Dey, P.K., 1987. Influence of long-term acute heat exposure on regional blood-brain barrier permeability, cerebral blood flow and 5-HT level in conscious normotensive young rats. Brain Res. 424 (1), 153–162. https://doi.org/10.1016/0006-8993(87)91205-4.

Sharma, H.S., Dey, P.K., 1988. EEG changes following increased blood-brain barrier permeability under long-term immobilization stress in young rats. Neurosci. Res. 5 (3), 224–239. https://doi.org/10.1016/0168-0102(88)90051-x.

Sharma, H.S., Olsson, Y., 1990. Edema formation and cellular alterations following spinal cord injury in the rat and their modification with p-chlorophenylalanine. Acta Neuropathol. 79 (6), 604–610. https://doi.org/10.1007/BF00294237.

Sharma, H.S., Sharma, A., 2011. New strategies for CNS injury and repair using stem cells, nanomedicine, neurotrophic factors and novel neuroprotective agents. Expert Rev. Neurother. 11 (8), 1121–1124. https://doi.org/10.1586/ern.11.100.

Sharma, A., Sharma, H.S., 2019. Nanoneuroprotection and nanoneurotoxicology. Prog. Brain Res. 245, 1–321. https://doi.org/10.1016/S0079-6123(19)30076-7.

Sharma, H.S., Sharma, A., 2020. Neuropharmacology of neuroprotection. Prog. Brain Res. 258, 1–491. https://doi.org/10.1016/S0079-6123(20)30227-2.

Sharma, H.S., Sjöquist, P.O., 2002. A new antioxidant compound H-290/51 modulates glutamate and GABA immunoreactivity in the rat spinal cord following trauma. Amino Acids 23 (1–3), 261–272. https://doi.org/10.1007/s00726-001-0137-z.

Sharma H.S., Westman J., (2004). The Blood-Spinal Cord and Brain Barriers in Health and Disease, Academic Press, San Diego, pp. 1-617 (Release date: Nov. 9, 2003).

Sharma, H.S., Olsson, Y., Dey, P.K., 1990a. Changes in blood-brain barrier and cerebral blood flow following elevation of circulating serotonin level in anesthetized rats. Brain Res. 517 (1–2), 215–223. https://doi.org/10.1016/0006-8993(90)91029-g.

Sharma, H.S., Olsson, Y., Dey, P.K., 1990b. Early accumulation of serotonin in rat spinal cord subjected to traumatic injury. Relation to edema and blood flow changes. Neuroscience 36 (3), 725–730. https://doi.org/10.1016/0306-4522(90)90014-u.

Sharma, H.S., Cervós-Navarro, J., Dey, P.K., 1991. Acute heat exposure causes cellular alteration in cerebral cortex of young rats. Neuroreport 2 (3), 155–158. https://doi.org/10.1097/00001756-199103000-00012.

Sharma, H.S., Zimmer, C., Westman, J., Cervós-Navarro, J., 1992a. Acute systemic heat stress increases glial fibrillary acidic protein immunoreactivity in brain: experimental observations in conscious normotensive young rats. Neuroscience 48 (4), 889–901. https://doi.org/10.1016/0306-4522(92)90277-9.

Sharma, H.S., Kretzschmar, R., Cervós-Navarro, J., Ermisch, A., Rühle, H.J., Dey, P.K., 1992b. Age-related pathophysiology of the blood-brain barrier in heat stress. Prog. Brain Res. 91, 189–196. https://doi.org/10.1016/s0079-6123(08)62334-1.

Sharma, H.S., Olsson, Y., Persson, S., Nyberg, F., 1995. Trauma-induced opening of the the blood-spinal cord barrier is reduced by indomethacin, an inhibitor of prostaglandin biosynthesis. Experimental observations in the rat using [131I]-sodium, Evans blue and lanthanum as tracers. Restor. Neurol. Neurosci. 17 (4), 207–215. https://doi.org/10.3233/RNN-1994-7403.

Sharma, H.S., Westman, J., Olsson, Y., Alm, P., 1996. Involvement of nitric oxide in acute spinal cord injury: an immunocytochemical study using light and electron microscopy in the rat. Neurosci. Res. 24 (4), 373–384. https://doi.org/10.1016/0168-0102(95)01015-7.

Sharma, H.S., Nyberg, F., Gordh, T., Alm, P., Westman, J., 1997a. Topical application of insulin like growth factor-1 reduces edema and upregulation of neuronal nitric oxide synthase following trauma to the rat spinal cord. Acta Neurochir. Suppl. 70, 130–133. https://doi.org/10.1007/978-3-7091-6837-0_40.

Sharma, H.S., Westman, J., Cervós-Navarro, J., Dey, P.K., Nyberg, F., 1997b. Opioid receptor antagonists attenuate heat stress-induced reduction in cerebral blood flow, increased blood-brain barrier permeability, vasogenic edema and cell changes in the rat. Ann. N. Y. Acad. Sci. 813, 559–571. https://doi.org/10.1111/j.1749-6632.1997.tb51747.x.

Sharma, H.S., Westman, J., Nyberg, F., 1998a. Pathophysiology of brain edema and cell changes following hyperthermic brain injury. Prog. Brain Res. 115, 351–412. https://doi.org/10.1016/s0079-6123(08)62043-9.

Sharma, H.S., Alm, P., Westman, J., 1998b. Nitric oxide and carbon monoxide in the brain pathology of heat stress. Prog. Brain Res. 115, 297–333. https://doi.org/10.1016/s0079-6123(08)62041-5.

Sharma, H.S., Drieu, K., Alm, P., Westman, J., 2000. Role of nitric oxide in blood-brain barrier permeability, brain edema and cell damage following hyperthermic brain injury. An experimental study using EGB-761 and Gingkolide B pretreatment in the rat. Acta Neurochir. Suppl. 76, 81–86. https://doi.org/10.1007/978-3-7091-6346-7_17.

Sharma, H.S., Winkler, T., Stålberg, E., Gordh, T., Alm, P., Westman, J., 2003. Topical application of TNF-alpha antiserum attenuates spinal cord trauma induced edema formation, microvascular permeability disturbances and cell injury in the rat. Acta Neurochir. Suppl. 86, 407–413. https://doi.org/10.1007/978-3-7091-0651-8_85.

Sharma, H.S., Patnaik, R., Ray, A.K., Dey, P.K., 2004. Blood-central nervous system barriers in morphine dependence and withdrawal. In: Sharma, H S, Westman, J (Eds.), The Blood-Spinal Cord and Brain Barriers in Health and Disease. Elsevier Academic Press, San Diego, pp. 299–328.

Sharma, H.S., Wiklund, L., Badgaiyan, R.D., Mohanty, S., Alm, P., 2006. Intracerebral administration of neuronal nitric oxide synthase antiserum attenuates traumatic brain injury-induced blood-brain barrier permeability, brain edema formation, and sensory motor disturbances in the rat. Acta Neurochir. Suppl. 96, 288–294. https://doi.org/10.1007/3-211-30714-1_62.

Sharma, H.S., Sjöquist, P.O., Ali, S.F., 2007. Drugs of abuse-induced hyperthermia, blood-brain barrier dysfunction and neurotoxicity: neuroprotective effects of a new antioxidant compound H-290/51. Curr. Pharm. Des. 13 (18), 1903–1923. https://doi.org/10.2174/138161207780858375.

Sharma, H.S., Ali, S.F., Dong, W., Tian, Z.R., Patnaik, R., Patnaik, S., Sharma, A., Boman, A., Lek, P., Seifert, E., Lundstedt, T., 2007a. Drug delivery to the spinal cord tagged with nanowire enhances neuroprotective efficacy and functional recovery following trauma to the rat spinal cord. Ann. N. Y. Acad. Sci. 1122, 197–218. https://doi.org/10.1196/annals.1403.014.

Sharma, H.S., Lundstedt, T., Flärdh, M., Skottner, A., Wiklund, L., 2007b. Neuroprotective effects of melanocortins in CNS injury. Curr. Pharm. Des. 13 (19), 1929–1941. https://doi.org/10.2174/138161207781039797.

Sharma, H.S., Ali, S., Tian, Z.R., Patnaik, R., Patnaik, S., Lek, P., Sharma, A., Lundstedt, T., 2009a. Nano-drug delivery and neuroprotection in spinal cord injury. J. Nanosci. Nanotechnol. 9 (8), 5014–5037. https://doi.org/10.1166/jnn.2009.gr04.

Sharma, H.S., Ali, S.F., Tian, Z.R., Hussain, S.M., Schlager, J.J., Sjoquist, P.O., Sharma, A., Muresanu, D.F., 2009b. Chronic treatment with nanoparticles exacerbate hyperthermia induced blood-brain barrier breakdown, cognitive dysfunction and brain pathology in the rat. Neuroprotective effects of nanowired-antioxidant compound H-290/51. J. Nanosci. Nanotechnol. 9 (8), 5073–5090. https://doi.org/10.1166/jnn.2009.gr10.

Sharma, H.S., Sharma, A., Hussain, S., Schlager, J., Sjöquist, P.O., Muresanu, D., 2010. A new antioxidant compound H-290/51 attenuates nanoparticle induced neurotoxicity and enhances neurorepair in hyperthermia. Acta Neurochir. Suppl. 106, 351–357. https://doi.org/10.1007/978-3-211-98811-4_64.

Sharma, A., Muresanu, D.F., Mossler, H., Sharma, H.S., 2012a. Superior neuroprotective effects of cerebrolysin in nanoparticle-induced exacerbation of hyperthermia-induced brain pathology. CNS Neurol. Disord. Drug Targets 11 (1), 7–25. https://doi.org/10.2174/187152712799960790.

Sharma, H.S., Castellani, R.J., Smith, M.A., Sharma, A., 2012b. The blood-brain barrier in Alzheimer's disease: novel therapeutic targets and nanodrug delivery. Int. Rev. Neurobiol. 102, 47–90. https://doi.org/10.1016/B978-0-12-386986-9.00003-X.

Sharma, H.S., Sharma, A., Mossler, H., Muresanu, D.F., 2012c. Neuroprotective effects of cerebrolysin, a combination of different active fragments of neurotrophic factors and peptides on the whole body hyperthermia-induced neurotoxicity: modulatory roles of co-morbidity factors and nanoparticle intoxication. Int. Rev. Neurobiol. 102, 249–276. https://doi.org/10.1016/B978-0-12-386986-9.00010-7.

Sharma, H.S., Muresanu, D.F., Sharma, A., 2013. Novel therapeutic strategies using nanodrug delivery, stem cells and combination therapy for CNS trauma and neurodegenerative disorders. Expert Rev. Neurother. 13 (10), 1085–1088. https://doi.org/10.1586/14737175.2013.836297.

Sharma, H.S., Muresanu, D.F., Lafuente, J.V., Nozari, A., Patnaik, R., Skaper, S.D., Sharma, A., 2016. Pathophysiology of blood-brain barrier in brain injury in cold and hot environments: novel drug targets for neuroprotection. CNS Neurol. Disord. Drug Targets 15 (9), 1045–1071. https://doi.org/10.2174/1871527315666160902145145.

Sharma, H.S., Muresanu, D.F., Sharma, A., 2016. Alzheimer's disease: cerebrolysin and nanotechnology as a therapeutic strategy. Neurodegener. Dis. Manag. 6 (6), 453–456. https://doi.org/10.2217/nmt-2016-0037.

Sharma, A., Muresanu, D.F., Lafuente, J.V., Sjoquist, P.O., Patnaik, R., Ryan Tian, Z., Ozkizilcik, A., Sharma, H.S., 2018a. Cold environment exacerbates brain pathology and oxidative stress following traumatic brain injuries: potential therapeutic effects of nanowired antioxidant compound H-290/51. Mol. Neurobiol. 55 (1), 276–285. https://doi.org/10.1007/s12035-017-0740-y.

Sharma, H.S., Muresanu, D.F., Lafuente, J.V., Patnaik, R., Tian, Z.R., Ozkizilcik, A., Castellani, R.J., Mossler, H., Sharma, A., 2018b. Co-administration of TiO2 nanowired mesenchymal stem cells with cerebrolysin potentiates neprilysin level and reduces brain pathology in Alzheimer's disease. Mol. Neurobiol. 55 (1), 300–311. https://doi.org/10.1007/s12035-017-0742-9.

Sharma, A., Castellani, R.J., Smith, M.A., Muresanu, D.F., Dey, P.K., Sharma, H.S., 2019a. 5-Hydroxytryptophan: a precursor of serotonin influences regional blood-brain barrier breakdown, cerebral blood flow, brain edema formation, and neuropathology. Int. Rev. Neurobiol. 146, 1–44. https://doi.org/10.1016/bs.irn.2019.06.005.

Sharma, H.S., Muresanu, D.F., Castellani, R.J., Nozari, A., Lafuente, J.V., Tian, Z.R., Ozkizilcik, A., Manzhulo, I., Mossler, H., Sharma, A., 2019b. Nanowired delivery of cerebrolysin with neprilysin and p-tau antibodies induces superior neuroprotection in Alzheimer's disease. Prog. Brain Res. 245, 145–200. https://doi.org/10.1016/bs.pbr.2019.03.009.

Sharma, A., Muresanu, D.F., Castellani, R.J., Nozari, A., Lafuente, J.V., Sahib, S., Tian, Z.R., Buzoianu, A.D., Patnaik, R., Wiklund, L., Sharma, H.S., 2020a. Mild traumatic brain injury exacerbates Parkinson's disease induced hemeoxygenase-2 expression and brain pathology: neuroprotective effects of co-administration of TiO2 nanowired mesenchymal stem cells and cerebrolysin. Prog. Brain Res. 258, 157–231. https://doi.org/10.1016/bs. pbr.2020.09.010.

Sharma, A., Muresanu, D.F., Sahib, S., Tian, Z.R., Castellani, R.J., Nozari, A., Lafuente, J.V., Buzoianu, A.D., Bryukhovetskiy, I., Manzhulo, I., Patnaik, R., Wiklund, L., Sharma, H.S., 2020b. Concussive head injury exacerbates neuropathology of sleep deprivation: superior neuroprotection by co-administration of TiO2-nanowired cerebrolysin, alpha-melanocyte-stimulating hormone, and mesenchymal stem cells. Prog. Brain Res. 258, 1–77. https://doi. org/10.1016/bs.pbr.2020.09.003.

Sharma, H.S., Muresanu, D.F., Castellani, R.J., Nozari, A., Lafuente, J.V., Tian, Z.R., Sahib, S., Bryukhovetskiy, I., Bryukhovetskiy, A., Buzoianu, A.D., Patnaik, R., Wiklund, L., Sharma, A., 2020c. Pathophysiology of blood-brain barrier in brain tumor. Novel therapeutic advances using nanomedicine. Int. Rev. Neurobiol. 151, 1–66. https:// doi.org/10.1016/bs.irn.2020.03.001.

Sharma, H.S., Sahib, S., Tian, Z.R., Muresanu, D.F., Nozari, A., Castellani, R.J., Lafuente, J.V., Wiklund, L., Sharma, A., 2020d. Protein kinase inhibitors in traumatic brain injury and repair: new roles of nanomedicine. Prog. Brain Res. 258, 233–283. https://doi.org/ 10.1016/bs.pbr.2020.09.009.

Shinagawa, S., 2016. Language symptoms of Alzheimer's disease. Brain Nerve 68 (5), 551–557. https://doi.org/10.11477/mf.1416200437.

Shiotsuki, H., Yoshimi, K., Shimo, Y., Funayama, M., Takamatsu, Y., Ikeda, K., Takahashi, R., Kitazawa, S., Hattori, N., 2010. A rotarod test for evaluation of motor skill learning. J. Neurosci. Methods 189 (2), 180–185. https://doi.org/10.1016/j.jneu-meth.2010.03.026 (Epub 2010 Mar 30).

Sibener, L., Zaganjor, I., Snyder, H.M., Bain, L.J., Egge, R., Carrillo, M.C., 2014. Alzheimer's Disease prevalence, costs, and prevention for military personnel and veterans. Alzheimers Dement. 10 (3), S105–110. https://doi.org/10.1016/j.jalz.2014.04.011.

Silva, M.C., Ferguson, F.M., Cai, Q., Donovan, K.A., Nandi, G., Patnaik, D., Zhang, T., Huang, H.T., Lucente, D.E., Dickerson, B.C., Mitchison, T.J., Fischer, E.S., Gray, N.S., Haggarty, S.J., 2019a. Targeted degradation of aberrant tau in frontotemporal dementia patient-derived neuronal cell models. Elife 8, e45457. https://doi.org/10.7554/eLife. 45457.

Silva, M.V.F., Loures, C.M.G., Alves, L.C.V., de Souza, L.C., Borges, K.B.G., Carvalho, M.D.G., 2019b. Alzheimer's disease: risk factors and potentially protective measures. J. Biomed. Sci. 26 (1), 33. https://doi.org/10.1186/s12929-019-0524-y.

Smith, D.H., Okiyama, K., Thomas, M.J., Claussen, B., McIntosh, T.K., 1991. Evaluation of memory dysfunction following experimental brain injury using the Morris water maze. J. Neurotrauma. 8 (4), 259–269. https://doi.org/10.1089/neu.1991.8.259.

Smith, D.H., Uryu, K., Saatman, K.E., Trojanowski, J.Q., McIntosh, T.K., 2003. Protein accumulation in traumatic brain injury. Neuromolecular Med. 4 (1–2), 59–72. https://doi. org/10.1385/NMM:4:1-2:59.

Solstrand Dahlberg L, Becerra L, Borsook D, Linnman C. 2018. Brain changes after spinal cord injury, a quantitative meta-analysis and review. Neurosci. Biobehav. Rev. 90: 272-293. doi: 10.1016/j.neubiorev.2018.04.018. Epub 2018 Apr 24.

Sontag, J.M., Sontag, E., 2014. Protein phosphatase 2A dysfunction in Alzheimer's disease. Front. Mol. Neurosci. 7, 16. https://doi.org/10.3389/fnmol.2014.00016.

Soria Lopez, J.A., Gonzalez, H.M., Leger, G.C., 2019. Alzheimer's disease. Handb. Clin. Neurol. 167, 231–255. https://doi.org/10.1016/B978-0-12-804766-8.00013-3.

Spignoli, G., Pepeu, G., 1986. Oxiracetam prevents electroshock-induced decrease in brain acetylcholine and amnesia. Eur. J. Pharmacol. 126 (3), 253–257. https://doi.org/10.1016/0014-2999(86)90055-5.

Spignoli, G., Pepeu, G., 1987. Interactions between oxiracetam, aniracetam and scopolamine on behavior and brain acetylcholine. Pharmacol. Biochem. Behav. 27 (3), 491–495. https://doi.org/10.1016/0091-3057(87)90353-4.

Spignoli, G., Pedata, F., Giovannelli, L., Banfi, S., Moroni, F., Pepeu, G., 1986. Effect of oxiracetam and piracetam on central cholinergic mechanisms and active-avoidance acquisition. Clin. Neuropharmacol. 9 (Suppl. 3), S39–S47. https://www.ncbi.nlm.nih.gov/pubmed/3594455.

Stamouli, E.C., Politis, A.M., 2016. Pro-inflammatory cytokines in Alzheimer's disease. Psychiatriki 27 (4), 264–275. https://doi.org/10.22365/jpsych.2016.274.264.

Stepler, K.E., Robinson, R.A.S., 2019. The potential of 'omics to link lipid metabolism and genetic and comorbidity risk factors of Alzheimer's disease in African Americans. Adv. Exp. Med. Biol. 1118, 1–28. https://doi.org/10.1007/978-3-030-05542-4_1.

Stoccoro, A., Coppede, F., 2018. Role of epigenetics in Alzheimer's disease pathogenesis. Neurodegener. Dis. Manag. 8 (3), 181–193. https://doi.org/10.2217/nmt-2018-0004.

Sun, X.Y., Tuo, Q.Z., Liuyang, Z.Y., Xie, A.J., Feng, X.L., Yan, X., Qiu, M., Li, S., Wang, X.L., Cao, F.Y., Wang, X.C., Wang, J.Z., Liu, R., 2016. Extrasynaptic NMDA receptor-induced tau overexpression mediates neuronal death through suppressing survival signaling ERK phosphorylation. Cell Death Dis. 7 (11), e2449. https://doi.org/10.1038/cddis.2016.329.

Taheri, P., Keshavarzi, S., Ebadi, M., Motaghinejad, M., Motevalian, M., 2018. Neuroprotective effects of forced exercise and bupropion on chronic methamphetamine-induced cognitive impairment via modulation of cAMP response element-binding protein/brain-derived neurotrophic factor signaling pathway, oxidative stress, and inflammatory biomarkers in rats. Adv. Biomed. Res. 7, 151. https://doi.org/10.4103/abr.abr_11_18.

Takahashi-Nakazato, A., Parajuli, L.K., Iwasaki, H., Tanaka, S., Okabe, S., 2019. Ultrastructural observation of glutamatergic synapses by focused ion beam scanning electron microscopy (FIB/SEM). Methods Mol. Biol. 1941, 17–27. https://doi.org/10.1007/978-1-4939-9077-1_2.

Takizawa, C., Thompson, P.L., van Walsem, A., Faure, C., Maier, W.C., 2015. Epidemiological and economic burden of Alzheimer's disease: a systematic literature review of data across Europe and the United States of America. J. Alzheimers Dis. 43 (4), 1271–1284. https://doi.org/10.3233/JAD-141134.

Tan, J.Z.A., Gleeson, P.A., 2019. The trans-Golgi network is a major site for alpha-secretase processing of amyloid precursor protein in primary neurons. J. Biol. Chem. 294 (5), 1618–1631. https://doi.org/10.1074/jbc.RA118.005222.

Tan, R., Lam, A.J., Tan, T., Han, J., Nowakowski, D.W., Vershinin, M., Simo, S., Ori-McKenney, K.M., McKenney, R.J., 2019. Microtubules gate tau condensation to spatially regulate microtubule functions. Nat. Cell Biol. 21 (9), 1078–1085. https://doi.org/10.1038/s41556-019-0375-5.

Tao, Y., Peters, M.E., Drye, L.T., Devanand, D.P., Mintzer, J.E., Pollock, B.G., Porsteinsson, A.P., Rosenberg, P.B., Schneider, L.S., Shade, D.M., Weintraub, D., Yesavage, J., Lyketsos, C.G., Munro, C.A., 2018. Sex differences in the neuropsychiatric symptoms of patients with Alzheimer's disease. Am. J. Alzheimers Dis. Other Demen. 33 (7), 450–457. https://doi.org/10.1177/1533317518783278.

Tcw, J., Goate, A.M., 2017. Genetics of beta-amyloid precursor protein in Alzheimer's disease. Cold Spring Harb. Perspect. Med. 7 (6), a024539. https://doi.org/10.1101/cshperspect.a024539.

Tian, Z.R., Sharma, A., Nozari, A., Subramaniam, R., Lundstedt, T., Sharma, H.S., 2012. Nanowired drug delivery to enhance neuroprotection in spinal cord injury. CNS Neurol. Disord. Drug Targets 11 (1), 86–95. https://doi.org/10.2174/187152712799960727.

Tian, M., Hou, D., Deng, Y., Li, W., Feng, X., 2013. Expression of STAT3 and P-STAT3 in the brain of a transgenic mouse model of Alzheimer's disease. Nan Fang Yi Ke Da Xue Xue Bao 33 (12), 1778–1782. https://www.ncbi.nlm.nih.gov/pubmed/24369244.

Tomiyama, T., 2010. Involvement of beta-amyloid in the etiology of Alzheimer's disease. Brain Nerve 62 (7), 691–699. https://www.ncbi.nlm.nih.gov/pubmed/20675873.

Tönnies, E., Trushina, E., 2017. Oxidative stress, synaptic dysfunction, and Alzheimer's disease. J. Alzheimers Dis. 57 (4), 1105–1121. https://doi.org/10.3233/jad-161088.

Troutman AD, Gallardo EJ, Brown MB, Coggan AR. 2018. Measurement of nitrate and nitrite in biopsy-sized muscle samples using HPLC. J. Appl. Physiol. (1985), 125(5):1475-1481. doi: 10.1152/japplphysiol.00625.2018. Epub 2018 Aug 16.

Truong, N.P., Whittaker, M.R., Mak, C.W., Davis, T.P., 2015. The importance of nanoparticle shape in cancer drug delivery. Expert Opin. Drug Deliv. 12 (1), 129–142. https://doi.org/10.1517/17425247.2014.950564.

Ueno, M., Tomimoto, H., Akiguchi, I., Wakita, H., Sakamoto, H., 2002. Blood-brain barrier disruption in white matter lesions in a rat model of chronic cerebral hypoperfusion. J. Cereb. Blood Flow Metab. 22 (1), 97–104. https://doi.org/10.1097/00004647-200201000-00012.

VanItallie, T.B., 2019. Traumatic brain injury (TBI) in collision sports: possible mechanisms of transformation into chronic traumatic encephalopathy (CTE). Metabolism 100S, 153943. https://doi.org/10.1016/j.metabol.2019.07.007.

Vannemreddy, P., Ray, A.K., Patnaik, R., Patnaik, S., Mohanty, S., Sharma, H.S., 2006. Zinc protoporphyrin IX attenuates closed head injury-induced edema formation, blood-brain barrier disruption, and serotonin levels in the rat. Acta Neurochir. Suppl. 96, 151–156. https://doi.org/10.1007/3-211-30714-1_34.

Vega, I.E., Cabrera, L.Y., Wygant, C.M., Velez-Ortiz, D., Counts, S.E., 2017. Alzheimer's disease in the Latino community: intersection of genetics and social determinants of health. J. Alzheimers Dis. 58 (4), 979–992. https://doi.org/10.3233/JAD-161261.

Venkatramani, A., Panda, D., 2019. Regulation of neuronal microtubule dynamics by tau: implications for tauopathies. Int. J. Biol. Macromol. 133, 473–483. https://doi.org/10.1016/j.ijbiomac.2019.04.120.

Wang, Y., Mandelkow, E., 2016. Tau in physiology and pathology. Nat. Rev. Neurosci. 17 (1), 5–21. https://doi.org/10.1038/nrn.2015.1.

Wang, R., Reddy, P.H., 2017. Role of glutamate and NMDA receptors in Alzheimer's disease. J. Alzheimers Dis. 57 (4), 1041–1048. https://doi.org/10.3233/jad-160763.

Wang, C., Li, F., Guan, Y., Zhu, L., Fei, Y., Zhang, J., Pan, Y., 2014a. Bone marrow stromal cells combined with oxiracetam influences the expression of B-cell lymphoma 2 in rats with ischemic stroke. J. Stroke Cerebrovasc. Dis. 23 (10), 2591–2597. https://doi.org/10.1016/j.jstrokecerebrovasdis.2014.05.035.

Wang, X., Wang, W., Li, L., Perry, G., Lee, H.G., Zhu, X., 2014b. Oxidative stress and mitochondrial dysfunction in Alzheimer's disease. Biochim. Biophys. Acta 1842 (8), 1240–1247. https://doi.org/10.1016/j.bbadis.2013.10.015.

Wang, J., Sun, R., Li, Z., Pan, Y., 2019. Combined bone marrow stromal cells and oxiracetam treatments ameliorates acute cerebral ischemia/reperfusion injury through TRPC6. Acta Biochim. Biophys. Sin. (Shanghai) 51 (8), 767–777. https://doi.org/10.1093/abbs/gmz059.

Weller, J., Budson, A., 2018. Current understanding of Alzheimer's disease diagnosis and treatment. F1000Res 7. F1000 Faculty Rev-1161 https://doi.org/10.12688/f1000research.14506.1.

Weston NM, Rolfe AT, Freelin AH, Reeves TM, Sun D. Traumatic brain injury modifies synaptic plasticity in newly-generated granule cells of the adult hippocampus. Exp. Neurol. 2021 Feb;336:113527. doi: 10.1016/j.expneurol.2020.113527. Epub 2020 Nov 11.

Wilkins, H.M., Swerdlow, R.H., 2016. Relationships between mitochondria and neuroinflammation: implications for Alzheimer's disease. Curr. Top. Med. Chem. 16 (8), 849–857. https://doi.org/10.2174/1568026615666150827095102.

Wilkins, H.M., Swerdlow, R.H., 2017. Amyloid precursor protein processing and bioenergetics. Brain Res. Bull. 133, 71–79. https://doi.org/10.1016/j.brainresbull.2016.08.009.

Wolf, J.A., Johnson, B.N., Johnson, V.E., Putt, M.E., Browne, K.D., Mietus, C.J., Brown, D.P., Wofford, K.L., Smith, D.H., Grady, M.S., Cohen, A.S., Cullen, D.K., 2017. Concussion induces hippocampal circuitry disruption in swine. J. Neurotrauma 34 (14), 2303–2314. https://doi.org/10.1089/neu.2016.4848.

Wolinsky, D., Drake, K., Bostwick, J., 2018. Diagnosis and management of neuropsychiatric symptoms in Alzheimer's disease. Curr. Psychiatry Rep. 20 (12), 117. https://doi.org/10.1007/s11920-018-0978-8.

Woodcock T., Morganti-Kossmann M.C., The role of markers of inflammation in traumatic brain injury, Front Neurol. 2013, 4:18. doi: 10.3389/fneur.2013.00018. eCollection 2013.

Wu, L., Rosa-Neto, P., Hsiung, G.Y., Sadovnick, A.D., Masellis, M., Black, S.E., Jia, J., Gauthier, S., 2012. Early-onset familial Alzheimer's disease (EOFAD). Can. J. Neurol. Sci. 39 (4), 436–445. https://doi.org/10.1017/s0317167100013949.

Xu, J., Zhang, H., Zhang, J., Liu, X., He, X., Xu, D., Qian, J., Liu, L., Sun, J., 2012. Hydrothermal synthesis of potassium/sodium titanate nanofibres and their ultraviolet properties. Micro Nano Lett. 7 (5), 407–411.

Xu, J., Qi, Q., Lv, P., Dong, Y., Jiang, X., Liu, Z., 2019. Oxiracetam ameliorates cognitive deficits in vascular dementia rats by regulating the expression of neuronal apoptosis/autophagy-related genes associated with the activation of the Akt/mTOR signaling pathway. Braz. J. Med. Biol. Res. 52 (11), e8371. https://doi.org/10.1590/1414-431X20198371.

Xue-Shan, Z., Juan, P., Qi, W., Zhong, R., Li-Hong, P., Zhi-Han, T., Zhi-Sheng, J., Gui-Xue, W., Lu-Shan, L., 2016. Imbalanced cholesterol metabolism in Alzheimer's disease. Clin. Chim. Acta 456, 107–114. https://doi.org/10.1016/j.cca.2016.02.024.

Yao, X.L., Yao, Z.H., Li, L., Nie, L., Zhang, S.F., 2016. Oxiracetam can improve cognitive impairment after chronic cerebral hypoperfusion in rats. Psychiatry Res. 246, 284–292. https://doi.org/10.1016/j.psychres.2016.10.006.

Yilmaz, U., 2015. Alzheimer's disease. Radiologe 55 (5), 386–388. https://doi.org/10.1007/s00117-014-2796-2. (Alzheimer-Demenz.).

Yuksel, M., Tacal, O., 2019. Trafficking and proteolytic processing of amyloid precursor protein and secretases in Alzheimer's disease development: an up-to-date review. Eur. J. Pharmacol. 856, 172415. https://doi.org/10.1016/j.ejphar.2019.172415.

Zhang, Y.W., Thompson, R., Zhang, H., Xu, H., 2011. APP processing in Alzheimer's disease. Mol. Brain 4, 3. https://doi.org/10.1186/1756-6606-4-3.

Zhang, T., Zhang, Y., Cui, M., Jin, L., Wang, Y., Lv, F., Liu, Y., Zheng, W., Shang, H., Zhang, J., Zhang, M., Wu, H., Guo, J., Zhang, X., Hu, X., Cao, C.M., Xiao, R.P., 2016. CaMKII is a RIP3 substrate mediating ischemia- and oxidative stress-induced myocardial necroptosis. Nat. Med. 22 (2), 175–182. https://doi.org/10.1038/nm.4017.

Zhang, W., Zhang, Y., Yu, L., Wu, N.-L., Huang, H., Wei, M., 2019. TiO_2-B nanowires via topological conversion with enhanced lithium-ion intercalation properties. J. Mater. Chem. A 7 (8), 3842–3847.

Zhong, J., 2016. RAS and downstream RAF-MEK and PI3K-AKT signaling in neuronal development, function and dysfunction. Biol. Chem. 397 (3), 215–222. https://doi.org/10.1515/hsz-2015-0270.

Zissimopoulos, J.M., Barthold, D., Brinton, R.D., Joyce, G., 2017. Sex and race differences in the association between statin use and the incidence of Alzheimer disease. JAMA Neurol. 74 (2), 225–232. https://doi.org/10.1001/jamaneurol.2016.3783.

Zysk, M., Clausen, F., Aguilar, X., Sehlin, D., Syvanen, S., Erlandsson, A., 2019. Long-term effects of traumatic brain injury in a mouse model of Alzheimer's Disease. J. Alzheimers Dis. 72 (1), 161–180. https://doi.org/10.3233/JAD-190572.

Clinical neurorestorative cell therapies for stroke

5

Hongyun Huang[a,b,*], Gengsheng Mao[a], Lin Chen[c], and Hari Shanker Sharma[d,*]

[a]*Beijing Hongtianji Neuroscience Academy, Beijing, People Republic of China*
[b]*Institute of Neurorestoratology, Third Medical Center of General Hospital of PLA, Beijing, People Republic of China*
[c]*Department of Neurosurgery, Dongzhimen Hospital of Beijing University of Traditional Chinese Medicine, Beijing, China*
[d]*International Experimental Central Nervous System Injury & Repair (IECNSIR), Department of Surgical Sciences, Anesthesiology & Intensive Care Medicine, Uppsala University Hospital, Uppsala University, Uppsala, Sweden*
*Corresponding authors: Tel.: +86-13910116608; Fax: +86-10-57976081 (Hongyun Huang); Tel.: +46 70 2011 801; Fax: +46 18 24 38 99 (Hari Shanker Sharma), e-mail address: huanghongyun001@126.com; sharma@surgsci.uu.se

Abstract

Clinical neurorestorative cell therapies for stroke have been explored for over 20 years. Majority cell therapies have shown neurorestorative effects for stroke on non-double-blind studies. In this review, we summarize types of cell transplantation, transplanted routes, therapeutic time windows, dosage, results of exploring trials or clinical studies, results of multicenter, double-blind or observing-blind, randomized, placebo-controlled clinical trials. The clinical application prospects of majority cell therapies for stroke need to prove their neurorestorative effects through trials with higher-level evidence-based medical evidence. Currently olfactory ensheathing cell is only one kind of cell to show neurorestorative effects through multicenter, double-blind, randomized, placebo-controlled clinical trials, which should be explored to optimize themselves effects and combination with others.

Keywords

Cell therapy, Clinical trial, Olfactory ensheathing cell, Neurorestoration, Stroke

1 Introduction

Clinical cell therapies (CTs) for stroke have been explored for over 20 years. Preliminary results for stroke, in the beginning, gave a big promising hope to patients, health care providers, and clinical scientific circle (Kondziolka et al., 2000). Following studies in varying kind CTs showed them to be able to restore neurological functions

Progress in Brain Research, Volume 265, ISSN 0079-6123, https://doi.org/10.1016/bs.pbr.2021.06.006

in a degree for stroke, but most of them do not prove their clinical application value through multicenter, randomized, double-blind, placebo-control clinical trials so far (Huang et al., 2020a,b). Excitingly, recently publishing a paper of olfactory ensheathing cell (OEC) clinical trial showed positive results of neurorestorative effects with higher-level evidence-based medical evidence (Wang et al., 2020). Here we introduce the current state of clinical CTs for stroke and discuss the issues related to cell effects.

2 Basic information of cell therapies

According to FDA cell categories (Guidance for Industry: Preclinical Assessment of Investigational Cellular and Gene Therapy Products) (http://www.fda.gov/Bio logicsBloodVaccines/GuidanceComplianceRegulatoryInformation/Guidances/de fault.htm), cells are divided into stem cell-derived CT products and mature/ functionally differentiated cell-derived CT products. Clinical CTs for stroke in mature/functional cell products include neuronal cells, mononuclear cells (MNCs), mesenchymal stromal cells (MSCs), olfactory ensheathing cells (OECs), Schwann cells (SCs), fetal brain cells and in stem cell-derived CT products include hematopoietic stem cells (HST), mesenchymal stem cell products, and neural stem/progenitor cell (NSPC) or products, and above combination. Cells were transplanted in varying dosage in acute, sub-acute or chronic phase. They were directly injected into brain parenchyma by stereotactic surgical technique or through intra-arterial, intravenous, intrathecal infusion and olfactory sub-mucosa injection. The basic cell therapy information of types, transplanting routes, dosage was briefly shown in Table 1.

Table 1 Cell type, transplanting route, therapeutic time window and dosage of cell therapy for stroke.

Cell type	Transplanting routes	Therapeutic time windows	Dosage
Neurons[2]	Inject cells into brain lesion area	Chronic phase (Kondziolka et al., 2004)	2 or 6×10^6 5 or 10×10^6
MNCs	Intra-arterial(IA) infusion Intravenous infusion Intrathecal injection	Acute phase (Mendonça et al., 2006) 5–9 days (Moniche et al., 2012) 8–15 days (Bhatia et al., 2018) Non-acute phase (Battistella et al., 2011) 3 months to 5 years (Hammadi and Alhimyari, 2019) Acute phase (24–72 h) (Savitz et al., 2011; Vahidy et al., 2019) 7–30 days (Prasad et al., 2012) 7–10 days (Taguchi et al., 2015) 3 months to 2 years (Bhasin et al., 2012, 2013) 3 months to 1.5 years (Bhasin et al., 2016) Chronic phase (Sharma et al., 2014a, 2014b) Non-acute phase (Chernykh et al., 2016)	30×10^7 1.59×10^8 6.1×10^8 $1–5 \times 10^8$ $5–6 \times 10^8$ $7–10 \times 10^6$/kg 8×10^7 2.5 or 3.4×10^8 $5–6 \times 10^7$ $6–7 \times 10^7$ 1×10^6/kg M2 macrophage 21.9×10^6

Table 1 Cell type, transplanting route, therapeutic time window and dosage of cell therapy for stroke.—cont'd

Cell type	Transplanting routes	Therapeutic time windows	Dosage
MSCs	Intravenous infusion IA infusion intraventricular injection	4 weeks (Bang et al., 2005; Lee et al., 2010) 36–133 days (Honmou et al., 2011) 8–24 weeks (Bhasin et al., 2011) 1 year (Tsang et al., 2017) Averaged 4.2 ± 4.6 years (Levy et al., 2019) 3 months (Jiang et al., 2013) >24 weeks (Al Fauzi et al., 2017)	1×10^8 or 5×10^7 $0.6–1.6 \times 10^8$ $5–6 \times 10^7$ $1.43–8.40 \times 10^7$ $0.5–1.5 \times 10^6$/kg 2×10^7 2×10^6, add 2 time/month
OECs	Inject cells to olfactory submucosa Inject cells to the adjacent lesion	4 months (Guo et al., 2019) Over 1 year (Wang et al., 2020) Chronic phase (Huang et al., 2009)	$2 \times 5 \times 10^6$ $2 \times 5 \times 10^6$ 2×10^6
HSCs or CD34 + cells	Intrathecal injection Inject cells to the adjacent infarct Intra-arterial infusion	1–7 years (Wang et al., 2013) 6 months to 5 years (Chen et al., 2014) Within 7 days (Banerjee et al., 2014) 8.6 ± 6.4 years (Sung et al., 2018)	$0.8–3.3 \times 10^7$ $3–8 \times 10^6$ 1×10^8 3.0×10^7
Mesenchymal stem cell products	Inject cells to the adjacent infarct	7–36 months (Steinberg et al., 2016, 2018)	2.5×10^6, 5.0×10^6, or 10×10^6
Neural stem cell products	Inject cells into putamen Intracerebral injection	6–60 months (Kalladka et al., 2016) 2–13 months (Muir et al., 2020) 3–24 months (Zhang et al., 2019)	2×10^6, 5×10^6, 10×10^6, 20×10^6 20×10^6 1.2×10^7, 1.2×10^7, 1.2×10^7
Endothelial progenitor cells	Intravenous infusion	7 days + 3–4 weeks (Fang et al., 2019)	2.5×10^6/kg

3 Results of exploring trials or studies

3.1 Neuron cells

Kondziolka et al. (2000) reported neuron transplantation in 12 patients with chronic basal ganglia stroke. Six patients improved 3–10 points of assessment score by European Stroke Scale (ESS) and showed improving fluorodeoxyglucose uptake at the implant site by PET scans after 6 months (Kondziolka et al., 2000). There was no adverse cell-related serologic or imaging-defined effects. Fetal porcine nervous cells after transplantation exhibited significant neurological improvements in 2/5 patients (Savitz et al., 2005), but the study was terminated by the US Federal Drug Administration.

3.2 Mononuclear cells (MNCs)

Unmanipulated MNCs come generally from bone marrow, cord blood, and peripheral blood. Mendonça reported intra-arterial autologous bone marrow mononuclear cells (BMMNC) transplantation for a patient with acute ischemic stroke (Mendonça et al., 2006). The procedure and cell itself was safe, and neurological functions in part patients got improvement (Bhatia et al., 2018; Moniche et al., 2012). Friedrich reported that intra-arterial infusion of autologous BMMNCs for acute ischemic stroke showed clinical improvement in 8 of 20 patients (Friedrich et al., 2012) or no significant improvement of motor, language disturbance, and infarction volume for sub-acute stroke patients (Ghali et al., 2016). The cell transplantation and procedure was also feasible and safe in patients with non-acute stroke (Battistella et al., 2011) and even improving some functions in chronic ischemic strokes (Hammadi and Alhimyari, 2019). Moniche reported that a higher dose of autologous BMMNC through intra-arterial infusion showed better outcome in patients with sub-acute stroke (Moniche et al., 2016).

Savitz et al. (2011) found that intravenous infusion of autologous BMMNCs for acute ischemic stroke showed some improvement of daily life quality in 7/10 patients 6 months after cell transplantation, safe and feasible in their phase 1 clinical trial (Vahidy et al., 2019); but it should be caution to do cell therapy at earlier (1–3 days) time (Vahidy et al., 2013). In magnetic resonance spectroscopy study, intravenous administration of autologous BMMNCs increased trends of N-acetylasperate concentration which significantly correlated with NIHSS improvement (Haque et al., 2019). Laskowitz reported that allogeneic umbilical cord blood intravenous infusion for adults with 3–9 day post-stroke was safe and could improve some neurological functions (Laskowitz et al., 2018). Prasad reported that 7/11 patients with sub-acute ischemic stroke had a favorable clinical outcome after autologous intravenous BMMNC infusion (Prasad et al., 2012). During this period, more clinical studies of BMMNCs were conducted through I.V. for patients with chronic ischemic stroke. Taguchi reported that a higher dose of intravenous autologous BMMNC transplantation for sub-acute stroke had a trend toward improved neurologic outcomes (Taguchi et al., 2015). Bhasin et al. (2012, 2013, 2016) reported that BMMNC transplantation studies showed increasing some scores of modified Barthel index (Bhasin et al., 2012, 2013) or no statistically significant clinical improvement between the groups (Bhasin et al., 2016). Vasconcelos-dos-Santos compared intravenous and intra-arterial administration of BMMNCs after focal cerebral ischemia and found that transplantation of BMMNCs by IA or IV routes might lead to similar brain homing and therapeutic efficacy (Vasconcelos-dos-Santos et al., 2012). MMNC labeling with technetium-99 m allowed imaging for up to 24 h after intra-arterial or intravenous injection and similar distribution for both routes in stroke patients (Rosado-de-Castro et al., 2013).

Sharma found that autologous BMMNCs through intrathecal transplantation could improve some neurological function in patients with chronic stroke (Sharma et al., 2014a,b). Chernykh et al. (2016) reported that intrathecal administration of M2 macrophages for patients with non-acute stroke could improve neurological functions (Chernykh et al., 2016).

3.3 Mesenchymal stromal cells (MSCs)

Before 2005, the plastic-adherent cells isolated from BM and other sources have come to be widely named as mesenchymal stem cells or mesenchymal stromal cells (MSCs). The name of mesenchymal stem cells and MSCs were mixed to identify these cells. The International Society for Cellular Therapy (2005) state that the recognized biologic properties of the unfractionated population of cells do not seem to meet generally accepted criteria for stem cell activity, rendering the name scientifically inaccurate and potentially misleading to the lay public (Horwitz et al., 2005). Therefore it suggests those cells should be named MSCs, and establish their criteria (Dominici et al., 2006; Galipeau et al., 2016). In this review paper, we use MSCs to replace mesenchymal stem cells except them identified by special stem cell markers or differenced into derivative cells.

Bang et al. (2005) reported that intravenous infusion of autologous MSCs appeared to be a feasible, safe and have some functional recovery for patients with 4 week post-stroke (Bang et al., 2005). A long-term follow-up study showed procedure and cell safety, but was not able to conclude the neurological improvement comparing treatment group with control group (Lee et al., 2010). Interestingly, intravenously delivering cells during 36–133 days post-stroke showed reducing >20% lesion volume by magnetic resonance imaging (MRI) examination at 1 week post-cell infusion (Honmou et al., 2011). Intravenous infusion of autologous MSCs was safe and feasible in patients with chronic (8–24 weeks) stroke (Bhasin et al., 2011). Four year follow-up results indicated safety, tolerance and applicability of MSCs, but the majority of impairment scales except modified Barthel Index did not show any significant improvement comparing with their previous published report (Bhasin et al., 2017). Later, more intravenous infusion studies of MSC for chronic stroke were reported to be safe and have the potential of improving neurological functions (Levy et al., 2019; Tsang et al., 2017).

Intra-artery delivery of MSCs *via* catheterization was a feasible and safe approach and might improve the neurological function of ischemic stroke patients (Jiang et al., 2013).

Repeated intraventricular transplanting MSCs showed safety and 5 of 8 patients improved their neurologic function (Al Fauzi et al., 2017).

MSCs from bone marrow could also be stereotactically transplanted into the brain of stroke patients with excellent tolerance and without complications (Suárez-Monteagudo et al., 2009).

3.4 Olfactory ensheathing cells (OECs)

Guo et al. (2019) reported that OEC transplantation through olfactory sub-mucosa (Guo et al., 2019) or injecting them to the adjacent lesion (Huang et al., 2009) could improve neurological functions and their quality of life for patient with chronic phase stroke.

3.5 Hemopoietic stem cells (HSTs)

Mobilizing bone marrow CD34+ hematopoietic stem cells by granulocyte-colony-stimulating factor (G-CSF) was safe, could increase the CD34+ cell count in peripheral blood circulation and potentially improved some neurological functions for acute or sub-acute stroke patients (Alasheev et al., 2011; Boy et al., 2011; England et al., 2012; Mizuma et al., 2016; Prasad et al., 2011; Ringelstein et al., 2013; Schäbitz et al., 2010; Shyu et al., 2006; Sprigg et al., 2006). Transplanting CD34+ stem cells mobilized by G-CSF or directly collected from the bone marrow through intrathecal injection (Wang et al., 2013), stereotactic surgical injection (Chen et al., 2014) and intra-arterial infusion (Banerjee et al., 2014; Sung et al., 2018) was safe and might offer some benefit to patients with stroke.

3.6 Mesenchymal stem cell products

Cells (SB623) derived mesenchymal stem cells were transplanted into sites surrounding the residual stroke through stereotactic technique for patients with chronic stroke. Most patients showed improvement in clinical outcome after 12 month and 2 year follow-up (Steinberg et al., 2016, 2018).

3.7 Neural stem cell products

Kalladka et al. reported to inject human neural stem cells (CTX0E03) into putamen in patients with chronic ischemic stroke. Cells and procedure were safe and associated with improved neurological function (Kalladka et al., 2016). In their following study, only a few (4/23) patients with residual upper limb movement at baseline in 2–13 months after ischemic stroke showed improvements in upper limb function (Muir et al., 2020). Zhang et al. reported to transplant NSI-566 from neural stem cell line through intracerebral injections for patients with chronic motor deficit stroke and showed some functional improvement (Zhang et al., 2019).

3.8 Combination cell therapy

Rabinovich et al. (2005) reported that transplanting cells from immature nervous (fetal brain) and hemopoietic tissues for patients with brain stroke consequences could improve their neurological function (Rabinovich et al., 2005). Multiple cell transplantation including OEC, neural progenitor cells, umbilical cord MSCs, and Schwann cells (SCs) for patients with chronic stroke showed neurorestorative effects (Chen et al., 2013). Cotransplantation of neural stem/progenitor cells and MSCs in patients with chronic stroke also improved their some functions (Qiao et al., 2014).

4 Results of multicenter, double-blind or observing-blind, randomized, placebo-controlled trials

Kondziolka reported a randomized, observer-blinded trial of neuron transplantation in 18 (5 million cells in 7, 10 million cells in 7 and 4 rehabilitation only) patients with chronic stroke; but there was no significant change in the European Stroke Scale motor score in patients who received cell implants compared with control or baseline values (Kondziolka et al., 2005). In a multicenter, randomized trial with blinded outcome assessment, autologous BMMNC therapy for subacute ischemic stroke did not show beneficial effects (Prasad et al., 2014). Hess reported that a randomized, double-blind, placebo-controlled, phase 2 trial of multipotent adult progenitor cells derived from bone marrow for patients with acute ischemic stroke didn't show the difference between the cell therapy group and placebo groups in neurological recovery at day 90 (Hess et al., 2017). Although SB623 transplantation for chronic stroke in a phase 2a trial showed some neurological improvement (Steinberg et al., 2016, 2018); unfortunately SB623 treatment groups in a phase 2b study did not demonstrate a statistically significant improvement compared to the sham surgery (control) group (SanBio Co. Ltd., 2019). In a two-center, randomized, placebo-controlled phase I/IIa endothelial progenitor cell trial with blinded outcome assessment on 18 patients with acute cerebral infarct, there was no significant difference in neurological or functional improvement observed among the three groups (two different dosage groups and sham surgery group) after 4 year follow-up (Fang et al., 2019). A multicenter, randomized, blinded assessment, sham-controlled trial of autologous bone marrow-derived ALD-401 cells through internal carotid artery infusion for patients with recovering ischemic stroke did not show the difference between the groups (Savitz et al., 2019).

Recently a multicenter, randomized, double-blind and placebo-controlled cell therapy trial showed significantly difference of functional assessments among OEC group, SC group and placebo group. OECs injecting into olfactory sub-mucosa had neurorestorative effects on improving quality of life for chronic ischemic stroke (Wang et al., 2020) (Figs. 1–4).

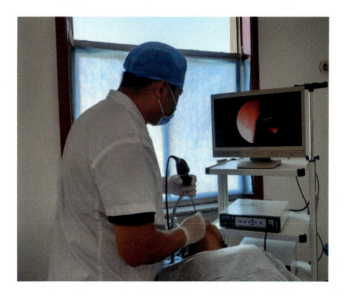

FIG. 1

A otolaryngologist was preparing to do injection.

Reprinted from Wang et al., (2020). Olfactory ensheathing cells in chronic ischemic stroke: a phase 2, double-blind, randomized, controlled trial. J. Neurorestoratol. 8 (3), 182–193 Copyright (2020), with permission from Journals of Tsinghua University Press.

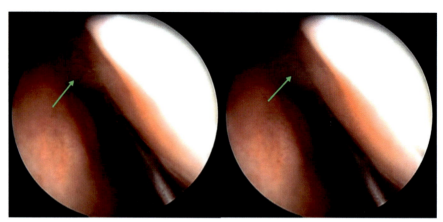

FIG. 2

Left: before injection, Right: after cell culture medium injecting into sub-mucosa, local tissue looks plump.

Reprinted from Wang et al., (2020). Olfactory ensheathing cells in chronic ischemic stroke: a phase 2, double-blind, randomized, controlled trial. J. Neurorestoratol. 8 (3), 182–193 Copyright (2020), with permission from Journals of Tsinghua University Press.

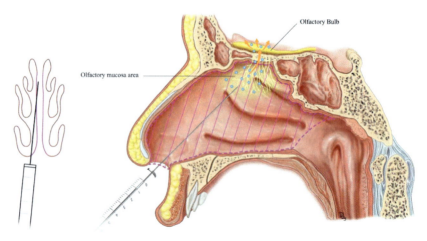

FIG. 3

Diagram of cell injection and migration: after cell being injected into sub-mucosa of olfactory area, they migrated into olfactory bulb and then to lesion area.

Reprinted from Wang et al., (2020). Olfactory ensheathing cells in chronic ischemic stroke: a phase 2, double-blind, randomized, controlled trial. J. Neurorestoratol. 8 (3), 182–193 Copyright (2020), with permission from Journals of Tsinghua University Press.

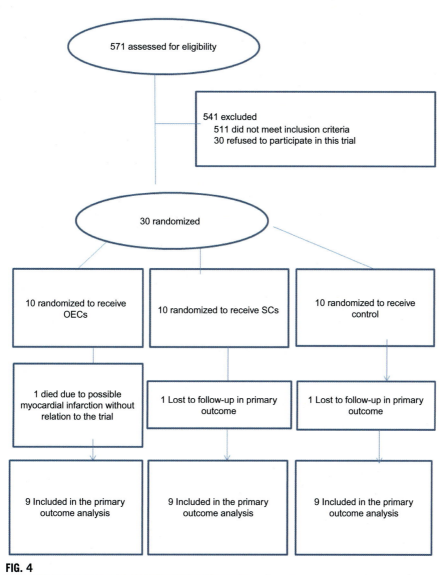

FIG. 4

Trial flow diagram.

5 Discussion

There are many contradictory or conflicting effects about cell therapy, even in same kind of cells. The reasons possibly are lack of uniform standard of cell culture and quality control, differences of dosage, therapeutic time window, transplanting route, and therapeutic times. Twenty years pasted with huge exploring clinical studies or trials, clinical physicians still failed to prove most kinds of cellular neurorestorative effects through multicenter, randomized, double-blind, placebo-control clinical trials.

Fortunately OECs had been proved their neurorestorative effect in higher level of evidence-based medical evidences (Wang et al., 2020). Human olfactory nervous system is the only nervous tissue which has the ability to automatically regenerate and restore itself in whole life (Holbrook et al., 2011). Olfactory neurons show the characteristic of photoreceptor cells *in vitro* with special medium (Lu et al., 2017). OECs sharing characteristic of Schwann cells (SCs), oligodendrocytes and astrocytes have strong ability to secrete neurotrophic factors which changed lesion area microenvironment to unmask or trigger the lively silent damaged neurons and play a restorative role through neurorestorative mechanisms (Huang et al., 2015). Those are potential reasons why OECs showed their positive neurorestorative effects through a multicenter, randomized, double-blind, placebo-control clinical trial. Therefore OECs and also olfactory neurons may be the most potential cells to restore damaged neurological structure/functions in CNS.

Currently there are many clinical trials or studies of CTs every year (Huang et al., 2019, 2020a,b), but the most important task for majority kinds of cells should be to prove their neurorestorative effects through multicenter, randomized, double-blind, placebo-control clinical trials. The most important task for OECs is to set up the uniform world standards of cell culture and quality control in current basis (Xiao et al., 2017); and then to explore the optimizing cell delivery methods, dosage, therapeutic time window, effects of repeatedly transplanting cells and combination with other kinds of cells or other neurorestorative therapies.

6 Summary

Varying clinical CTs have been explored in stroke for over two decades. Majority of CTs showed positive results; but they were small, single-center, phase 1–2 non-double-blind observing trials or retrospective clinical studies. At present, most CTs for stroke through multicenter, randomized, double-blind placebo-control or observing blind clinical trials did not show positive results except OECs. In future clinical trials of CTs for stroke with higher level evidence-based medical evidences should be done to prove their affirmative neurorestorative effects. OECs should be explored to optimize themselves neurorestorative effects or combination with others.

Disclosure

The authors declare that they have no competing interests.

References

Al Fauzi, A., Sumorejo, P., Suroto, N.S., Parenrengi, M.A., Wahyuhadi, J., Turchan, A., Mahyudin, F., Suroto, H., Rantam, F.A., Machfoed, M.H., Bajamal, A.H., Lumenta, C.B., 2017. Clinical outcomes of repeated intraventricular transplantation of autologous bone marrow mesenchymal stem cells in chronic haemorrhagic stroke. A one-year follow up. Open Neurol. J. 11, 74–83.

Alasheev, A.M., Belkin, A.A., Leiderman, I.N., Ivanov, R.A., Isakova, T.M., 2011. Granulocyte-colony-stimulating factor for acute ischemic stroke: a randomized controlled trial (STEMTHER). Transl. Stroke Res. 2, 358–365.

Banerjee, S., Bentley, P., Hamady, M., Marley, S., Davis, J., Shlebak, A., Nicholls, J., Williamson, D.A., Jensen, S.L., Gordon, M., Habib, N., Chataway, J., 2014. Intra-arterial immunoselected CD34+ stem cells for acute ischemic stroke. Stem Cells Transl. Med. 3, 1322–1330.

Bang, O.Y., Lee, J.S., Lee, P.H., Lee, G., 2005. Autologous mesenchymal stem cell transplantation in stroke patients. Ann. Neurol. 57 (6), 874–882.

Battistella, V., de Freitas, G.R., da Fonseca, L.M., Mercante, D., Gutfilen, B., Goldenberg, R.C., Dias, J.V., Kasai-Brunswick, T.H., Wajnberg, E., Rosado-de-Castro, P.H., Alves-Leon, S.V., Mendez-Otero, R., Andre, C., 2011. Safety of autologous bone marrow mononuclear cell transplantation in patients with nonacute ischemic stroke. Regen. Med. 61, 45–52.

Bhasin, A., Srivastava, M.V., Kumaran, S.S., Mohanty, S., Bhatia, R., Bose, S., Gaikwad, S., Garg, A., Airan, B., 2011. Autologous mesenchymal stem cells in chronic stroke. Cerebrovasc. Dis. Extra 1 (1), 93–104.

Bhasin, A., Srivastava, M., Bhatia, R., Mohanty, S., Kumaran, S., Bose, S., 2012. Autologous intravenous mononuclear stem cell therapy in chronic ischemic stroke. J. Stem Cells Regen. Med. 8 (3), 181–189.

Bhasin, A., Srivastava, M.V., Mohanty, S., Bhatia, R., Kumaran, S.S., Bose, S., 2013. Stem cell therapy: a clinical trial of stroke. Clin. Neurol. Neurosurg. 115 (7), 1003–1008.

Bhasin, A., Srivastava, M.V.P., Mohanty, S., Vivekanandhan, S., Sharma, S., Kumaran, S., Bhatia, R., 2016. Paracrine mechanisms of intravenous bone marrow-derived mononuclear stem cells in chronic ischemic stroke. Cerebrovasc. Dis. Extra 6 (3), 107–119.

Bhasin, A., Kumaran, S.S., Bhatia, R., Mohanty, S., Srivastava, M.V.P., 2017. Safety and feasibility of autologous mesenchymal stem cell transplantation in chronic stroke in Indian patients. A four-year follow up. J. Stem Cells Regen. Med. 13 (1), 14–19.

Bhatia, V., Gupta, V., Khurana, D., Sharma, R.R., Khandelwal, N., 2018. Randomized assessment of the safety and efficacy of intra-arterial infusion of autologous stem cells in subacute ischemic stroke. Am. J. Neuroradiol. 39 (5), 899–904.

Boy, S., Sauerbruch, S., Kraemer, M., Schormann, T., Schlachetzki, F., Schuierer, G., Luerding, R., Hennemann, B., Orso, E., Dabringhaus, A., Winkler, J., Bogdahn, U., RAIS (Regeneration in Acute Ischemic stroke) Study Group, 2011. Mobilisation of hematopoietic CD34+ precursor cells in patients with acute stroke is safe—results of an open-labeled non randomized phase I/II trial. PLoS One 6 (8), e23099.

Chen, L., Xi, H., Huang, H., Zhang, F., Liu, Y., Chen, D., Xiao, J., 2013. Multiple cell transplantation based on an intraparenchymal approach for patients with chronic phase stroke. Cell Transplant. 22 (Suppl. 1), S83–S91.

Chen, D.C., Lin, S.Z., Fan, J.R., Lin, C.H., Lee, W., Lin, C.C., Liu, Y.J., Tsai, C.H., Chen, J.C., Cho, D.Y., Lee, C.C., Shyu, W.C., 2014. Intracerebral implantation of autologous peripheral blood stem cells in stroke patients: a randomized phase II study. Cell Transplant. 23 (12), 1599–1612.

Chernykh, E.R., Shevela, E.Y., Starostina, N.M., Morozov, S.A., Davydova, M.N., Menyaeva, E.V., Ostanin, A.A., 2016. Safety and therapeutic potential of M2 macrophages in stroke treatment. Cell Transplant. 25 (8), 1461–1471.

Dominici, M., Le Blanc, K., Mueller, I., Slaper-Cortenbach, I., Marini, F., Krause, D., Deans, R., Keating, A., Prockop, D.J., Horwitz, E., 2006. Minimal criteria for defining multipotent mesenchymal stromal cells. The international society for cellular therapy position statement. Cytotherapy 8 (4), 315–317.

England, T.J., Abaei, M., Auer, D.P., Lowe, J., Jones, D.R., Sare, G., Walker, M., Bath, P.M., 2012. Granulocyte-colony stimulating factor for mobilizing bone marrow stem cells in subacute stroke: the stem cell trial of recovery enhancement after stroke 2 randomized controlled trial. Stroke 43, 405–411.

Fang, J., Guo, Y., Tan, S., Li, Z., Xie, H., Chen, P., Wang, K., He, Z., He, P., Ke, Y., Jiang, X., Chen, Z., 2019. Autologous endothelial progenitor cells transplantation for acute ischemic stroke: a 4-year follow-up study. Stem Cells Transl. Med. 8 (1), 14–21.

Friedrich, M.A., Martins, M.P., Araújo, M.D., Klamt, C., Vedolin, L., Garicochea, B., Raupp, E.F., Sartori, E.L., Ammar, J., Machado, D.C., Costa, J.C., Nogueira, R.G., Rosado-de-Castro, P.H., Mendez-Otero, R., Freitas, G.R., 2012. Intra-arterial infusion of autologous bone marrow mononuclear cells in patients with moderate to severe middle cerebral artery acute ischemic stroke. Cell Transplant. 21 (Suppl. 1), S13–S21.

Galipeau, J., Krampera, M., Barrett, J., Dazzi, F., Deans, R.J., DeBruijn, J., Dominici, M., Fibbe, W.E., Gee, A.P., Gimble, J.M., Hematti, P., Koh, M.B., LeBlanc, K., Martin, I., McNiece, I.K., Mendicino, M., Oh, S., Ortiz, L., Phinney, D.G., Planat, V., Shi, Y., Stroncek, D.F., Viswanathan, S., Weiss, D.J., Sensebe, L., 2016. International Society for Cellular Therapy perspective on immune functional assays for mesenchymal stromal cells as potency release criterion for advanced phase clinical trials. Cytotherapy 18 (2), 151–159.

Ghali, A.A., Yousef, M.K., Ragab, O.A., ElZamarany, E.A., 2016. Intra-arterial infusion of autologous bone marrow mononuclear stem cells in subacute ischemic stroke patients. Front. Neurol. 7, 228.

Guo, X., Wang, X., Li, Y., Zhou, B., Chen, W., Ren, L., 2019. Olfactory ensheathing cell transplantation improving cerebral infarction sequela: a case report and literature review. J. Neurorestoratol. 7 (2), 82–88.

Hammadi, A.M.A., Alhimyari, F., 2019. Intra-arterial injection of autologous bone marrow-derived mononuclear cells in ischemic stroke patients. Exp. Clin. Transplant. 17 (Suppl. 1), 239–241.

Haque, M.E., Gabr, R.E., George, S.D., Boren, S.B., Vahidy, F.S., Zhang, X., Arevalo, O.D., Alderman, S., Narayana, P.A., Hasan, K.M., Friedman, E.R., Sitton, C.W., Savitz, S.I., 2019. Serial cerebral metabolic changes in patients with ischemic stroke treated with autologous bone marrow derived mononuclear cells. Front. Neurol. 10, 141.

Hess, D.C., Wechsler, L.R., Clark, W.M., Savitz, S.I., Ford, G.A., Chiu, D., Yavagal, D.R., Uchino, K., Liebeskind, D.S., Auchus, A.P., Sen, S., Sila, C.A., Vest, J.D., Mays, R.W., 2017. Safety and efficacy of multipotent adult progenitor cells in acute ischaemic stroke (MASTERS): a randomised, double-blind, placebo-controlled, phase 2 trial. Lancet Neurol. 16 (5), 360–368.

Holbrook, E.H., Wu, E., Curry, W.T., Lin, D.T., Schwob, J.E., 2011. Immunohistochemical characterization of human olfactory tissue. Laryngoscope 121 (8), 1687–1701.

Honmou, O., Houkin, K., Matsunaga, T., Niitsu, Y., Ishiai, S., Onodera, R., Waxman, S.G., Kocsis, J.D., 2011. Intravenous administration of auto serum-expanded autologous mesenchymal stem cells in stroke. Brain 134 (Pt. 6), 1790–1807.

Horwitz, E.M., Le Blanc, K., Dominici, M., Mueller, I., Slaper-Cortenbach, I., Marini, F.C., Deans, R.J., Krause, D.S., Keating, A., International Society for Cellular Therapy, 2005. Clarification of the nomenclature for MSC: the International Society for Cellular Therapy position statement. Cytotherapy 7 (5), 393–395.

Huang, H., Chen, L., Xi, H., Wang, Q., Zhang, J., Liu, Y., Zhang, F., 2009. Olfactory ensheathing cells transplantation for central nervous system diseases in 1,255 patients. Zhongguo Xiu Fu Chong Jian Wai Ke Za Zhi 23 (1), 14–20.

Huang, H., Raisman, G., Sanberg, P.R., Sharma, H., Chen, L., 2015. Neurorestoratology. vol. 1 Nova Biomedical, New York, pp. 68–71.

Huang, H., Sharma, H.S., Chen, L., Saberi, H., Mao, G., 2019. 2018 yearbook of neurorestoratology. J. Neurorestoratol. 7 (1), 11–20.

Huang, H., Chen, L., Mao, G., Bach, J., Xue, Q., Han, F., Guo, X., Otom, A., Chernykh, E., Alvarez, E., Bryukhovetskiy, A., Sarnowaska, A., He, X., Dimitrijevic, M., Shanti, I., von Wild, K., Ramón-Cueto, A., Alzoubi, Z., Moviglia, G., Mobasheri, H., Alzoubi, A., Zhang, W., 2020a. The 2019 yearbook of neurorestoratology. J. Neurorestoratol. 8 (1), 1–11.

Huang, H., Chen, L., Mao, G., Sharma, H., 2020b. Clinical neurorestorative cell therapies: developmental process, current state, and future prospective. J. Neurorestoratol. 8 (2), 61–82.

Jiang, Y., Zhu, W., Zhu, J., Wu, L., Xu, G., Liu, X., 2013. Feasibility of delivering mesenchymal stem cells via catheter to the proximal end of the lesion artery in patients with stroke in the territory of the middle cerebral artery. Cell Transplant. 22 (12), 2291–2298.

Kalladka, D., Sinden, J., Pollock, K., Haig, C., McLean, J., Smith, W., McConnachie, A., Santosh, C., Bath, P.M., Dunn, L., Muir, K.W., 2016. Human neural stem cells in patients with chronic ischaemic stroke (PISCES): a phase 1, first-in-man study. Lancet 388 (10046), 787–796.

Kondziolka, D., Wechsler, L., Goldstein, S., Meltzer, C., Thulborn, K.R., Gebel, J., Jannetta, P., DeCesare, S., Elder, E.M., McGrogan, M., Reitman, M.A., Bynum, L., 2000. Transplantation of cultured human neuronal cells for patients with stroke. Neurology 55 (4), 565–569.

Kondziolka, D., Steinberg, G.K., Cullen, S.B., McGrogan, M., 2004. Evaluation of surgical techniques for neuronal cell transplantation used in patients with stroke. Cell Transplant. 13 (7–8), 749–754.

Kondziolka, D., Steinberg, G.K., Wechsler, L., Meltzer, C.C., Elder, E., Gebel, J., Decesare, S., Jovin, T., Zafonte, R., Lebowitz, J., Flickinger, J.C., Tong, D., Marks, M.P., Jamieson, C., Luu, D., Bell-Stephens, T., Teraoka, J., 2005. Neurotransplantation for patients with subcortical motor stroke: a phase 2 randomized trial. J. Neurosurg. 103 (1), 38–45.

Laskowitz, D.T., Bennett, E.R., Durham, R.J., Volpi, J.J., Wiese, J.R., Frankel, M., Shpall, E., Wilson, J.M., Troy, J., Kurtzberg, J., 2018. Allogeneic umbilical cord blood infusion for adults with ischemic stroke: clinical outcomes from a phase I safety study. Stem Cells Transl. Med. 7 (7), 521–529.

Lee, J.S., Hong, J.M., Moon, G.J., Lee, P.H., Ahn, Y.H., Bang, O.Y., Starting Collaborators, 2010. A long-term follow-up study of intravenous autologous mesenchymal stem cell transplantation in patients with ischemic stroke. Stem Cells 28 (6), 1099–1106.

Levy, M.L., Crawford, J.R., Dib, N., Verkh, L., Tankovich, N., Cramer, S.C., 2019. Phase I/II study of safety and preliminary efficacy of intravenous allogeneic mesenchymal stem cells in chronic stroke. Stroke 50 (10), 2835–2841.

Lu, W., Duan, D., Ackbarkhan, Z., Lu, M., Huang, M.L., 2017. Differentiation of human olfactory mucosa mesenchymal stem cells into photoreceptor cells in vitro. Int. J. Ophthalmol. 10 (10), 1504–1509.

Mendonça, M.L., Freitas, G.R., Silva, S.A., Manfrim, A., Falcão, C.H., Gonzáles, C., André, C., Dohmann, H.F., Borojevic, R., Otero, R.M., 2006. Safety of intra-arterial autologous bone marrow mononuclear cell transplantation for acute ischemic stroke. Arq. Bras. Cardiol. 86 (1), 52–55.1.

Mizuma, A., Yamashita, T., Kono, S., Nakayama, T., Baba, Y., Itoh, S., Asakura, K., Niimi, Y., Asahi, T., Kanemaru, K., Mutoh, T., Kuroda, S., Kinouchi, H., Abe, K., Takizawa, S., 2016. Phase II trial of intravenous low-dose granulocyte colony-stimulating factor in acute ischemic stroke. J. Stroke Cerebrovasc. Dis. 25, 1451–1457.

Moniche, F., Gonzalez, A., Gonzalez-Marcos, J.R., Carmona, M., Piñero, P., Espigado, I., Garcia-Solis, D., Cayuela, A., Montaner, J., Boada, C., Rosell, A., Jimenez, M.D., Mayol, A., Gil-Peralta, A., 2012. Intra-arterial bone marrow mononuclear cells in ischemic stroke: a pilot clinical trial. Stroke 43 (8), 2242–2244.

Moniche, F., Rosado-de-Castro, P.H., Escudero, I., Zapata, E., de la Torre Laviana, F.J., Mendez-Otero, R., Carmona, M., Piñero, P., Bustamante, A., Lebrato, L., Cabezas, J.A., Gonzalez, A., de Freitas, G.R., Montaner, J., 2016. Increasing dose of autologous bone marrow mononuclear cells transplantation is related to stroke outcome: results from a pooled analysis of two clinical trials. Stem Cells Int. 2016, 8657173. https://doi.org/10.1155/2016/8657173.

Muir, K.W., Bulters, D., Willmot, M., Sprigg, N., Dixit, A., Ward, N., Tyrrell, P., Majid, A., Dunn, L., Bath, P., Howell, J., Stroemer, P., Pollock, K., Sinden, J., 2020. Intracerebral implantation of human neural stem cells and motor recovery after stroke: multicentre prospective single-arm study (PISCES-2). J. Neurol. Neurosurg. Psychiatry 91 (4), 396–401.

Prasad, K., Kumar, A., Sahu, J.K., Srivastava, M.V., Mohanty, S., Bhatia, R., Gaikwad, S.B., Srivastava, A., Goyal, V., Tripathi, M., Bal, C., Mishra, N.K., 2011. Mobilization of stem cells using G-CSF for acute ischemic stroke: a randomized controlled, pilot study. Stroke Res. Treat. 2011, 283473. https://doi.org/10.4061/2011/283473.

Prasad, K., Mohanty, S., Bhatia, R., Srivastava, M.V., Garg, A., Srivastava, A., Goyal, V., Tripathi, M., Kumar, A., Bal, C., Vij, A., Mishra, N.K., 2012. Autologous intravenous bone marrow mononuclear cell therapy for patients with subacute ischaemic stroke: a pilot study. Indian J. Med. Res. 136 (2), 221–228.

Prasad, K., Sharma, A., Garg, A., Mohanty, S., Bhatnagar, S., Johri, S., Singh, K.K., Nair, V., Sarkar, R.S., Gorthi, S.P., Hassan, K.M., Prabhakar, S., Marwaha, N., Khandelwal, N., Misra, U.K., Kalita, J., Nityanand, S., InveST Study Group, 2014. Intravenous autologous bone marrow mononuclear stem cell therapy for ischemic stroke: a multicentric, randomized trial. Stroke 45 (12), 3618–3624.

Qiao, L.Y., Huang, F.J., Zhao, M., Xie, J.H., Shi, J., Wang, J., Lin, X.Z., Zuo, H., Wang, Y.L., Geng, T.C., 2014. A two-year follow-up study of cotransplantation with neural stem/progenitor cells and mesenchymal stromal cells in ischemic stroke patients. Cell Transplant. 23 (1_suppl), 65–72.

Rabinovich, S.S., Seledtsov, V.I., Banul, N.V., Poveshchenko, O.V., Senyukov, V.V., Astrakov, S.V., Samarin, D.M., Taraban, V.Y., 2005. Cell therapy of brain stroke. Bull. Exp. Biol. Med. 139 (1), 126–128.

Ringelstein, E.B., Thijs, V., Norrving, B., Chamorro, A., Aichner, F., Grond, M., Saver, J., Laage, R., Schneider, A., Rathgeb, F., Vogt, G., Charissé, G., Fiebach, J.B., Schwab, S., Schäbitz, W.R., Kollmar, R., Fisher, M., Brozman, M., Skoloudik, D., Gruber, F., Serena Leal, J., Veltkamp, R., Köhrmann, M., Berrouschot, J., 2 Investigators, A.X.I.S., 2013. Granulocyte colony-stimulating factor in patients with acute ischemic stroke results of the Ax200 for ischemic stroke trial. Stroke 44, 2681–2687.

Rosado-de-Castro, P.H., Schmidt Fda, R., Battistella, V., Lopes de Souza, S.A., Gutfilen, B., Goldenberg, R.C., Kasai-Brunswick, T.H., Vairo, L., Silva, R.M., Wajnberg, E., Alvarenga Americano do Brasil, P.E., Gasparetto, E.L., Maiolino, A., Alves-Leon, S.V., Andre, C., Mendez-Otero, R., Rodriguez de Freitas, G., Barbosa da Fonseca, L.M., 2013. Biodistribution of bone marrow mononuclear cells after intra-arterial or intravenous transplantation in subacute stroke patients. Regen. Med. 8 (2), 145–155.

SanBio Co. Ltd., 2019. SanBio and Sumitomo Dainippon Pharma Announce Topline Results From a Phase 2b Study in the U.S. Evaluating SB623, a Regenerative Cell Medicine for the Treatment of Patients With Chronic Stroke. Available at https://www.ds-pharma.com/ir/news/pdf/ene20190129.1.pdf.

Savitz, S.I., Dinsmore, J., Wu, J., Henderson, G.V., Stieg, P., Caplan, L.R., 2005. Neurotransplantation of fetal porcine cells in patients with basal ganglia infarcts: a preliminary safety and feasibility study. Cerebrovasc. Dis. 20 (2), 101–107.

Savitz, S.I., Misra, V., Kasam, M., Juneja, H., Cox Jr., C.S., Alderman, S., Aisiku, I., Kar, S., Gee, A., Grotta, J.C., 2011. Intravenous autologous bone marrow mononuclear cells for ischemic stroke. Ann. Neurol. 70 (1), 59–69.

Savitz, S.I., Yavagal, D., Rappard, G., Likosky, W., Rutledge, N., Graffagnino, C., Alderazi, Y., Elder, J.A., Chen, P.R., Budzik Jr., R.F., Tarre, L.R., Huang, D.Y., Hinson Jr., J.M., 2019. A phase 2 randomized, sham-controlled trial of internal carotid artery infusion of autologous bone marrow derived ALD-401 cells in patients with recent stable ischemic stroke (RECOVER-Stroke). Circulation 139 (2), 192–205.

Schäbitz, W.R., Laage, R., Vogt, G., Koch, W., Kollmar, R., Schwab, S., Schneider, D., Hamann, G.F., Rosenkranz, M., Veltkamp, R., Fiebach, J.B., Hacke, W., Grotta, J.C., Fisher, M., Schneider, A., 2010. AXIS: a trial of intravenous granulocyte colony-stimulating factor in acute ischemic stroke. Stroke 41, 2545–2551.

Sharma, A., Sane, H., Gokulchandran, N., Khopkar, D., Paranjape, A., Sundaram, J., Gandhi, S., Badhe, P., 2014a. Autologous bone marrow mononuclear cells intrathecal transplantation in chronic stroke. Stroke Res. Treat. 2014, 234095.

Sharma, A., Sane, H., Nagrajan, A., Gokulchandran, N., Badhe, P., Paranjape, A., Biju, H., 2014b. Autologous bone marrow mononuclear cells in ischemic cerebrovascular accident paves way for neurorestoration: a case report. Case Rep. Med. 2014, 530239.

Shyu, W.C., Lin, S.Z., Lee, C.C., Liu, D.D., Li, H., 2006. Granulocyte colony-stimulating factor for acute ischemic stroke: a randomized controlled trail. CMAJ 174, 927–933.

Sprigg, N., Bath, P.M., Zhao, L., Willmot, M.R., Gray, L.J., Walker, M.F., Dennis, M.S., Russell, N., 2006. Granulocyte-colony-stimulating factor mobilizes bone marrow stem cells in patients with subacute ischemic stroke: the stem cell trial of recovery enhancement after stroke (STEMS) pilot randomized, controlled trial. Stroke 37, 2979–2983.

Steinberg, G.K., Kondziolka, D., Wechsler, L.R., Lunsford, L.D., Coburn, M.L., Billigen, J.B., Kim, A.S., Johnson, J.N., Bates, D., King, B., Case, C., McGrogan, M., Yankee, E.W., Schwartz, N.E., 2016. Clinical outcomes of transplanted modified bone marrow-derived mesenchymal stem cells in stroke: a phase 1/2a study. Stroke 47 (7), 1817–1824.

Steinberg, G.K., Kondziolka, D., Wechsler, L.R., Lunsford, L.D., Kim, A.S., Johnson, J.N., Bates, D., Poggio, G., Case, C., McGrogan, M., Yankee, E.W., Schwartz, N.E., 2018. Two-year safety and clinical outcomes in chronic ischemic stroke patients after implantation of modified bone marrow-derived mesenchymal stem cells (SB623): a phase 1/2a study. J. Neurosurg. 1, 1–11.

Suárez-Monteagudo, C., Hernández-Ramírez, P., Alvarez-González, L., García-Maeso, I., de la Cuétara-Bernal, K., Castillo-Díaz, L., Bringas-Vega, M.L., Martínez-Aching, G., Morales-Chacón, L.M., Báez-Martín, M.M., Sánchez-Catasús, C., Carballo-Barreda, M., Rodríguez-Rojas, R., Gómez-Fernández, L., Alberti-Amador, E., Macías-Abraham, C., Balea, E.D., Rosales, L.C., Del Valle Pérez, L., Ferrer, B.B., González, R.M., Bergado, J.A., 2009. Autologous bone marrow stem cell neurotransplantation in stroke patients. An open study. Restor. Neurol. Neurosci. 27 (3), 151–161.

Sung, P.H., Lin, H.S., Lin, W.C., Chang, C.C., Pei, S.N., Ma, M.C., Chen, K.H., Chiang, J.Y., Chang, H.W., Lee, F.Y., Lee, M.S., Yip, H.K., 2018. Intra-carotid arterial transfusion of autologous circulatory derived CD34+ cells for old ischemic stroke patients—a phase I clinical trial to evaluate safety and tolerability. Am. J. Transl. Res. 10 (9), 2975–2989.

Taguchi, A., Sakai, C., Soma, T., Kasahara, Y., Stern, D.M., Kajimoto, K., Ihara, M., Daimon, T., Yamahara, K., Doi, K., Kohara, N., Nishimura, H., Matsuyama, T., Naritomi, H., Sakai, N., Nagatsuka, K., 2015. Intravenous autologous bone marrow mononuclear cell transplantation for stroke: phase1/2a clinical trial in a homogeneous group of stroke patients. Stem Cells Dev. 24 (19), 2207–2218.

Tsang, K.S., Ng, C.P.S., Zhu, X.L., Wong, G.K.C., Lu, G., Ahuja, A.T., Wong, K.S.L., Ng, H.K., Poon, W.S., 2017. Phase I/II randomized controlled trial of autologous bone marrow-derived mesenchymal stem cell therapy for chronic stroke. World J. Stem Cells 9 (8), 133–143.

Vahidy, F.S., Alderman, S., Savitz, S.I., 2013. Challenges enrolling patients with acute ischemic stroke into cell therapy trials. Stem Cells Dev. 22 (1), 27–30.

Vahidy, F.S., Haque, M.E., Rahbar, M.H., Zhu, H., Rowan, P., Aisiku, I.P., Lee, D.A., Juneja, H.S., Alderman, S., Barreto, A.D., Suarez, J.I., Bambhroliya, A., Hasan, K.M., Kassam, M.R., Aronowski, J., Gee, A., Cox Jr., C.S., Grotta, J.C., Savitz, S.I., 2019. Intravenous bone marrow mononuclear cells for acute ischemic stroke: safety, feasibility, and effect size from a phase I clinical trial. Stem Cells 37 (11), 1481–1491.

Vasconcelos-dos-Santos, A., Rosado-de-Castro, P.H., Lopes de Souza, S.A., da Costa Silva, J., Ramos, A.B., Rodriguez de Freitas, G., Barbosa da Fonseca, L.M., Gutfilen, B., Mendez-Otero, R., 2012. Intravenous and intra-arterial administration of bone marrow mononuclear cells after focal cerebral ischemia: is there a difference in biodistribution and efficacy? Stem Cell Res. 9 (1), 1–8.

Wang, L., Ji, H., Li, M., Zhou, J., Bai, W., Zhong, Z., Li, N., Zhu, D., Zhang, Z., Liu, Y., Wu, M., 2013. Intrathecal Administration of Autologous CD34 positive cells in patients with past cerebral infarction: a safety study. ISRN Neurol. 2013, 128591. https://doi.org/10.1155/2013/128591.

Wang, Y., Guo, X., Liu, J., Zheng, Z., Ying Liu, Y., Gao, W., Xiao, J., Liu, Y.Q., Li, Y., Tang, M., Wang, L., Chen, L., Chen, D., Guo, D., Liu, F., Chen, W., Chan, B., Zhou, B.O., Liu, A., Mao, G., Huang, H., 2020. Olfactory ensheathing cells in chronic ischemic stroke: a phase 2 double-blind, randomized, controlled trial. J. Neurorestoratol. 8 (3), 182–193.

Xiao, J., Chen, L., Mao, G., Gao, W., Lu, M., He, X., Huang, H., 2017. Neurorestorative clinical application standards for the culture and quality control of olfactory ensheathing cells. J. Neurorestoratol. 5 (1), 175–179.

Zhang, G., Li, Y., Reuss, J.L., Liu, N., Wu, C., Li, J., Xu, S., Wang, F., Hazel, T.G., Cunningham, M., Zhang, H., Dai, Y., Hong, P., Zhang, P., He, J., Feng, H., Lu, X., Ulmer, J.L., Johe, K.K., Xu, R., 2019. Stable intracerebral transplantation of neural stem cells for the treatment of paralysis due to ischemic stroke. Stem Cells Transl. Med. 8 (10), 999–1007.

Nanodelivery of traditional Chinese Gingko Biloba extract EGb-761 and bilobalide BN-52021 induces superior neuroprotective effects on pathophysiology of heat stroke

Seaab Sahib[a], Aruna Sharma[b],*, Dafin F. Muresanu[c,d], Zhiqiang Zhang[e], Cong Li[e], Z. Ryan Tian[a], Anca D. Buzoianu[f], José Vicente Lafuente[g], Rudy J. Castellani[h], Ala Nozari[i], Ranjana Patnaik[j], Preeti K. Menon[k], Lars Wiklund[b], and Hari Shanker Sharma[b],*

[a]*Department of Chemistry & Biochemistry, University of Arkansas, Fayetteville, AR, United States*
[b]*International Experimental Central Nervous System Injury & Repair (IECNSIR), Department of Surgical Sciences, Anesthesiology & Intensive Care Medicine, Uppsala University Hospital, Uppsala University, Uppsala, Sweden*
[c]*Department of Clinical Neurosciences, University of Medicine & Pharmacy, Cluj-Napoca, Romania*
[d]*"RoNeuro" Institute for Neurological Research and Diagnostic, Cluj-Napoca, Romania*
[e]*Department of Neurosurgery, Chinese Medicine Hospital of Guangdong Province, The Second Affiliated Hospital, Guangzhou University of Chinese Medicine, Yuexiu, Guangzhou, China*
[f]*Department of Clinical Pharmacology and Toxicology, "Iuliu Hatieganu" University of Medicine and Pharmacy, Cluj-Napoca, Romania*
[g]*LaNCE, Department of Neuroscience, University of the Basque Country (UPV/EHU), Leioa, Bizkaia, Spain*
[h]*Department of Pathology, University of Maryland, Baltimore, MD, United States*
[i]*Anesthesiology & Intensive Care, Massachusetts General Hospital, Boston, MA, United States*
[j]*Department of Biomaterials, School of Biomedical Engineering, Indian Institute of Technology, Banaras Hindu University, Varanasi, India*
[k]*Department of Biochemistry and Biophysics, Stockholm University, Stockholm, Sweden*
*Corresponding authors: Tel.: +46-70-21-95-963; Fax: +46-18-24-38-99 (Aruna Sharma); Tel.: +46-70-20-11-801; Fax: +46-18-24-38-99 (Hari Shanker Sharma), e-mail address: Aruna.sharma@surgsci.uu.se; Sharma@surgsci.uu.se

Progress in Brain Research, Volume 265, ISSN 0079-6123, https://doi.org/10.1016/bs.pbr.2021.06.007

Abstract

Military personnel often exposed to high summer heat are vulnerable to heat stroke (HS) resulting in abnormal brain function and mental anomalies. There are reasons to believe that leakage of the blood-brain barrier (BBB) due to hyperthermia and development of brain edema could result in brain pathology. Thus, exploration of suitable therapeutic strategies is needed to induce neuroprotection in HS. Extracts of Gingko Biloba (EGb-761) is traditionally used in a variety of mental disorders in Chinese traditional medicine since ages. In this chapter, effects of TiO2 nanowired EGb-761 and BN-52021 delivery to treat brain pathologies in HS is discussed based on our own investigations. We observed that TiO2 nanowired delivery of EGb-761 or TiO2 BN-52021 is able to attenuate more that 80% reduction in the brain pathology in HS as compared to conventional drug delivery. The functional outcome after HS is also significantly improved by nanowired delivery of EGb-761 and BN-52021. These observations are the first to suggest that nanowired delivery of EGb-761 and BN-52021 has superior therapeutic effects in HS not reported earlier. The clinical significance in relation to the military medicine is discussed.

Keywords

Heat stroke, Brain edema, Hyperthermia, Gingko Biloba, EGb-761, Nanowired delivery, Neuroprotection, Blood-brain barrier, Brain pathology

1 Introduction

Heat wave and heat stress induced alterations in brain function and death is known since Biblical times (Judith 8:2,3). Effect of heat or sunstroke was well described affecting the brain and central nervous system as early as 1911 in Philadelphia followed by more than 150 cases in Chicago county hospital in 1917 (Gauss & Meyer, 1917; Weisenburg, 1912). In these cases, central nervous system was severely affected. At that time this was suspected that a lesion in brain stem or hypothalamic lesion or tumors is affecting the CNS function that looks like similar to ischemic stroke (Sharma & Hoopes, 2003). In 1944 few reports and autopsy finding in cerebellum neuropathology was reported in heat stroke (Freeman and Dumoff, 1944). The cerebellum pathology showed sponginess, edema, softness of the cerebellar tissues, neuronal damage and death of both Purkinje and granule cells. However, the first detailed account of human brain pathology following heat stress was described in detail in 1946 in 125 fatal cases (Malamud et al., 1946). Human brain pathology in heat stress clearly shows extensive cellular changes in the cortex, hippocampus, hypothalamus, thalamus, cerebellum, and brain stem. In addition, the endothelial cells deformity, swelling of brain areas and in some cases softening of brain tissue was also observed. This shows that endothelial cells are vulnerable to heat stress induced brain pathology (Haymaker et al., 1947; Malamud et al., 1946). Similar fatal cases were reported in US Army and heat exhaustion cases in Iraq (Schickele, 1947; Waterlow, 1947).

The severity of heat stress associated with heat waves and mortality was first greatly emphasized in 1995 when in Chicago heat wave more than 739 deaths were heat related (Semenza et al., 1996; Whitman et al., 1997). Similar heat wave in 1980 in Kansas City more than hundred deaths are confirmed as heat-related mortality (Jones et al., 1982). During heat wave of 2003 about 35 k mortality was assigned to heat-related death in Europe (Filleul et al., 2006). These incidences show that heat-related deaths are affecting large number of populations during excessive high heat waves across the continent throughout years (CDC 1999, 2005). Excess mortality due to heat wave is recorded in Istanbul in 2017 (Can et al., 2019). An estimated 70 k death is attributed in Europe during 2019 heat wave where in France alone more than 15 k people died due to heat related injuries (Roye et al., 2020).

Thus, heat stress and heat waves are serious environmental threat to many lives in the World. Heat stress or heat waves leads to heat exhaustion and heat Stroke when the body temperature reaches above 39–40 °C (Sharma and Hoopes, 2003; Sharma et al., 1998b). At this high temperature if uncontrolled, neuronal damage and vital center brain pathology could lead to instant death. Available data show that heat stroke results in death of more than 50% of the individuals (Sharma, 2005a, 2006).

The pathophysiology of heat stress is further aggravated in individuals after consumption of alcohol, substance abuse and/or drugs producing hyperthermia (Bongers et al., 2020; Martin-Latry et al., 2007; Tacke and Venalainen, 1987). Apart from these factors, individuals with co-morbidity factors such as hypertension or diabetes are more susceptible to heat stress induced neurological dysfunction and death (Muresanu and Sharma, 2007; Muresanu et al., 2010, 2019, 2020). However, the detailed mechanisms of heat-induced brain pathology and death are still not well known.

In 2005 neuropathological study was done in three patients who died in 2003 heat wave in France showed severe loss of Purkinje cell associated with heat shock protein 70 (HSP-70) expression in Bergmann glia (Bazille et al., 2005). These observations clearly show that cerebellum is quite vulnerable to heat-induced injury. The other areas showed degeneration include dentate nucleus, cerebellar peduncle and associated projections in the thalamus. Axonal damages and myelin vesiculation were also prominent (Bazille et al., 2005).

Taken together these observations suggest that heat stress induced brain damage and death is very serious problems Worldwide and exploration of suitable therapeutic strategies are the need of the hour (Sharma and Hoopes, 2003).

Experiments carried out in our laboratory in a heat stress model for the first time showed that when young rats were exposed to summer heat (34–36 °C) from 8 AM to 6 PM leakage of Evans blue albumin was seen in their brain in more than 90% of cases (Sharma and Dey, 1984, 1986). Whole cerebral cortex on dorsal surface and piriform cortex on the ventral surface was stained deep blue and the dye entered into the cerebrospinal fled (CSF) staining the dorsal surface of the hippocampus, caudate nucleus and colliculi. This suggests that heat exposure induces breakdown of the blood-brain barrier (BBB) and blood-CSF-barrier (BCSFB) to Evans blue albumin (68 kDa) (Sharma and Dey, 1986, 1987). Hemisecting of the brain showed significant deep staining of cerebellar vermis, cerebellar cortex, thalamus and hypothalamus. However, this effect was not seen in old rats exposed to same environment at identical conditions of heat stress (Sharma et al., 1992a,b).

Further experiments from our laboratory showed that heat stress induced BBB breakdown was associated with edema formation and brain pathology (Sharma et al., 1991a, 1994). Treatment with antioxidant drugs like H-290/51 (a chain breaking antioxidant) or extracts of Gingko biloba EGb-761 and bilobalide (BN-52021) are able to thwart heat stress induced brain pathology and upregulation of nitric oxide synthase (NOS) in model experiments (Sharma et al., 1997a, 2000a, 2003a, 2009b). Our study was thus the first to point out a neuroprotective role of EGB-761 and BN-52021 in the brain pathology of heat stress (Sharma et al., 2000a, 2003a; Westman et al., 2000).

Gingko biloba of the Ginkgoaceae family a native tree from Eastern China originated about 200 million years ago and considered as "living fossil" (Major, 1967). Its first record in Chinese Materia Medica appears about 2000 years ago (Major, 1967). In 1964 the extract of Gingko leaves (EGb-761) is introduced to Western Medical Practice and thereafter used Worldwide for age-associated memory disorders, dementia, and peripheral diseases (Chan et al., 2018; Evans, 2000). Now-a-days EGB-761 and its counterpart bilobalide is used for the pre-clinical and clinical treatment of ischemic stroke, Alzheimer's disease (AD), Parkinson's Disease (PD) and other neurodegenerative diseases (Chong et al., 2020; Feng et al., 2019; Kim et al., 2004; Liu et al., 2019a; Moradi et al., 2020; Rojas et al., 2009; Vellas et al., 2012; Yang et al., 2016; Zhao et al., 2020). The EGB-761 appears to be very potent in improving behavioral functions and prevention of cellular injuries and neurodegeneration in wide variety of neurological diseases (DeKosky et al., 2008; Singh et al., 2019).

In 2018 we are the first to nanowire EGb-761 and BN-52021 for further enhancing its efficacy on neuroprotection in heat stress induced brain pathology (Sharma et al., 2000a, 2003a; Westman et al., 2000). This review deals with the effect of nanowired EGb-761 and BN-52021 on heat-stress induced pathophysiology of brain injury based on our own investigation. We found that nanowired delivery of Gingko biloba extract has superior and longtime neuroprotective effects in heat stress induced neurodegeneration. The functional significance of our findings in relation to its clinical use in heat waves and medical practice is discussed in the light of the current literature.

1.1 Heat stress and heat wave induced mortality

Endogenous hyperthermia and exogenous heat strokes are well-recognized problems since several decades. Extreme hyperthermia caused by fever, chemotherapy or strenuous heat exposure induces serious CNS damage in populations exposed for longer periods. Military personnel often exposed to hyperthermia at hot environments are victims of heat stroke (Bazille et al., 2005; Bensimhon et al., 2004). The symptoms of heat stress associated with the exercise in high heat include excessive loss of electrolytes that could be reversed by replacement of electrolytes (Derrick, 1934; Edsall, 1908; Glover, 1931; Haldane, 1929). Prolonged training and/ or exercise in high environmental heat leads to heat stroke affecting CNS damage and altering hypothalamic thermoregulation set point (Nakai et al., 1999; Simon, 1994). The severity of heat stress is exemplified with the fact that in 1995 heat wave about thousand people died due to the heat stroke in Chicago (Changnon et al., 1996; Dematte et al., 1998; Naughton et al., 2002; Semenza et al., 1999) and similar episode happened in the 1980 heat wave in St Louis and Kansas City in 1980 (Jones et al., 1982). The neurological damages accompanying with hyperthermia over 40 °C are common symptoms with severe head injury culminating in death of individuals. Those survived this heat wave showed mild symptoms including diarrhea, vomiting, nausea, vertigo, locomotor disturbances, skin rash, and hyperthermia less than 40 °C. In some severe cases symptoms of heat exposure showed shock, respiratory distress, kidney failure, muscles cramps, tachycardia, blood coagulation, alterations in enzymes and electrolyte imbalance that leads to coma and eventually death (al-Mashhadani et al., 1994; Barrow and Clark, 1998; Bouchama and Knochel, 2002; Ferris et al., 1938; Knochel, 1989; Mirabelli et al., 2010; Yeo, 2004). These observations show that heat stroke following hat stress is a serious condition and caused brain damage. Thus, efforts are needed to expand our knowledge to treat these dreaded conditions in clinics for the benefit of heat-injured victims.

1.2 Heat shock proteins (HSPs) synthesis in heat stress

A large number of heat stroke survived patients exhibit serious neurological problems for long-tine (Yeo, 2004). It appears that these neurological dysfunctions are associated with protein denaturation, necrotic and apoptotic neurons, degeneration of axons altered synaptic functions (Jardine, 2007). Protein denaturation occurs due to inability in heat transfer or exchange to the surrounding environment leading to heat shock response (Jardine, 2007). This heat shock response induces cell damage and triggers synthesis of specific class of proteins called heat shock proteins (HSPs) (Ashburner and Bonner, 1979; Craig and Schlesinger, 1985). HSP response is universal to every cell and it can be upregulated due to any kind of stress response (Ritchie et al., 1994). Cells that survive hypothermia or heat shock response upregulate HSPs representing the state of thermotolerance against future heat elevation (Gerner and Schneider, 1975). Development of thermotolerance is proportional to the amount of the HSPS synthesized within the cells (Landry et al., 1982; Li and Werb, 1982; Subjeck et al., 1982).

Heat shock proteins are categorized based on their different molecular weights to six different proteins namely HSP/HSP B (12–43 kDa), HSP 40 (DnaJ), HSP 60 (HSP D), HSP 70 (HSP A), HSP 90 (HSP C), and HSP110 (HSP H) (Kampinga et al., 2009). HSPs as chaperones, play major roles in protein synthesis and folding and protect unfolded proteins or deteriorated proteins under stressful cellular conditions (Hartl et al., 2011; Treweek et al., 2015). HSP proteins determine the fate of several tissues under stressful or pathological conditions by triggering specific mechanism for self-digestion, stimulating secretory vesicles, signal transduction, and/or cell death (Muchowski and Wacker, 2005). Moreover, HSP also regulate cellular signaling pathways to alter cell metabolism and function (Basu and Srivastava, 2000; Mehlen et al., 1996; Murashov et al., 2001; Stokoe et al., 1992). Thus, HSP70 significantly influence apoptosis by diminishing c-Jun N-terminal kinase (JNK) and p38 mitogen-activated protein kinase (MAPK) dependent signaling pathways (Gabai et al., 1997; Park et al., 2001). In addition, under stress conditions, HSP attenuate the release of inflammatory mediators, nitric oxide, immunologic nitric oxide synthase (iNOS), reactive oxygen species (ROS), TNFα, IL-1, and IL-12 by inhibiting NFκB stimulation (Ding et al., 2001; Feinstein et al., 1996; Heneka et al., 2000; Tang et al., 2007; Van Molle et al., 2002; Zheng et al., 2008). Downregulation of NFκB occurs through inhibiting IκB phosphorylation through IκB kinase (IKK) activity. Reduction in NFκB results by inhibition of IKKγ activity that in turn attenuate NFκB activity (Ran et al., 2004; Zheng et al., 2008). Heat shock transcription factor 1 (HSF1) plays major roles in cell death. Stimulation of HSF1 induced Fas/Fas ligand initiates cell death (Xia et al., 2000). Overexpression of HSP70 regulates TCR/CD3- and Fas/Apo-1/CD95 that is enhancing apoptosis (Liossis et al., 1997).

Besides triggering apoptosis HSP also regulate the cell survival and repair in stressful conditions. Induction of HSP70 production following brain ischemia attenuates brain damage and reduces neuronal and glial cells injuries (Giffard et al., 2004; Hoehn et al., 2001; Rajdev et al., 2000). This effect of HSP70 is related to repairing of moderately denatured proteins, reducing neuroinflammation, proteins extravasation in brain, and downregulating apoptotic cell death (Garrido et al., 2001; Giffard et al., 2004; Mosser and Morimoto, 2004). Elevation of antiapoptotic protein Bcl-2 expression, inhibition of microglial/monocyte activation, and reduction in matrix metalloproteinases (MMPs) activity by HSP induction is likely to induce neuroprotection in ischemic injury (Yenari et al., 2005). Several isoforms of HSP such as HSP70, HSP90, and HSP27 regulate cell apoptosis during development via Akt kinase pathways (Beere, 2001; Murashov et al., 2001; Rane et al., 2003; Sato et al., 2000; Stetler et al., 2010). HSPs by regulating apoptosis-inducing factor (AIF) and inhibiting caspase signaling balance the cell death mechanism in health and disease (Beere, 2001; Beere et al., 2000; Matsumori et al., 2005). These observations suggest that HSP is an endogenous mechanism to protect neuronal injury and regulate CNS activity during stress and hyperthermia (Brown, 2007).

2 Neurological dysfunction in hyperthermia

Severe hyperthermia greater than 40 °C or 41 °C induces brain injuries that appears to be irreversible (Walter and Carraretto, 2016). After heat exposure, mild increase in body temperature greater than 38.5 °C results in neurological dysfunction (Lee et al., 2012). For mild hyperthermia induced neurological functions are reduced after cooling as standard protocol (Pease et al., 2009). Decreasing body temperature within 1 h after heatstroke below 38.5 °C improved brain function and other heat injury symptoms (Vicario et al., 1986). When hyperthermia exceeds 41 °C after heat exposure, vital centers in the brain stem are affected leading to permanent disability or even death in more than 50% of heat stress victims (Jimenez-Mejias et al., 1990; Pease et al., 2009; Trujillo et al., 2009). Hyperthermia induced neurological deficits such as cognitive decline, confusion, speech and locomotion difficulty, hyper excitability, seizures, or coma is seen more than 60% of patients when hyperthermia exceeds more than 40 °C (Albukrek et al., 1997; Flournoy et al., 2003; Lee et al., 2015; Lefkowitz et al., 1983; Walsh et al., 1997; Webster et al., 1985; Yaqub et al., 1987). These observations suggest that heat-induced hyperpyrexia is dangerous for brain function.

The cell membranes are particularly sensitive to heat induced hyperthermia. Thus, excessive hyperthermia greater than 40 °C lead to structural changes to cells membranes and alter signal transduction (White et al., 2007). Mitochondria and plasma membrane are the most heat-sensitive cellular structures and hyperthermia above 40 °C induces deleterious and irreparable damage to them in short period of time (Kiyatkin, 2007). Increase in body temperature over 41.6 °C induces apoptosis and cell death within few hours (Sakaguchi et al., 1995). In vitro studies suggest that, if the temperature reaches 49 °C cell necrosis and death occur within 300 s (Buckley, 1972). These studies suggest that heat is the sensitive indicator for cell and tissue damage.

In vivo situation, necrosis or apoptosis occur when the body temperature reaches above 40 °C (Peng et al., 2010). This hyperthermic effect is seen on both non-neuronal and neuronal cells (Khan and Brown, 2002). Heat injury also affects endothelial cells and causes their degeneration (Bouchama et al., 1996; Brinton et al., 2012; Roberts et al., 2008). Heat induced denatured proteins triggers cell necrosis or apoptosis (Bellmann et al., 2010) probably through caspase-mediated pathways (Peng et al., 2010). Taken together heat injury leads to neuronal dysfunction and cellular damage. Thus, further studies are needed to explore possible therapeutic strategies to rescue them in time.

2.1 Hyperthermia and cell signaling

Hyperthermia induces phosphorylation of several protein kinases including c-Jun NH2-terminal protein kinase (JNK) leading to differentiation and stimulation of apoptosis (Walter and Carraretto, 2016). Hyperthermia also activates of JNK 2 and JNK 3 pathways (Schiaffonati et al., 2001) causing intellectual dysfunctions (Xiong et al., 2014). Phosphorylation of specific Threonine groups (Thr) within the Calcium/calmodulin-dependent protein kinase II (CaMKII) results in cognitive dysfunction, learning and memory impairment and synaptic plasticity (Irvine et al., 2006; Uchinoumi et al., 2016; Xiong et al., 2014). Hyperthermia stimulates CaMKIIα Thr in presence of brain injury (305) whereas CaMKIIα Thr (286) phosphorylation is inhibited in hypothermia (Xiong et al., 2014).

3 Cytokines release in hyperthermia

Increased release of cytokines following heat stroke induces neuronal damage (Chang, 1993). However, pro- and anti-inflammatory cytokines are overexpressed immediately or during hyperthermia after heat stroke (Leon, 2007). In particular, tumor necrosis factor (TNF)-α, anti-inflammatory cytokines IL-1RA, IL-10, interleukin (IL)-1, IL-6, are released after heat injury (Bouchama et al., 1991, 1993, 2000; Hammami et al., 1997; Hashim et al., 1997; Kao et al., 1994). Thus, using inhibitors of cytokine release in heat stress appears to be beneficial treatment strategies (Leon et al., 2006; Walter and Carraretto, 2016).

3.1 Amino acid Neurotransmitters in hyperthermia

Several neurotransmitters are altered or released following hyperthermia that may contribute to cell damage. However, release of excitatory amino acid transmitter glutamate and inhibitory neurotransmitter glycine is also released in hyperthermia. Glutamate is a potent excitatory neurotransmitter plays significant role in cell and tissue injury. Increased glutamate level in neuronal tissue after hyperthermia enhances calcium influx, depletes ATP generation, changes ionic balance, and induces caspase-dependent apoptosis (Niu et al., 2020; Sharma et al., 2020a,d). On the other hand, glycine is an inhibitory neurotransmitter essential for stimulating NMDA receptors together with glutamate. These neurotransmitters are involved in several pathological conditions such as acute ischemia (Castillo et al., 1999), traumatic CNS injury (Richards et al., 2003), multiple sclerosis (Klivenyi et al., 1997; Pitt et al., 2003), and Alzheimer's disease (Jimenez-Jimenez et al., 1998). This suggests that neuronal damages in hyperthermia is related to amino acid neurotransmitters as well (Walter and Carraretto, 2016). However, further studies are needed to clarify these points.

3.2 Blood-brain barrier damage in hyperthermia

Hyperthermia disrupts the blood-brain barrier (BBB), and perivascular astrocytes (Sharma et al., 1992b, 1997a, 1998b). The BBB maintains the homeostasis of the brain fluid microenvironment within narrow limits thus prevents harmful molecules to enter into the brain (Cardoso et al., 2010; Kun, 2009). However, hyperthermia influences the selective permeability of BBB when the blood or body temperature rises above 39°C in vivo. The BBB that is composed of endothelial cell membranes of the brain connected with the tight junctions starts start losing its tightness, allowing diffusion of unwanted molecules, and serum proteins (Kiyatkin and Sharma, 2009). Entry of plasma proteins into the brain fluid microenvironment disruption leads to vasogenic cerebral edema after heat stroke. Excessive brain swelling could result in instant death in hyperthermic patients with a core temperature greater than 40°C (Goldstein et al., 2003; Sharma et al., 1998b). However, the degree of BBB damage depends on the intensity and duration of heat exposure (Sharma et al., 1992b,

1998b). Rats exposed to heat stress at 38 °C for 2 h does not develop BBB breakdown. Whereas continued exposure of heat at 38 °C for 4 h resulted in massive BBB breakdown to Evans blue albumin associated with edema formation (Kiyatkin and Sharma, 2009; Sharma et al., 1998b). On the other hand, hyperthermia in anesthetized rats exhibited significant BBB breakdown together with edema formation the temperature range of 38.5–39 °C (Kiyatkin and Sharma, 2009). These observations suggest that hyperthermia induced heat stress is crucial in BBB disruption, edema formation and cellular damages in the brain.

3.3 Cerebral blood flow and metabolism disruption in hyperthermia

Heat stroke causes severe alterations in the cerebral blood flow (CBF) and metabolism (Dietrich et al., 1990; Shiozaki et al., 1993). A previous About 1 °C increase in temperature results in about 25% elevation in the CBF (Busija et al., 1988). Development of cerebral edema, and intracranial hypertension following hyperthermia causes reduction in the CBF (Cremer and Kalkman, 2007). Temperature above 40 °C after heat stress induces hypotension and significant reduction in the CBF (Qian et al., 2014). However, the mechanism underlying hypotension following hyperthermia is still unknown. Decreased CBF and ischemia is associated with BBB breakdown and edema formation (Berhouma et al., 2019; Sharma et al., 1998b). Changes in the CBF in hyperthermia could be due to loss of cerebral autoregulation resulting in hemodynamic changes and blood pressure (Rangel-Castillo et al., 2008; Rapoport, 1976; Stefanits et al., 2019). This high blood pressure or intracranial hypertension also disrupts BBB and brain swelling in hyperthermia (Gasser et al., 2003; Karibe et al., 1994; Muresanu and Sharma, 2007; Patnaik et al., 2000; Sharma et al., 2006a; Shiozaki et al., 1993). However, various other factors such as capillary filtration pressure, blood osmotic pressure or hydrostatic pressure, and tissue compliance during heat injury could all influence the BBB function and brain edema formation (Cherian et al., 2018; Clement et al., 2020; Lent-Schochet and Jialal, 2021; Mokhtarudin et al., 2018; Rapoport, 1976; Wurzer et al., 2018).

Alterations in cerebral glucose and oxygen metabolism is also altered with the CBF changes as well in hyperthermia (Cremer and Kalkman, 2007). When the body temperature reached 40 °C mitochondrial oxygen consumption declines significantly (Cremer and Kalkman, 2007). However, alteration in cerebral metabolism varies in different part of the brain after heat exposure. Bran metabolic rate during hyperthermia appears to be related with the rise in body temperature. Thus, at body temperature elevation in adult males by 1.5–2 °C above the basal level increases the cerebral metabolic rate by 25% (Saxton, 1981). Further increase in body temperature by additional 2 °C increases brain metabolism by additional 10% (South, 1958). Previous studies on different animal models of induced hyperthermia indicated that for every 1 °C rise in body temperature results in about 5–10% increase in the cerebral metabolic rate and glucose consumption (Carlsson et al., 1976; McCulloch et al., 1982; Mickley et al., 1997; Nemoto and Frankel, 1970).

Hyperthermia induced increase in rectal temperature to 42.1 °C in dogs increases the cerebral metabolic rate by 21%. However, this increase in the cerebral metabolic rate (CMR) to normal level when the body temperature reaches 43 °C. (Nemoto and Frankel, 1970). This reduction in the CMR at extremely high body temperature appears to be due to the nucleotide degradation and BBB disintegration (Bynum et al., 1978). In humans increase in core body temperature to 39 °C significantly increase cerebral metabolism for glucose consumption in corpus callosum, hypothalamus, thalamus, cerebellum, and cingulate gyrus (Nunneley et al., 2002). At the same time the CMR for glucose consumption decreased in the caudate, putamen, insula, and posterior cingulum shows a remarkable decrease in the glucose metabolic rate (Nunneley et al., 2002). These observations suggest that hyperthermia induces selective changes in the CMR for oxygen and glucose consumption (Bain et al., 2014).

3.4 Oxidative stress and brain pathology in hyperthermia

Hyperthermia induces severe oxidative stress that is largely responsible for the BBB breakdown and brain damage. In any biological system, increase in free radical production results in oxidative stress. Reactive oxygen species (ROS) or reactive nitrogen species (NOS) and other related molecules such as hydrogen peroxide, nitric oxide, carbon monoxide and peroxynitrite have the potential to cause significant damage to the cell and cell membranes by denaturing lipids, proteins, and the nucleic acids (Ferrari, 2000; Gilgun-Sherki et al., 2001). As a result, increase in harmful molecules such as aldehydes, isoprostanes, protein carbonyls, and base adducts occur within the cell leading to further cellular injuries (Markesbery, 1999). Brain tissues contain large portions of proteins and lipids with rapid metabolic rate and iron content they are highly vulnerable to oxidative stress leading to brain pathology (Fernandez-Sanchez et al., 2011; Furukawa et al., 2017; Manna and Jain, 2015).

Oxidative stress is more effective among the elderly (Lenaz, 1998). With advancing age and incidences of Alzheimer's disease (AD) enhanced mitochondrial electron transport chain system is associated with fatigue (Cunnane et al., 2011; Reddy et al., 2012) and leads to accumulation of ROS damaging most of the cell components (Trifunovic and Larsson, 2008). Thus, mitochondrial dysfunction is the major factor in development of Alzheimer's disease and other neurodegenerative disease (Wang et al., 2014). Further evidence supporting this idea is evident from the findings that antioxidants significantly improve the rain function in Alzheimer's disease (AD), Parkinson's disease (PD), Multiple Sclerosis (MS), amyotrophic lateral sclerosis (ALS), and dementia (Commenges et al., 2000; Engelhart et al., 2002; Morris et al., 2002; Uttara et al., 2009). Oxidative stress induce lipid peroxidation (Lovell et al., 1995), triggers protein and nucleic acid oxidation (Gabbita et al., 1998, 1999; Hensley et al., 1995; Mecocci et al., 1994), reduces the membrane polyunsaturated fatty acids (Prasad et al., 1998; Svennerholm and Gottfries, 1994), and elevate the trace elements in the brain tissues including iron and aluminum (Cornett et al., 1998; Good et al., 1992; Smith et al., 1997). Increased neurotoxins and lipid peroxidation with elevation of 4-hydroxynonenal (HNE), F4-neuroprostanes, and

F2-isoprostanes further contributes to the cell injury in oxidative stress (Lovell et al., 1997; Markesbery and Lovell, 1998; Montine et al., 1998).

Lipid peroxidation also results in production of several deleterious byproducts such as acrolein and 4-hydroxy-2-nonenal (HNE), and malondialdehyde (Ayala et al., 2014; Draper and Hadley, 1990; Uchida, 1999). Liberation of these toxic lipid aldehyde species result in damaging the DNA and proteins resulting in genotoxicity and cytotoxicity together wish inhibition of gene expression and apoptosis (Erejuwa et al., 2013; Taso et al., 2019).

Acrolein is one of the lipid peroxidation reaction products generated during metal-catalyzed oxidation of polyunsaturated fatty acids such as arachidonic acids (Uchida et al., 1998). Acrolein is a highly reactive electrophile that reacts with sulfhydryl groups of specific amino acid residues such as Cystine, lysine and Histidine in the nucleic acid structure (Calingasan et al., 1999; Erejuwa et al., 2013; Ian, 2008). Overproduction of acrolein during oxidative stress inhibits metabolic pathways such as glutamate and glucose uptake leading to cell damage (Haenen et al., 1988; Keller et al., 1997).

Hyperthermia significant elevates interleukin-1β, dopamine, and glutamate in ischemic stroke (Chiu et al., 1995; Kao et al., 1994; Lin, 1997; Lin et al., 1995) associated with increased reactive oxygen or nitrogen species generation (Halliwell, 1991; Ikeda and Long, 1990). Cerebral ischemia in heat stroke causes neuronal depolarization leading to tissue hypoxia (Chang et al., 2007). Free radical's generation increases intracellular calcium influx activating glutamate receptors and triggers cell death (Ermak and Davies, 2002; Mattson, 2003).

Administration of magnolol, a-tocopherol, or mannitol in hyperthermia induced significant neuroprotection (Chang et al., 2007) and reduce the generation of hydroxyl radicals (Suzuki et al., 1985). These observations suggest that cerebral ischemia, hypoxia, reperfusion, inflammation, and hyperthermia are associated with significant elevation in ROS or/and RNS levels (Babior, 2000; Cao et al., 1988; Chan, 2001; Riedel and Maulik, 1999; Suzuki et al., 1985). Thus, antioxidants therapy in hyperthermia is capable of reducing brain pathology and functional disabilities. This aspect requires further investigations to enhance clinical practice in hyperthermia induced brain pathology in future.

3.5 Chinese traditional medicine extract of Gingko Biloba (EGb-761)

Ginkgo biloba is a Chinese plant extract used for the treatment of several diseases due to its high contents of antioxidants and pharmaceutical compounds. These compounds include flavonoid fraction and nonflavonoid fraction (6%). The flavonoid fraction comprises 24% of the total ingredients including quercetin, kaempferol and isorhamnetin glycosides. The non-flavonoid fraction, also called terpenoid fraction, consists of trilactonic diterpenes (ginkgolide A-C, ginkgolide J-M), and trilactonic sesquiterpene (Bilobalide) (Nasr et al., 1986, 1987; Stewart, 1992). Other compounds in Gingko biloba extract are kynurenic acid (KYNA),

6-hydroxykynurenic acid (6-HKA), glucuronic acid, shikimic acid, vanillic acid, proanthocyanidins, rhamnose, and glucose (Stewart, 1992).

Ginkgo biloba extract (EGb-761) has the potential to reduce oxidative stress damage and prevents blood clot formation. In hypobaric hypoxia model EGb-761 extends the survival time for 30 min as compared to 5 min in the untreated group (Karcher et al., 1984). Treatment with EGb 761 in neonatal hypoxia significantly minimizes the symptoms and enhances neuroprotection (Chatterjee and Gabard, 1981). The non-flavone ingredients of EGb 761 reduces hypoxic brain damage in the rat and increases cerebral blood flow (CBF) in thalamus, hypothalamus, cortex, substantia nigra, and nucleus accumbens (Ahlemeyer and Krieglstein, 2003).

In 1997, Klein et al. reported that the non-flavonoid fraction (bilobalide) of EGb 761 has superior neuroprotective activity against lipid peroxidation and choline release in hypoxia (Klein et al., 1997). Pretreatment with EGb 761 in hydrogen peroxide induced oxidative brain damage is significantly reduced (Chen et al., 1999; Ni et al., 1996; Oyama et al., 1996). Moreover, glutamate neurotoxicity is also reduced by EGb 761 extract (Krieglstein et al., 1995; Zhu et al., 1997). EGb 761 is a potent neuroprotective agent against β-amyloid induced neuronal cell injury (Bastianetto et al., 2000a; Zhou et al., 2000). EGb 761 extract reduced Parkinson's disease induced brain damage in 1-methyl-4-phenyl-1,2,3,6-tetrahydropyridine (MPTP) intoxicated animal model (Yang et al., 2001). These neuroprotective effects of EGb-761 is related to its antioxidant properties (Yang et al., 2001).

The neuroprotective activity of flavonoid fraction of EGb-761 is due to its free radical scavenging capability together with inhibiting the nitric oxide (NO)-induced protein kinase C activity (Bastianetto et al., 2000b; Sharma et al., 2000a). Hyperthermia significantly induces nitric oxide synthase (NOS) of neuronal (nNOS) and inducible (iNOS) isoforms upregulation in cerebral cortex, hippocampus, cerebellum, thalamus, hypothalamus, and brain stem (Sharma et al., 2000a). This expression of nNOS and iNOS following hyperthermia is significantly reduced by EGB-761 or bilobalide BN-52021 treatments (Sharma et al., 2000a). In addition, the BBB breakdown and brain edema formation in hyperthermia is also reduced by EGB-762 and BN-52021 treated hyperthermic group (Sharma et al., 2000a). These observations suggest that EBG-761 and BN 52021 exert powerful neuroprotective effects in hyperthermia. It would be interesting to see whether nanodelivery of EGb-761 and BN-52021 has further superior neuroprotective effects in heat stress induced brain damage.

3.6 Nanowires drug delivery to CNS

Recent advancement in drug delivery techniques opened new vista to overcome the challenges associated with brain targeting. Due to presence of the blood-brain barrier (BBB) and blood-spinal cord barrier (BSCB) drug delivery to the neuronal tissues is severely restricted (Emerich and Thanos, 2006; Sahib et al., 2019, 2020; Sharma, 2005b). Drug Nanodelivery techniques use lipophilic analog attachment (Pardridge, 1988), chemical modified compounds (Bodor and Buchwald, 1999), vector-mediated drug delivery (Kang et al., 1994), colloidal drug carrier (Wong et al., 2012), intracerebral implants (DiMeco et al., 2002), and micro-infusion pump (Cunningham et al., 2008) may increase drug content in the brain or spinal cord. Neuroprotective efficiency of several drugs and antibodies are also enhanced by their topical application on injured brain or spinal cord for effective delivery as another approach (Sahib et al., 2019, 2020; Sharma, 1994, 2000, 2005b; Sharma and Dey, 1982; Sharma et al., 1990b, 1995a, 1997a, 1998b, 2000a).

Topical application of therapeutic agents results in superior pharmacological effects at a lower concentration as compared to their systemic administration in higher doses (Sahib et al., 2019, 2020; Sharma, 1994, 2004; Sharma et al., 1990b, 1996, 1997b, 2003b, 2006b). Recently, nanocarriers for drug delivery using biocompatible nanomaterials is the preferred mode of drug development in neurological diseases (Li et al., 2015; Sahib et al., 2019, 2020; Zhang et al., 2014). Disruption of BBB during CNS jury could enhance drug delivery to the neuronal tissues (Chen and Liu, 2012; Lo et al., 2001; Niu et al., 2020; Sharma et al., 2020b,c,d). However, BBB breakdown through damage also results in edema and tissue swelling that could further restricts the drug penetration into the neural tissues (Sandoval and Witt, 2008; Shlosberg et al., 2010).

Delivering neuroprotective agents across the BBB can be accomplished efficiently using nano-delivery techniques (Sahib et al., 2019). Biomolecules anchored to the surface of biocompatible nanowires could be a promising approach for drug delivery to the CNS (Sahib et al., 2019; Sharma et al., 2007). Delivering drugs via biocompatible nanocarriers revealed significant improvement in therapeutic efficiency (Sahib et al., 2019; Sharma, 2004; Sharma and Sharma, 2007). Polysorbate 80—enveloped poly butyl cyanoacrylate nanostructures target the neuronal tissue efficiently and enhances the efficacy of the loaded Dalargin (Kreuter, 1995). This confirmed the superior effect of the nanomaterial-loading as a drug vehicle for targeting CNS delivery (Alyautdin et al., 1997; Friese et al., 2000). Drug-loaded titanate nanowires significantly potentiate the drug effect across the BSCB and enhances neuroprotective efficacy of the spinal cord (Niu et al., 2020; Sahib et al., 2019, 2020; Sharma et al., 2007).

The stability and biocompatibility of Titanate nanomaterials are greatly influenced by the method of synthesis (Kulkarni et al., 2015; Linsebigler et al., 1995; Sahib et al., 2019, 2020; Yin et al., 2013). Several types of titanate nanostructures have been developed for biomedical and industrial use. Titanate whiskers are used for delivering anticancer drugs (Li et al., 2009). Titanate nanowires are used to control the release of proteins via ultraviolet light stimulation (Sua et al., 2011). Injection of titanate nanotubes improves bone regeneration due to its ability to adsorb and concentrate FITC-BSA proteins, collagen, and fibronectin at the site of bone fracture (Ding et al., 2019). Different sizes of titanate nanotubes reduce oxidative stress and enhance osteocytes differentiation (Yu et al., 2018).

Titanate nanofibers as drug carrier potentiates neuroprotective effect of cerebrolysin in traumatic brain injury (Ozkizilcik et al., 2018; Sahib et al., 2020). Titanate nanowires loaded cerebrolysin also significantly enhance its neuroprotective ability in spinal cord injury (Sahib et al., 2020; Sharma et al., 2009a). Delivery of

DL-3-n-butylphthalide using titanate nanowires in concussive head injury potentiates neuronal survival and motor functions (Sahib et al., 2019). Titanate nanowires etched implants enhances spinal cord conduction and improve moor function (Sahib et al., 2019). Structural based conductivity of titanate nanostructures on biomolecules and tissue behavior is examined in several studies. Lager titanate nanotubes have superior conductive properties as compared to the smaller ones (Ding et al., 2019; Yang et al., 2014). Based on the 3D shape of these structures several techniques are applied to load the drugs either as scaffolds or within the hollow spheres (Bertrand and Leroux, 2012; Yin et al., 2013). Accordingly, TiO2 nanospheres or nanowires used for drug entrapment and release of sodium phenytoin (Heredia-Cervera et al., 2009), temozolomide (Lopez et al., 2006), Valproic acid (Uddin et al., 2011), daunorubicin (Wu et al., 2011), SCI1, SCI2, SCI5 (Sharma et al., 2007), and DL-3-n-butylphthalide (Sahib et al., 2019) for their in vivo treatment with superior results.

TiO2 nanofibers conductivity, piezoelectricity, polarization, or stability, and their influence on biological molecules make them one of the superior materials for further use in drug delivery (Choi et al., 2016; Othman et al., 2014; Wu et al., 2017). The ability to produce different morphological nanostructures of titanium using various synthesis procedures makes them an ideal choice for diverse drug delivery applications (Adams et al., 2006; Gittens et al., 2011; Ren et al., 2018). In present investigation Gingko biloba extract EGb-761 and BN-52021 are nanowired using TiO2 synthesis for the first time to potentiate their neuroprotective effects in heat stress induced brain pathology.

4 Our observations on EGb-761 and BN-52021 on heat stress induced brain pathology

We examined in detail on brain pathology in heat stress following nanowired delivery of EGb-761 or BN-52021 in the rat and compared with conventional EGb-761 and BN-52021 delivery using standard protocol (see below).

5 Nanowired preparation of Gingko biloba extract
5.1 Materials

Titanium dioxide TiO2 powder (Degussa, P25) (98+%, VERTEC® TVBT, Alfa Aesar), Sodium hydroxide pellets (Macron Fine Chemicals), Extracts of Gingko Biloba EGb-761, Bilobalide BN-52021 (IPSEN Beaufour, Paris, France; Dr. Willmar Schwabe GmbH, Karlsruhe, Germany), deionized water prepared by ion exchange in our laboratory.

5.2 **Fabrication of titanate nanowires**

The following hydrothermal synthesis process was conducted based on previous reports (Jitputti et al., 2008; Kolen'ko et al., 2006; Mund et al., 2014; Yoshida et al., 2005) with some modifications. The reaction started by mixing a 0.375 g of Titanium Dioxide nanopowder (Degussa P25) with 50 mL of 10 M NaOH solution and stirring for 24 h. Then, sonication for 15 min was done to make sure no aggregation is left. Following this step, the mixture was heated in a sealed stainless-steel Teflon linear at 240 °C for 72 h. After that the product was centrifuged to obtain TiO_2 nanowires. The nanowires were then washed with DI water until pH dropped to (pH 7.2). Nanowires were then dried in the autoclave at 45 °C for 1 day to collect a white dry material ready for the drug loading process (Sahib et al., 2019, 2020).

5.3 **Characterization**

Scanning electron microscope (SEM, Tescan VEGA II SBH) was used to observe the morphologies of titanate nanowires. X-ray Diffractometer (XRD, Rigaku MiniFlex II) was used in this study to analyze the crystal structure of titanate nanowires. For the elemental composition analysis, Energy dispersed X-ray diffraction (EDX) technique (Bruker AXS Microanalysis GmbH Berlin, Germany, EDX system) was used. All the results for the above characterization parameters (SEM, XRD, and EDX) have been discussed in detail earlier (Sahib et al., 2019, 2020) and shown in Fig. 1.

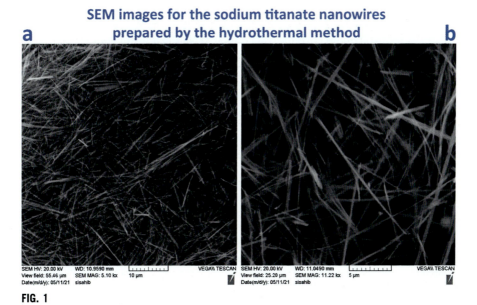

FIG. 1

Scanning electron micrograph (ESM) of titanium nanowires (A, 10 μm) and (B, 5 μm) for nanowiring of EGb-761 and BN 52021 in heat stress. For details see text.

5.4 Drug loading and delivery

Both EGb-761 and BN-52021 were loaded on the titanate nanowires separately following a standard loading technique describe in detail earlier (Sahib et al., 2019; Sharma et al., 2000a; Westman et al., 2000) and shown in Fig. 2. The TiO2 loaded drug (EGb-761 and BN-52021) were administered according to the standard protocol in heat stress and brain pathology is examined (see below).

6 Heat Stress Model of hyperthermic brain injury

6.1 Animals

Experiments were carried out on inbred Charles Foster male rats (100–150 g body weight; age 10–12 weeks) housed at controlled ambient temperature ($21 \pm 1 \,^{\circ}\text{C}$) with 12 h light and 12 h dark schedule. Rat feed and tap water were supplied ad libitum.

6.2 Exposure to heat stress

Rats were exposed to heat stress in a Biological Oxygen Demand Incubator (BOD) maintained at $38 \,^{\circ}\text{C}$ with relative humidity 45–47% and wind velocity 20–25 cm/s kept constant. The animals were exposed to 1–2 h daily for 8 days in chronic experiments and maximum 4 h for acute experiments (Sharma and Dey, 1984, 1986, 1987).

Control group was placed in identical conditions at $21 \,^{\circ}\text{C}$ either at the BOD chamber or at room temperature for identical periods (Sharma et al., 1991a).

All experiments were conducted according to National Institute of Health (NIH) Guidelines for Care and Handling of Experimental Animals and approved by Local Institutional Ethics Committee.

7 Heat Stress symptoms

The following symptoms were recorded and observed during heat exposure as stress symptoms.

(i) *Rectal temperature*

The rectal temperature was recorded using rodent thermistor probes (Yellow Springfield, NJ, USA) connected to a 12-channel telethermometer (Harvard Apparatus, Holliston, USA). The thermistor probes were inserted into the rectum to record deep visceral temperature as describe earlier (Sharma and Dey, 1987). For this purpose, rats were handled 1 week before the experiments once daily to record their body temperature so that the measurement of rectal temperature either before or after heat exposure was not affected due to handling stress (Sharma and Dey, 1986, 1987).

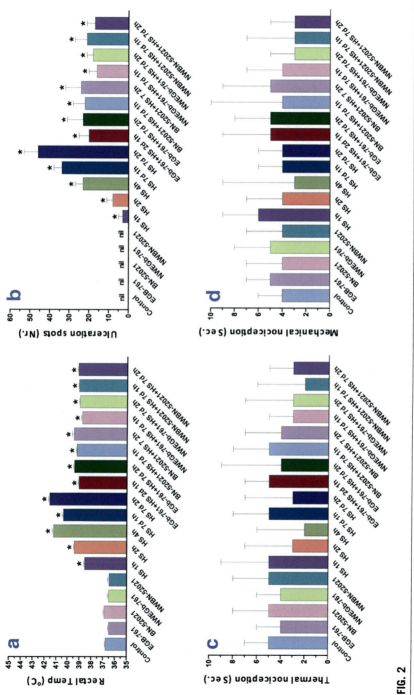

FIG. 2

Effect of EGb-761 and BN-52021 on stress symptoms including rectal temperature (A), gastric ulceration (B) thermal nociception (C) and mechanical nociception (D) following acute and chronic heat stress. Animals were subjected to heat stress in a biological oxygen demand (BOD) incubator maintained at 38°C with food and water ad libitum (For details, see text). Values are Mean ± SD of 6–8 rats at one point. * $P < 0.05$ from control group. # $P < 0.05$ from respective chronic heat stress group. ANOVA followed by Dunnett's test for multiple group comparison using one control.

(ii) *Behavioral salivation*

When rats are exposed to high ambient heat exposure the heat dissipation occurs through salivation, as the animals do not have enough sweat glands. Thus, rats spread their saliva over the snout to dissipate heat and this behavioral is referred to salivation (Sharma and Dey, 1984, 1986, 1987). The spread of saliva over the snout was measured at the end of the experiment. The spread of saliva is proportional to the degree of heat stress (Sharma and Dey, 1984, 1986, 1987).

(iii) *Behavioral prostration*

Rats during heat exposure and heat dissipation efforts became lethargic and often lay prostrate on the floors of cage. Sometimes even after gentle pushing, they do not move or walk. This behavior is known as prostration. When the animals are just lying on the floor is graded as one, when they do not move even after hearing a knock on the cage or other sound this may be stage 2 and when they do not move even after gentle pushing, they have entered into the grade 3 of prostration due to heat stress (Sharma and Dey, 1984, 1987). These different grades of prostration are related with the severity of heat stress (Sharma and Dey, 1987).

(iv) *Gastric hemorrhages*

Another parameter of stress is the development of peptic ulcers. The magnitude and severity of stress can thus be evaluated by presence of stress ulcers in the stomach (Sharma and Dey, 1986, 1987). Thus, at the end of the experiments stress ulcers were examined in the stomach at postmortem examination. The stomach was cleaned and seen for any ulcerations or spots using a hand help magnifying lens (x4) and number of hemorrhagic spots or microhemorrhages were identified and counted manually.

8 Physiological parameters

The following physiological parameters were assessed using standard protocol as described earlier (Rojas et al., 2009; Sharma et al., 2009a,b).

(i) *Thermal nociception*

In order to test pain perception in heat-exposed rats, thermal nociception induced pain response was examined using Tail-Flick analgesiometer (LE7106, 76-0187 Rat, Harvard Apparatus, Holliston, MA, USA) against a radiant heat stimulus. The rat-tail was placed on the cliff and radiant heat stimulus applied through a light beam. The latency by which rat flick its tail was counted manually (Sharma et al., 2020b). The cut off time was maintained for 30 s to avoid any injury or sensitivity to the tail skin.

(ii) *Mechanical nociception*

In this model the rat's tail was pinched using a mechanical blunt pressure and the time to react the tail flick response was recorded manually. The cut off time was maintained 20 s (Sharma et al., 2004). The animal tail after heat exposure was exposed to the mechanical pressure unit for pain management (LE7306, Harvard Apparatus, Holliston, MA; USA).

(iii) *Mean arterial blood pressure*

Mean arterial blood pressure (MABP) was recorded using an indwelling polythene cannula PE10 implanted into the left carotid artery retrogradely toward the heart aseptically 1 week before the experiment. At the time of the experiment, the carotid artery catheter was connected to a Strain Gauge Pressure Transducer (Statham, P23, Holliston, MA, USA) attached to a chart recorder (Electromed, London, UK) (Sharma, 1987).

(iv) *Arterial pH*

Before carotid artery cannula was connected to the Strain Gauge Pressure transducer about 1 mL of arterial blood as withdrawn for later determination of arterial pH using Radiometer Apparatus (ALB 800 Flex, Copenhagen, Denmark) (Sharma et al., 2009c).

(v) *Arterial PaO2*

The arterial PaO2 was also recorded on Radiometer Apparatus (ALB 800 Flex, Copenhagen, Denmark).

(vi) *Arterial PaCO2*

Arterial PaCO2 from the carotid cannula blood was analyzed in the Radiometer Apparatus (ALB 800 Flex, Copenhagen, Denmark) as described earlier (Sharma and Dey, 1986, 1987).

9 Pathophysiological parameters

(i) *Blood-brain barrier permeability*

The blood-brain barrier (BBB) permeability was measured using Evans blue albumin (EBA) and radioiodine ([131]-I-Na) as described earlier (Sharma and Dey, 1986, 1987). In brief, 2% solution of EBA in sterile physiological saline (pH 7.4) was administered in a dose of 3 mL/kg intravenously either through needle puncture in the right femoral vein under anesthesia (Equithesin 3 mL/kg, i.p.) or through an indwelling polythene catheter implanted into the right jugular vein 1 week before the experiment aseptically. Radioiodine was administered (100 μCi/kg, i.v.) either mixed with EBA or immediately after EBA administration. The tracers were allowed to circulate 5–10 min in the systemic circulation (Sharma, 1987; Sharma and Dey, 1987).

The animals were then deeply anesthetized, and chest was rapidly opened to exposed heart and right auricle was cut. A butterfly needle (21G) was inserted into the left ventricle of the heart connected with a peristaltic pump (Harvard Apparatus, Holliston, MA; USA) attached with a cold 0.9% saline bottle. The intravascular tracer was washed out by saline perfusion at 90 Torr for about 3–5 min (50–80 mL saline). Immediately before saline perfusion, about 1 mL of whole blood was withdrawn from the left cardiac ventricle for later determination of EBA or radioiodine (Sharma and Dey, 1986).

Immediately after perfusion, the brains were dissected out and examined for EBA penetration over the dorsal and ventral surfaces of the brain. After that, a mid-sagittal section was made to examine leakage of dye across the choroid plexus in the cerebral ventricle and subcortical structures. After that, the desired portions of the brains were dissected out, weighed immediately and the radioactivity counted in a 3-in well type Gamma counter at energy window 500–800 KeV (Packard, Ramsey, MN, USA). The extravasation of radioactivity in the brain was calculated as percentage increase over the blood radioactivity (Sharma and Dey, 1986, 1987).

After counting the radioactivity, the sample were homogenized in a mixture of 0.5 M Sodium phosphate (pH 7.0) and ultrapure analytical grade acetone (BDH, UK) to extract EVA dye entered brain tissue. The samples were centrifuged at $900 \times g$ for 10 min and the supernatant were measured for dye calorimetrically by Spectrophotometer (Harvard Apparatus, Holliston, MA, USA) at 620 nm. The dye entered into the brain was calculated from the EBA standard curve prepared earlier (Sharma and Dey, 1986, 1987). The dye entered was measured as mg % in brain tissue (Sharma, 1987).

(ii) *Brain edema*

Brain edema formation was assessed using brain water content measurement as described earlier (Sharma and Cervos-Navarro, 1990; Sharma and Olsson, 1990; Sharma et al., 1991a). In brief, immediately after termination of experiments, the brains were dissected out after decapitation and placed over cold saline wetted filter paper. The large superficial blood vessels and blood clot if any was removed and weighed on a preweighed filter paper (No. 1, Whatman, Sigma Aldrich, St. Louis, MO, USA) in a sensitive digital electronic balance (Mettler Toledo, Columbus, OH, USA, sensitivity 0.1 mg). After weighing the brain samples the tissues were placed in an incubator maintained at 90 °C to evaporate brain water for 72 h. When the dry weights of the samples became constant after 3-consecutive measurements on the same electronic balance the water content was calculated from the difference between wet and dry weight as described earlier (Sharma and Cervos-Navarro, 1990).

(iii) *Volume swelling*

The volume swelling was calculated from the differences between brain water content from the control and experimental group using the formula of Elliott and Jasper (1949) as described earlier (Sharma and Cervos-Navarro, 1990). In general, about 1% of brain water increase is approximately equal to 4% of volume swelling (Sharma et al., 1990a).

(iv) *Cerebral blood flow*

The cerebral blood flow (CBF) was measured using carbonized microsphere (o.d. $15 \pm 0.6 \,\mu m$) labeled with [125]-I-Na as described earlier (Sharma and Cervos-Navarro, 1990; Sharma and Dey, 1986, 1987). In brief, about 90k microspheres were administered as bolus into the heart through an indwelling cannula (PE10) retrogradely implanted toward heart aseptically 1 week before the experiments. Serial samples of blood from the femoral artery via another indwelling cannula (PE10) was withdrawn at a rate of 0.8 mL/min starting from 30 s before microsphere injection and continued at every 30 s up to 90 s after completion of microsphere administration.

After that the brains were removed by decapitation and placed over filter paper (Whatman, Sigma-Aldrich, St. Louis, MO, USA = wetted with cold saline. The superficial vessels and blood clots if any were removed and the brain tissue was dissected into several brain regions. The tissue samples were weighed immediately, and the radioactivity counted in a 3-in well type Gamma Counter at energy window 25–50 KeV (Packard, Ramsey, MN, USA). The whole blood radioactivity was also counted in each sample. The CBF (mL/g/min) was calculated according to formula as describe earlier (Sharma and Dey, 1987) where CBF = CPM. Brain × RBF/CPM. Blood whereas CPM (counts per min) brain in grams, reference blood flow (RBF 0.8 mL/min) and CPM. Blood = counts in total blood samples collected (Sharma and Dey, 1987).

(v) *Neuropathology*

Brain pathology was examined using standard protocol for histological investigation as describe earlier (Sharma et al., 1991b, 1992b, 1993, 1995b). For neuropathological examination, after the experiments rats were perfused in situ with either Somogyi Fixative containing 0.25% picric acid (Sharma and Cervos-Navarro, 1990) or 4% buffered paraformaldehyde (Sharma et al., 1992b). For this purpose, under Equithesin anesthesia (3 mL/kg, i.p.) the chest was rapidly opened, and the right auricle cut, and a butterfly stainless steel needle (21G) was inserted into the left cardiac ventricle. The intravascular blood was washed either by 50–80 mL of cold physiological saline of by cold phosphate buffer (0.1 M, pH 7.0) through a peristaltic pump (Harvard Apparatus, Holliston, USA) at 90 Torr. This was followed by about 150 mL of cold Somogyi or paraformaldehyde fixative under identical condition (Sharma and Cervos-Navarro, 1990; Sharma et al., 1995b).

After perfusion, the animals were wrapped in an aluminum foil and kept overnight in a refrigerator at 4 °C. On the next day, the brains were dissected, and several coronal sections were cut and embedded in paraffin (Sakura FineTek, Torrance, CA, USA).

(a) *Light microscopy*

About 3-μm thick paraffin sections were cut and stained with Haematoxylin and Eosin (H&E) or Nissl stain according to commercial protocol (Sharma and Cervos-Navarro, 1990). The sections were examined under a Carl-Zeiss Bright field light microscope (Jena, Germany) and images were captured with the attached digital camera (Zeiss Axiocam 506 color, White Plains, NY, USA) and stored in a Macintosh Apple Power Book (Mc Os El Capitan 10.11.6). These controls, experimental and drug treated images were all processed using commercial software Adobe Photoshop CS 12.0.4 (San José, CA, USA) using identical brightness and color filters (Sharma and Sjoquist, 2002).

(b) *Transmission electron microscopy*

For ultrastructural studies, small tissue pieces were postfixed in osmium tetraoxide (OsO4) and embedded in plastic (Epon 812). About 1-μm semi thick sections were cut and stained with Toluidine blue and examined under a bench microscope for identifying the areas for ultrastructural investigation (Sharma and Olsson, 1990). After identifying the specified area, the block ware trimmed and ultrathin section about 500 nm were cut on an Ultramicrotome (LKB, Stockholm, Sweden) using diamond knife (Diatome, Hatfield, PA, USA). The sections were collected on single hole (600 μm) copper grid (Stansted, Essex, UK) and counter stained with lead citrate and uranyl acetate for viewing under a Phillips 400 Transmission Electron Microscope (Eindhoven, The Netherlands) and images were captured on a Gatan digital camera (Pleasanton, CA, USA) at the original magnification of 4–12 k as described earlier (Sharma and Cervos-Navarro, 1990; Sharma and Olsson, 1990).

(vi) *Drug treatments*

Chinese EGBb-761 and BN-52021 was administered in heat stress as follows (Sharma, 2000; Sharma et al., 2000a).

(a) *EGb-761*

EGb-761 (IPSEN, Paris, France) was administered orally (50 mg/kg) in water administered orally by gavage daily for 1 week. In some cases, EGb-761 was administered intraperitoneally once daily for 1 week. Administration of EGb-761 was either given 1-day after heat stress or in some cases after the heat stress session on the 1st day.

(b) *BN-52021*

The bilobalide (BN-52021, IPSEN, Paris, France) was administered in a dose of 2 mg/kg orally by gavage for 1 week under identical condition either 1 day after heat stress session or on the same day after the heat stress treatment. In some animals, BN-52021 was administered intraperitoneally instead of gavage using identical protocol.

(vii) *Nanowired delivery*

Nanodelivery of EGb-761 and BN-52021 was done in heat stress and all the parameters were evaluated using standard protocol as follows (Sharma, 2018; Sharma et al., 2018c).

(a) *NWEGB-761*

In separate group of rats TiO2-nanowired EGb-761 (NWEGb-761) was administered (50 mg/kg, i.p.) 48 h after the onset of the heat stress session once daily for 5 days.

(b) *NWBN-52021*

TiO2-nanowired BN-52021 (NWBN-52021) was administered (2 mg/kg, i.p.) 48 h after the onset of the heat stress session once daily for 5 days.

10 Statistical analyses

The data were analyzed using commercial software StatView 5 (Abacus concept Inc., USA) on a Macintosh computer (Classic Environment) employing ANOVA followed by Dunnett's test for multiple group comparison with one control. For semiquantitative data Chi Square test was used to analyze significance of the data. A p-value of 0.05 was considered significant.

11 Our findings in heat stress

11.1 Stress symptoms

Subjection of rats to heat stress resulted in stress symptoms as evident from the increase in rectal temperature (Fig. 2). This rise in rectal temperature was linear up to 4 h acute heat stress exposure. Thus, the rectal temperature increased from control group of 37.80 ± 0.05 °C to 38.56 ± 0.08 °C ($P < 0.05$) after 1 h heat exposure at 38 °C that was further increased to 39.43 ± 0.06 °C ($P < 0.05$) after 2 h heat exposure. After 4 h heat exposure the rectal temperature was increased further to 41.23 ± 0.12 °C ($P < 0.05$) (Fig. 2).

When animals were subjected to 1 h heat exposure at 38 °C daily for 7 days their rectal temperature was increased to 40.36 ± 0.14 °C ($P < 0.05$). Whereas 2 h daily heat exposure at 38 °C resulted in rise in rectal temperature on the 7th day was 41.56 ± 0.08 ($P < 0.05$) (Fig. 2).

Salivation following heat exposure at 38 °C also showed a linear increase from 1 h acute heat exposure up to 4 h periods of heat stress (Sharma and Dey 1986, 1987). Animals exposed to 1 h or 2 h heat stress daily for 7 days resulted in salivation similar to 4 h acute heat exposure (Sharma and Hoopes, 2003).

The degree of prostration as seen in acute heat exposure was increased from 1 h period to 4 h period of heat exposure linearly. However, 1 h or 2 h daily heat exposure for 7 days shows prostration similar to that of 4 h acute heat exposure (Sharma, 1999; Sharma et al., 1998a,b).

On the other hand, gastric hemorrhages in stomach at postmortem examination revealed liner and gradual increase in microhemorrhages in the stomach from 1 h (3 ± 2 spots) to 2 h (8 ± 3 spots) and at 4 h (23 ± 4 spots) associated with microhemorrhages. Exposure of 1 h or 2 h heat exposure daily resulted in increase of gastric hemorrhagic spots in stomach to greater than 4 h acute heat exposure. Thus, 1 h heat exposure resected in gastric hemorrhagic spots 33 ± 4 in stomach while 2 h heat exposure daily caused an average of 46 ± 7 spots in stomach associated with microhemorrhages (Fig. 2). This shows that chronic heat stress for 1 h or 2 h daily induces severe stress in terms of microhemorrhages in stomach as compared to the acute heat exposure (Fig. 2).

12 Nociception

Subjection of acute heat stress from 1 h to 4 h at 38 °C in the BOD incubator or chronic heat stress for 7 days of 1 h or 2 h did not affect nociception responses to either thermal or mechanical noxious stimulus (Fig. 2). Thus, the tail flick responses to thermal or mechanical nociception were ranged from 2 ± 4 s to 5 ± 4 s for thermal and 3 ± 6 s to 6 ± 3 s to mechanical noxious stimulus in heat exposure. Exposure to acute heat stress from 1 h to 4 h or 7 days exposure of 1 h or 2 h heat at 38 °C did not influence thermal or mechanical noxious stimulation (Fig. 2). Thus, suggests that heat exposure to animals did not affect nociceptive response significantly.

13 Physiological variables

Rats subjected to acute heat exposure at 38 °C exhibited increase in MABP after 1 h (120 ± 6 Torr, $P < 0.05$) and 2 h (132 ± 8 Torr, $P < 0.05$) from control group at room temperature (21 ± 1 °C MABP 110 ± 6 Torr) (Fig. 3). However, after 4 h acute heat exposure the MABP significantly reduced to 80 ± 7 Torr ($P < 0.05$). Exposure of heat for 7 days 1 h daily at 38 °C resulted in significantly lower MABP (78 ± 8 Torr, $P < 0.05$) while 2 h daily exposure caused MABP to 82 ± 5 Torr ($P < 0.05$) from control group (Fig. 3).

The arterial pH showed a declined gradually from 1 h to 4 h acute heat exposure but the changes is not significant. Also 1 h or 2 h daily heat exposure for 7 days did not alter the arterial pH significantly although the trend was a little decline in arterial pH (Fig. 3). On the other hand, arterial PaO2 declined significantly after 4 h cute heat exposure (78.35 ± 0.07 Torr, $P < 0.05$) from control group (79.36 ± 0.04 Torr). The PaO2 values are also decreased significantly after daily exposure for 1 h (78.67 ± 0.07 Torr, $P < 0.05$) or 2 h (78.56 ± 0.04 Torr, $P < 0.05$) for 7 days indicates that chronic exposure to heat for short periods affects PaO2 values significantly (Fig. 3). Regarding PaCO2 values they are increased after 4 h acute heat exposure (35.78 ± 0.05, $P < 0.05$ from control 34.36 ± 0.06 Torr). Chronic exposure of heat

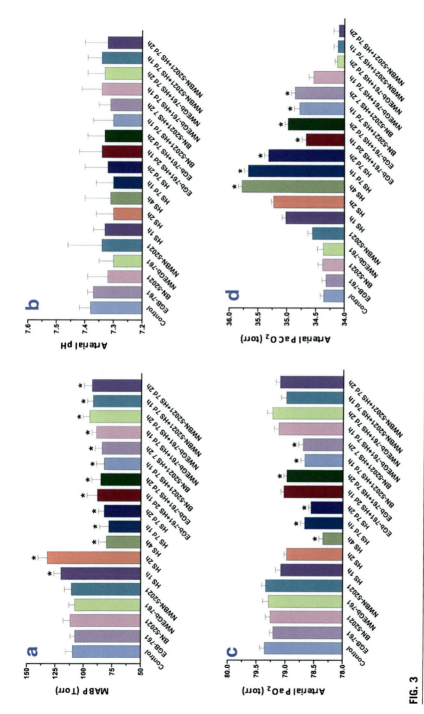

FIG. 3

Effect of EGb-761 and BN-52021 on physiological variables including mean arterial blood pressure (A), arterial pH (B) arterial PaO2 (C) and arterial PaCO2 (D) following acute and chronic heat stress. Animals were subjected to heat stress in a biological oxygen demand (BOD) incubator maintained at 38°C with food and water ad libitum (For details, see text). Values are Mean ± SD of 6–8 rats at one point. * $P < 0.05$ from control group. # $P < 0.05$ from respective chronic heat stress group. ANOVA followed by Dunnett's test for multiple group comparison using one control.

for 1 h (35.67 ± 0.07 Torr, $P < 0.05$) or 2 h (35.32 ± 0.07 Torr, $P < 0.05$) also declined PaCO2 values significantly (Fig. 3).

14 Blood-brain barrier permeability

The BBB permeability to Evans blue albumin (EBA) and radioiodine was significantly increased linearly following 1 h to 4 h acute heat exposure (Fig. 4). Thus, the BBB to EBA was 0.56 ± 0.08 mg % ($P < 0.05$) after 1 h heat stress that increased further to 0.76 ± 0.04 mg % ($P < 0.05$) after 2 h and increased tremendously to 1.89 ± 0.11 mg % ($P < 0.05$) after 4 h exposure as compared to the control group (0.18 ± 0.04 mg %). Extravasation of radioiodine was greater than EBA after heat exposure of 1 h (0.68 ± 0.05%, $P < 0.05$), 2 h (0.83 ± 0.08%, $P < 0.05$) and at 4 h period (2.35 ± 0.14%, $P < 0.05$) as compared to the control group (0.23 ± 0.06%) (Fig. 4).

Chronic heat exposure for 1 h daily showed leakage of EBA 1.23 ± 0.12 mg % $P < 0.05$) and following 2 h daily was 1.64 ± 0.14 mg % ($P < 0.05$) on the 7th day exposure (Fig. 4). For radioiodine significantly higher values in the brain was observed after 1 h daily (2.13 ± 0.10%, $P < 0.05$) and 2 h daily (2.10 ± 0.12%, $P < 0.05$) on the 7th day of heat exposure (Fig. 4).

Regional BBB permeability was examined using radioiodine leakage during heat exposure of either 1 h or 2 h daily on the 7th day (Fig. 4). Our results showed significant increase in cerebral cortex, hippocampus, and cerebellum after chronic heat exposure of 1 h or 2 h daily on the 7th day (Fig. 6). Thus, the cerebral cortex exhibited about 2.34 ± 0.12% ($P < 0.05$) after 1 h daily heat exposure and for 2 h daily the values are 2.56 ± 0.13% ($P < 0.05$) increases on the 7th day as compared to the control group (0.23 ± 0.04%). In the hippocampus the value for 1 h daily heat exposure was 1.34 ± 0.14% ($P < 0.05$) and after 2 h daily was 1.67 ± 0.10% ($P < 0.05$) on the 7th day as compared to the control group (0.18 ± 0.03%). In the cerebellum, 1 h daily heat exposure the BBB leakage to radioiodine was 1.18 ± 0.08% ($P < 005$) and after 2 h daily to 1.56 ± 0.18% ($P < 0.05$) on the 7th day as compared to the control value of cerebellum (0.08 ± 0.02%) (Figs. 4 and 6).

15 Cerebral blood flow

The cerebral blood flow (CBF) significantly decreased following acute heat exposure at 38 °C for 1–4 h periods. This decrease in the CBF was progressive (Fig. 4). Thus, the CBF decreased to 0.98 ± 0.04 mL/g/min ($P < 0.05$) after 1 h heat exposure that decreased to declined further at 2 h 0.89 ± 0.05 mL/g/min ($P < 0.05$) (Table 1). After 4 h the CBF decreased to 0.76 ± 0.10 mL/g/min ($P < 0.05$) as compared to the control group (1.10 ± 0.04 mL/g/min).

Exposure to 1 h or 2 h chronic heat stress also decreased the CBF on the 7th day (Fig. 4). Thus, 1 h daily heat stress resulted in decline of CBF to 0.80 ± 0.08 mL/g/min

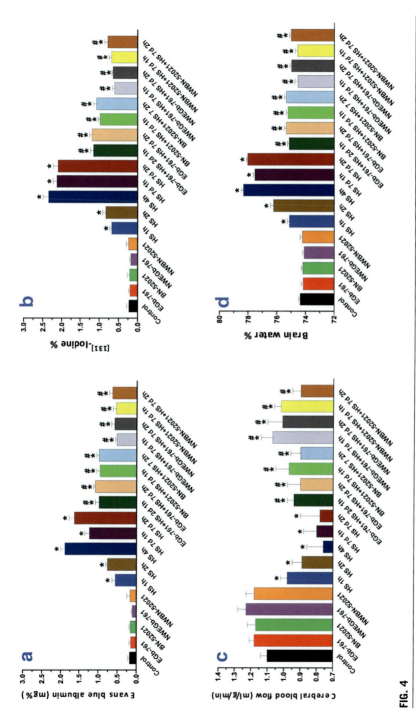

FIG. 4

Effect of EGb-761 and BN-52021 on Evans blue albumin (A), radioiodine leakage (B) cerebral blood flow (C) and brain water content (D) following acute and chronic heat stress. Animals were subjected to heat stress in a biological oxygen demand (BOD) incubator maintained at 38°C with food and water ad libitum (For details, see text). Values are Mean ± SD of 6–8 rats at one point. * $P < 0.05$ from control group. # $P < 0.05$ from respective chronic heat stress group. ANOVA followed by Dunnett's test for multiple group comparison using one control.

Table 1 Effect of EGb-761 and BN-52021 on blood-brain barrier, brain edema and cerebral blood flow and brain pathology following heat stress.

Type of experiment	Stress symptoms		Brain pathology		
	Salivation	Prostration	Neuronal distortion nr.	Myelin Vesiculation nr.	Endothelial cell distortion nr.
(A) Control group					
Control	Nil	Nil	2 ± 3	3 ± 4	2 ± 3
EGB-761	Nil	Nil	2 ± 4	2 ± 1	1 ± 3
BN-52021	Nil	Nil	2 ± 3	3 ± 2	2 ± 3
NWEGb-761	Nil	Nil	2 ± 2	1 ± 3	1 ± 2
NWBN-52021	Nil	Nil	3 ± 2	3 ± 4	2 ± 3
(B) Heat stress					
HS 1h	++	+	28 ± 8^a	18 ± 4^a	23 ± 7^a
HS 2h	+++	++	43 ± 6^a	28 ± 3^a	41 ± 9^a
HS 4h	++++	+++	187 ± 12^a	98 ± 10^a	156 ± 12^a
HS 7d 1h	++++	++++	165 ± 18^a	102 ± 12^a	108 ± 18^a
HS 7d 2h	++++	++++	178 ± 21^a	150 ± 13^a	124 ± 16^a
(C) Drug treatments					
EGb-761+HS 7d 1h	++++	++	$76\pm4^{a,b}$	$70\pm9^{a,b}$	$78\pm7^{a,b}$
EGb-761+HS 7d 2h	+++	+++	$84\pm8^{a,b}$	$89\pm6^{a,b}$	$87\pm4^{a,b}$
BN-52021+HS 7d 1h	+++	+++	$83\pm8^{a,b}$	$75\pm4^{a,b}$	$85\pm6^{a,b}$
BN-52021+HS 7 2h	+++	+++	$93\pm6^{a,b}$	$87\pm8^{a,b}$	$94\pm9^{a,b}$
(D) Nanowired drug delivery					
NWEGb-761+HS 7d 1h	+++	++	$29\pm8^{a,b}$	$18\pm6^{a,b}$	$14\pm6^{a,b}$
NWEGb-761+HS 7d 2h	+++	++	$34\pm12^{a,b}$	$28\pm10^{a,b}$	$18\pm8^{a,b}$
NWBN-52021+HS 7d1h	+++	+++	$40\pm8^{a,b}$	$30\pm12^{a,b}$	$20\pm7^{a,b}$
NWBN-52021+HS 7d2h	+++	+++	$48\pm8^{a,b}$	$39\pm9^{a,b}$	$24\pm10^{a,b}$

Animals were subjected to heat stress in a biological oxygen demand (BOD) incubator maintained at 38°C with food and water ad libitum (For details, see text).
Values are Mean ± SD of 6–8 rats at one point.
+ = mild, ++ = moderate, +++ = severe, ++++ = extensive.
[a]P < 0.05 from control group.
[b]P < 0.05 from respective chronic heat stress group. ANOVA followed by Dunnett's test for multiple group comparison using one control.

($P < 0.05$) and 2 h daily heat stress resulted in further decline in CBF to 0.78 ± 0.12 mL/g/min ($P < 0.05$) from the control value (Fig. 4).

Regional CBF also declined after chronic heat stress in the cerebral cortex, hippocampus and in cerebellum (Fig. 6). Thus, 1 h daily heat stress resulted in decline in CBF in cerebral cortex to 0.78 ± 0.04 mL/g/min ($P < 0.05$) and 2 h daily declined the CBF to 0.75 ± 0.09 ($P < 0.05$) as compared to the control group (1.26 ± 0.08 mL/g/min). In the hippocampus the decrease in CBF was 0.70 ± 0.05 mL/g/min after 1 h daily heat exposure and following 2 h exposure daily resulted in CNF decline in hippocampus to 0.68 ± 0.08 mL/g/min ($P < 0.05$) from control value (0.89 ± 0.06 mL/g/min). In cerebellum also CBF declined significantly after 1 h daily heat exposure to 0.72 ± 0.08 mL/g/min ($P < 0.05$) and following 2 h daily heat stress the CBF values show further reduction to 0.68 ± 0.08 mL/g/min ($P < 0.05$) on the 7th day as compared to the control group (1.28 ± 0.04 mL/g/min) (Fig. 6).

16 Brain edema formation and volume swelling

Exposure to heat acute heat stress resulted in a linear and gradual increase in brain edema formation over the hours up to 4 h period. Thus, about 4% volume swelling was seen in animals after 1 h heat exposure at 38 °C. The brain water content rose to $75.13 \pm 0.14\%$ ($P < 0.05$) after 1 h heat exposure as compared to the control group ($74.38 \pm 0.12\%$, Fig. 4). This was further increased to $76.23 \pm 0.21\%$ ($P < 0.05$) after 2 h heat exposure equal to 8% volume swelling and at 4 h there was tremendous increase in brain water up to $78.34 \pm 0.15\%$ that amounts to about 16% volume swelling (Figs. 4 and 5).

Short-term heat exposure daily for 1 h or 2 h also resulted in brain water content increase on 7th day (Fig. 4). Thus, 1 h heat exposure daily lead to an increase in brain water $77.54 \pm 0.11\%$ ($P < 005$) amounting to 13% volume swelling from control value. When 2 h heat stress was given daily at 38 °C the brain water rose to 78.04 ± 0.08 ($P < 0.05$) equal to 15% volume swelling on the 7th day as compared to the controls (Figs. 4 and 5).

Regional brain edema measured in the cerebral cortex, hippocampus and cerebellum also showed profound increases after daily chronic heat exposure for 1 h or 2 h period for 7 days (Fig. 6). Thus, the cerebral cortex showed significant increase in brain water after 1 h daily to $78.34 \pm 0.12\%$ ($P < 0.05$) and 2 h daily exhibited a rise in brain water content to $78.79 \pm 0.21\%$ ($P < 0.05$) as compared to the control values (75.45 ± 0.13). In hippocampus the brain water increased to $84.34 \pm 0.12\%$ ($P < 0.05$) after 1 h daily heat exposure and 2 h daily heat stress resulted in hippocampus brain water increase to $84.76 \pm 0.10\%$ ($P < 0.05$) as compared to the control value ($80.34 \pm 0.15\%$) on the 7th day (Fig. 6). In cerebellum the brain water content increased after 1 h daily heat stress to $80.23 \pm 0.14\%$ ($P < 0.05$) and after 2 h daily heat exposer to $80.65 \pm 0.12\%$ ($P < 0.05$) as compared to control group ($78.34 \pm 0.23\%$) on 7th day (Fig. 6).

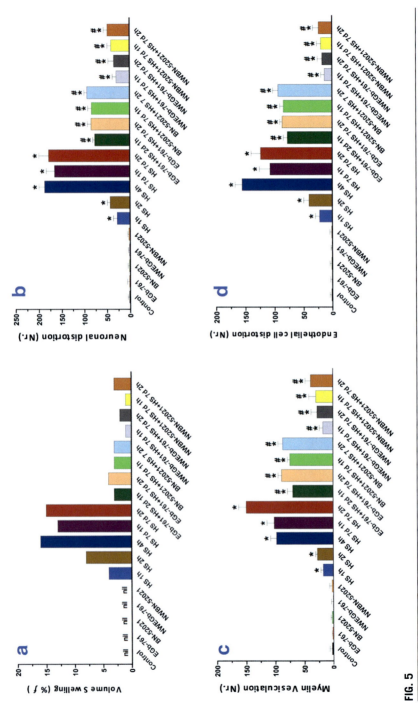

FIG. 5

Effect of EGb-761 and BN-52021 on volume swelling (A), neuronal damages (B) myelin vesiculation (C) and endothelial deformity (D) following acute and chronic heat stress. Animals were subjected to heat stress in a biological oxygen demand (BOD) incubator maintained at 38°C with food and water ad libitum (For details, see text). Values are Mean ± SD of 6–8 rats at one point. * $P < 0.05$ from control group. # $P < 0.05$ from respective chronic heat stress group. ANOVA followed by Dunnett's test for multiple group comparison using one control.

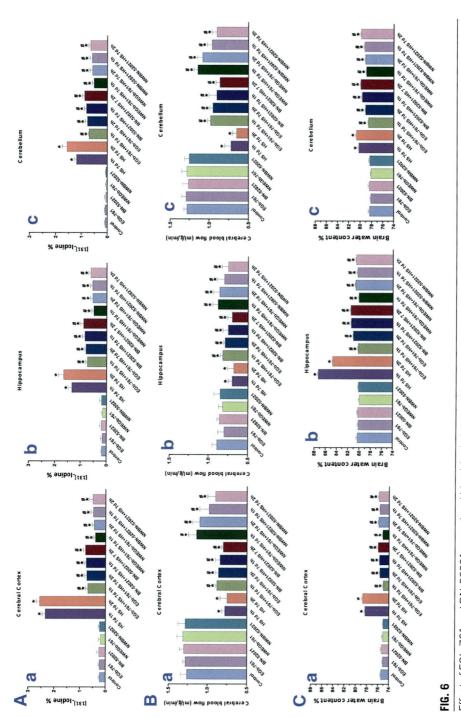

FIG. 6

Effect of EGb-761 and BN-52021 on regional blood-brain barrier (A), regional cerebral blood flow (B) and regional brain edema and (C) in cerebral cortex (a), hippocampus (b) and in cerebellum (c) following chronic heat stress. Animals were subjected to heat stress in a biological oxygen demand (BOD) incubator maintained at 38°C with food and water ad libitum (For details, see text). Values are Mean ± SD of 6–8 rats at one point. * $P < 0.05$ from control group. # $P < 0.05$ from respective chronic heat stress group. ANOVA followed by Dunnett's test for multiple group comparison using one control.

17 Brain pathology after heat exposure

At light microcopy, cell changes occur after acute heat exposure from 1 h to 4 h period. Neuronal changes examined either by H&E or Nissl staining showed neuronal damage, chromatolysis, swollen or shrunken neurons located with the areas of BBB permeability within the neuropil (Sharma and Sharma 2007; Sharma et al., 1991a,b, 1998a,b. Semiquantitative data showed a progressive increase in neuronal injury starting from 1 h acute heat stress (28 ± 8 cells, $P < 0.05$) that was increased to 43 ± 6 cells ($P < 0.05$) after 2 h and following 4 h heat exposure there was tremendous increase in cell damage (187 ± 12 cells, $P < 0.05$) as compared to the control group (2 ± 3 cells).

Chronic exposure of 1 h daily showed 165 ± 18 distorted nerve cells ($P < 0.05$) and 2 h daily exhibited further increase in the number of damaged nerve cells (178 ± 21, $P < 0.05$) in several brain areas on the 7th day as compared to the control group. These neuronal injuries are located within the expanded ad spongy neuropil showing BBB breakdown to EBA and radioiodine (Figs. 5–8).

At the ultrastructural level, myelin damage and vesiculation is evident in acute heat stress group starting from 1 h after exposure (28 ± 8 axons, $P < 0.05$) and this was increased at 2 h after heat stress (43 ± 6 axons, $P < 0.05$) as compared to the control group of myelin vesiculation (3 ± 4 axons). Alter 4 h of acute heat stress the number of axonal damages and myelin vesiculation increased tremendously to 187 ± 12 axons ($P < 0.05$) as compared to the controls (Figs. 5, 9 and 10).

Chronic exposure to heat stress for 1 h daily resulted in myelin vesiculation and damage in more than 165 ± 18 axons ($P < 0.05$) and 2 h daily heat exposure further extended to myelin vesiculation and damage to 178 ± 21 axons ($P < 0.05$) on the 7th day as compared to the control group (Figs. 5 and 10).

Study of endothelial cells at ultrastructural level exhibited deformation and altered shape of capillaries or cerebral microvessels starting from 1 h of heat exposure at 38 °C to 23 ± 7 vessels ($P < 0.05$) that was increased after 2 h exposure to 41 ± 9 microvessels ($P < 0.05$) as compared to the control group (2 ± 3 vessels). After 4 h heat exposure severe endothelial cell deformities with collapse of many microvessels were seen (156 ± 12 vessels, $P < 0.05$) (Figs. 5, 9 and 10).

Chronic heat exposure for 1 h daily also showed endothelial cells deformation (108 ± 18 vessels, $P < 0.05$) and 2 h daily heat exposure resulted in more than 124 ± 16 vessels deformed or collapsed ($P < 0.05$) on the 7th day as compared to the control group (Fig. 7).

18 EGb-761 treatment on heat exposure induced pathophysiology

Effects of EGb-761 on brain pathophysiology were examined in chronic heat exposure for 1 h or 2 h daily heat stress for 7 days. Acute heat stress effects of 4 h exposure and EGb-761 were reported earlier (Sharma et al., 2000a,b; Westman et al., 2000).

HS 38°C 2h 7d

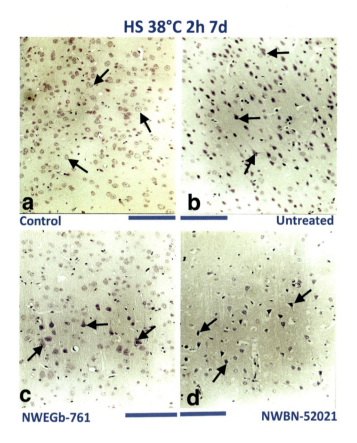

FIG. 7

Light micrograph of parietal cerebral cortex in control (A) heat stress (2 h daily for 7 days) (B) and its modification with treatment with titanium nanowired EGb-761 (C) or nanowired BN-52021 (D). Cerebral cortical cells in control (A) are healthy showing clear cytoplasm and nucleus and evenly distributed in the neuropil (arrows). In heat stress the density of nerve cells is decreased due to expansion or loss of nerve cells and also showing dark in appearance (arrows) (B). On the other hand, in NWEGb-761 treatment in heat stress only few dark and distorted cells could be seen (arrows, C) in the neuropil and most other cells appear near normal (C). In NWBN-52021 treated heat stressed group several cells are in dark appearance (arrows, D) and the nerve cells appear distorted in the neuropil (D). However, in NWBN-52021 treated group (D) the nerve cells appear to be better preserved than the untreated heat stress group (B). Nissl Stain on Paraffin 3-µm section, Bar = 35 µm.

EGb-761 treatment after 1 h or 2 h daily significantly attenuated rectal temperature, salivation and prostration symptoms as seen on 7th day after heat exposure. The occurrence of gastric hemorrhage was also significantly reduced on 7th day (Fig. 2). However, EGb-761 did not influence thermal or mechanical nociception behavior. The MABP lowering effect of chronic heat stress of 1 h or 2 h heat exposure on

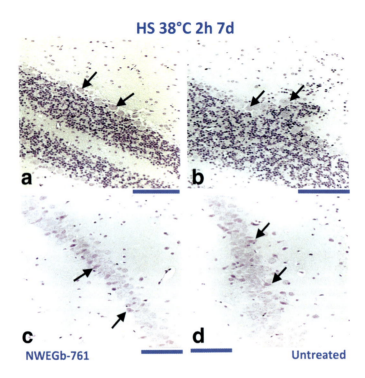

HS 38°C 2h 7d

a

b

c — NWEGb-761

d — Untreated

FIG. 8

Light micrograph of cerebellum (A and B) and hippocampus (C and D) in heat stress (2 h daily for 7 days) (B and D) and its modification with titanium nanowired EGb-761 treatment (A and C). Bothe Purkinje cells and granule cells in the cerebellum shows swelling, loss of cells in an edematous expansion of the neuropil (B, arrow) in heat stress and treatment with NWEGb-761 markedly reduced these cellular changes in the cerebellum (A, arrows). Likewise, hippocampal CA1 area several dark cells (arrows, D) are seen in the edematous neuropil (D) in heat stress that are markedly reduced in NWEGb-761 treat rat (arrows). The hippocampus CA 1 cells look compact and the number of dark cells are much less evident in the hippocampus of NWEGb-761 treated heat stressed rat Ⓒ as compared to the untreated heat stressed group (D). Nissl Stain on Paraffin 3-μm section, Bar=35μm.

the 7th day was slightly improved as compared to the untreated heat stressed group (Fig. 3). The arterial pH was not significantly influenced by the EGb-761 treatment in 1 h or 2 h daily heat stress for 7th day. The arterial PaO2 was not greatly affected by EGb-761 treatment but PaCO2 values were significantly controlled in 1 h or 2 h daily heat exposure on the 7th day (Fig. 3).

The BBB permeability in chronic heat exposure for 1 h or 2 h daily for 7 days was significantly reduced in EGb-761 treated rats to EBA and radioiodine as compared to the untreated heat stressed groups (Fig. 4). Regional changes in the BBB breakdown to radioiodine in the cerebral cortex, hippocampus and the cerebellum was

HS 38°C 2h 7d

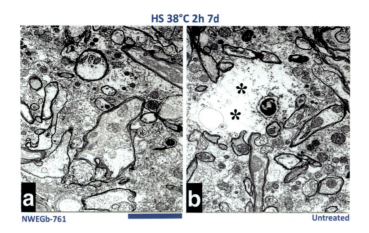

FIG. 9

High power transmission electron micrograph showing neuropil from parietal cerebral cortex in heat stress (2 h daily for 7 days) (B) and its modification with titanium nanowired EGb-761 treatment (A). Heat stress resulted in membrane vacuolation (**), sponginess and edema (B). Treatment with titanium nanowired EGb-761 (NWEGb-761) markedly reduced the membrane vacuolation, sponginess and edema in the neuropil after heat stress (A). Bar = 2 μm.

significantly attenuated by EGb-761 treatment following 1 h or 2 h heat exposure daily for 7 days (Fig. 4).

The CBF was maintained at higher level in EGb-761 treated group after either 1 h or 2 h daily heat exposure for 7 days as compared to the untreated heat exposed group (Table 1). Regional CBF in EGb-761 treated group was also maintained at higher level in the cerebral cortex, hippocampus and cerebellum as compared to the untreated 1 h or 2 h heat exposed group for 7 days (Fig. 4).

Brain edema formation in EGb-761 treated rats after 1 h or 2 h daily for 7 days exhibited significant reduction as compared to the untreated heat exposed group (Fig. 4). The regional measurement of brain edema in the cerebral cortex, hippocampus and cerebellum also showed significant reduction in EGb-761 treated group as compared to the untreated heat exposed rats for 7 days (Fig. 4).

EGb-761 was also able to reduce brain pathology at light or electron microscopy (Figs. 5 and 6). Thus, after 1 h or 2 h daily heat exposure induced brain pathology on the 7th day in EGb-761 treated rats reduced neuronal injury and the number of axons showing myelin vesiculation or damage was reduced. (Figs. 7–10). The number of endothelial cell deformity was in EGb-761 treated chronic heat stressed rats whether for 1 h or 2 h showed significantly less changes on the 7th day as compared to the untreated heat exposed group (Fig. 5).

HS 38°C 2h 7d

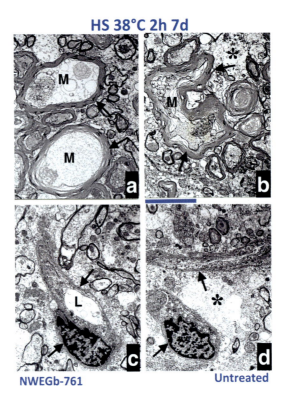

NWEGb-761 Untreated

FIG. 10

High power transmission electron micrograph showing myelin vesiculation (B) and endothelial cell deformity (D) in heat stress (2 h daily for 7 days) and its modification with titanium nanowired EGb-761 (A and C). Treatment with NWEGb-761 markedly induced neuroprotection in myelin (M) vesiculation (A) and cerebral capillary collapse (C) as compared to untreated heat stressed rat showing myelin collapse and vesiculation (B) and almost complete obliteration of capillary lumen (L) (D). Morphology of pericytes (arrow) is also altered in heat stress (D) and NWEGb-761 protected the pericytes shape changes (arrow) and lumen (L) is clearly seen (C). Better preservation of myelin (M) structure is also seen in NWEGb-761 treated rat in heat stress (A, arrow) as compared to the untreated group of myelin (M, arrow). Expansion and edema (*) is also much less seen in NWEGb-761 treated heat stressed rat (A) as compared to the untreated group (B). Bar = 2 μm.

19 BN-52021 treatment on heat exposure induced pathophysiology

Treatment with BN-52021 in chronic heat stress either for 1 h or 2 h daily significantly attenuated rectal temperature elevation, salivation, and prostration symptoms and reduced the occurrence of gastric hemorrhagic spots and microhemorrhages in stomach after 7th day of heat stress as compared to the untreated heat exposed group. However, BN-52021 did affect thermal or mechanical nociception behaviors of chronic heat exposed rats of 1 h or 2 h daily on the 7th day (Fig. 2).

Regarding physiological variables, BN-52021 treated heat exposed group of either 1 h or 2 h daily on the 7th day did not affect lowering of blood pressure and arterial pH as compared to the untreated heat exposed group (Fig. 3). This drug treatment (NB-52021) did not affect greatly PaO_2 but maintained slightly lower $PaCO_2$ values in chronic heat exposed rats either for 1 h or 2 h daily for 7 day (Fig. 3).

The BBB permeability to EBA and radioiodine was significantly reduced by BN-52021 given to either 1 h or 2 h daily heat exposure on the 7th day (Fig. 4) as compared to the untreated chronic heat exposed rats (Fig. 4). Regional changes in the BBB breakdown to radioiodine in the cerebral cortex, hippocampus and cerebellum was also significantly attenuated in BN-52021 treated heat-exposed rats subjected to either 1 h or 2 h daily for 7 days (Fig. 6).

Cerebral blood flow (CBF) studies indicate an improved level of circulation to brain in heat stressed group subjected to either 1 h or 2 h daily heat exposure for 7 days (Fig. 4). Regional CBF examined in the cerebral cortex, hippocampus and cerebellum also showed significantly higher CBF in BN-52021 treated heat exposed group for either 1 h or 2 h daily for 7 days as compared to the untreated heat exposed rats (Fig. 6).

Brain edema showed a significant reduction in brain water and volume swellings in BN-52021 treated heat-exposed rats subjected to either 1 h or 2 h daily for 7 days (Fig. 4). Regional brain water in the cerebral cortex, hippocampus, and cerebellum (Fig. 6) was also reduced in chronic heat exposed rats either 1 h or 2 h daily for 7 days as compared to the untreated chronic heat exposed rats (Fig. 6).

The brain pathology at light and electron microscopy was significantly reduced in BN-52021 treated rats following chronic heat exposure of either 1 h or 2 h daily for 7 days. The nerve cell injuries were significantly reduced as seen on light microscopy in chronic heat exposed rats either for 1 h or 2 h daily for 7 days (Figs. 7–10) as compared to the untreated chronic heat exposed animals. At electron microscopy, the number of axonal damages and myelin vesiculation were significantly reduced in BN-52021 treated heat exposed rats for either 1 h or 2 h daily for 7 days as compared to untreated heat exposed group under identical conditions (Figs. 9 and 10). The number of endothelial cell deformity is also reduced significantly in BN-52021 treated chromic heat stressed group either for 1 h or 2 h daily for 7 days (Fig. 6).

20 NWEGb-761 treatment on chronic heat stressed induced pathophysiology

Treatment with NWEGb-761 induced superior neuroprotection as compared to EGb-761 on all parameters examined in chronic heat stress group either exposed to 1 h or 2 h daily for 7 days. Thus, the effects of NWEGb-761 significantly reduced rectal temperature limited to 38.78 ± 0.08 °C ($P < 0.05$) after 1 h daily and 38.98 ± 0.05 °C ($P < 0.05$) after 2 h daily heat exposure on the 7th day as compared to the untreated heat stressed group for 1 h (40.36 ± 0.14 °C) and 2 h (41.56 ± 0.08 °C) after 7 days (Fig. 2). The other symptom of heat stress such as salivation and prostration together with gastric hemorrhagic spots in the stomach were reduced in NWEGb-761 treatment in chronic heat stress (Fig. 2). The MABP was maintained at significantly higher level in NWEGb-761 treated chronic heat stressed rats either for 1 h or 2 h daily exposure for 7 days (Fig. 3). The arterial pH was maintained near normal levels and the PaO2 and PaCO2 levels were almost restored near normal values in NWEGb-761 treated chronic heat stressed group of either 1 h or 2 h daily for 7 days (Fig. 3).

The BBB permeability was significantly reduced to EBA 0.52 ± 0.04 mg % ($P < 0.05$) for 1 h and 0.58 ± 0.08 mg % ($P < 0.05$) for 2 h daily and for radioiodine $0.60 \pm 0.06\%$ ($P < 0.05$) for 1 h and $0.64 \pm 0.03\%$ ($P < 0.05$) for 2 h daily on the 7th day in NWEGb-761 treated group as compared to the untreated chronic heat stressed groups (Fig. 4). The regional BBB permeability in the cerebral cortex, hippocampus and cerebellum for radioiodine was also significantly reduced in NWEGb-761 treated rats after 1 h or 2 h daily heat exposure for 7 day (Fig. 6).

The CBF values were also restored near normal values in NWEGb-761 treated chronic heat stressed rats for either 1 h or 2 h daily exposer for 7 day. Thus, the values of CBF in NWEGb-761 treated heat stressed rats were 1.07 ± 0.07 mL/g/min ($P < 0.05$) for 1 h and 1.01 ± 0.09 mL/g/min ($P < 0.05$) after 2 h daily exposure on the 7th day as compared to the untreated chronic heat stressed rats for 1 h (0.80 ± 0.08 mL/g/min) and for 2 h (0.78 ± 0.06 mL/g/min) on the 7th day (Fig. 4). The regional CBF measured in the cerebral cortex, hippocampus and in cerebellum in NWEGb-761 treated chronic heat stressed group was also significantly elevated near normal values after 1 h or 2 h daily heat exposure for 7th day (Fig. 6).

The brain water content was also reduced together with the volume swelling in NWEGb-761 treated chronically heat stressed group either for 1 h or 2 h daily for 7 day (Fig. 2). This was also seen in regional brain edema measurement in the cerebral cortex, hippocampus and in cerebellum in chronic heat stressed group treated with NWEGb-761 either for 1 h or 2 h daily form 7 day heat exposure (Fig. 6). Thus, the brain water content is reduced in NWEGb-761 treated chronic heat stressed rats for 1 h ($74.54 \pm 0.21\%$, $P < 0.05$) and 2 h ($74.98 \pm 0.10\%$, $P < 0.05$) at 7th day as compared to the untreated heat stressed rats for 1 h ($77.54 \pm 0.11\%$) and for 2 h ($78.04 \pm 0.08\%$) on the 7th day (Figs. 4 and 6).

Brain pathology seen at light microscopy regarding neuronal injuries were considerably reduced in NWEGb-761 treated chronic heat stressed rats either for 1 h or 2 h daily on the 7th day (Figs. 7 and 8). Thus, only 29 ± 9 nerve cells ($P < 0.05$) after 1 h and 34 ± 12 nerve cells ($P < 0.05$) after 2 h daily heat stress were seen on 7th day in NWEGb-761 treated rats as compared to the untreated chronic heat stressed rats for 1 h (165 ± 18 cells) and for 2 h (178 ± 21 cells) on the 7th day. At electron microscope the number of myelin damage and vesiculation was reduced in NWEGb-761 treated chronic heat stressed rats for 1 h (18 ± 6 axons, $P < 0.05$) and for 2 h (28 ± 10 axons, $P < 0.05$) on the 7th day as compared to the untreated heat stressed rats. Likewise, the number of endothelial cells deformity was significantly reduced in both 1 h (14 ± 6 vessels, $P < 0.05$) and after 2 h (18 ± 8 vessels, $P < 0.05$) in NWEGb-761 treated rats chronic heat stressed rats on the 7th day as compared to the untreated chronic heat stressed group (Fig. 5).

21 NWBN-52021 treatment on chronic heat stressed induced pathophysiology

NWBN-52021 treatment also induced many superior effects in chronic heat stress either for 1 h or 2 h daily for 7 days as compared to the untreated chronic heat stressed rats for 7 days. Thus, BN-52021 treatment reduced rise of rectal temperature and attenuated stress response of salivation and prostration. The occurrence of gastric hemorrhages and ulcerations were also significantly reduced as compared to the conventional BN-52021 treated chronic heat stressed rats (Fig. 2).

However, BN-52021 treatment dud does not affect nociceptive response to either thermal or mechanical stress in rats subjected to 1 h or 2 h daily chronic heat stress for 7 days. The physiological variables were better managed in chronic heat stressed by NWBN-52021 as compared to the conventional delivery of BN-52021 in chronic heat stressed rats. The reduction in MABP after 7 days of heat stress in 1 h or 2 h daily was significantly reduced by NWBN-52021 treatment. The arterial pH, PaO2 and PaCO2 values were much superior in BN-52021 retreated chronic heat stressed group for 1 h or 2 h daily for 7 days (Fig. 2).

The BBB breakdown to EBA and radioiodine is reduced significantly that is much superior in NWBN-52021 treated heat stressed group for 1 h or 2 h daily for 7 days exposure than the conventional BN-52021 treatment (Fig. 4). The regional BBB changes in the cerebral cortex, hippocampus and in cerebellum was also significantly reduced after 1 h or 2 h daily heat exposure up to 7 days in NWBN-52021 treated group. This effect was much more superior than conventional BN-52021 treatment in chronic heat stressed animals for 7 days (Figs. 4 and 6).

The CBF was maintained much higher in brain in NWBN-52021 treated 1 h or 2 h daily heat exposed rats for 7 days as compared to the conventional treatment if BN-52021 treatment under identical conation (Fig. 4). The regional changes in CBF in cerebral cortex, hippocampus and in cerebellum was also maintained higher in NWBN-52021 treated heat exposed animals for either 1 h or 2 h daily for 7 days as compared to the conventional BN-52021 treated rats under identical conditions (Figs. 4 and 6).

The reduction in brain water content and volume swelling in NWBN-52021 treated group was seen after 1 h or 2 h heat exposed daily in rats for 7 days (Fig. 4). The regional brain water content in the cerebral cortex, hippocampus and in cerebellum was also reduced by NWBN-52021 treatment in chronic heat exposed rats for 1 h or 2 h daily for 7 days. These reduction in brain edema was much superior by NWBN-52021 treatment as compared to the conventional treatment with BN-52021 in chronic heat stress (Figs. 4 and 6).

Brain pathology was also reduced in NWBN-52021 treated chronic heat stressed rats. Thus, at light microscope, the neuronal cell changes were significantly greater reduced in rats exposed to either 1 h or 2 h heat stress daily for 7 days as compared to the conventional BN-52021 treatment under identical conditions (Figs. 7 and 8). At electron microscope, changes in myelin vesiculation and damage was significantly reduced by NWBN-52021 treated in chronic heat stressed rats either for 1 h or 2 h daily for 7 days as compared to the conventional treatment with BN-52021 (Figs. 9 and 10). The number of endothelial cell deformity was also reduced by NWBN-52021 treatment in 1 h or 2 h daily heat exposed rats for 7 days as compared to the conventional BN-52021 treated group under identical condition (Figs. 9 and 10).

22 Discussion

The salient new findings of this investigation shows that chronic heat stress for 7 days either for 1 h or 2 h exposure at 38 °C induces profound brain pathology. This suggests that repeated heat exposure even for short durations could induce brain pathology and related symptoms. That means even slight heat stress is not suitable for the normal brain function. This is further evident from the findings that during heat waves several mortalities and/or hospital admissions are recorded in the literature (Bazille et al., 2005; CDC 1993, 1994a,b; Davis et al., 2016; Ghumman and Horney, 2016; Liu et al., 2020; Mirchandani et al., 1996; Odhiambo Sewe et al., 2018; Weinberger et al., 2018). These facts support the idea that heat stroke or heat stress is bad for human health and one should avoid the situations of heat stress to protect CNS structure and function.

Another important point of this investigation shows that treatment with nanodelivery of EGb-761 or BN-52021 has superior is neuroprotective ability in repeated heat stress as compared to the conventional therapy using EGb-761 or BN-52021. This observation suggests that nanowired therapy of EGb-761 or BN-52021 induces better neuroprotection in repeated heat stress induced brain pathology. We found that 5 days administration of 50 mg/kg dose of NWEGb-761 induces far more superior neuroprotective effects on brain pathology in heat stress as compared to the 2 mg/kg of NWBN-52021 for 5 days treatment in identical conditions of repeated heat exposure of short 1 h or 2 h durations. This suggests that enhancement of doses of NWBN-52021 is needed to further enhance the neuroprotective effects in repeated heat exposure. Currently, our laboratory is engaged in finding out a dose related neuroprotective effects of NWEGb-761 and NWBN-52021 in repeated heat exposure induced brain pathology.

It appears than nanowired delivery of bilobalide NWBN-52021 is very potent in inducing neuroprotection in heat stress induced brain pathology. This is apparent from the fining that only a small dose of NWBN-52021 is quite effective in repeated heat stress induced brain dysfunction as compared to NWEGb-761 therapy. This suggests that increased dose of NWBN-52021 could have the potential of reducing brain pathology in heat stress. This is a feature that requires additional investigation.

We observed that convectional doses of EGb-761 and BN-52021 also induce neuroprotection in repeated heat stress induced brain pathology. This effect is potentiated by nanowired delivery of EGb-761 and BN-52021 in identical condition. The exact possible mechanisms are still unclear for a superior neuroprotection by nanowired delivery of drugs in heat stress. However, available evidence suggests that nanowired delivery of EGb-761 or NWBN-52021 could induce superior neuroprotection than their conventional delivery due to increased bioavailability of the active compounds in the CNS. This is further evident that nanowired delivery could work as a slow release of active compounds in the brain for long time period without being catabolized quickly by endogenous cerebral enzymes for detoxification (Sahib et al., 2019; Sharma et al., 2015, 2016a,b,c,d, 2017, 2018a,b, 2019b, 2020c,d). Furthermore, drugs labeled with nanowired technology could easily reach central nervous system by piercing endothelial cell membrane of the cerebral capillaries by making a tiny hole without damaging the cell membrane and enter brain fluid compartments easily than convention delivery (Sahib et al., 2019, 2020; Sharma et al., 2016a,b, 2020a,b,c,d). Thus, this is likely that nanowired delivery of EGb-761 or BN-52021 could induce superior neuroprotection because of their active and fast presence into the brain for longer periods of time.

Nanodelivery of drugs could induce superior neuroprotection because of their inherent properties for exerting beneficial effects in the nervous system (Niu et al., 2020; Sahib et al., 2019, 2020; Sharma et al., 2020a,b,c,d). The EGb-761 and BN-52021 are potent antioxidative compounds and thus they are able to prevent lipid peroxidation and production of free radicals in the brain (Liu et al., 2019b; Makarewicz-Plonska et al., 1998; Prochazkova et al., 2018; Sarikcioglu et al., 2004; Sharma et al., 2000a, 2003a; Tian et al., 2017; Westman et al., 2000; Zhou et al., 2017; Zhu et al., 2019). Thus, the EGb-761 and BN-52021 treatment reduce oxidative stress and induce neuroprotection in heat stress induced brain pathology. To prove this hypothesis measurement of oxidative stress parameter in heat stress is needed. However, this requires additional investigations.

Heat stress can induce BBB opening though alterations in neurochemicals within the brain (Sharma and Dey, 1984, 1986, 1987; Sharma et al., 1991a, 1992b). Thus, increased serotonin levels in brain and hyperthermia of greater than 40°C is likely to affect the BBB dysfunction. In addition, prostaglandins released during hyperthermia also contribute to the BBB disruption (Sharma and Johanson, 2007; Sharma et al., 1997a, 1994). After that opioids, nitric oxide, amino acids and histamine is associated with BBB breakdown in heat stress induced hyperthermia (Alm et al., 1998, 2000; Patnaik et al., 2000; Sharma, 2006; Sharma et al., 1991a, 1992a,b, 1997a, 2000a, 2006a; Westman et al., 2000). When the BBB is opened to macromolecules such as EBA and other serum proteins enter into the brain (Brightman et al., 1970; Hultstrom et al., 1984; Lien et al., 2012; Michinaga and Koyama, 2017; Sahib et al., 2019, 2020; Sharma et al., 2019a,b, 2020b,d). Extravasation of serum proteins within the brain extracellular or intracellular fluid environment induces vasogenic brain edema formation (Sharma and Olsson, 1990; Sharma et al., 1995b, 1998b). Spread of edema fluid along with other vasoactive substances, immunological and neuro humoral agents induce cellular reaction within the microfluid environment of the brain causing brain pathology (Sharma et al., 1998a,b).

Hyperthermia is associated with oxidative stress in the brain (Aghazadeh et al., 2020; Bain et al., 2014; Chang et al., 2007; Crespo et al., 2018). Oxidative stress induces lipid peroxidation and oxygen radical formation inducing damage to membranes of endothelial cells, neurons, and glial cells (Kochevar et al., 1996). As results BBB damage, neuronal injury, axonal alterations, and endothelial cell deformation could occur in heat stress. NWEGb-761 and NWBN-52021 is thus able to attenuate cell damage by preventing or reducing the BBB breakdown. It would be interesting to examine ultrastructural changes of lanthanum transport across the BBB in BWEGb-761 and NWBN-52021 treated heat stressed brains. This is a feature currently being examined in our laboratory.

Previous reports from our laboratory showed that lanthanum is transported across the endothelial cell using increased pinocytosis at the transcellular membranes to reach basal lamina in heat stress (Sharma et al., 1998b). In all our studies we did not observe opening of the tight junctions in heat stress for the lanthanum tracers (Sharma, 2003, 2004). Thus, it is possible that oxidative stress, neurochemical stress, and other neuro humoral and immunologic factors potentiate pinocytosis of the cerebral endothelial cells causing breakdown of the BBB resulting in brain edema and cellular changes (Muresanu et al., 2020; Niu et al., 2020; Sahib et al., 2019, 2020; Sharma, 2003, 2004; Sharma et al., 2020a,b,c,d). These factors caused by oxidative stress are largely prevented by the NWEGb-761 and NWBN-52021 treatment in heat stress. This is further evident from the findings that NWEGb-761 and NWBN-52021 in heat stress attenuated the BBB breakdown and resulting brain edema formation. As a result, the cell changes are also reduced in these drug treatments in heat stress. This observation is supported by our precious experiments in heat stress that prevented the structural changes using another antioxidant compound H-290/51 treatment (Alm et al., 1998, 2000; Muresanu et al., 2020, 2012).

23 Conclusion and future perspectives

In summary, our results show that nanowired delivery of EGb-761 or BN-52021 has superior neuroprotective effects in repeated heat stress. This indicates that nanodelivery of compounds potentiate the neuroprotective efficacy of suitable therapeutic agents in heat stress. It appears that EGb-761 and BN-52021 is the potent antioxidant compound and this could be one of the reasons to induce neuroprotective effects in chronic heat stress. To prove these point further studies on oxidative stress parameters in repeated heat stress is needed. Furthermore, it would be interesting to examine the role of EGb-761 and BN-52021 in heat stress induced nitric oxide synthase upregulation. Also, a dose related effects of EGb-761 and BN-52021 in repeated heat stress is needed to expand our knowledge for development of suitable therapeutic strategies using nanowired delivery. Further studies are needed to understand the effects of long-term heat exposure in inducing brain pathology and different therapeutic windows for NWEGb-761 and NWBN-52021 in heat stress. These features are currently being examined in our laboratory.

Acknowledgments

Authors (HSS/AS) are grateful to Dr. Mrs. Katy Drieu (IPSEN-Beaufour, Paris, France) and Dr. Shyam S Chatterjee (Dr. Willmar Schwabe GmbH, Karlsruhe, Germany) for supply of EGb-761 and BN-52021 compounds and their support and advice on heat stress research. This investigation is supported by grants from the Air Force Office of Scientific Research (EOARD, London, UK), and Air Force Material Command, USAF, under grant number FA8655-05-1-3065; Grants from the Alzheimer's Association (IIRG-09-132087), the National Institutes of Health (R01 AG028679) and the Dr. Robert M. Kohrman Memorial Fund (RJC); Swedish Medical Research Council (Nr 2710-HSS), Göran Gustafsson Foundation, Stockholm, Sweden (HSS), Astra Zeneca, Mölndal, Sweden (HSS/AS), The University Grants Commission, New Delhi, India (HSS/AS), Ministry of Science & Technology, Govt. of India (HSS/AS), Indian Medical Research Council, New Delhi, India (HSS/AS) and India-EU Cooperation Program (RP/AS/HSS) and IT-901/16 (JVL), Government of Basque Country and PPG 17/51 (JVL), JVL thanks to the support of the University of the Basque Country (UPV/EHU) PPG 17/51 and 14/08, the Basque Government (IT-901/16 and CS-2203) Basque Country, Spain; and Foundation for Nanoneuroscience and Nanoneuroprotection (FSNN), Romania. Technical and human support provided by Dr. Ricardo Andrade from SGIker (UPV/EHU) is gratefully acknowledged. Dr. Seaab Sahib is supported by Research Fellowship at University of Arkansas Fayetteville AR by Department of Community Health; Middle Technical University; Wassit; Iraq, and The Higher Committee for Education Development in Iraq; Baghdad; Iraq. We thank Suraj Sharma, Blekinge Inst. Technology, Karlskrona, Sweden and Dr. Saja Alshafeay, University of Arkansas Fayetteville, Fayetteville AR, USA for computer and graphic support. The U.S. Government is authorized to reproduce and distribute reprints for Government purpose notwithstanding any copyright notation thereon. The views and conclusions contained herein are those of the authors and should not be interpreted as necessarily representing the official policies or endorsements, either expressed or implied, of the Air Force Office of Scientific Research or the U.S. Government.

Conflict of interest

The authors declare no conflict of interest with any entity mentioned here.

References

Adams, L.K., Lyon, D.Y., Alvarez, P.J., 2006. Comparative eco-toxicity of nanoscale TiO2, SiO2, and ZnO water suspensions. Water Res. 40 (19), 3527–3532. https://doi.org/10.1016/j.watres.2006.08.004.

Aghazadeh, A., Feizi, M.A.H., Fanid, L.M., Ghanbari, M., Roshangar, L., 2020. Effects of hyperthermia on TRPV1 and TRPV4 channels expression and oxidative markers in mouse brain. Cell. Mol. Neurobiol. https://doi.org/10.1007/s10571-020-00909-z. PMID: 32661579.

Ahlemeyer, B., Krieglstein, J., 2003. Pharmacological studies supporting the therapeutic use of Ginkgo biloba extract for Alzheimer's disease. Pharmacopsychiatry 36 (S 1), 8–14.

Albukrek, D., Bakon, M., Moran, D.S., Faibel, M., Epstein, Y., 1997. Heat-stroke-induced cerebellar atrophy: clinical course, CT and MRI findings. Neuroradiology 39 (3), 195–197. https://doi.org/10.1007/s002340050392.

Alm, P., Sharma, H.S., Hedlund, S., Sjoquist, P.O., Westman, J., 1998. Nitric oxide in the pathophysiology of hyperthermic brain injury. Influence of a new anti-oxidant compound H-290/51. A pharmacological study using immunohistochemistry in the rat. Amino Acids 14 (1f3), 95–103. https://doi.org/10.1007/BF01345249.

Alm, P., Sharma, H.S., Sjoquist, P.O., Westman, J., 2000. A new antioxidant compound H-290/51 attenuates nitric oxide synthase and heme oxygenase expression following hyperthermic brain injury. An experimental study using immunohistochemistry in the rat. Amino Acids 19 (1), 383–394. https://doi.org/10.1007/s007260070069.

al-Mashhadani, S.A., Gader, A.G., al Harthi, S.S., Kangav, D., Shaheen, F.A., Bogus, F., 1994. The coagulopathy of heat stroke: alterations in coagulation and fibrinolysis in heat stroke patients during the pilgrimage (Haj) to Makkah. Blood Coagul. Fibrinolysis 5 (5), 731–736. https://doi.org/10.1097/00001721-199410000-00009.

Alyautdin, R.N., Petrov, V.E., Langer, K., Berthold, A., Kharkevich, D.A., Kreuter, J., 1997. Delivery of loperamide across the blood-brain barrier with polysorbate 80-coated polybutylcyanoacrylate nanoparticles. Pharm. Res. 14 (3), 325–328. https://doi.org/10.1023/a:1012098005098.

Ashburner, M., Bonner, J., 1979. The induction of gene activity in Drosophila by heat shock. Cell 17 (2), 241–254.

Ayala, A., Munoz, M.F., Arguelles, S., 2014. Lipid peroxidation: production, metabolism, and signaling mechanisms of malondialdehyde and 4-hydroxy-2-nonenal. Oxid. Med. Cell. Longev. 2014. https://doi.org/10.1155/2014/360438, 360438.

Babior, B.M., 2000. Phagocytes and oxidative stress. Am. J. Med. 109 (1), 33–44. https://doi.org/10.1016/s0002-9343(00)00481-2.

Bain, A.R., Morrison, S.A., Ainslie, P.N., 2014. Cerebral oxygenation and hyperthermia. Front. Physiol. 5, 92. https://doi.org/10.3389/fphys.2014.00092.

Barrow, M.W., Clark, K.A., 1998. Heat-related illnesses. Am. Fam. Physician 58 (3), 749–756. 759. Retrieved from https://www.ncbi.nlm.nih.gov/pubmed/9750542.

Bastianetto, S., Ramassamy, C., Dore, S., Christen, Y., Poirier, J., Quirion, R., 2000a. The Ginkgo biloba extract (EGb 761) protects hippocampal neurons against cell death induced by beta-amyloid. Eur. J. Neurosci. 12 (6), 1882–1890. https://doi.org/10.1046/j.1460-9568.2000.00069.x.

Bastianetto, S., Zheng, W.H., Quirion, R., 2000b. The Ginkgo biloba extract (EGb 761) protects and rescues hippocampal cells against nitric oxide-induced toxicity: involvement of its flavonoid constituents and protein kinase C. J. Neurochem. 74 (6), 2268–2277. https://doi.org/10.1046/j.1471-4159.2000.0742268.x.

Basu, S., Srivastava, P.K., 2000. Heat shock proteins: the fountainhead of innate and adaptive immune responses. Cell Stress Chaperones 5 (5), 443–451. https://doi.org/10.1379/1466-1268(2000)005<0443:hsptfo>2.0.co;2.

Bazille, C., Megarbane, B., Bensimhon, D., Lavergne-Slove, A., Baglin, A.C., Loirat, P., Woimant, F., Mikol, J., Gray, F., 2005. Brain damage after heat stroke. J. Neuropathol. Exp. Neurol. 64 (11), 970–975. https://doi.org/10.1097/01.jnen.0000186924.88333.0d.

Beere, H.M., 2001. Stressed to death: regulation of apoptotic signaling pathways by the heat shock proteins. Sci. STKE 2001 (93), re1. https://doi.org/10.1126/stke.2001.93.re1.

Beere, H.M., Wolf, B.B., Cain, K., Mosser, D.D., Mahboubi, A., Kuwana, T., Tailor, P., Morimoto, R.I., Cohen, G.M., Green, D.R., 2000. Heat-shock protein 70 inhibits apoptosis by preventing recruitment of procaspase-9 to the Apaf-1 apoptosome. Nat. Cell Biol. 2 (8), 469–475. https://doi.org/10.1038/35019501.

Bellmann, K., Charette, S.J., Nadeau, P.J., Poirier, D.J., Loranger, A., Landry, J., 2010. The mechanism whereby heat shock induces apoptosis depends on the innate sensitivity of cells to stress. Cell Stress Chaperones 15 (1), 101–113. https://doi.org/10.1007/s12192-009-0126-9.

Bensimhon, D., Bazille, C., Baglin, C., Mikol, J., Gray, F., 2004. Cerebral damage following heat stroke. J. Neuropathol. Exp. Neurol. 63 (5), 534.

Berhouma, M., Jacquesson, T., Jouanneau, E., Cotton, F., 2019. Pathogenesis of peri-tumoral edema in intracranial meningiomas. Neurosurg. Rev. 42 (1), 59–71. https://doi.org/10.1007/s10143-017-0897-x.

Bertrand, N., Leroux, J.C., 2012. The journey of a drug-carrier in the body: an anatomo-physiological perspective. J. Control. Release 161 (2), 152–163. https://doi.org/10.1016/j.jconrel.2011.09.098.

Bodor, N., Buchwald, P., 1999. Recent advances in the brain targeting of neuropharmaceuti-cals by chemical delivery systems. Adv. Drug Deliv. Rev. 36 (2–3), 229–254. https://doi.org/10.1016/s0169-409x(98)00090-8.

Bongers, K.S., Salahudeen, M.S., Peterson, G.M., 2020. Drug-associated hyperthermia: a lon-gitudinal analysis of hospital presentations. J. Clin. Pharm. Ther. 45 (3), 477–487. https://doi.org/10.1111/jcpt.13090.

Bouchama, A., Knochel, J.P., 2002. Heat stroke. N. Engl. J. Med. 346 (25), 1978–1988. https://doi.org/10.1056/NEJMra011089.

Bouchama, A., Parhar, R.S., el-Yazigi, A., Sheth, K., al-Sedairy, S., 1991. Endotoxemia and release of tumor necrosis factor and interleukin 1 alpha in acute heatstroke. J. Appl. Physiol. (1985) 70 (6), 2640–2644. https://doi.org/10.1152/jappl.1991.70.6.2640.

Bouchama, A., al-Sedairy, S., Siddiqui, S., Shail, E., Rezeig, M., 1993. Elevated pyrogenic cytokines in heatstroke. Chest 104 (5), 1498–1502. https://doi.org/10.1378/chest.104.5.1498.

Bouchama, A., Hammami, M.M., Haq, A., Jackson, J., al-Sedairy, S., 1996. Evidence for en-dothelial cell activation/injury in heatstroke. Crit. Care Med. 24 (7), 1173–1178. https://doi.org/10.1097/00003246-199607000-00018.

Bouchama, A., Hammami, M.M., Al Shail, E., De Vol, E., 2000. Differential effects of in vitro and in vivo hyperthermia on the production of interleukin-10. Intensive Care Med. 26 (11), 1646–1651. https://doi.org/10.1007/s001340000665.

Brightman, M.W., Klatzo, I., Olsson, Y., Reese, T.S., 1970. The blood-brain barrier to proteins under normal and pathological conditions. J. Neurol. Sci. 10 (3), 215–239. https://doi.org/10.1016/0022-510x(70)90151-6.

Brinton, M.R., Tagge, C.A., Stewart, R.J., Cheung, A.K., Shiu, Y.T., Christensen, D.A., 2012. Thermal sensitivity of endothelial cells on synthetic vascular graft material. Int. J. Hyper-thermia 28 (2), 163–174. https://doi.org/10.3109/02656736.2011.638963.

Brown, I.R., 2007. Heat shock proteins and protection of the nervous system. Ann. N. Y. Acad. Sci. 1113 (1), 147–158. https://doi.org/10.1196/annals.1391.032.

Buckley, I.K., 1972. A light and electron microscopic study of thermally injured cultured cells. Lab. Invest. 26 (2), 201–209. Retrieved from https://www.ncbi.nlm.nih.gov/pubmed/4551120.

Busija, D.W., Leffler, C.W., Pourcyrous, M., 1988. Hyperthermia increases cerebral metabolic rate and blood flow in neonatal pigs. Am. J. Physiol. 255 (2 Pt. 2), H343–H346. https://doi.org/10.1152/ajpheart.1988.255.2.H343.

Bynum, G.D., Pandolf, K.B., Schuette, W.H., Goldman, R.F., Lees, D.E., Whang-Peng, J., Atkinson, E.R., Bull, J.M., 1978. Induced hyperthermia in sedated humans and the concept of critical thermal maximum. Am. J. Physiol. 235 (5), R228–R236. https://doi.org/10.1152/ajpregu.1978.235.5.R228.

Calingasan, N.Y., Uchida, K., Gibson, G.E., 1999. Protein-bound acrolein: a novel marker of oxidative stress in Alzheimer's disease. J. Neurochem. 72 (2), 751–756. https://doi.org/10.1046/j.1471-4159.1999.0720751.x.

Can, G., Şahin, U., Sayılı, U., Dubé, M., Kara, B., Acar, H.C., İnan, B., Aksu Sayman, O., Lebel, G., Bustinza, R., Küçükali, H., Güven, U., Gosselin, P., 2019. Excess mortality in istanbul during extreme heat waves between 2013 and 2017. Int. J. Environ. Res. Public Health 16 (22), 4348. https://doi.org/10.3390/ijerph16224348. PMID: 31703402, PMCID: PMC6887774.

Cao, W., Carney, J.M., Duchon, A., Floyd, R.A., Chevion, M., 1988. Oxygen free radical involvement in ischemia and reperfusion injury to brain. Neurosci. Lett. 88 (2), 233–238. https://doi.org/10.1016/0304-3940(88)90132-2.

Cardoso, F.L., Brites, D., Brito, M.A., 2010. Looking at the blood–brain barrier: molecular anatomy and possible investigation approaches. Brain Res. Rev. 64 (2), 328–363.

Carlsson, C., Hägerdal, M., Siesjö, B.K., 1976. The effect of hyperthermia upon oxygen consumption and upon organic phosphates, glycolytic metabolites, citric acid cycle intermediates and associated amino acids in rat cerebral cortex. J. Neurochem. 26 (5), 1001–1006.

Castillo, J., Dávalos, A., Noya, M., 1999. Aggravation of acute ischemic stroke by hyperthermia is related to an excitotoxic mechanism. Cerebrovasc. Dis. 9 (1), 22–27.

CDC, 1993. Centers for Disease, Control Prevention, Heat-related deaths—United States, 1993. MMWR Morb. Mortal. Wkly Rep. 42 (28), 558–560. Retrieved from https://www.ncbi.nlm.nih.gov/pubmed/8326949.

CDC, 1994a. From the centers for disease control and prevention. Heat-related deaths—Philadelphia and United States, 1993-1994. JAMA 272 (3), 197. Retrieved from https://www.ncbi.nlm.nih.gov/pubmed/8022030.

CDC, 1994b. Heat-related deaths—Philadelphia and United States, 1993-1994. MMWR Morb. Mortal. Wkly Rep. 43 (25), 453–455.

CDC, 1999. Centers for disease, control prevention, heat-related illnesses and deaths—Missouri, 1998, and United States, 1979-1996. MMWR Morb. Mortal. Wkly Rep. 48 (22), 469–473. Retrieved from https://www.ncbi.nlm.nih.gov/pubmed/10428101.

CDC, 2005. Centers for disease, control prevention, heat-related mortality—Arizona, 1993-2002, and United States, 1979-2002. MMWR Morb. Mortal. Wkly Rep. 54 (25), 628–630. Retrieved from https://www.ncbi.nlm.nih.gov/pubmed/15988407.

Chan, P.H., 2001. Reactive oxygen radicals in signaling and damage in the ischemic brain. J. Cereb. Blood Flow Metab. 21 (1), 2–14. https://doi.org/10.1097/00004647-200101000-00002.

Chan, E.S., Bautista, D.T., Zhu, Y., You, Y., Long, J.T., Li, W., Chen, C., 2018. Traditional Chinese herbal medicine for vascular dementia. Cochrane Database Syst. Rev. 12 (12). https://doi.org/10.1002/14651858.CD010284.pub2, CD010284.

Chang, D.M., 1993. The role of cytokines in heat stroke. Immunol. Invest. 22 (8), 553–561. https://doi.org/10.3109/08820139309084183.

Chang, C.K., Chang, C.P., Liu, S.Y., Lin, M.T., 2007. Oxidative stress and ischemic injuries in heat stroke. Prog. Brain Res. 162, 525–546. https://doi.org/10.1016/S0079-6123(06)62025-6.

Changnon, S.A., Kunkel, K.E., Reinke, B.C., 1996. Impacts and responses to the 1995 heat wave: a call to action. Bull. Am. Meteorol. Soc. 77 (7), 1497–1506.

Chatterjee, S., Gabard, B., 1981. Protective Effect of an Extract of Ginkgo Biloba and Other Hydroxyl Radical Scavengers Against Hypoxia. Paper presented at the Eighth International Congress of Pharmacology.

Chen, Y., Liu, L., 2012. Modern methods for delivery of drugs across the blood-brain barrier. Adv. Drug Deliv. Rev. 64 (7), 640–665. https://doi.org/10.1016/j.addr.2011.11.010.

Chen, C., Wei, T., Gao, Z., Zhao, B., Hou, J., Xu, H., Xin, W., Packer, L., 1999. Different effects of the constituents of EGb761 on apoptosis in rat cerebellar granule cells induced by hydroxyl radicals. Biochem. Mol. Biol. Int. 47 (3), 397–405. https://doi.org/10.1080/15216549900201423.

Cherian, I., Beltran, M., Landi, A., Alafaci, C., Torregrossa, F., Grasso, G., 2018. Introducing the concept of "CSF-shift edema" in traumatic brain injury. J. Neurosci. Res. 96 (4), 744–752.

Chiu, W.T., Kao, T.Y., Lin, M.T., 1995. Interleukin-1 receptor antagonist increases survival in rat heatstroke by reducing hypothalamic serotonin release. Neurosci. Lett. 202 (1–2), 33–36. https://doi.org/10.1016/0304-3940(95)12203-6.

Choi, W., Choi, K., Yang, G., Kim, J.C., Yu, C., 2016. Improving piezoelectric performance of lead-free polymer composites with high aspect ratio BaTiO3 nanowires. Polym. Test. 53, 143–148.

Chong, P.Z., Ng, H.Y., Tai, J.T., Lee, S.W.H., 2020. Efficacy and safety of ginkgo biloba in patients with acute ischemic stroke: a systematic review and meta-analysis. Am. J. Chin. Med. 48 (3), 513–534. https://doi.org/10.1142/S0192415X20500263.

Clement, T., Rodriguez-Grande, B., Badaut, J., 2020. Aquaporins in brain edema. J. Neurosci. Res. 98 (1), 9–18. https://doi.org/10.1002/jnr.24354.

Commenges, D., Scotet, V., Renaud, S., Jacqmin-Gadda, H., Barberger-Gateau, P., Dartigues, J.F., 2000. Intake of flavonoids and risk of dementia. Eur. J. Epidemiol. 16 (4), 357–363. https://doi.org/10.1023/a:1007614613771.

Cornett, C.R., Markesbery, W.R., Ehmann, W.D., 1998. Imbalances of trace elements related to oxidative damage in Alzheimer's disease brain. Neurotoxicology 19 (3), 339–345. Retrieved from https://www.ncbi.nlm.nih.gov/pubmed/9621340.

Craig, E.A., Schlesinger, M.J., 1985. The heat shock respons. Crit. Rev. Biochem. 18 (3), 239–280.

Cremer, O.L., Kalkman, C.J., 2007. Cerebral pathophysiology and clinical neurology of hyperthermia in humans. Prog. Brain Res. 162, 153–169. https://doi.org/10.1016/S0079-6123(06)62009-8.

Crespo, M., Leon-Navarro, D.A., Martin, M., 2018. Cerebellar oxidative stress and fine motor impairment in adolescent rats exposed to hyperthermia-induced seizures is prevented by maternal caffeine intake during gestation and lactation. Eur. J. Pharmacol. 822, 186–198. https://doi.org/10.1016/j.ejphar.2018.01.023.

Cunnane, S., Nugent, S., Roy, M., Courchesne-Loyer, A., Croteau, E., Tremblay, S., Castellano, A., Pifferi, F., Bocti, C., Paquet, N., Begdouri, H., Bentourkia, M., Turcotte, E., Allard, M., Barberger-Gateau, P., Fulop, T., Rapoport, S.I., 2011. Brain fuel metabolism, aging, and Alzheimer's disease. Nutrition 27 (1), 3–20. https://doi.org/10.1016/j.nut.2010.07.021.

Cunningham, J., Pivirotto, P., Bringas, J., Suzuki, B., Vijay, S., Sanftner, L., Kitamura, M., Chan, C., Bankiewicz, K.S., 2008. Biodistribution of adeno-associated virus type-2 in non-human primates after convection-enhanced delivery to brain. Mol. Ther. 16 (7), 1267–1275. https://doi.org/10.1038/mt.2008.111.

Davis, R.E., Hondula, D.M., Patel, A.P., 2016. Temperature observation time and type influence estimates of heat-related mortality in seven U.S. Cities. Environ. Health Perspect. 124 (6), 795–804. https://doi.org/10.1289/ehp.1509946.

DeKosky, S.T., Williamson, J.D., Fitzpatrick, A.L., Kronmal, R.A., Ives, D.G., Saxton, J.A., Lopez, O.L., Burke, G., Carlson, M.C., Fried, L.P., Kuller, L.H., Robbins, J.A., Tracy, R.P., Woolard, N.F., Dunn, L., Snitz, B.E., Nahin, R.L., Furberg, C.D., Ginkgo Evaluation of Memory Study Investigators, 2008. Ginkgo biloba for prevention of dementia: a randomized controlled trial. JAMA 300 (19), 2253–2262. https://doi.org/10.1001/jama.2008.683.

Dematte, J.E., O'Mara, K., Buescher, J., Whitney, C.G., Forsythe, S., McNamee, T., Adiga, R.B., Ndukwu, I.M., 1998. Near-fatal heat stroke during the 1995 heat wave in Chicago. Ann. Intern. Med. 129 (3), 173–181. https://doi.org/10.7326/0003-4819-129-3-199808010-00001.

Derrick, E.H., 1934. Heat cramps and uraemic cramps, with special reference to their treatment with sodium chloride. Med. J. Aust. 2, 612–616.

Dietrich, W.D., Busto, R., Halley, M., Valdes, I., 1990. The importance of brain temperature in alterations of the blood-brain barrier following cerebral ischemia. J. Neuropathol. Exp. Neurol. 49 (5), 486–497. https://doi.org/10.1097/00005072-199009000-00004.

DiMeco, F., Li, K.W., Tyler, B.M., Wolf, A.S., Brem, H., Olivi, A., 2002. Local delivery of mitoxantrone for the treatment of malignant brain tumors in rats. J. Neurosurg. 97 (5), 1173–1178. https://doi.org/10.3171/jns.2002.97.5.1173.

Ding, X.Z., Fernandez-Prada, C.M., Bhattacharjee, A.K., Hoover, D.L., 2001. Over-expression of hsp-70 inhibits bacterial lipopolysaccharide-induced production of cytokines in human monocyte-derived macrophages. Cytokine 16 (6), 210–219. https://doi.org/10.1006/cyto.2001.0959.

Ding, X., Wang, Y., Xu, L., Zhang, H., Deng, Z., Cai, L., Wu, Z., Yao, L., Wu, X., Liu, J., Shen, X., 2019. Stability and osteogenic potential evaluation of micro-patterned titania mesoporous-nanotube structures. Int. J. Nanomedicine 14, 4133–4144. https://doi.org/10.2147/IJN.S199610.

Draper, H., Hadley, M., 1990. [43] Malondialdehyde determination as index of lipid Peroxidation. In: Methods in Enzymology. vol. 186. Elsevier, pp. 421–431.

Edsall, D.L., 1908. A disorder due to exposure to intense heat, characterized clinically chiefly by violent muscular spasms and excessive irritability of the muscles. Preliminary note. JAMA 51 (23), 1969–1971.

Elliott, K.A., Jasper, H., 1949. Measurement of experimentally induced brain swelling and shrinkage. Am. J. Physiol. 157 (1), 122–129. https://doi.org/10.1152/ajplegacy.1949.157.1.122.

Emerich, D.F., Thanos, C.G., 2006. The pinpoint promise of nanoparticle-based drug delivery and molecular diagnosis. Biomol. Eng. 23 (4), 171–184. https://doi.org/10.1016/j.bioeng.2006.05.026.

Engelhart, M.J., Geerlings, M.I., Ruitenberg, A., van Swieten, J.C., Hofman, A., Witteman, J.C., Breteler, M.M., 2002. Dietary intake of antioxidants and risk of Alzheimer disease. JAMA 287 (24), 3223–3229. https://doi.org/10.1001/jama.287.24.3223.

Erejuwa, O.O., Sulaiman, S.A., Ab Wahab, M.S., 2013. Evidence in support of potential applications of lipid peroxidation products in cancer treatment. Oxid. Med. Cell. Longev. 2013, 931251. https://doi.org/10.1155/2013/931251.

Ermak, G., Davies, K.J., 2002. Calcium and oxidative stress: from cell signaling to cell death. Mol. Immunol. 38 (10), 713–721. https://doi.org/10.1016/s0161-5890(01)00108-0.

Evans, V., 2000. Herbs and the brain: friend or foe? The effects of ginkgo and garlic on warfarin use. J. Neurosci. Nurs. 32 (4), 229–232. https://doi.org/10.1097/01376517-200008000-00007.

Feinstein, D.L., Galea, E., Aquino, D.A., Li, G.C., Xu, H., Reis, D.J., 1996. Heat shock protein 70 suppresses astroglial-inducible nitric-oxide synthase expression by decreasing NFkappaB activation. J. Biol. Chem. 271 (30), 17724–17732. https://doi.org/10.1074/jbc.271.30.17724.

Feng, Z., Sun, Q., Chen, W., Bai, Y., Hu, D., Xie, X., 2019. The neuroprotective mechanisms of ginkgolides and bilobalide in cerebral ischemic injury: a literature review. Mol. Med. 25 (1), 57. https://doi.org/10.1186/s10020-019-0125-y.

Fernandez-Sanchez, A., Madrigal-Santillan, E., Bautista, M., Esquivel-Soto, J., Morales-Gonzalez, A., Esquivel-Chirino, C., Durante-Montiel, I., Sanchez-Rivera, G., Valadez-Vega, C., Morales-Gonzalez, J.A., 2011. Inflammation, oxidative stress, and obesity. Int. J. Mol. Sci. 12 (5), 3117–3132. https://doi.org/10.3390/ijms12053117.

Ferrari, C.K.B., 2000. Free radicals, lipid peroxidation and antioxidants in apoptosis: implications in cancer, cardiovascular and neurological diseases. Biologia 55 (6), 581–590.

Ferris, E.B., Blankenhorn, M.A., Robinson, H.W., Cullen, G.E., 1938. Heat stroke: clinical and chemical observations on 44 cases. J. Clin. Invest. 17 (3), 249–262. https://doi.org/10.1172/JCI100949.

Filleul, L., Cassadou, S., Medina, S., Fabres, P., Lefranc, A., Eilstein, D., Le Tertre, A., Pascal, L., Chardon, B., Blanchard, M., Declercq, C., Jusot, J.F., Prouvost, H., Ledrans, M., 2006. The relation between temperature, ozone, and mortality in nine French cities during the heat wave of 2003. Environ. Health Perspect. 114 (9), 1344–1347. https://doi.org/10.1289/ehp.8328.

Flournoy, W.S., Wohl, J.S., Macintire, D.K., 2003. Heatstroke in dogs: pathophysiology and predisposing factors. Compendium 25 (6), 410–418.

Freeman, W., Dumoff, E., 1944. Cerebellar syndrome following heat stroke. Arch. Neurol. Psychiatry 51 (1), 67–72.

Friese, A., Seiller, E., Quack, G., Lorenz, B., Kreuter, J., Biopharmaceutics, 2000. Increase of the duration of the anticonvulsive activity of a novel NMDA receptor antagonist using poly (butylcyanoacrylate) nanoparticles as a parenteral controlled release system. Euro. J. Pharm. 49 (2), 103–109.

Furukawa, S., Fujita, T., Shimabukuro, M., Iwaki, M., Yamada, Y., Nakajima, Y., Nakayama, O., Makishima, M., Matsuda, M., Shimomura, I., 2017. Increased oxidative stress in obesity and its impact on metabolic syndrome. J. Clin. Invest. 114 (12), 1752–1761. https://doi.org/10.1172/JCI21625.

Gabai, V.L., Meriin, A.B., Mosser, D.D., Caron, A.W., Rits, S., Shifrin, V.I., Sherman, M.Y., 1997. Hsp70 prevents activation of stress kinases. A novel pathway of cellular thermotolerance. J. Biol. Chem. 272 (29), 18033–18037. https://doi.org/10.1074/jbc.272.29.18033.

Gabbita, S.P., Lovell, M.A., Markesbery, W.R., 1998. Increased nuclear DNA oxidation in the brain in Alzheimer's disease. J. Neurochem. 71 (5), 2034–2040. https://doi.org/10.1046/j.1471-4159.1998.71052034.x.

Gabbita, S.P., Aksenov, M.Y., Lovell, M.A., Markesbery, W.R., 1999. Decrease in peptide methionine sulfoxide reductase in Alzheimer's disease brain. J. Neurochem. 73 (4), 1660–1666. https://doi.org/10.1046/j.1471-4159.1999.0731660.x.

Garrido, C., Gurbuxani, S., Ravagnan, L., Kroemer, G., 2001. Heat shock proteins: endogenous modulators of apoptotic cell death. Biochem. Biophys. Res. Commun. 286 (3), 433–442. https://doi.org/10.1006/bbrc.2001.5427.

Gasser, S., Khan, N., Yonekawa, Y., Imhof, H.G., Keller, E., 2003. Long-term hypothermia in patients with severe brain edema after poor-grade subarachnoid hemorrhage: feasibility and intensive care complications. J. Neurosurg. Anesthesiol. 15 (3), 240–248. https://doi.org/10.1097/00008506-200307000-00012.

Gauss, H., Meyer, K., 1917. Heat stroke: report of one hundred and fifty-eight cases from Cook County hospital, Chicago. Am. J. Med. Sci. *154* (4), 554.

Gerner, E.W., Schneider, M.J., 1975. Induced thermal resistance in HeLa cells. Nature 256 (5517), 500–502. https://doi.org/10.1038/256500a0.

Ghumman, U., Horney, J., 2016. Characterizing the impact of extreme heat on mortality, Karachi, Pakistan, June 2015. Prehosp. Disaster Med. 31 (3), 263–266. https://doi.org/10.1017/S1049023X16000273.

Giffard, R.G., Xu, L., Zhao, H., Carrico, W., Ouyang, Y., Qiao, Y., Sapolsky, R., Steinberg, G., Hu, B., Yenari, M.A., 2004. Chaperones, protein aggregation, and brain protection from hypoxic/ischemic injury. J. Exp. Biol. 207 (Pt. 18), 3213–3220. https://doi.org/10.1242/jeb.01034.

Gilgun-Sherki, Y., Melamed, E., Offen, D., 2001. Oxidative stress induced-neurodegenerative diseases: the need for antioxidants that penetrate the blood brain barrier. Neuropharmacology 40 (8), 959–975.

Gittens, R.A., McLachlan, T., Olivares-Navarrete, R., Cai, Y., Berner, S., Tannenbaum, R., Schwartz, Z., Sandhage, K.H., Boyan, B.D., 2011. The effects of combined micron-/submicron-scale surface roughness and nanoscale features on cell proliferation and differentiation. Biomaterials 32 (13), 3395–3403. https://doi.org/10.1016/j.biomaterials.2011.01.029.

Glover, D.M., 1931. Heat cramps in industry: their treatment and prevention by means of sodium chloride. J. Ind. Hyg. 13, 347–360.

Goldstein, L.S., Dewhirst, M.W., Repacholi, M., Kheifets, L., 2003. Summary, conclusions and recommendations: adverse temperature levels in the human body. Int. J. Hyperthermia 19 (3), 373–384. https://doi.org/10.1080/0265673031000090701.

Good, P.F., Perl, D.P., Bierer, L.M., Schmeidler, J., 1992. Selective accumulation of aluminum and iron in the neurofibrillary tangles of Alzheimer's disease: a laser microprobe (LAMMA) study. Ann. Neurol. 31 (3), 286–292. https://doi.org/10.1002/ana.410310310.

Haenen, G.R., Vermeulen, N.P., Tai Tin Tsoi, J.N., Ragetli, H.M., Timmerman, H., Blast, A., 1988. Activation of the microsomal glutathione-S-transferase and reduction of the glutathione dependent protection against lipid peroxidation by acrolein. Biochem. Pharmacol. 37 (10), 1933–1938. https://doi.org/10.1016/0006-2952(88)90539-4.

Haldane, J.S., 1929. Salt depletion by sweating. Br. Med. J. 3, 469.

Halliwell, B., 1991. Hydroxylation of salicylate as an assay for hydroxyl radicals: a cautionary note. Free Radic. Biol. Med. 10, 439–441.

Hammami, M.M., Bouchama, A., Al-Sedairy, S., Shail, E., AlOhaly, Y., Mohamed, G.E., 1997. Concentrations of soluble tumor necrosis factor and interleukin-6 receptors in heatstroke and heatstress. Crit. Care Med. 25 (8), 1314–1319. https://doi.org/10.1097/00003246-199708000-00017.

Hartl, F.U., Bracher, A., Hayer-Hartl, M., 2011. Molecular chaperones in protein folding and proteostasis. Nature 475 (7356), 324–332. https://doi.org/10.1038/nature10317.

Hashim, I.A., Al-Zeer, A., Al-Shohaib, S., Al-Ahwal, M., Shenkin, A., 1997. Cytokine changes in patients with heatstroke during pilgrimage to Makkah. Mediators Inflamm. 6 (2), 135–139. https://doi.org/10.1080/09629359791839.

Haymaker, W., Malamud, N., Custer, R.P., 1947. Heat stroke; a clinic-pathologic study of 125 fatal cases. J. Neuropathol. Exp. Neurol. 6 (2), 209–211. Retrieved from https://www.ncbi.nlm.nih.gov/pubmed/20340260.

Heneka, M.T., Sharp, A., Klockgether, T., Gavrilyuk, V., Feinstein, D.L., 2000. The heat shock response inhibits NF-kappaB activation, nitric oxide synthase type 2 expression, and macrophage/microglial activation in brain. J. Cereb. Blood Flow Metab. 20 (5), 800–811. https://doi.org/10.1097/00004647-200005000-00006.

Hensley, K., Hall, N., Subramaniam, R., Cole, P., Harris, M., Aksenov, M., Aksenova, M., Gabbita, S.P., Wu, J.F., Carney, J.M., 1995. Brain regional correspondence between Alzheimer's disease histopathology and biomarkers of protein oxidation. J. Neurochem. 65 (5), 2146–2156.

Heredia-Cervera, B., González-Azcorra, S., Rodríguez-Gattorno, G., López, T., Ortiz-Islas, E., Oskam, G., 2009. Controlled release of phenytoin from nanostructured TiO2 reservoirs. Sci. Adv. Mater. 1 (1), 63–68.

Hoehn, B., Ringer, T.M., Xu, L., Giffard, R.G., Sapolsky, R.M., Steinberg, G.K., Yenari, M.A., 2001. Overexpression of HSP72 after induction of experimental stroke protects neurons from ischemic damage. J. Cereb. Blood Flow Metab. 21 (11), 1303–1309. https://doi.org/10.1097/00004647-200111000-00006.

Hultstrom, D., Tengvar, C., Forssen, M., Olsson, Y., 1984. Distribution of exudated FITC-dextrans in experimental vasogenic brain edema produced by a focal cryogenic injury. Acta Neuropathol. 63 (1), 13–17. https://doi.org/10.1007/BF00688465.

Ian, H., 2008. Enols and Enolates-the Michael Additions. University of Calgary. Available at: http://en.wikipedia.org/wiki/Michael_addition#cite_note-0.

Ikeda, Y., Long, D.M., 1990. The molecular basis of brain injury and brain edema: the role of oxygen free radicals. Neurosurgery 27 (1), 1–11. https://doi.org/10.1097/00006123-199007000-00001.

Irvine, E.E., von Hertzen, L.S., Plattner, F., Giese, K.P., 2006. alphaCaMKII autophosphorylation: a fast track to memory. Trends Neurosci. 29 (8), 459–465. https://doi.org/10.1016/j.tins.2006.06.009.

Jardine, D.S., 2007. Heat illness and heat stroke. Pediatr. Rev. 28 (7), 249–258. https://doi.org/10.1542/pir.28-7-249.

Jimenez-Jimenez, F.J., Molina, J.A., Gomez, P., Vargas, C., de Bustos, F., Benito-Leon, J., Tallon-Barranco, A., Orti-Pareja, M., Gasalla, T., Arenas, J., 1998. Neurotransmitter amino acids in cerebrospinal fluid of patients with Alzheimer's disease. J. Neural Transm. (Vienna) 105 (2–3), 269–277. https://doi.org/10.1007/s007020050056.

Jimenez-Mejias, M.E., Montano Diaz, M., Villalonga, J., Bollain Tienda, E., Lopez Pardo, F., Pineda, J.A., Gonzalez de la Puente, M.A., 1990. Classical heatstroke in Spain. Analysis of a series 78 cases. Med. Clin. (Barc.) 94 (13), 481–486. Retrieved from https://www.ncbi.nlm.nih.gov/pubmed/2355761.

Jitputti, J., Suzuki, Y., Yoshikawa, S., 2008. Synthesis of TiO2 nanowires and their photocatalytic activity for hydrogen evolution. Cat. Com. 9 (6), 1265–1271.

Jones, T.S., Liang, A.P., Kilbourne, E.M., Griffin, M.R., Patriarca, P.A., Wassilak, S.G., Mullan, R.J., Herrick, R.F., Donnell Jr., H.D., Choi, K., Thacker, S.B., 1982. Morbidity and mortality associated with the July 1980 heat wave in St Louis and Kansas City, Mo. JAMA 247 (24), 3327–3331. Retrieved from https://www.ncbi.nlm.nih.gov/pubmed/7087075.

Kampinga, H.H., Hageman, J., Vos, M.J., Kubota, H., Tanguay, R.M., Bruford, E.A., Cheetham, M.E., Chen, B., Hightower, L.E., 2009. Guidelines for the nomenclature of the human heat shock proteins. Cell Stress Chaperones 14 (1), 105–111. https://doi.org/10.1007/s12192-008-0068-7.

Kang, Y.S., Bickel, U., Pardridge, W.M., 1994. Pharmacokinetics and saturable blood-brain barrier transport of biotin bound to a conjugate of avidin and a monoclonal antibody to the transferrin receptor. Drug Metab. Dispos. 22 (1), 99–105. Retrieved from https://www.ncbi.nlm.nih.gov/pubmed/8149897.

Kao, T.Y., Chio, C.C., Lin, M.T., 1994. Hypothalamic dopamine release and local cerebral blood flow during onset of heatstroke in rats. Stroke 25 (12), 2483–2486, (discussion 2486–2487). https://doi.org/10.1161/01.str.25.12.2483.

Karcher, L., Zagermann, P., Krieglstein, J., 1984. Effect of an extract of Ginkgo biloba on rat brain energy metabolism in hypoxia. Naunyn Schmiedebergs Arch. Pharmacol. 327 (1), 31–35. https://doi.org/10.1007/BF00504988.

Karibe, H., Zarow, G.J., Graham, S.H., Weinstein, P.R., 1994. Mild intraischemic hypothermia reduces postischemic hyperperfusion, delayed postischemic hypoperfusion, blood-brain barrier disruption, brain edema, and neuronal damage volume after temporary focal cerebral ischemia in rats. J. Cereb. Blood Flow Metab. 14 (4), 620–627. https://doi.org/10.1038/jcbfm.1994.77.

Keller, J.N., Pang, Z., Geddes, J.W., Begley, J.G., Germeyer, A., Waeg, G., Mattson, M.P., 1997. Impairment of glucose and glutamate transport and induction of mitochondrial oxidative stress and dysfunction in synaptosomes by amyloid beta-peptide: role of the lipid peroxidation product 4-hydroxynonenal. J. Neurochem. 69 (1), 273–284. https://doi.org/10.1046/j.1471-4159.1997.69010273.x.

Khan, V.R., Brown, I.R., 2002. The effect of hyperthermia on the induction of cell death in brain, testis, and thymus of the adult and developing rat. Cell Stress Chaperones 7 (1), 73–90. https://doi.org/10.1379/1466-1268(2002)007<0073:teohot>2.0.co;2.

Kim, M.S., Lee, J.I., Lee, W.Y., Kim, S.E., 2004. Neuroprotective effect of Ginkgo biloba L. extract in a rat model of Parkinson's disease. Phytother. Res. 18 (8), 663–666. https://doi.org/10.1002/ptr.1486.

Kiyatkin, E.A., 2007. Physiological and pathological brain hyperthermia. Prog. Brain Res. 162, 219–243. https://doi.org/10.1016/S0079-6123(06)62012-8.

Kiyatkin, E.A., Sharma, H.S., 2009. Permeability of the blood-brain barrier depends on brain temperature. Neuroscience 161 (3), 926–939. https://doi.org/10.1016/j.neuroscience.2009.04.004.

Klein, J., Chatterjee, S.S., Loffelholz, K., 1997. Phospholipid breakdown and choline release under hypoxic conditions: inhibition by bilobalide, a constituent of Ginkgo biloba. Brain Res. 755 (2), 347–350. https://doi.org/10.1016/s0006-8993(97)00239-4.

Klivenyi, P., Kekesi, K., Juhasz, G., Vecsei, L., 1997. Amino acid concentrations in cerebrospinal fluid of patients with multiple sclerosis. Acta Neurol. Scand. 95 (2), 96–98. https://doi.org/10.1111/j.1600-0404.1997.tb00076.x.

Knochel, J.P., 1989. Heat stroke and related heat stress disorders. Dis. Mon. 35 (5), 301–377. Retrieved from https://www.ncbi.nlm.nih.gov/pubmed/2653754.

Kochevar, I.E., Lambert, C.R., Lynch, M.C., Tedesco, A.C., 1996. Comparison of photosensitized plasma membrane damage caused by singlet oxygen and free radicals. Biochim. Biophys. Acta 1280 (2), 223–230. https://doi.org/10.1016/0005-2736(95)00297-9.

Kolen'ko, Y.V., Kovnir, K.A., Gavrilov, A.I., Garshev, A.V., Frantti, J., Lebedev, O.I., Churagulov, B.R., Van Tendeloo, G., Yoshimura, M., 2006. Hydrothermal synthesis and characterization of nanorods of various titanates and titanium dioxide. J. Phys. Chem. B 110 (9), 4030–4038.

Kreuter, J., 1995. Nanoparticulate Systems in Drug Delivery and Targeting. Taylor & Francis.

Krieglstein, J., Ausmeier, F., El-Abhar, H., Lippert, K., Welsch, M., Rupalla, K., Henrich-Noack, P., 1995. Neuroprotective effects of Ginkgo biloba constituents. Eur. J. Pharm. Sci. 3 (1), 39–48.

Kulkarni, M., Mazare, A., Gongadze, E., Perutkova, S., Kralj-Iglic, V., Milosev, I., Schmuki, P., Iglic, A., Mozetic, M., 2015. Titanium nanostructures for biomedical applications. Nanotechnology 26 (6). https://doi.org/10.1088/0957-4484/26/6/062002, 062002.

Kun, G., 2009. Application of Biologically Active Micelles in Drug Delivery Across the Blood-Brain Barrier. Ph.D. Thesis (Open Access), ScholarBank@NUS Repository. http://scholarbank.nus.edu.sg/handle/10635/28240.

Landry, J., Chretien, P., Bernier, D., Nicole, L.M., Marceau, N., Tanguay, R.M., 1982. Thermotolerance and heat shock proteins induced by hyperthermia in rat liver cells. Int. J. Radiat. Oncol. Biol. Phys. 8 (1), 59–62. https://doi.org/10.1016/0360-3016(82)90385-6.

Lee, B.H., Inui, D., Suh, G.Y., Kim, J.Y., Kwon, J.Y., Park, J., Tada, K., Tanaka, K., Ietsugu, K., Uehara, K., 2012. Association of body temperature and antipyretic treatments with mortality of critically ill patients with and without sepsis: multi-centered prospective observational study. Crit. Care 16 (1), R33.

Lee, W., Moon, M., Kim, H.G., Lee, T.H., Oh, M.S., 2015. Heat stress-induced memory impairment is associated with neuroinflammation in mice. J. Neuroinflammation 12 (1), 102.

Lefkowitz, D., Ford, C.S., Rich, C., Biller, J., McHenry Jr., L.C., 1983. Cerebellar syndrome following neuroleptic induced heat stroke. J. Neurol. Neurosurg. Psychiatry 46 (2), 183–185. https://doi.org/10.1136/jnnp.46.2.183.

Lenaz, G., 1998. Role of mitochondria in oxidative stress and ageing. Biochim. Biophys. Acta 1366 (1–2), 53–67. https://doi.org/10.1016/s0005-2728(98)00120-0.

Lent-Schochet, D., Jialal, I., 2021. Physiology, edema. In: StatPearls. StatPearls Publishing, Treasure Island, FL.

Leon, L.R., 2007. Heat stroke and cytokines. Prog. Brain Res. 162, 481–524. https://doi.org/10.1016/S0079-6123(06)62024-4.

Leon, L.R., Blaha, M.D., DuBose, D.A., 2006. Time course of cytokine, corticosterone, and tissue injury responses in mice during heat strain recovery. J. Appl. Physiol. (1985) 100 (4), 1400–1409. https://doi.org/10.1152/japplphysiol.01040.2005.

Li, G.C., Werb, Z., 1982. Correlation between synthesis of heat shock proteins and development of thermotolerance in Chinese hamster fibroblasts. Proc. Natl. Acad. Sci. U. S. A. 79 (10), 3218–3222.

Li, Q., Wang, X., Lu, X., Tian, H., Jiang, H., Lv, G., Guo, D., Wu, C., Chen, B., 2009. The incorporation of daunorubicin in cancer cells through the use of titanium dioxide whiskers. Biomaterials 30 (27), 4708–4715.

Li, J., Wang, Y., Liang, R., An, X., Wang, K., Shen, G., Tu, Y., Zhu, J., Tao, J., 2015. Recent advances in targeted nanoparticles drug delivery to melanoma. Nanomedicine 11 (3), 769–794. https://doi.org/10.1016/j.nano.2014.11.006.

Lien, C.F., Mohanta, S.K., Frontczak-Baniewicz, M., Swinny, J.D., Zablocka, B., Gorecki, D.C., 2012. Absence of glial alpha-dystrobrevin causes abnormalities of the blood-brain barrier and progressive brain edema. J. Biol. Chem. 287 (49), 41374–41385. https://doi.org/10.1074/jbc.M112.400044.

Lin, M.T., 1997. Heatstroke-induced cerebral ischemia and neuronal damage. Involvement of cytokines and monoamines. Ann. N. Y. Acad. Sci. 813 (1), 572–580. https://doi.org/10.1111/j.1749-6632.1997.tb51748.x.

Lin, M.T., Kao, T.Y., Chio, C.C., Jin, Y.T., 1995. Dopamine depletion protects striatal neurons from heatstroke-induced ischemia and cell death in rats. Am. J. Physiol. 269 (2 Pt. 2), H487–H490. https://doi.org/10.1152/ajpheart.1995.269.2.H487.

Linsebigler, A.L., Lu, G., Yates, J.R., John, T., 1995. Photocatalysis on TiO2 surfaces: principles, mechanisms, and selected results. Chem. Rev. 95 (3), 735–758.

Liossis, S., Ding, X.Z., Kiang, J.G., Tsokos, G.C., 1997. Overexpression of the heat shock protein 70 enhances the TCR/CD3-and Fas/Apo-1/CD95-mediated apoptotic cell death in Jurkat T cells. J. Immunol. 158 (12), 5668–5675.

Liu, H., Ye, M., Guo, H., 2019a. An updated review of randomized clinical trials testing the improvement of cognitive function of ginkgo biloba extract in healthy people and Alzheimer's patients. Front. Pharmacol. 10, 1688. https://doi.org/10.3389/fphar.2019.01688.

Liu, Q., Jin, Z., Xu, Z., Yang, H., Li, L., Li, G., Li, F., Gu, S., Zong, S., Zhou, J., Cao, L., Wang, Z., Xiao, W., 2019b. Antioxidant effects of ginkgolides and bilobalide against cerebral ischemia injury by activating the Akt/Nrf2 pathway in vitro and in vivo. Cell Stress Chaperones 24 (2), 441–452. https://doi.org/10.1007/s12192-019-00977-1.

Liu, S.Y., Song, J.C., Mao, H.D., Zhao, J.B., Song, Q., Expert Group of Heat Stroke Prevention and Treatment of the People's Liberation Army, and People's Liberation Army Professional Committee of Critical Care Medicine, 2020. Expert consensus on the diagnosis and treatment of heat stroke in China. Mil. Med. Res. 7 (1), 1. https://doi.org/10.1186/s40779-019-0229-2.

Lo, E.H., Singhal, A.B., Torchilin, V.P., Abbott, N.J., 2001. Drug delivery to damaged brain. Brain Res. Brain Res. Rev. 38 (1–2), 140–148. https://doi.org/10.1016/s0165-0173(01)00083-2.

Lopez, T., Sotelo, J., Navarrete, J., Ascencio, J.A., 2006. Synthesis of TiO2 nanostructured reservoir with temozolomide: structural evolution of the occluded drug. Opt. Mater. 29 (1), 88–94.

Lovell, M.A., Ehmann, W.D., Butler, S.M., Markesbery, W.R., 1995. Elevated thiobarbituric acid-reactive substances and antioxidant enzyme activity in the brain in Alzheimer's disease. Neurology 45 (8), 1594–1601. https://doi.org/10.1212/wnl.45.8.1594.

Lovell, M.A., Ehmann, W.D., Mattson, M.P., Markesbery, W.R., 1997. Elevated 4-hydroxynonenal in ventricular fluid in Alzheimer's disease. Neurobiol. Aging 18 (5), 457–461. https://doi.org/10.1016/s0197-4580(97)00108-5.

Major, R.T., 1967. The ginkgo, the most ancient living tree. The resistance of Ginkgo biloba L. to pests accounts in part for the longevity of this species. Science 157 (3794), 1270–1273. https://doi.org/10.1126/science.157.3794.1270.

Makarewicz-Plonska, M., Witek, A., Farbiszewski, R., 1998. BN 52021, PAF-receptor antagonist, improves diminished antioxidant defense system of lungs in experimentally induced haemorrhagic shock. Pol. J. Pharmacol. 50 (3), 265–269. Retrieved from https://www.ncbi.nlm.nih.gov/pubmed/9861635.

Malamud, N., Haymaker, W., Custer, R.P., 1946. Heat stroke; a clinico-pathologic study of 125 fatal cases. Mil. Surg. 99 (5), 397–449. Retrieved from https://www.ncbi.nlm.nih.gov/pubmed/20276794.

Manna, P., Jain, S.K., 2015. Obesity, oxidative stress, adipose tissue dysfunction, and the associated health risks: causes and therapeutic strategies. Metab. Syndr. Relat. Disord. 13 (10), 423–444. https://doi.org/10.1089/met.2015.0095.

Markesbery, W.R., 1999. The role of oxidative stress in Alzheimer disease. Arch. Neurol. 56 (12), 1449–1452. https://doi.org/10.1001/archneur.56.12.1449.

Markesbery, W., Lovell, M., 1998. Four-hydroxynonenal, a product of lipid peroxidation, is increased in the brain in Alzheimer's disease. Neurobiol. Aging 19 (1), 33–36.

Martin-Latry, K., Goumy, M.P., Latry, P., Gabinski, C., Begaud, B., Faure, I., Verdoux, H., 2007. Psychotropic drugs use and risk of heat-related hospitalisation. Eur. Psychiatry 22 (6), 335–338. https://doi.org/10.1016/j.eurpsy.2007.03.007.

Matsumori, Y., Hong, S.M., Aoyama, K., Fan, Y., Kayama, T., Sheldon, R.A., Vexler, Z.S., Ferriero, D.M., Weinstein, P.R., Liu, J., 2005. Hsp70 overexpression sequesters AIF and reduces neonatal hypoxic/ischemic brain injury. J. Cereb. Blood Flow Metab. 25 (7), 899–910. https://doi.org/10.1038/sj.jcbfm.9600080.

Mattson, M.P., 2003. Excitotoxic and excitoprotective mechanisms: abundant targets for the prevention and treatment of neurodegenerative disorders. Neuromolecular Med. 3 (2), 65–94. https://doi.org/10.1385/NMM:3:2:65.

McCulloch, J., Savaki, H.E., Jehle, J., Sokoloff, L., 1982. Local cerebral glucose utilization in hypothermic and hyperthermic rats. J. Neurochem. 39 (1), 255–258. https://doi.org/10.1111/j.1471-4159.1982.tb04729.x.

Mecocci, P., MacGarvey, U., Beal, M.F., 1994. Oxidative damage to mitochondrial DNA is increased in Alzheimer's disease. Ann. Neurol. 36 (5), 747–751.

Mehlen, P., Schulze-Osthoff, K., Arrigo, A.-P., 1996. Small stress proteins as novel regulators of apoptosis heat shock protein 27 blocks Fas/APO-1-and staurosporine-induced cell death. J. Biol. Chem. 271 (28), 16510–16514.

Michinaga, S., Koyama, Y., 2017. Protection of the blood-brain barrier as a therapeutic strategy for brain damage. Biol. Pharm. Bull. 40 (5), 569–575. https://doi.org/10.1248/bpb.b16-00991.

Mickley, G.A., Cobb, B.L., Farrell, S.T., 1997. Brain hyperthermia alters local cerebral glucose utilization: a comparison of hyperthermic agents. Int. J. Hyperthermia 13 (1), 99–114. https://doi.org/10.3109/02656739709056434.

Mirabelli, M.C., Quandt, S.A., Crain, R., Grzywacz, J.G., Robinson, E.N., Vallejos, Q.M., Arcury, T.A., 2010. Symptoms of heat illness among Latino farm workers in North Carolina. Am. J. Prev. Med. 39 (5), 468–471.

Mirchandani, H.G., McDonald, G., Hood, I.C., Fonseca, C., 1996. Heat-related deaths in Philadelphia—1993. Am. J. Forensic Med. Pathol. 17 (2), 106–108. https://doi.org/10.1097/00000433-199606000-00004.

Mokhtarudin, M.J.M., Shabudin, A., Payne, S.J., 2018. Effects of brain tissue mechanical and fluid transport properties during ischaemic brain oedema: a poroelastic finite element analysis. In: Paper Presented at the 2018 IEEE-EMBS Conference on Biomedical Engineering and Sciences (IECBES).

Montine, T.J., Markesbery, W.R., Morrow, J.D., Roberts 2nd., L.J., 1998. Cerebrospinal fluid F2-isoprostane levels are increased in Alzheimer's disease. Ann. Neurol. 44 (3), 410–413. https://doi.org/10.1002/ana.410440322.

Moradi, S.Z., Momtaz, S., Bayrami, Z., Farzaei, M.H., Abdollahi, M., 2020. Nanoformulations of herbal extracts in treatment of neurodegenerative disorders. Front. Bioeng. Biotechnol. 8, 238. https://doi.org/10.3389/fbioe.2020.00238.

Morris, M.C., Evans, D.A., Bienias, J.L., Tangney, C.C., Bennett, D.A., Aggarwal, N., Wilson, R.S., Scherr, P.A., 2002. Dietary intake of antioxidant nutrients and the risk of incident Alzheimer disease in a biracial community study. JAMA 287 (24), 3230–3237. https://doi.org/10.1001/jama.287.24.3230.

Mosser, D.D., Morimoto, R.I., 2004. Molecular chaperones and the stress of oncogenesis. Oncogene 23 (16), 2907–2918. https://doi.org/10.1038/sj.onc.1207529.

Muchowski, P.J., Wacker, J.L., 2005. Modulation of neurodegeneration by molecular chaperones. Nat. Rev. Neurosci. 6 (1), 11–22. https://doi.org/10.1038/nrn1587.

Mund, R., Panda, N., Nimesh, S., Biswas, A., 2014. Novel titanium oxide nanoparticles for effective delivery of paclitaxel to human breast cancer cells. J. Nanopart. Res. 16, 2739. https://doi.org/10.1007/s11051-014-2739-x.

Murashov, A.K., Haq, I.U., Hill, C., Park, E., Smith, M., Wang, X., Wang, X., Goldberg, D.J., Wolgemuth, D.J., 2001. Crosstalk between p38, Hsp25 and Akt in spinal motor neurons after sciatic nerve injury. Mol. Brain Res. 93 (2), 199–208.

Muresanu, D.F., Sharma, H.S., 2007. Chronic hypertension aggravates heat stress–induced cognitive dysfunction and brain pathology: an experimental study in the rat, using growth hormone therapy for possible neuroprotection. Ann. N. Y. Acad. Sci. 1122 (1), 1–22.

Muresanu, D.F., Sharma, A., Sharma, H.S., 2010. Diabetes aggravates heat stress-induced blood-brain barrier breakdown, reduction in cerebral blood flow, edema formation, and brain pathology: possible neuroprotection with growth hormone. Ann. N. Y. Acad. Sci. 1199, 15–26. https://doi.org/10.1111/j.1749-6632.2009.05328.x.

Muresanu, D.F., Sharma, A., Tian, Z.R., Smith, M.A., Sharma, H.S., 2012. Nanowired drug delivery of antioxidant compound H-290/51 enhances neuroprotection in hyperthermia-induced neurotoxicity. CNS Neurol. Disord. Drug Targets 11 (1), 50–64.

Muresanu, D.F., Sharma, A., Patnaik, R., Menon, P.K., Mossler, H., Sharma, H.S., 2019. Exacerbation of blood-brain barrier breakdown, edema formation, nitric oxide synthase upregulation and brain pathology after heat stroke in diabetic and hypertensive rats. Potential neuroprotection with cerebrolysin treatment. Int. Rev. Neurobiol. 146, 83–102. https://doi.org/10.1016/bs.irn.2019.06.007.

Muresanu, D.F., Sharma, A., Sahib, S., Tian, Z.R., Feng, L., Castellani, R.J., Nozari, A., Lafuente, J.V., Buzoianu, A.D., Sjoquist, P.O., Patnaik, R., Wiklund, L., Sharma, H.S., 2020. Diabetes exacerbates brain pathology following a focal blast brain injury: new role of a multimodal drug cerebrolysin and nanomedicine. Prog. Brain Res. 258, 285–367. https://doi.org/10.1016/bs.pbr.2020.09.004.

Nakai, S., Itoh, T., Morimoto, T., 1999. Deaths from heat-stroke in Japan: 1968-1994. Int. J. Biometeorol. 43 (3), 124–127. https://doi.org/10.1007/s004840050127.

Nasr, C., Haag-Berrurier, M., Lobstein-Guth, A., Anton, R., 1986. Kaempferol coumaroyl glucorhamnoside from Ginkgo biloba. Phytochemistry 25 (3), 770–771.

Nasr, C., Lobstein-Guth, A., Haag-Berrurier, M., Anton, R., 1987. Quercetin coumaroyl glucorhamnoside from Ginkgo biloba. Phytochemistry 26 (10), 2869–2870.

Naughton, M.P., Henderson, A., Mirabelli, M.C., Kaiser, R., Wilhelm, J.L., Kieszak, S.M., Rubin, C.H., McGeehin, M.A., 2002. Heat-related mortality during a 1999 heat wave in Chicago. Am. J. Prev. Med. 22 (4), 221–227. https://doi.org/10.1016/s0749-3797(02)00421-x.

Nemoto, E.M., Frankel, H.M., 1970. Cerebral oxygenation and metabolism during progressive hyperthermia. Am. J. Physiol. 219 (6), 1784–1788. https://doi.org/10.1152/ajplegacy.1970.219.6.1784.

Ni, Y., Zhao, B., Hou, J., Xin, W., 1996. Preventive effect of Ginkgo biloba extract on apoptosis in rat cerebellar neuronal cells induced by hydroxyl radicals. Neurosci. Lett. 214 (2–3), 115–118. https://doi.org/10.1016/0304-3940(96)12897-4.

Niu, F., Sharma, A., Wang, Z., Feng, L., Muresanu, D.F., Sahib, S., Tian, Z.R., Lafuente, J.V., Buzoianu, A.D., Castellani, R.J., Nozari, A., Patnaik, R., Wiklund, L., Sharma, H.S., 2020. Co-administration of TiO2-nanowired dl-3-n-butylphthalide (dl-NBP) and mesenchymal stem cells enhanced neuroprotection in Parkinson's disease exacerbated by concussive head injury. Prog. Brain Res. 258, 101–155. https://doi.org/10.1016/bs.pbr.2020.09.011.

Nunneley, S.A., Martin, C.C., Slauson, J.W., Hearon, C.M., Nickerson, L.D., Mason, P.A., 2002. Changes in regional cerebral metabolism during systemic hyperthermia in humans. J. Appl. Physiol. (1985) 92 (2), 846–851. https://doi.org/10.1152/japplphysiol.00072.2001.

Odhiambo Sewe, M., Bunker, A., Ingole, V., Egondi, T., Oudin Astrom, D., Hondula, D.M., Rocklov, J., Schumann, B., 2018. Estimated effect of temperature on years of life lost: a retrospective time-series study of low-, middle-, and high-income regions. Environ. Health Perspect. 126 (1), 017004. https://doi.org/10.1289/EHP1745.

Othman, M.A., Amat, N.F., Ahmad, B.H., Rajan, J., 2014. Electrical Conductivity Characteristic of TiO2 Nanowires From Hydrothermal Method. Paper Presented at the Journal of Physics: Conference Series.

Oyama, Y., Chikahisa, L., Ueha, T., Kanemaru, K., Noda, K., 1996. Ginkgo biloba extract protects brain neurons against oxidative stress induced by hydrogen peroxide. Brain Res. 712 (2), 349–352. https://doi.org/10.1016/0006-8993(95)01440-3.

Ozkizilcik, A., Williams, R., Tian, Z.R., Muresanu, D.F., Sharma, A., Sharma, H.S., 2018. Synthesis of biocompatible titanate nanofibers for effective delivery of neuroprotective agents. In: Neurotrophic Factors. Springer, pp. 433–442.

Pardridge, W.M., 1988. Recent advances in blood-brain barrier transport. Annu. Rev. Pharmacol. Toxicol. 28 (1), 25–39. https://doi.org/10.1146/annurev.pa.28.040188.000325.

Park, H.S., Lee, J.S., Huh, S.H., Seo, J.S., Choi, E.J., 2001. Hsp72 functions as a natural inhibitory protein of c-Jun N-terminal kinase. EMBO J. 20 (3), 446–456.

Patnaik, R., Mohanty, S., Sharma, H.S., 2000. Blockade of histamine H 2 receptors attenuate blood-brain barrier permeability, cerebral blood flow disturbances, edema formation and cell reactions following hyperthermic brain injury in the rat. In: Brain Edema XI. Springer, pp. 535–539.

Pease, S., Bouadma, L., Kermarrec, N., Schortgen, F., Regnier, B., Wolff, M., 2009. Early organ dysfunction course, cooling time and outcome in classic heatstroke. Intensive Care Med. 35 (8), 1454–1458. https://doi.org/10.1007/s00134-009-1500-x.

Peng, H., Sola, A., Moore, J., Wen, T., 2010. Caspase inhibition by cardiotrophin-1 prevents neuronal death in vivo and in vitro. J. Neurosci. Res. 88 (5), 1041–1051. https://doi.org/10.1002/jnr.22269.

Pitt, D., Nagelmeier, I.E., Wilson, H.C., Raine, C.S., 2003. Glutamate uptake by oligodendrocytes: implications for excitotoxicity in multiple sclerosis. Neurology 61 (8), 1113–1120. https://doi.org/10.1212/01.wnl.0000090564.88719.37.

Prasad, M.R., Lovell, M.A., Yatin, M., Dhillon, H., Markesbery, W.R., 1998. Regional membrane phospholipid alterations in Alzheimer's disease. Neurochem. Res. 23 (1), 81–88. https://doi.org/10.1023/a:1022457605436.

Prochazkova, K., Sejna, I., Skutil, J., Hahn, A., 2018. Ginkgo biloba extract EGb 761((R)) versus pentoxifylline in chronic tinnitus: a randomized, double-blind clinical trial. Int. J. Clin. Pharmacol. 40 (5), 1335–1341. https://doi.org/10.1007/s11096-018-0654-4.

Qian, S., Jiang, Q., Liu, K., Li, B., Li, M., Li, L., Yang, X., Yang, Z., Sun, G., 2014. Effects of short-term environmental hyperthermia on patterns of cerebral blood flow. Physiol. Behav. 128, 99–107. https://doi.org/10.1016/j.physbeh.2014.01.028.

Rajdev, S., Hara, K., Kokubo, Y., Mestril, R., Dillmann, W., Weinstein, P.R., Sharp, F.R., 2000. Mice overexpressing rat heat shock protein 70 are protected against cerebral infarction. Ann. Neurol. 47 (6), 782–791. Retrieved from https://www.ncbi.nlm.nih.gov/pubmed/10852544.

Ran, R., Lu, A., Zhang, L., Tang, Y., Zhu, H., Xu, H., Feng, Y., Han, C., Zhou, G., Rigby, A.C., 2004. Hsp70 promotes TNF-mediated apoptosis by binding IKKγ and impairing NF-κB survival signaling. Genes Dev. 18 (12), 1466–1481.

Rane, M.J., Pan, Y., Singh, S., Powell, D.W., Wu, R., Cummins, T., Chen, Q., McLeish, K.R., Klein, J.B., 2003. Heat shock protein 27 controls apoptosis by regulating Akt activation. J. Biol. Chem. 278 (30), 27828–27835. https://doi.org/10.1074/jbc.M303417200.

Rangel-Castillo, L., Gopinath, S., Robertson, C.S., 2008. Management of intracranial hypertension. Neurol. Clin. 26 (2), 521–541.

Rapoport, S.I., 1976. Blood-Brain Barrier in Physiology and Medicine. Raven Pr.

Reddy, P.H., Tripathi, R., Troung, Q., Tirumala, K., Reddy, T.P., Anekonda, V., Shirendeb, U.P., Calkins, M.J., Reddy, A.P., Mao, P., 2012. Abnormal mitochondrial dynamics and synaptic degeneration as early events in Alzheimer's disease: implications to mitochondria-targeted antioxidant therapeutics. Biochim. Biophys. Acta Mol. Basis Dis. 1822 (5), 639–649.

Ren, B., Wan, Y., Wang, G., Liu, Z., Huang, Y., Wang, H., 2018. Morphologically modified surface with hierarchical micro-/nano-structures for enhanced bioactivity of titanium implants. J. Mater. Sci. 53, 12679–12691. https://doi.org/10.1007/s10853-018-2554-3.

Richards, D.A., Tolias, C.M., Sgouros, S., Bowery, N.G., 2003. Extracellular glutamine to glutamate ratio may predict outcome in the injured brain: a clinical microdialysis study in children. Pharmacol. Res. 48 (1), 101–109.

Riedel, W., Maulik, G., 1999. Fever: an integrated response of the central nervous system to oxidative stress. In: Stress Adaptation, Prophylaxis and Treatment. Springer, pp. 125–132.

Ritchie, K.P., Keller, B.M., Syed, K.M., Lepock, J.R., 1994. Hyperthermia (heat shock)-induced protein denaturation in liver, muscle and lens tissue as determined by differential scanning calorimetry. Int. J. Hyperthermia 10 (5), 605–618.

Roberts, G.T., Ghebeh, H., Chishti, M.A., Al-Mohanna, F., El-Sayed, R., Al-Mohanna, F., Bouchama, A., 2008. Microvascular injury, thrombosis, inflammation, and apoptosis in the pathogenesis of heatstroke: a study in baboon model. Arterioscler. Thromb. Vasc. Biol. 28 (6), 1130–1136. https://doi.org/10.1161/ATVBAHA.107.158709.

Rojas, P., Montes, S., Serrano-Garcia, N., Rojas-Castaneda, J., 2009. Effect of EGb761 supplementation on the content of copper in mouse brain in an animal model of Parkinson's disease. Nutrition 25 (4), 482–485. https://doi.org/10.1016/j.nut.2008.10.013.

Roye, D., Codesido, R., Tobias, A., Taracido, M., 2020. Heat wave intensity and daily mortality in four of the largest cities of Spain. Environ. Res. 182, 109027. https://doi.org/10.1016/j.envres.2019.109027.

Sahib, S., Niu, F., Sharma, A., Feng, L., Tian, Z.R., Muresanu, D.F., Nozari, A., Sharma, H.S., 2019. Potentiation of spinal cord conduction and neuroprotection following nanodelivery of DL-3-n-butylphthalide in titanium implanted nanomaterial in a focal spinal cord injury induced functional outcome, blood-spinal cord barrier breakdown and edema formation. Int. Rev. Neurobiol. 146, 153–188. https://doi.org/10.1016/bs.irn.2019.06.009.

Sahib, S., Sharma, A., Menon, P.K., Muresanu, D.F., Castellani, R.J., Nozari, A., Lafuente, J.V., Bryukhovetskiy, I., Tian, Z.R., Patnaik, R., Buzoianu, A.D., Wiklund, L., Sharma, H.S., 2020. Cerebrolysin enhances spinal cord conduction and reduces blood-spinal cord barrier breakdown, edema formation, immediate early gene expression and cord pathology after injury. Prog. Brain Res. 258, 397–438. https://doi.org/10.1016/bs.pbr.2020.09.012.

Sakaguchi, Y., Stephens, L.C., Makino, M., Kaneko, T., Strebel, F.R., Danhauser, L.L., Jenkins, G.N., Bull, J.M., 1995. Apoptosis in tumors and normal tissues induced by whole body hyperthermia in rats. Cancer Res. 55 (22), 5459–5464. Retrieved from https://www.ncbi.nlm.nih.gov/pubmed/7585616.

Sandoval, K.E., Witt, K.A., 2008. Blood-brain barrier tight junction permeability and ischemic stroke. Neurobiol. Dis. 32 (2), 200–219. https://doi.org/10.1016/j.nbd.2008.08.005.

Sarikcioglu, S.B., Oner, G., Tercan, E., 2004. Antioxidant effect of EGb 761 on hydrogen peroxide-induced lipoperoxidation of G-6-PD deficient erythrocytes. Phytother. Res. 18 (10), 837–840. https://doi.org/10.1002/ptr.1544.

Sato, S., Fujita, N., Tsuruo, T., 2000. Modulation of Akt kinase activity by binding to Hsp90. Proc. Natl. Acad. Sci. U. S. A. 97 (20), 10832–10837. https://doi.org/10.1073/pnas.170276797.

Saxton, C., 1981. Effects of severe heat stress on respiration and metabolic rate in resting man. Aviat. Space Environ. Med. 52 (5), 281–286. Retrieved from https://www.ncbi.nlm.nih.gov/pubmed/7247898.

Schiaffonati, L., Maroni, P., Bendinelli, P., Tiberio, L., Piccoletti, R., 2001. Hyperthermia induces gene expression of heat shock protein 70 and phosphorylation of mitogen activated protein kinases in the rat cerebellum. Neurosci. Lett. 312 (2), 75–78. https://doi.org/10.1016/s0304-3940(01)02182-6.

Schickele, E., 1947. Environment and fatal heat stroke; an analysis of 157 cases occurring in the Army in the U.S. during World War II. Mil. Surg. 100 (3), 235–256. Retrieved from https://www.ncbi.nlm.nih.gov/pubmed/20287512.

Semenza, J.C., Rubin, C.H., Falter, K.H., Selanikio, J.D., Flanders, W.D., Howe, H.L., Wilhelm, J.L., 1996. Heat-related deaths during the July 1995 heat wave in Chicago. N. Engl. J. Med. 335 (2), 84–90. https://doi.org/10.1056/NEJM199607113350203.

Semenza, J.C., McCullough, J.E., Flanders, W.D., McGeehin, M.A., Lumpkin, J.R., 1999. Excess hospital admissions during the July 1995 heat wave in Chicago. Am. J. Prev. Med. 16 (4), 269–277. https://doi.org/10.1016/s0749-3797(99)00025-2.

Sharma, H.S., 1987. Effect of captopril (a converting enzyme inhibitor) on blood-brain barrier permeability and cerebral blood flow in normotensive rats. Neuropharmacology 26 (1), 85–92. https://doi.org/10.1016/0028-3908(87)90049-9.

Sharma, H.S., 1994. Topical application of dynorphin-A antibodies reduces edema and cell changes in traumatise tar spinal cord. Regul. Pept. Suppl. 1, S92.

Sharma, H.S., 1999. Pathophysiology of Blood-Brain Barrier, Brain Edema and Cell Injury Following Hyperthermia: New Role of Heat Shock Protein, Nitric Oxide and Carbon Monoxide. An Experimental Study in the Rat using Light and Electron Microscopy, vol. 830 Acta Universitatis Upsaliensis, pp. 1–94.

Sharma, H.S., 2000. A bradykinin BK 2 receptor antagonist HOE-140 attenuates blood-spinal cord barrier permeability following a focal trauma to the rat spinal cord. In: Brain Edema XI. Springer, pp. 159–163.

Sharma, H.S., 2003. Blood-Spinal Cord and Brain Barriers in Health and Disease. Elsevier.

Sharma, H.S., 2004. Pathophysiology of the blood–spinal cord barrier in traumatic injury. In: Blood-Spinal Cord and Brain Barriers in Health and Disease. Elsevier, pp. 437–518.

Sharma, H.S., 2005a. Heat-related deaths are largely due to brain damage. Indian J. Med. Res. 121 (5), 621–623. Retrieved from https://www.ncbi.nlm.nih.gov/pubmed/15937362.

Sharma, H.S., 2005b. Neuroprotective effects of neurotrophins and melanocortins in spinal cord injury: an experimental study in the rat using pharmacological and morphological approaches. Ann. N. Y. Acad. Sci. 1053 (1), 407–421.

Sharma, H.S., 2006. Hyperthermia induced brain oedema: current status and future perspectives. Indian J. Med. Res. 123 (5), 629–652. Retrieved from https://www.ncbi.nlm.nih.gov/pubmed/16873906.

Sharma, A., 2018. Nanodelivery of Chinese traditional medicine extract of Gingko Biloba (EGb-761) induces superior neuroprotection in heat stroke. Am. J. Ethnomed. 5, 42. https://doi.org/10.21767/2348-9502-C1-004. 6th Edition of International Conference on Pharmacognosy and Medicinal Plants, April 16-17, 2018 Amsterdam, Netherlands.

Sharma, H.S., Cervos-Navarro, J., 1990. Brain oedema and cellular changes induced by acute heat stress in young rats. Acta Neurochir. Suppl. (Wien) 51, 383–386. https://doi.org/10.1007/978-3-7091-9115-6_129.

Sharma, H.S., Dey, P.K., 1982. Correlation of spinal cord tissue 5-HT with edema development following surgical spinal cord trauma in rats. Indian J. Physiol. Pharmacol. 26, 8–9.

Sharma, H.S., Dey, P.K., 1984. Role of 5-HT on increased permeability of blood-brain barrier under heat stress. Indian J. Physiol. Pharmacol. 28 (4), 259–267. Retrieved from https://www.ncbi.nlm.nih.gov/pubmed/6242051.

Sharma, H.S., Dey, P.K., 1986. Probable involvement of 5-hydroxytryptamine in increased permeability of blood-brain barrier under heat stress in young rats. Neuropharmacology 25 (2), 161–167. https://doi.org/10.1016/0028-3908(86)90037-7.

Sharma, H.S., Dey, P.K., 1987. Influence of long-term acute heat exposure on regional blood-brain barrier permeability, cerebral blood flow and 5-HT level in conscious normotensive young rats. Brain Res. 424 (1), 153–162. https://doi.org/10.1016/0006-8993(87)91205-4.

Sharma, H.S., Hoopes, P.J., 2003. Hyperthermia induced pathophysiology of the central nervous system. Int. J. Hyperthermia 19 (3), 325–354. https://doi.org/10.1080/0265673 021000054621.

Sharma, H.S., Johanson, C.E., 2007. Blood-cerebrospinal fluid barrier in hyperthermia. Prog. Brain Res. 162, 459–478. https://doi.org/10.1016/S0079-6123(06)62023-2.

Sharma, H.S., Olsson, Y., 1990. Edema formation and cellular alterations following spinal cord injury in the rat and their modification with p-chlorophenylalanine. Acta Neuropathol. 79 (6), 604–610. https://doi.org/10.1007/BF00294237.

Sharma, H.S., Sharma, A., 2007. Nanoparticles aggravate heat stress induced cognitive deficits, blood-brain barrier disruption, edema formation and brain pathology. Prog. Brain Res. 162, 245–273. https://doi.org/10.1016/S0079-6123(06)62013-X.

Sharma, H.S., Sjoquist, P.O., 2002. A new antioxidant compound H-290/51 modulates glutamate and GABA immunoreactivity in the rat spinal cord following trauma. Amino Acids 23 (1–3), 261–272. https://doi.org/10.1007/s00726-001-0137-z.

Sharma, H.S., Olsson, Y., Dey, P.K., 1990a. Changes in blood-brain barrier and cerebral blood flow following elevation of circulating serotonin level in anesthetized rats. Brain Res. 517 (1–2), 215–223. https://doi.org/10.1016/0006-8993(90)91029-g.

Sharma, H.S., Westman, J., Olsson, Y., Johansson, O., Dey, P.K., 1990b. Increased 5-hydroxytryptamine immunoreactivity in traumatized spinal cord. An experimental study in the rat. Acta Neuropathol. 80 (1), 12–17. https://doi.org/10.1007/BF00294216.

Sharma, H.S., Cervos-Navarro, J., Dey, P.K., 1991a. Acute heat exposure causes cellular alteration in cerebral cortex of young rats. Neuroreport 2 (3), 155–158. https://doi.org/10.1097/00001756-199103000-00012.

Sharma, H.S., Winkler, T., Stålberg, E., Olsson, Y., Dey, P.K., 1991b. Evaluation of traumatic spinal cord edema using evoked potentials recorded from the spinal epidural space: an experimental study in the rat. J. Neurol. Sci. 102 (2), 150–162.

Sharma, H.S., Kretzschmar, R., Cervos-Navarro, J., Ermisch, A., Ruhle, H.J., Dey, P.K., 1992a. Age-related pathophysiology of the blood-brain barrier in heat stress. Prog. Brain Res. 91, 189–196. https://doi.org/10.1016/s0079-6123(08)62334-1.

Sharma, H.S., Zimmer, C., Westman, J., Cervos-Navarro, J., 1992b. Acute systemic heat stress increases glial fibrillary acidic protein immunoreactivity in brain: experimental observations in conscious normotensive young rats. Neuroscience 48 (4), 889–901. https://doi.org/10.1016/0306-4522(92)90277-9.

Sharma, H.S., Olsson, Y., Cervos-Navarro, J., 1993. Early perifocal cell changes and edema in traumatic injury of the spinal cord are reduced by indomethacin, an inhibitor of prostaglandin synthesis. Experimental study in the rat. Acta Neuropathol. 85 (2), 145–153. https://doi.org/10.1007/BF00227761.

Sharma, H.S., Westman, J., Nyberg, F., Cervos-Navarro, J., Dey, P.K., 1994. Role of serotonin and prostaglandins in brain edema induced by heat stress. An experimental study in the young rat. Acta Neurochir. Suppl. (Wien) 60, 65–70. https://doi.org/10.1007/978-3-7091-9334-1_17.

Sharma, H.S., Olsson, Y., Nyberg, F., 1995a. Influence of dynorphin A antibodies on the formation of edema and cell changes in spinal cord trauma. In: Progress in Brain Research. vol. 104. Elsevier, pp. 401–416.

Sharma, H.S., Olsson, Y., Persson, S., Nyberg, F., 1995b. Trauma-induced opening of the blood-spinal cord barrier is reduced by indomethacin, an inhibitor of prostaglandin biosynthesis. Experimental observations in the rat using [131I]-sodium, Evans blue and lanthanum as tracers. Restor. Neurol. Neurosci. 7 (4), 207–215. https://doi.org/10.3233/RNN-1994-7403.

Sharma, H.S., Westman, J., Olsson, Y., Alm, P., 1996. Involvement of nitric oxide in acute spinal cord injury: an immunocytochemical study using light and electron microscopy in the rat. Neurosci. Res. 24 (4), 373–384.

Sharma, H.S., Westman, J., Alm, P., Sjoquist, P.O., Cervos-Navarro, J., Nyberg, F., 1997a. Involvement of nitric oxide in the pathophysiology of acute heat stress in the rat. Influence of a new antioxidant compound H-290/51. Ann. N. Y. Acad. Sci. 813 (1), 581–590. https://doi.org/10.1111/j.1749-6632.1997.tb51749.x.

Sharma, H.S., Westman, J., Nyberg, F., 1997b. Topical application of 5-HT antibodies reduces edema and cell changes following trauma of the rat spinal cord. In: Brain Edema X. Springer, pp. 155–158.

Sharma, H.S., Alm, P., Westman, J., 1998a. Nitric oxide and carbon monoxide in the brain pathology of heat stress. Prog. Brain Res. 115, 297–333. https://doi.org/10.1016/s0079-6123(08)62041-5.

Sharma, H.S., Westman, J., Nyberg, F., 1998b. Pathophysiology of brain edema and cell changes following hyperthermic brain injury. In: Progress in Brain Research. vol. 115. Elsevier, pp. 351–412.

Sharma, H.S., Drieu, K., Alm, P., Westman, J., 2000a. Role of nitric oxide in blood-brain barrier permeability, brain edema and cell damage following hyperthermic brain injury. An experimental study using EGB-761 and Gingkolide B pretreatment in the rat. In: Brain Edema XI. Springer, pp. 81–86.

Sharma, H.S., Drieu, K., Alm, P., Westman, J., 2000b. Role of nitric oxide in blood-brain barrier permeability, brain edema and cell damage following hyperthermic brain injury. An experimental study using EGB-761 and Gingkolide B pretreatment in the rat. Acta Neurochir. Suppl. 76, 81–86. https://doi.org/10.1007/978-3-7091-6346-7_17.

Sharma, H.S., Drieu, K., Westman, J., 2003a. Antioxidant compounds EGB-761 and BN-52021 attenuate brain edema formation and hemeoxygenase expression following hyperthermic brain injury in the rat. Acta Neurochir. Suppl. 86, 313–319. https://doi.org/10.1007/978-3-7091-0651-8_68.

Sharma, H.S., Winkler, T., Stålberg, E., Gordh, T., Alm, P., Westman, J., 2003b. Topical application of TNF-α antiserum attenuates spinal cord trauma induced edema formation, microvascular permeability disturbances and cell injury in the rat. In: Brain Edema XII. Springer, pp. 407–413.

Sharma, H.S., Patnaik, R., Ray, A.K., Dey, P.K., 2004. Blood-central nervous system barriers in morphine dependence and withdrawal. In: Blood-Spinal Cord and Brain Barriers in Health and Disease. Elsevier, pp. 299–328.

Sharma, H.S., Duncan, J.A., Johanson, C.E., 2006a. Whole-body hyperthermia in the rat disrupts the blood-cerebrospinal fluid barrier and induces brain edema. In: Brain Edema XIII. Springer, pp. 426–431.

Sharma, H.S., Nyberg, F., Gordh, T., Alm, P., 2006b. Topical application of dynorphin A (1-17) antibodies attenuates neuronal nitric oxide synthase up-regulation, edema formation, and cell injury following focal trauma to the rat spinal cord. In: Brain Edema XIII. Springer, pp. 309–315.

Sharma, H.S., Ali, S.F., Dong, W., Tian, Z.R., Patnaik, R., Patnaik, S., Sharma, A., Boman, A., Lek, P., Seifert, E., Lundstedt, T., 2007. Drug delivery to the spinal cord tagged with nanowire enhances neuroprotective efficacy and functional recovery following trauma to the rat spinal cord. Ann. N. Y. Acad. Sci. 1122 (1), 197–218. https://doi.org/10.1196/annals.1403.014.

Sharma, H.S., Ali, S., Tian, Z.R., Patnaik, R., Patnaik, S., Lek, P., Sharma, A., Lundstedt, T., 2009a. Nano-drug delivery and neuroprotection in spinal cord injury. J. Nanosci. Nanotechnol. 9 (8), 5014–5037. https://doi.org/10.1166/jnn.2009.gr04.

Sharma, H.S., Ali, S.F., Tian, Z.R., Hussain, S.M., Schlager, J.J., Sjoquist, P.O., Sharma, A., Muresanu, D.F., 2009b. Chronic treatment with nanoparticles exacerbate hyperthermia induced blood-brain barrier breakdown, cognitive dysfunction and brain pathology in the rat. Neuroprotective effects of nanowired-antioxidant compound H-290/51. J. Nanosci. Nanotechnol. 9 (8), 5073–5090. https://doi.org/10.1166/jnn.2009.gr10.

Sharma, H.S., Muresanu, D., Sharma, A., Patnaik, R., 2009c. Cocaine-induced breakdown of the blood-brain barrier and neurotoxicity. Int. Rev. Neurobiol. 88, 297–334. https://doi.org/10.1016/S0074-7742(09)88011-2.

Sharma, H.S., Feng, L., Lafuente, J.V., Muresanu, D.F., Tian, Z.R., Patnaik, R., Sharma, A., 2015. TiO2-nanowired delivery of mesenchymal stem cells thwarts diabetes-induced exacerbation of brain pathology in heat stroke: an experimental study in the rat using morphological and biochemical approaches. CNS Neurol. Disord. Drug Targets 14 (3), 386–399. https://doi.org/10.2174/1871527314666150318114335.

Sharma, A., Menon, P., Muresanu, D.F., Ozkizilcik, A., Tian, Z.R., Lafuente, J.V., Sharma, H.S., 2016a. Nanowired drug delivery across the blood-brain barrier in central nervous system injury and repair. CNS Neurol. Disord. Drug Targets 15 (9), 1092–1117. https://doi.org/10.2174/1871527315666160819123059.

Sharma, H.S., Muresanu, D.F., Lafuente, J.V., Nozari, A., Patnaik, R., Skaper, S.D., Sharma, A., 2016b. Pathophysiology of blood-brain barrier in brain injury in cold and hot environments: novel drug targets for neuroprotection. CNS Neurol. Disord. Drug Targets 15 (9), 1045–1071. https://doi.org/10.2174/1871527315666160902145145.

Sharma, H.S., Muresanu, D.F., Sharma, A., 2016c. Alzheimer's disease: cerebrolysin and nanotechnology as a therapeutic strategy. Neurodegener. Dis. Manag. 6 (6), 453–456. https://doi.org/10.2217/nmt-2016-0037.

Sharma, H.S., Skaper, S.D., Sharma, A., 2016d. Editorial: (Thematic Issue) novel concepts on the blood-brain barrier and brain pathology. New therapeutic approaches. CNS Neurol. Disord. Drug Targets 15 (9), 1014–1015. https://doi.org/10.2174/18715273150916 1007123931.

Sharma, A., Menon, P.K., Patnaik, R., Muresanu, D.F., Lafuente, J.V., Tian, Z.R., Ozkizilcik, A., Castellani, R.J., Mossler, H., Sharma, H.S., 2017. Novel treatment strategies using TiO2-nanowired delivery of histaminergic drugs and antibodies to tau with cerebrolysin for superior neuroprotection in the pathophysiology of Alzheimer's disease. Int. Rev. Neurobiol. 137, 123–165. https://doi.org/10.1016/bs.irn.2017.09.002.

Sharma, A., Muresanu, D.F., Lafuente, J.V., Sjoquist, P.O., Patnaik, R., Ryan Tian, Z., Ozkizilcik, A., Sharma, H.S., 2018a. Cold environment exacerbates brain pathology and oxidative stress following traumatic brain injuries: potential therapeutic effects of nanowired antioxidant compound H-290/51. Mol. Neurobiol. 55 (1), 276–285. https://doi.org/10.1007/s12035-017-0740-y.

Sharma, H.S., Muresanu, D.F., Lafuente, J.V., Patnaik, R., Tian, Z.R., Ozkizilcik, A., Castellani, R.J., Mossler, H., Sharma, A., 2018b. Co-administration of TiO2 nanowired mesenchymal stem cells with cerebrolysin potentiates neprilysin level and reduces brain pathology in Alzheimer's disease. Mol. Neurobiol. 55 (1), 300–311. https://doi.org/10.1007/s12035-017-0742-9.

Sharma, A., Muresanu, D.F., Lafuente, J.V., Zhang, Z.Q., Li, C., Patnaik, R., Tuian, Z.R., Ozkizilcik, A., Sharma, H.S., 2018c. Nanodelivery of Chinese traditional medicine extract of Gigko Biloba (EGb-761) induces superior neuroprotection following traumatic brain injury in heat stroke. In: Biotech, Biomaterials and Biomedical: TechConnect Briefs 2018c. TechConnect.org, pp. 89–92. ISBN:978-0-9988782-4-9.

Sharma, A., Castellani, R.J., Smith, M.A., Muresanu, D.F., Dey, P.K., Sharma, H.S., 2019a. 5-Hydroxytryptophan: a precursor of serotonin influences regional blood-brain barrier breakdown, cerebral blood flow, brain edema formation, and neuropathology. Int. Rev. Neurobiol. 146, 1–44. https://doi.org/10.1016/bs.irn.2019.06.005.

Sharma, A., Muresanu, D.F., Ozkizilcik, A., Tian, Z.R., Lafuente, J.V., Manzhulo, I., Mossler, H., Sharma, H.S., 2019b. Sleep deprivation exacerbates concussive head injury induced brain pathology: neuroprotective effects of nanowired delivery of cerebrolysin with alpha-melanocyte-stimulating hormone. Prog. Brain Res. 245, 1–55. https://doi.org/10.1016/bs.pbr.2019.03.002.

Sharma, A., Muresanu, D.F., Castellani, R.J., Nozari, A., Lafuente, J.V., Sahib, S., Tian, Z.R., Buzoianu, A.D., Patnaik, R., Wiklund, L., Sharma, H.S., 2020a. Mild traumatic brain injury exacerbates Parkinson's disease induced hemeoxygenase-2 expression and brain pathology: neuroprotective effects of co-administration of TiO2 nanowired mesenchymal stem cells and cerebrolysin. Prog. Brain Res. 258, 157–231. https://doi.org/10.1016/bs.pbr.2020.09.010.

Sharma, A., Muresanu, D.F., Sahib, S., Tian, Z.R., Castellani, R.J., Nozari, A., Lafuente, J.V., Buzoianu, A.D., Bryukhovetskiy, I., Manzhulo, I., Patnaik, R., Wiklund, L., Sharma, H.S., 2020b. Concussive head injury exacerbates neuropathology of sleep deprivation: superior neuroprotection by co-administration of TiO2-nanowired cerebrolysin, alpha-melanocyte-stimulating hormone, and mesenchymal stem cells. Prog. Brain Res. 258, 1–77. https://doi.org/10.1016/bs.pbr.2020.09.003.

Sharma, H.S., Muresanu, D.F., Castellani, R.J., Nozari, A., Lafuente, J.V., Tian, Z.R., Sahib, S., Bryukhovetskiy, I., Bryukhovetskiy, A., Buzoianu, A.D., Patnaik, R., Wiklund, L., Sharma, A., 2020c. Pathophysiology of blood-brain barrier in brain tumor. Novel therapeutic advances using nanomedicine. Int. Rev. Neurobiol. 151, 1–66. https://doi.org/10.1016/bs.irn.2020.03.001.

Sharma, H.S., Sahib, S., Tian, Z.R., Muresanu, D.F., Nozari, A., Castellani, R.J., Lafuente, J.V., Wiklund, L., Sharma, A., 2020d. Protein kinase inhibitors in traumatic brain injury and repair: new roles of nanomedicine. Prog. Brain Res. 258, 233–283. https://doi.org/10.1016/bs.pbr.2020.09.009.

Shiozaki, T., Sugimoto, H., Taneda, M., Yoshida, H., Iwai, A., Yoshioka, T., Sugimoto, T., 1993. Effect of mild hypothermia on uncontrollable intracranial hypertension after severe head injury. J. Neurosurg. 79 (3), 363–368. https://doi.org/10.3171/jns.1993.79.3.0363.

Shlosberg, D., Benifla, M., Kaufer, D., Friedman, A., 2010. Blood–brain barrier breakdown as a therapeutic target in traumatic brain injury. Nat. Rev. Neurol. 6 (7), 393.

Simon, H.B., 1994. Hyperthermia and heatstroke. Hosp. Pract. (Off. Ed) 29 (8), 65–68 (73, 78-80) https://doi.org/10.1080/21548331.1994.11443062.

Singh, S.K., Srivastav, S., Castellani, R.J., Plascencia-Villa, G., Perry, G., 2019. Neuroprotective and antioxidant effect of ginkgo biloba extract against AD and other neurological disorders. Neurotherapeutics 16 (3), 666–674. https://doi.org/10.1007/s13311-019-00767-8.

Smith, M.A., Harris, P.L., Sayre, L.M., Perry, G., 1997. Iron accumulation in Alzheimer disease is a source of redox-generated free radicals. Proc. Natl. Acad. Sci. 94 (18), 9866–9868.

South Jr., F.E., 1958. Rates of oxygen consumption and glycolysis of ventricle and brain slices, obtained from hibernating and non-hibernating mammals, as function of temperature. Physiol. Zool. 31 (1), 6–15.

Stefanits, H., Reinprecht, A., Klein, K.U., 2019. Intracranial pressure. In: Mashour, G.A., Engelhard, K. (Eds.), Oxford Textbook of Neuroscience Anaesthesiology. Oxford University Press, https://doi.org/10.1093/med/9780198746645.003.0002.

Stetler, R.A., Gan, Y., Zhang, W., Liou, A.K., Gao, Y., Cao, G., Chen, J., 2010. Heat shock proteins: cellular and molecular mechanisms in the central nervous system. Prog. Neurobiol. 92 (2), 184–211. https://doi.org/10.1016/j.pneurobio.2010.05.002.

Stewart, A., 1992. Gingko biloba extract (EGb 761): pharmacological activities and clinical applications: FV DeFeudis, Elsevier, 1991. FFr250. 00 in France, FFr280. 00 elsewhere (xii+ 187 pages) ISBN 2 906077 24 0 hbk, 2 906077 21 6 pbk. Trends Pharmacol. Sci. 13, 334–335.

Stokoe, D., Engel, K., Campbell, D.G., Cohen, P., Gaestel, M., 1992. Identification of MAP-KAP kinase 2 as a major enzyme responsible for the phosphorylation of the small mammalian heat shock proteins. FEBS Lett. 313 (3), 307–313.

Sua, R., Konga, T., Zhanga, B., Songa, Q., Chenga, G., 2011. TiO2/Si core/shell nanowires array as molecule carriers. Procedia Environ. Sci. 8, 492–497.

Subjeck, J.R., Sciandra, J.J., Johnson, R.J., 1982. Heat shock proteins and thermotolerance; a comparison of induction kinetics. Br. J. Radiol. 55 (656), 579–584. https://doi.org/10.1259/0007-1285-55-656-579.

Suzuki, J., Imaizumi, S., Kayama, T., Yoshimoto, T., 1985. Chemiluminescence in hypoxic brain—the second report: cerebral protective effect of mannitol, vitamin E and glucocorticoid. Stroke 16 (4), 695–700. https://doi.org/10.1161/01.str.16.4.695.

Svennerholm, L., Gottfries, C.G., 1994. Membrane lipids, selectively diminished in Alzheimer brains, suggest synapse loss as a primary event in early-onset form (type I) and demyelination in late-onset form (type II). J. Neurochem. 62 (3), 1039–1047.

Tacke, U., Venalainen, E., 1987. Heat stress and neuroleptic drugs. J. Neurol. Neurosurg. Psychiatry 50 (7), 937–938. https://doi.org/10.1136/jnnp.50.7.937-a.

Tang, D., Kang, R., Xiao, W., Jiang, L., Liu, M., Shi, Y., Wang, K., Wang, H., Xiao, X., 2007. Nuclear heat shock protein 72 as a negative regulator of oxidative stress (hydrogen peroxide)-induced HMGB1 cytoplasmic translocation and release. J. Immunol. 178 (11), 7376–7384.

Taso, O.V., Philippou, A., Moustogiannis, A., Zevolis, E., Koutsilieris, M., 2019. Lipid peroxidation products and their role in neurodegenerative diseases. Ann. Res. Hosp. 3 (2), 1–10. https://doi.org/10.21037/arch.2018.12.02.

Tian, J., Liu, Y., Chen, K., 2017. Ginkgo biloba extract in vascular protection: molecular mechanisms and clinical applications. Curr. Vasc. Pharmacol. 15 (6), 532–548. https://doi.org/10.2174/1570161115666170713095545.

Treweek, T.M., Meehan, S., Ecroyd, H., Carver, J.A., 2015. Small heat-shock proteins: important players in regulating cellular proteostasis. Cell. Mol. Life Sci. 72 (3), 429–451. https://doi.org/10.1007/s00018-014-1754-5.

Trifunovic, A., Larsson, N.G., 2008. Mitochondrial dysfunction as a cause of ageing. J. Intern. Med. 263 (2), 167–178.

Trujillo, M.H., Bellorin-Font, E., Fragachan, C.F., Perret-Gentil, R., 2009. Multiple organ failure following near fatal exertional heat stroke. J. Intensive Care Med. 24 (1), 72–78. https://doi.org/10.1177/0885066608327122.

Uchida, K., 1999. Current status of acrolein as a lipid peroxidation product. Trends Cardiovasc. Med. 9 (5), 109–113.

Uchida, K., Kanematsu, M., Morimitsu, Y., Osawa, T., Noguchi, N., Niki, E., 1998. Acrolein is a product of lipid peroxidation reaction formation of free acrolein and its conjugate with lysine residues in oxidized low density lipoproteins. J. Biol. Chem. 273 (26), 16058–16066.

Uchinoumi, H., Yang, Y., Oda, T., Li, N., Alsina, K.M., Puglisi, J.L., Chen-Izu, Y., Cornea, R.L., Wehrens, X.H.T., Bers, D.M., 2016. CaMKII-dependent phosphorylation of RyR2 promotes targetable pathological RyR2 conformational shift. J. Mol. Cell. Cardiol. 98, 62–72. https://doi.org/10.1016/j.yjmcc.2016.06.007.

Uddin, M., Mondal, D., Morris, C., Lopez, T., Diebold, U., Gonzalez, R., 2011. An in vitro controlled release study of valproic acid encapsulated in a titania ceramic matrix. Appl. Surf. Sci. 257 (18), 7920–7927.

Uttara, B., Singh, A.V., Zamboni, P., Mahajan, R.T., 2009. Oxidative stress and neurodegenerative diseases: a review of upstream and downstream antioxidant therapeutic options. Curr. Neuropharmacol. 7 (1), 65–74. https://doi.org/10.2174/157015909787602823.

Van Molle, W., Wielockx, B., Mahieu, T., Takada, M., Taniguchi, T., Sekikawa, K., Libert, C., 2002. HSP70 protects against TNF-induced lethal inflammatory shock. Immunity 16 (5), 685–695. https://doi.org/10.1016/s1074-7613(02)00310-2.

Vellas, B., Coley, N., Ousset, P.J., Berrut, G., Dartigues, J.F., Dubois, B., Grandjean, H., Pasquier, F., Piette, F., Robert, P., Touchon, J., Garnier, P., Mathiex-Fortunet, H., Andrieu, S., GuidAge Study, G., 2012. Long-term use of standardised Ginkgo biloba extract for the prevention of Alzheimer's disease (GuidAge): a randomised placebo-controlled trial. Lancet Neurol. 11 (10), 851–859. https://doi.org/10.1016/S1474-4422(12)70206-5.

Vicario, S.J., Okabajue, R., Haltom, T., 1986. Rapid cooling in classic heatstroke: effect on mortality rates. Am. J. Emerg. Med. 4 (5), 394–398. https://doi.org/10.1016/0735-6757(86)90185-3.

Walsh, D., Li, Z., Wu, Y., Nagata, K., 1997. Heat shock and the role of the HSPs during neural plate induction in early mammalian CNS and brain development. Cell. Mol. Life Sci. 53 (2), 198–211. https://doi.org/10.1007/pl00000592.

Walter, E.J., Carraretto, M., 2016. The neurological and cognitive consequences of hyperthermia. Crit. Care 20 (1), 199. https://doi.org/10.1186/s13054-016-1376-4.

Wang, X., Wang, W., Li, L., Perry, G., Lee, H.-g., Zhu, X.D., 2014. Oxidative stress and mitochondrial dysfunction in Alzheimer's disease. Biomed. Biochim. Acta 1842 (8), 1240–1247.

Waterlow, J.C., 1947. Heat-stroke and heat-exhaustion in Iraq. Br. Med. Bull. 5 (1), 3. https://doi.org/10.1093/oxfordjournals.bmb.a073045.

Webster, W.S., Germain, M.A., Edwards, M.J., 1985. The induction of microphthalmia, encephalocele, and other head defects following hyperthermia during the gastrulation process in the rat. Teratology 31 (1), 73–82. https://doi.org/10.1002/tera.1420310109.

Weinberger, K.R., Zanobetti, A., Schwartz, J., Wellenius, G.A., 2018. Effectiveness of National Weather Service heat alerts in preventing mortality in 20 US cities. Environ. Int. 116, 30–38. https://doi.org/10.1016/j.envint.2018.03.028.

Weisenburg, T., 1912. Nervous symptoms following sunstroke. JAMA 58 (26), 2015–2017.

Westman, J., Drieu, K., Sharma, H.S., 2000. Antioxidant compounds EGB-761 and BN-520 21 attenuate heat shock protein (HSP 72 kD) response, edema and cell changes following hyperthermic brain injury. An experimental study using immunohistochemistry in the rat. Amino Acids 19 (1), 339–350. https://doi.org/10.1007/s007260070065.

White, M.G., Luca, L.E., Nonner, D., Saleh, O., Hu, B., Barrett, E.F., Barrett, J.N., 2007. Cellular mechanisms of neuronal damage from hyperthermia. Prog. Brain Res. 162, 347–371. https://doi.org/10.1016/S0079-6123(06)62017-7.

Whitman, S., Good, G., Donoghue, E.R., Benbow, N., Shou, W., Mou, S., 1997. Mortality in Chicago attributed to the July 1995 heat wave. Am. J. Public Health 87 (9), 1515–1518. https://doi.org/10.2105/ajph.87.9.1515.

Wong, H.L., Wu, X.Y., Bendayan, R., 2012. Nanotechnological advances for the delivery of CNS therapeutics. Adv. Drug Deliv. Rev. 64 (7), 686–700. https://doi.org/10.1016/j.addr.2011.10.007.

Wu, K.C., Yamauchi, Y., Hong, C.Y., Yang, Y.H., Liang, Y.H., Funatsu, T., Tsunoda, M., 2011. Biocompatible, surface functionalized mesoporous titania nanoparticles for intracellular imaging and anticancer drug delivery. Chem. Commun. (Camb.) 47 (18), 5232–5234. https://doi.org/10.1039/c1cc10659g.

Wu, F., Yu, Y., Yang, H., German, L.N., Li, Z., Chen, J., Yang, W., Huang, L., Shi, W., Wang, L., Wang, X., 2017. Simultaneous enhancement of charge separation and hole transportation in a TiO2-SrTiO3 core-shell nanowire photoelectrochemical system. Adv. Mater. 29 (28). https://doi.org/10.1002/adma.201701432, 1701432.

Wurzer, P., Culnan, D., Cancio, L.C., Kramer, G.C., 2018. Pathophysiology of burn shock and burn edema. In: Total Burn Care. Elsevier, pp. 66–76. e63.

Xia, W., Voellmy, R., Spector, N.L., 2000. Sensitization of tumor cells to Fas killing through overexpression of heat-shock transcription factor 1. J. Cell. Physiol. 183 (3), 425–431.

Xiong, Y., Zhou, H., Zhang, L., 2014. Influences of hyperthermia-induced seizures on learning, memory and phosphorylative state of CaMKIIalpha in rat hippocampus. Brain Res. 1557, 190–200. https://doi.org/10.1016/j.brainres.2014.02.026.

Yang, S.F., Wu, Q., Sun, A.S., Huang, X.N., Shi, J.S., 2001. Protective effect and mechanism of Ginkgo biloba leaf extracts for Parkinson disease induced by 1-methyl-4-phenyl-1,2,3,6-tetrahydropyridine. Acta Pharmacol. Sin. 22 (12), 1089–1093. Retrieved from https://www.ncbi.nlm.nih.gov/pubmed/11749805.

Yang, W., Xi, X., Shen, X., Liu, P., Hu, Y., Cai, K., 2014. Titania nanotubes dimensions-dependent protein adsorption and its effect on the growth of osteoblasts. J. Biomed. Mater. Res. A 102 (10), 3598–3608. https://doi.org/10.1002/jbm.a.35021.

Yang, G., Wang, Y., Sun, J., Zhang, K., Liu, J., 2016. Ginkgo biloba for mild cognitive impairment and Alzheimer's Disease: a systematic review and meta-analysis of randomized controlled trials. Curr. Top. Med. Chem. 16 (5), 520–528. https://doi.org/10.2174/1568026615666150813143520.

Yaqub, B.A., Daif, A.K., Panayiotopoulos, C.P., 1987. Pancerebellar syndrome in heat stroke: clinical course and CT scan findings. Neuroradiology 29 (3), 294–296. https://doi.org/10.1007/BF00451771.

Yenari, M.A., Liu, J., Zheng, Z., Vexler, Z.S., Lee, J.E., Giffard, R.G., 2005. Antiapoptotic and anti-inflammatory mechanisms of heat-shock protein protection. Ann. N. Y. Acad. Sci. 1053 (1), 74–83. https://doi.org/10.1196/annals.1344.007.

Yeo, T.P., 2004. Heat stroke: a comprehensive review. AACN Clin. Issues 15 (2), 280–293. https://doi.org/10.1097/00044067-200404000-00013.

Yin, Z.F., Wu, L., Yang, H.G., Su, Y.H., 2013. Recent progress in biomedical applications of titanium dioxide. Phys. Chem. Chem. Phys. 15 (14), 4844–4858. https://doi.org/10.1039/c3cp43938k.

Yoshida, R., Suzuki, Y., Yoshikawa, S., 2005. Syntheses of TiO2 (B) nanowires and TiO2 anatase nanowires by hydrothermal and post-heat treatments. J. Solid State Chem. 178 (7), 2179–2185.

Yu, Y., Shen, X., Luo, Z., Hu, Y., Li, M., Ma, P., Ran, Q., Dai, L., He, Y., Cai, K., 2018. Osteogenesis potential of different titania nanotubes in oxidative stress microenvironment. Biomaterials 167, 44–57. https://doi.org/10.1016/j.biomaterials.2018.03.024.

Zhang, Q., Wang, X., Li, P.Z., Nguyen, K.T., Wang, X.J., Luo, Z., Zhang, H., Tan, N.S., Zhao, Y., 2014. Biocompatible, uniform, and redispersible mesoporous silica nanoparticles for cancer-targeted drug delivery in vivo. Adv. Funct. Mater. 24 (17), 2450–2461.

Zhao, Y., Xiong, S., Liu, P., Liu, W., Wang, Q., Liu, Y., Tan, H., Chen, X., Shi, X., Wang, Q., Chen, T., 2020. Polymeric nanoparticles-based brain delivery with improved therapeutic efficacy of ginkgolide B in Parkinson's disease. Int. J. Nanomedicine 15, 10453–10467. https://doi.org/10.2147/IJN.S272831.

Zheng, Z., Kim, J.Y., Ma, H., Lee, J.E., Yenari, M.A., 2008. Anti-inflammatory effects of the 70 kDa heat shock protein in experimental stroke. J. Cereb. Blood Flow Metab. 28 (1), 53–63. https://doi.org/10.1038/sj.jcbfm.9600502.

Zhou, L.J., Song, W., Zhu, X.Z., Chen, Z.L., Yin, M.L., Cheng, X.F., 2000. Protective effects of bilobalide on amyloid beta-peptide 25-35-induced PC12 cell cytotoxicity. Acta Pharmacol. Sin. 21 (1), 75–79. Retrieved from https://www.ncbi.nlm.nih.gov/pubmed/11263252.

Zhou, X., Qi, Y., Chen, T., 2017. Long-term pre-treatment of antioxidant Ginkgo biloba extract EGb-761 attenuates cerebral-ischemia-induced neuronal damage in aged mice. Biomed. Pharmacother. 85, 256–263. https://doi.org/10.1016/j.biopha.2016.11.013.

Zhu, L., Gao, J., Wang, Y., Zhao, X.N., Zhang, Z.X., 1997. Neuron degeneration induced by verapamil and attenuated by EGb761. J. Basic Clin. Physiol. Pharmacol. 8 (4), 301–314. https://doi.org/10.1515/jbcpp.1997.8.4.301.

Zhu, P.C., Tong, Q., Zhuang, Z., Wang, Z.H., Deng, L.H., Zheng, G.Q., Wang, Y., 2019. Ginkgolide B for myocardial ischemia/reperfusion injury: a preclinical systematic review and meta-analysis. Front. Physiol. 10, 1292. https://doi.org/10.3389/fphys.2019.01292.

Upregulation of hemeoxygenase enzymes HO-1 and HO-2 following ischemia-reperfusion injury in connection with experimental cardiac arrest and cardiopulmonary resuscitation: Neuroprotective effects of methylene blue

Lars Wiklund[a],*, Aruna Sharma[a],*, Ranjana Patnaik[b], Dafin F. Muresanu[c,d],
Seaab Sahib[e], Z. Ryan Tian[e], Rudy J. Castellani[f], Ala Nozari[g],
José Vicente Lafuente[h], and Hari Shanker Sharma[a],*

[a]*International Experimental Central Nervous System Injury & Repair (IECNSIR), Department of Surgical Sciences, Anesthesiology & Intensive Care Medicine, Uppsala University Hospital, Uppsala University, Uppsala, Sweden*
[b]*Department of Biomaterials, School of Biomedical Engineering, Indian Institute of Technology, Banaras Hindu University, Varanasi, India*
[c]*Department of Clinical Neurosciences, University of Medicine & Pharmacy, Cluj-Napoca, Romania*
[d]*"RoNeuro" Institute for Neurological Research and Diagnostic, Cluj-Napoca, Romania*
[e]*Department of Chemistry & Biochemistry, University of Arkansas, Fayetteville, AR, United States*
[f]*Department of Pathology, University of Maryland, Baltimore, MD, United States*
[g]*Anesthesiology & Intensive Care, Massachusetts General Hospital, Boston, MA, United States*

Progress in Brain Research, Volume 265, ISSN 0079-6123, https://doi.org/10.1016/bs.pbr.2021.06.009

[h]*LaNCE, Department of Neuroscience, University of the Basque Country (UPV/EHU),*
Leioa, Bizkaia, Spain
*Corresponding authors: Tel.: +46-70-2195963; Fax: +46-18-243899, e-mail address: lars.
wiklund@surgsci.uu.se; aruna.sharma@surgsci.uu.se; sharma@surgsci.uu.se;
harishanker_sharma55@icloud.com; hssharma@aol.com

Abstract

Oxidative stress plays an important role in neuronal injuries after cardiac arrest. Increased production of carbon monoxide (CO) by the enzyme hemeoxygenase (HO) in the brain is induced by the oxidative stress. HO is present in the CNS in two isoforms, namely the inducible HO-1 and the constitutive HO-2. Elevated levels of serum HO-1 occurs in cardiac arrest patients and upregulation of HO-1 in cardiac arrest is seen in the neurons. However, the role of HO-2 in cardiac arrest is not well known. In this review involvement of HO-1 and HO-2 enzymes in the porcine brain following cardiac arrest and resuscitation is discussed based on our own observations. In addition, neuroprotective role of methylene blue- an antioxidant dye on alterations in HO under in cardiac arrest is also presented. The biochemical findings of HO-1 and HO-2 enzymes using ELISA were further confirmed by immunocytochemical approach to localize selective regional alterations in cardiac arrest. Our observations are the first to show that cardiac arrest followed by successful cardiopulmonary resuscitation results in significant alteration in cerebral concentrations of HO-1 and HO-2 levels indicating a prominent role of CO in brain pathology and methylene blue during CPR followed by induced hypothermia leading to superior neuroprotection after return of spontaneous circulation (ROSC), not reported earlier.

Keywords

Cardiac arrest, Reestablishment of spontaneous circulation, Global ischemia, Hemeoxygenase enzyme, Carbon monoxide, Methylene blue, Oxidative stress

1 Introduction

Cardiac arrest (CA) and its consequences on brain diseases are devastating (Hawkes and Rabinstein, 2019; Lopez Soto et al., 2020; Rossetti et al., 2016). More than 356k cases of CA are recorded annually in the United States out of which the survival rates are only between 10% and 11% (Andersen et al., 2019; Madder and Reynolds, 2018; Panchal et al., 2019). The survivors often show serious neurological disorders and lifetime disability (Roberts et al., 2018; Torbey et al., 2004). This is largely due to neuronal damages following CA induced global ischemia (Chambers and Fallouh, 2010; Frisch et al., 2017; Reis et al., 2017; Weil et al., 1988). Human postmortem studies in CA showed severe neuronal damage in the hippocampus, thalamus, cerebellum, cerebral cortex and brain stem (Bjorklund et al., 2014; Cao et al., 2020; Endisch et al., 2020). The magnitude and intensities of neuronal damages are largely responsible for disability and death in CA victims (Binks and Nolan, 2010; Haglund et al., 2019; Keijzer et al., 2018; Uray et al., 2018). Thus, there is an urgent need to expand our knowledge regarding brain pathology in CA. It appears that CA induced secondary injury cascade of events lead to neuronal damage (Elmer and Callaway, 2017; Jakkula et al., 2018; Sekhon et al., 2017). These include breakdown of the blood-brain barrier (BBB) (Kim et al., 2020; Park et al., 2019; Semenas et al., 2014), brain edema (Esdaille et al., 2020; Hayman et al., 2018; Rud et al., 2020), generation of oxidative stress (Ottolenghi et al., 2020; Tamura et al., 2020; Wang et al., 2019), formation of free radicals (Dey et al., 2018; Li et al., 2017; Zhang et al., 2018), production of nitric oxide (Kajimoto et al., 2020; Miyazaki and Ichinose, 2020; Wu et al., 2020), and carbon monoxide (Wollborn et al., 2019, 2020; Wu et al., 2019) that are deleterious to neuronal health.

Previous studies from our laboratory show that CA followed by return of spontaneous circulation (ROSC) induced BBB breakdown to Evans blue as early as within 5 min and continued for 180 min observation period (Sharma et al., 2011, 2015). Severe neuronal damages occur after 180 min of ROSC affecting serous cellular damages in the cerebral cortex, cerebellum, hippocampus, thalamus, and hypothalamus in a porcine model. The pathophysiology of brain injury in CA followed by ROSC includes upregulation of nitric oxide synthase (NOS) and production of oxidative stress leading to cellular injuries. These changes were attenuated by methylene blue alone or in combination with hypothermia (Miclescu et al., 2010; Sharma et al., 2015; Wiklund et al., 2007, 2013, 2018).

Since hemeoxygenase (HO) enzyme is responsible for synthesizing carbon monoxide (CO) (Sharma et al., 1998e, 2000b,c, 2003, 2020a; Sharma and Westman, 2004), it would be interesting to explore their involvement in pathophysiology of CA. In this chapter we explore the role of HO in CA induced brain pathology based on our own investigation. In addition, we used ethylene blue (MB) to explore its neuroprotective role on HO in a porcine model of CA. The functional significance of our findings in relation to clinical settings is discussed.

2 Cerebral ischemia and reperfusion

Several reports focus on the importance of the immediate treatment of acute cerebral artery occlusion because any delay in therapeutic protocol result in the development of cerebral ischemia after reperfusion (Costantini and Lenti, 2002; Little, 1978; Liu et al., 2001; Schaller and Graf, 2004). Cardiac arrest is one of the main leading factors in cerebral injuries (Wiklund et al., 2018). Following cardiac arrest, the most serious mechanism that could be initially activated is the generation of free radicals such as reactive oxygen species (ROS) and reactive nitrogen species (RNS) leading to cell damage.

Based on previous animal studies, it has been reported that cerebral damage following ischemia and reperfusion might occur within less than 15 min of cardiac arrest and 8 min after cardiopulmonary resuscitation (Miclescu et al., 2010; Mortberg et al., 2009; Sharma et al., 2011). Understanding the mechanisms and biochemical alterations following the initial insult is crucial in designing effective therapeutic strategy for the treatment of brain pathology (Schaller and Graf, 2004). Accordingly, therapeutic strategies including short term therapy for functional and neuronal improvement, long term therapy for neuroprotection, and extended-term therapy for prevention of oxidative stress and apoptosis are needed to reduce mortality following the injury (Schaller and Graf, 2004). Adenosine triphosphate (ATP) is a major determining factor for the survival of the living cells (Rosenthal et al., 1987). Under aerobic metabolism, mitochondrial oxidative phosphorylation is responsible for about 80% of the ATP synthesized in the brain (Fiskum, 1983, 1985). However, during cerebral ischemia or hypoxia, the amount of ATP synthesized in the brain will be depleted due to the disruption of the mitochondrial metabolism. As a result or diminution of O_2 supply the ATP storage will be consumed within a few minutes (Sato et al., 1984).

Several events and injury cascades are initiated following cerebral ischemia and could be responsible for neuronal cell injuries, apoptosis, necrosis, and/or cell death. Hypoxia results in increase in the intracellular Ca^{2+} levels (Dirnagl et al., 1999; Farber, 1982; Mergenthaler et al., 2004). Diminishing ATP supply results in escalating the cytosolic concentration of calcium ion level from about $0.1\,\mu M$ to about $1\,mM$. This excessive elevation in the intracellular Ca^{2+} level could be a key factor in stimulating degenerative enzymes such as phospholipases (Fiskum, 1985), inducing membrane lipid peroxidation (McCord, 1985), prompting vascular contraction (Borgers et al., 1983), diminishing regular mitochondrial metabolism (Mukherjee et al., 1979), and initiating necrosis (Farber, 1982) that collectively leads to apoptosis (Annunziato et al., 2003; Dirnagl et al., 1999; Rizzuto et al., 2003; Yu et al., 2001).

Animal studies reveal that the degree of cerebral damage is markedly influenced by the hypoxic event in a time-dependent manner. For instance, continuous hypoxia in neuronal tissue for half an hour leads to lower the mitochondrial respiration rate to about 50% in the isolated neuronal cells (Hamud and Fiskum, 1985; Mela, 1979; Rehncrona et al., 1979; Siesjo, 1981). Broughton and his collages demonstrate that the incidence of neuronal tissue necrosis could be observed immediately following cerebral ischemia that is associated with the decrease in blood supply (Broughton et al., 2009). The necrotic area becomes neurologically inactive and the surrounding tissues also forms a temporary nonfunctional region, the ischemic penumbra, that might respond efficiently to appropriate treatment (Broughton et al., 2009). The initiation of apoptosis in ischemic penumbra tissue is also seen within a few hours following the manifestation of the primary necrotic core (Broughton et al., 2009; Nakase et al., 2004; Rami and Kogel, 2008). Therefore, attention should be paid towards finding a suitable therapeutic strategy to recover ischemic penumbra tissue as it appears to about the half of volume of the entire injured area (Broughton et al., 2009).

Another factor that leading to programmed cell death in neuronal tissue following ischemic stroke is the release of pro-apoptotic factors such as BID, a Bcl-2 family protein (Li et al., 1998). After the ischemic event, Calpains mediated signal induces the activation of the BID pro-apoptotic factor to its active form tBID, the neuronal death ligand a key agent of neuronal apoptosis and cell death (Culmsee et al., 2005;

Love, 2003; Sugawara et al., 2004). Since calpains are Ca^{2+} dependent proteases, an increase in the intracellular Ca^{2+} level following ischemic injury result in the activation of calpains (Norberg et al., 2008). A previous report found that diminishing the action of BID via enzymatic degradation in rodent results in reduction in the ischemic volume lesion (Plesnila et al., 2002). Once BID activated to tBID, the later one mediates the activation of other pro-apoptotic factors such as Bak, Bax, Bad, and Bcl-XS that could further facilitate the programmed cell death via different mechanisms including the liberation of cytochrome c into the cytosol (Plesnila et al., 2002). Numerous factors thus contribute to the deleterious secondary damage of the neuronal tissue following the primary insult. Other factors inducing cell death include neurotransmitters such as glutamate, oxidative stress, membrane phospholipids disruption, and lipid or protein peroxidation products (Anderson et al., 1985; Balentine, 1985; Hall and Braughler, 1986; Hall et al., 1992; Young, 1988).

3 Mechanisms of neuronal cell death following ischemia/reperfusion

3.1 Caspase dependent cell death

Activation of caspase dependent pathway is another cascade that plays a major role in neuronal cell death triggered following ischemic brain stroke (Chan, 2004; Oshitari et al., 2008; Yakovlev and Faden, 2001). It has been reported that stimulating the release of some pro-apoptotic factors such as cytochrome c, Smac/DIABLO, and the serine protease HtrA2/Omi into the cytosol following neuronal tissue hypoxia activates a caspase-dependent mitochondrial cascade that is greatly attributed to the active form of caspase-9 and is responsible for neuronal cell death (Elmore, 2007; Guerrero et al., 2008; Logue and Martin, 2008). Accumulating evidences suggest that one of the major markers following cerebral ischemia is the remarkable elevation in cytochrome c level (Babu et al., 2000; Fujimura et al., 1998, 1999; Pérez-Pinzón et al., 1999; Sugawara et al., 1999). Cao and his coworkers report that Apaf-1/caspase-9 complex assembly significantly appears in the hippocampus within one day following global ischemia (Cao et al., 2004). This finding is further supported by subjecting ischemic rat neurons to Apaf-1 inhibitor resulting in neuronal survival and cellular protection (Cao et al., 2004; Cho et al., 2004). Furthermore, caspase-3 also mediates other developmental events during cerebral ischemia. Experimental evidence shows that during neuronal tissue ischemia, increased expression of caspase-3 is another indicator of cell termination (Dodel et al., 1999; Hartmann et al., 2000; Kim et al., 2005). Caspase-3 mRNA remarkably elevates after 60 min of cerebral ischemia, and similar increase is seen 120 min after middle cerebral artery occlusion in rodents (Asahi et al., 1997; Namura et al., 1998). The genetic redistribution of some caspases also occurs in humans following CNS ischemic injury (Rami et al., 2003). Thus, using caspase inhibitors including genetic expression suppressors would be an effective strategy to ameliorate neuronal injury and induce neuroprotection (Endres et al., 1998; Le et al., 2002; Ma et al., 1998; Mouw et al., 2002).

3.2 Caspase independent cell death

Activation of the caspase-independent pathway is another key feature of the ischemic brain pathology. Unlike caspase-dependent apoptosis, caspase-independent cell death does not require energy to take place since the latter one is physiologically designed to terminate the cells at the last stage of the energy depletion (Cho and Toledo-Pereyra, 2008). This pathway seems more approachable when cells under anoxia and a shortage of ATP production (Ferrer et al., 2003). Stressed mitochondria produces several apoptotic factors such as apoptosis-inducing factor (AIF), endonuclease G and Bcl-2 interacting protein (BNIP3) that plays major role in caspase-independent apoptosis (Cho and Toledo-Pereyra, 2008; Elmore, 2007). Several studies focus on the action of the AIF following ischemia-induced pathologies on hypoxia and energy exhaustion (Daugas et al., 2000). Tissue ischemia signals the translocation of AIF to the nucleus and consequently initiates a programmed intracellular pathway that makes them different from caspase-dependent apoptosis (Cho and Toledo-Pereyra, 2008). As an apoptotic factor, AIF induces destruction of the DNA (karyorrhexis) under the ischemic condition that in turn leads to condensation of nuclear chromatin (Pyknosis) in the periphery. Inhibiting the translocation of AIF via the deactivation of PARP-1 remarkably improves neuronal survival following ischemic stroke (Hong et al., 2004). Endonuclease G works in parallel with AIF during the caspase-independent apoptosis cell death in cerebral ischemia. Significant increase in the Endonuclease G level within less than one-day form the onset of ischemic stroke is in line with this idea. Also, translocation of both AIF and endonuclease G from the mitochondria to the nucleus occurs at the same time following the neuronal tissue hypoxia (Lee et al., 2005). However, the relationship between the AIF and endonuclease G in not fully understood yet. Finally, BNIP3 is a cytochrome c inducer that triggers cell death in a caspase-independent manner during specific abnormal conditions (Guo et al., 2001; Quinsay et al., 2010). The lack of oxygen supply and insufficient ATP production under some pathological condition, such as in hypoxia, stimulates the activation of hypoxia-inducible factor-1 (HIF-1) that upregulates the production of BNIP3 and subsequently triggers apoptosis (Althaus et al., 2006; Chen et al., 1997; Greijer and van der Wall, 2004; Sowter et al., 2001). However, the mechanisms underlying the BNIP3-induced apoptosis are unrelated to any of the death factors such as Apaf-1, cytochrome c, caspase introduction, and apoptosis-inducing factor AIF (Guo et al., 2001; Hamacher-Brady et al., 2007; Ray et al., 2000; Velde et al., 2000; Wan et al., 2003; Zhang and Ney, 2009).

3.3 Free radical production following cerebral ischemia

Several studies show pronounced impact of free radical overproduction and oxidative stress system during post cerebral ischemia and reperfusion (Higgins et al., 2010; Poh Loh et al., 2006; Schmidley, 1990). Although superoxide (O_2^-) and hydroxyl ($OH^{\cdot}$) are essential for some metabolic pathways signaling, the amount of the produced radicals becomes intolerable by the cellular antioxidant system under hypoxic conditions (Poh Loh et al., 2006). Moreover, superoxide (O_2^-) radical undergoes protonation state to form (OH_2^-) that is more potent from the superoxide radicals and contribute to severe damage to the cell components (Schmidley, 1990). Hydrogen peroxide induces considerable cell damage after it reacts with superoxide radicals in the presence of transition metal catalysts to produce hydroxyl ($OH^{\cdot}$) radicals (Morano et al., 2012; Sato et al., 1992; Sies, 2014). Apart from that, the mitochondrial electron transport chain (ETC) (Hoek et al., 1997; Shin et al., 2008), arachidonic acids (Cocco et al., 1999; Rao et al., 1983), xanthine oxidase (George and Struthers, 2009; Shin et al., 2008), and NADPH oxidases (Sagi and Fluhr, 2006) further contributes to the production of the free radicals.

The deleterious impact of the free radicals on the neuronal tissue following the ischemic stroke has been demonstrated in several reports (Adibhatla and Hatcher, 2006; Clemens, 2000; Miclescu et al., 2010; Niizuma et al., 2009; Olmez and Ozyurt, 2012; Pandey et al., 2012; Sharma et al., 2007). Excessive production of ROS and RNS provokes the neuronal damage together with antioxidants insufficiency and the weakness of the endogenous oxidative stress defense system further potentiate the effect of the free radical damages during the hypoxic event (Poh Loh et al., 2006). Free radicals damage the cellular system either directly through involvement in the demolition of proteins, fat, nucleic acid, or indirectly via interrupting important cellular pathways and signals that in turn disrupt cellular hemostasis (Chen et al., 2012; Horton, 2003; Wolff et al., 1986). The produced free radicals following tissue hypoxia could further cause more damage via the interaction with nitric oxide to form a very virulent molecule, peroxynitrite, that consequently induces programmed cell death (Dirnagl et al., 1999; Sisein, 2014). Moreover, free radical generated following cerebral ischemia not only be stimulated by the hypoxic state but also from other ROS and RNS inducing factors and disturbance in calcium ion hemostasis and glutamate overproduction (Mergenthaler et al., 2004). Mitochondrial oxidative stress and free radical production are at the maximum levels following cerebral reperfusion which in turn induce cell death via different mechanisms such as cytochrome c liberation and translocation of apoptotic factors to the neuronal cell cytoplasm (Kirkland et al., 2002; Sugawara et al., 2002). The ROS induced damage is strongly related to the anoxic pathology. The high lipid content of the neuronal tissue makes it very vulnerable to the attack of free radicals (Butterfield and Lauderback, 2002). Although free radicals cause damage to different compartments in the cell such as peptides, DNA, and RNS, the most affected molecules are the polyunsaturated fatty acids and the sulfur-containing amino acids (Chan and Fishman, 1980; Rokyta et al., 1996; Schmidley, 1990). Thus, damage to lipids and related molecules are sensible marker for oxidative stress following ischemic stroke (Schmidley, 1990). Perturbing the membrane phospholipids by the free radical attack results in rapid generation of free fatty acids that contributes to serious damage to the whole biological system (Hall and Braughler, 1986). Also, the generation of arachidonic acid metabolites during the oxidative stress could further aggravate due to generation of superoxide anion radicals, hydroxyl radicals, and hydrogen peroxide (Hall and Braughler, 1986; Kontos et al., 1984, 1985). Free radical attack on the cell membrane also disturbs the function and transport of proteins that in turn alters the ion hemostasis of the surface triggering osmotic lysis (Dean, 1987; Tesoriere et al., 1999).

Endogenous antioxidants play major role in eliminating the excessive amount of free radicals and their effects on the biological systems (Rizzo et al., 2010; Yuan and Kitts, 1997). Numerous molecules including enzymatic or non-enzymatic metabolites such as superoxide dismutase, catalase, glutathione peroxidase, uric acid, bilirubin, albumin, and metallothioneins are potent endogenous antioxidants (Jadon et al., 2017; Rizzo et al., 2010).

4 Hemeoxygenase enzymes HO-1 and HO-2 following ischemia-reperfusion injury

Hemeoxygenase (HO) is an enzyme responsible for catabolism of heme. It is also a member of the heat shock protein (HSP) known as (HSP32). When hemeoxygenase involved in the heme catabolism, the final products of the reaction are biliverdin, iron, and carbon monoxide (CO) (Kikuchi et al., 2005; Ryter et al., 2006). The endoplasmic reticulum, mitochondria, cell nucleus, and plasma membrane are the main location of the hemeoxygenase within the cell (Hopper et al., 2018). Whereas, the extracellular sources of hemeoxygenase are the hemoglobin and myoglobin (Hill-Kapturczak et al., 2002). Stress and injuries are the main factors inducing the activation of hemeoxygenase (Hill-Kapturczak et al., 2002; Nath et al., 1992). Injuries to the CNS are associated with the induction and release of several neurotransmitters, enzymes, biomolecules, and ions that are involved in the secondary cell injury pathways (Winkler et al., 1998). Also, free radical generation results into activation of hemeoxygenase enzymes. The HO activation upregulates several cellular pathways leading to oxidative stress injuries, protein oxidation, and lipid peroxidation in brain and spinal cord tissues (Schwab and Bartholdi, 1996; Sharma et al., 1998a,d; Tator and Fehlings, 1991). Hemeoxygenase contributes to the production of carbon monoxide CO gas, a free radical that further signals free radical generation and release (Piantadosi, 2008).

Administration of hemeoxygenase inhibitors following CNS injury caused by trauma or heat significantly reduced the development of brain swelling and cell injuries (Ewing and Maines, 1992). The elevated level of carbon monoxide in neuronal tissue modulates synapses and alters signal transductions (Abraham et al., 1996; Applegate et al., 1991). Application of a direct fluid pressure pulse on to the intact dura of rat brain leads to upregulate heme oxygenase-1 (HO-1) production in glial cells (Fukuda et al., 1996). Increase in heme oxygenase-2 (HO-2) activity in rat brains occur after 4 h following hyperthermia (Sharma et al., 1998b). These findings indicate that hemeoxygenase upregulation is associated with the development of neuronal pathology and increased CO production (Verma et al., 1993). A focal spinal cord injury in the rat model could also upregulate hemeoxygenase-2 around the lesion area (Sharma et al., 2000b). Topical application of brain-derived neurotrophic factor (BDNF) or insulin-like growth factor-1 (IGF-1) over the traumatized cord area induced marked reduction in HO-2 levels, spinal cord edema cord pathology (Sharma et al., 2000b). This indicates that upregulation of HO-2 is harmful for brain and spinal cord.

It seems that several factors such as free radicals, high body temperature or transition metals can trigger the activation and production of hemeoxygenase enzyme. A previous report demonstrates in heart tissue following the resuscitation of a preconditioned cardiac arrest in that hypoxia does not induce hemeoxygenase, but reperfusion does (Maulik et al., 1996). Thus increase in the hemeoxygenase activity is directly proportional to the reperfusion term length (Maulik et al., 1996). Interestingly, the administration of antioxidants before the reperfusion could substantially lower the hemeoxygenase induction in the heart muscle (Maulik et al., 1996). Likewise, injection of ROS in heart tissue reveals a remarkable elevation in hemeoxygenase activity that is suppressed by superoxide dismutase (SOD) and catalase administration (Maulik et al., 1996). Studies in myocardium following reperfusion showed a significant rise in hemeoxygenase-1 expression in the cardiomyocyte-rich perivascular tissue (Maulik et al., 1996; Yet et al., 1999). These findings suggest that reperfusion-included oxidative stress upregulates the production and the release of the hemeoxygenase in the heart muscle (Maulik et al., 1996). Another report demonstrates that elevation in HO-1 can be significantly increased in response to an ischemic pathology in perivascular cells of the heart and enhanced free radical scavenging pathways (Sharma et al., 1999).

4.1 Neuronal dysfunction following cardiac arrest and cardiopulmonary resuscitation

Several reports show that severity of cardiac arrest induced development of cerebral ischemia and reperfusion results in subsequent neuronal pathologies. Clinical observations suggest several CNS dysfunction including cognitive malfunction, tiredness, memory weakness, deterioration in thermoregulatory mechanisms, emotional disturbance, and neuroinflammation following cardiac arrest resuscitation (Fichet et al., 2010; Middelkamp et al., 2007; Moulaert et al., 2009, 2010; Teschendorf et al., 2008). There are quite similarities in brain pathologies following cardiac arrest and CNS insults following trauma or heat injury (Cantor et al., 2008; Hochstenbach et al., 2001; van de Port et al., 2007; Visser-Meily et al., 2009).

Most of the reports focus on the role of free radicals causing neuronal tissue injury following hypoxia in CA and ROSC. Ischemia-induced cardiopulmonary resuscitation (CPR) triggers a cascade of events in CNS leading to neuronal damages after 12 min of cardiac arrest (CA) and 8 min after cardiopulmonary resuscitation (Miclescu et al., 2010; Mortberg et al., 2009; Sharma et al., 2011). Increased lipid peroxidation markers such as 8-isoPGF2α and 15-keto-PGF2α as well as fatty acid metabolites seen within hours from the return of the spontaneous circulation (ROSC) (Basu et al., 2000, 2003; Liu et al., 2003; Miclescu et al., 2006). Such markers indicate oxidative stress is responsible for neuronal injury following CA and CPR (Idris et al., 2005). Study in canine models of cardiac arrest show lowering of mitochondrial respiration state to half within an hour following complete brain hypoxia (White et al., 1985). However, following 4 min of cardiac arrest, resuscitation helps the cardiac muscle to provide minimum amount of oxygenated blood to the brain and other vital organs within the first half hour and minimize neuronal damage (Ohomoto et al., 1976; Ralston et al., 1982; Voorhees et al., 1984; Voorhees et al., 1983; Walker et al., 1984; White et al., 1983).

Energy depletion occurs immediately in the brain and neuronal tissue duo to the circulatory arrest results in disruption of membrane ion pumps following cardiac arrest (Geocadin et al., 2008). This disruption in membrane activity causes further cellular damage via release of excitatory neurotransmitters such as glutamate. Increasing the cellular glutamate release causes glutamate-mediated excitotoxicity (Vaagenes et al., 1996). Several neurological pathologies such as Alzheimer's disease, Parkinson's disease, ischemia, and epilepsy causes hyperactivation of N-methyl-D-aspartate (NMDA) receptors due to the elevated glutamate level (Biegon et al., 2004; Gascon et al., 2008; Le Greves et al., 1997; Lin et al., 2012b; Mehta et al., 2013). However, the state of glutamate elevation is usually associated with a remarkable decrease in some inhibitory neurotransmitters such as glycine, serotonin, and gamma-aminobutyric acid (GABA) (Globus et al., 1991). Overstimulation of NMDA initiates a cascade of abnormal cellular events including disruption in the calcium ion hemostasis that triggers the generation of free radicals and apoptosis (de Baaij et al., 2015). Studies demonstrate that reperfusion potentiate the oxidative stress in neuronal tissue since it provides oxygen to some active enzymes which convert this oxygen to free radicals via specific enzymatic oxidation (Chan, 1996).

Blood-brain barrier (BBB) destruction is another common pathological feature of cardiac arrest and reperfusion. Disruption of BBB integrity following cardiac arrest induced ischemia results in abnormal deposition of serum proteins in brain tissue (Liu et al., 2010; Muresanu et al., 2010; Sharma and Sharma, 2010; Shlosberg et al., 2010). The influx of serum protein into the brain through the disrupted BBB induces a series of immunological and biological reactions causing brain swelling (Sharma and Ali, 2008; Sharma and Kiyatkin, 2009; Sharma et al., 2009). Accumulation of serum proteins during brain edema initiates pressure on neuronal tissue that subsequently disrupts neuronal network, synapses, cellular integrity and diminishes neuronal conductivity (Castejón, 2009; Castejon et al., 1995, 2002; Go, 1980; Reed et al., 1964; Sharma et al., 1998e, 2009). The most affected brain compartments following brain edematous swelling are the axonal and glial cells besides neurons affecting brain functions (Sharma and Kiyatkin, 2009; Wagner et al., 2005; Zador et al., 2009). Memory distortion following cardiac arrest appears to be a result of the neuronal injuries of the hippocampus due to the CA induced hypoxia (Wiltgen et al., 2010). As circulatory arrest causing hypothermia, several reports show that brains in hypothermic animals also reveal neuronal injury and dysfunction after ROSC (Alpers, 1940; Kiyatkin and Sharma, 2009; Wiklund et al., 2012). Neuronal injuries in CA are most prominent in the cerebral cortex and in the basal ganglia and thalamocortical centers (Alpers, 1940; Jones, 2009; Xiao et al., 2009). Several findings from cardiopulmonary resuscitation show that applying therapeutic hypothermia after CPR improves neurological functions and survival. Maintaining body temperature at normal or below normal values protects heat-sensitive metabolic enzymes from further damage (Geocadin et al., 2017). Our research suggests that immediately lowering of brain temperature following cardiac arrest or cardiopulmonary resuscitation using intravenous fluid results in a significant reduction in the CNS pathology and oxidative damage (Wiklund et al., 2018).

4.2 Neuroprotective effects of methylene blue

Methylene blue (MB) or methylthioninium chloride is a dye and drug that was first discovered in 1876 by Heinrich Caro and used to stain fabrics (Cooksey, 2017). This drug is approved by FDA to use for the treatment of methemoglobinemia, and as an effective suppressor for Tau protein aggregation (Hosokawa et al., 2012). Recently, methylene blue (MB) has been employed intensively for the treatment of various diseases such as malaria (Meissner et al., 2006), septic shock (Harvel et al., 2015), carbon monoxide poisoning (Haggard and Greenberg, 1933), cyanide poisoning (Chen et al., 1933), ifosfamide neurotoxicity (Alici-Evcimen and Breitbart, 2007), and Alzheimer's disease (Oz et al., 2009). Therapeutic activity of the methylene blue is due to activation of a unique type of reversible reduction-oxidation reaction by its auto-oxidizing capacity leading to maintain a productive mitochondrial electron transport pathway (Gonzalez-Lima and Auchter, 2015; Huang et al., 2013; Rojas et al., 2012). MB enhances the mitochondrial electron transport chain reaction via the direct transferring of electrons from nicotinamide adenine dinucleotide (NADH) to cytochrome c (Clifton and Leikin, 2003). Using MB for the treatment of neurological disorders improves cognitive behaviors and enhances memory function. This is because of its ability to maintain a constant ATP level and regulate oxygen utilization in tissues (Riha et al., 2005; Scott and Hunter, 1966; Zhang et al., 2006). Ameliorating the mitochondrial respiration state pathway is advantageous in reducing free radical accumulation and regulation of oxidative stress-induced pathological reaction (Atamna et al., 2012). Several studies used methylene blue to combat free radicals-related neurological dysfunctions including Alzheimer's disease, dementia, and neurotoxicity (Oz et al., 2009; Stack et al., 2014; Tsai and Boxer, 2014). Memory enhancement and neuroprotective effects of the MB could be due to its antioxidative capability (Heydrick et al., 2007; Ohlow and Moosmann, 2011; Rojas et al., 2012; Zhang et al., 2006). Administration of methylene blue significantly enhances the skill learning techniques in animal models.

Several reports show that methylene blue is able to minimize the symptoms of the cardia arrest and associated brain pathology (Deutsch et al., 1997; Kelner et al., 1988; Naylor et al., 1986; Pelgrims et al., 2000; Sharma et al., 2011; Wainwright and Crossley, 2002). The mechanism underlying the neuroprotective effect of MB appears to be related to its electron cycling property and reduction-oxidation capacity. MB also enhances the mitochondrial electron transport chain pathways (Bruchey and Gonzalez-Lima, 2008). Facilitating the release of O_2 from the methemoglobin in the case of methemoglobinemia provide another explanation for the positive effect of MB in neuroprotection following injuries-induced hypoxia (Wendel, 1939; Wright et al., 1999). MB also ameliorates the neuronal tissue metabolism, blood supply, potentiates synaptic functions and neuronal conduction through regulating glucose and oxygen utilization in ischemic neuronal tissues (Huang et al., 2013; Lin et al., 2012a). In rats preconditioned with mild TBI, treatment with MB induced a significant reduction in the lesion size, improved the cognitive behaviors, and attenuated in the neuronal damages (Talley Watts et al., 2014). Using MB alone or in combination with other therapies for the treatment of neurodegenerative disease such as Parkinson's disease or Alzheimer's disease resulted in significant improvements in behavior and attenuation in brain pathology (Medina et al., 2011; Rojas et al., 2012; Shen et al., 2013; Wen et al., 2011; Yoo et al., 2013).

Our studies show that administration of MB in CA significantly reduces the BBB leakage of albumin, tissue nitrite/nitrate content, and brain edema (Miclescu et al., 2010). Histochemical investigation shows marked neuroprotection and axonal damages in CA arrest in MB treated animals (Miclescu et al., 2010). Ischemia and reperfusion causes a deleterious effect on BBB integrity due to the generation of free radicals and nitric oxide (NO) during the hypoxic event (Kaur and Ling, 2008; Kubes and Granger, 1992). Schleien and his colleagues indicate that inhibiting nitric oxide using a nonspecific NO synthase inhibitor, L-NAME attenuates BBB disruption following cerebral ischemia associated with cardiac arrest and cardiopulmonary resuscitation (Schleien et al., 1998). Interestingly, MB injection causes a significant reduction in the cortical NOS level, lesion volume and induced neuroprotection (Schleien et al., 1998). These findings support the concept that MB achieves its neuroprotective effects through direct or indirect free radicals scavenging mechanisms (Mayer et al., 1993; Miclescu et al., 2006, 2007, 2010; Salaris et al., 1991; Wiklund et al., 2018).

Another mechanism for the neuroprotective effect of MB is its capability to reduce hypoxanthine accumulation during ischemia (Salaris et al., 1991). During the hypoxic event, anaerobic glycolysis results in an elevated level of hypoxanthine and xanthine substrates in the blood that induces xanthine oxidase production (Freeman and Crapo, 1982; McCord, 1985; Salaris et al., 1991; Saugstad et al., 1984). Xanthine oxidase catalyzes the conversion of hypoxanthine to xanthine and to uric acid (Cantu-Medellin and Kelley, 2013; Pasalic et al., 2012). Furthermore, xanthine oxidase also mediates the generation of reactive oxygen species and superoxide radicals (McCord, 1985). Superoxide radical directly damages cell components through oxidative stress, and formation of hydroxyl radical ($HO^{\cdot}$), by Fenton reaction in the presence of suitable iron complex (Deby and Goutier, 1990; Kuppusamy and Zweier, 1989; Kvietys et al., 1989). Several reports demonstrate the deleterious effects of hydroxyl radical on the biological systems (Josephson et al., 1991; Khalid and Ashraf, 1993; Riley, 1994; Ward et al., 1983). Protein is one of the most important system that extensively damaged in the presence of hydroxyl and superoxide radicals (Wolff et al., 1986). Such alterations in the protein structure include fragmentation and crosslinking disrupting enzyme functions (Hunt et al., 1988; Prinsze et al., 1990; Wolff et al., 1986). MB reduces cellular stress and improves neuronal cell functions by potentiating the mitochondrial electron transfer chain, enhancing mitochondria respiration, increasing the rate of cytochrome c reduction, and suppressing free radicals formation (Atamna et al., 2008; Callaway et al., 2004; Poteet et al., 2012; Riha et al., 2005; Wen et al., 2011).

Neuroprotective and free radicals scavenging activity of the MB is also due to the formation of leuco-MB, the reduced form of the MB, from NADH through mitochondrial complex I activation (Miclescu et al., 2006; Poteet et al., 2012). Formation of leuco-MB leads to bypass the formation of mitochondrial complex I-III after receiving an electron from NADH through complex I and then transfers that electron to cytochrome c (Das and Mandal, 1982). Significant increase in the formation of mitochondrial complex IV but not complexes I-III after cells being treated frequently with MB supports the idea (Atamna et al., 2008; Wrubel et al., 2007). These observations suggest that MB attenuates formation of superoxide radicals as a renewable antioxidant molecule and is complementary to the mitochondrial electron transfer chain (Poteet et al., 2012). The MB-mediated neuroprotective mechanism in Alzheimer's disease indicates its ability in penetrating the BBB and reaching the dysfunctional brain tissues (Oz et al., 2009). The concentration of MB in the cytoplasm and extracellular matrix of the neuronal tissue is 10 folds higher than in plasma within 1 h our administration supports this idea. (Oz et al., 2009). The charge distribution pattern of the MB molecule particularly the nitrogen and sulfur atoms accelerate its delivery through the BBB (Oz et al., 2009; Wagner et al., 1998). Discrepancy in the solubility and ionization state between the MB and its reduced form leuco-MB contributes to the kinetic of MB across the cell membranes (DiSanto and Wagner, 1972; Peter et al., 2000; Rengelshausen et al., 2004; Wagner et al., 1998). Thus, it seems likely that MB, due to its pharmacokinetic and dynamic activity, opens novel opportunity for an effective therapeutic strategy of the neurodegenerative diseases in future.

5 Our investigations on cardiac arrest and reperfusion induced brain pathology

5.1 Methodological consideration

5.1.1 Animals

All experiments were carried out on 12 to 14 weeks old piglets of triple breed and free from any apparent diseases (Federal authorized Pig Farm, Uppsala, Sweden) as described earlier (Sharma et al., 2011). The experimental conditions and laboratory use of piglets was approved by the Uppsala Institutional Review Board for Animal Experimentation and prospective randomized animal study.

5.1.2 Experimental cardiac arrest

The following procedures were used for cardiac arrest (CA) and return of spontaneous circulation (ROSC) in our experimental set-up (Sharma et al., 2011, 2015; Wiklund et al., 2013).

5.1.3 Anesthetic procedures and maintenance

The animals in the laboratory were injected intramuscular Zoletil (6 mg/kg; Tiletamine and zolazepam, Virbac, Carros cedex, France) in combination with Rompun (2.2 mg/kg; Xylazine, Bayer, Frankfurt, Germany) and Atropine (Atropin 0.04/kg; NM Pharma, Stockholm, Sweden).

After that intravenous administration of Morphine (20 mg, Morfin, Bioglan, Malmö, Sweden) and Ketamine (100 mg Ketaminol, Veterinaria AG, Pharmacia, Uppsala, Sweden) was done followed by infusion of sodium pentobarbital (8 mg/kg/h; Apoteket, Uppsala, Sweden) together with morphine (0.5 mg/kg/h, Morphine, Pharmacia, Uppsala, Sweden) and Pavulon (0.25 mg/kg/h; Panacuronium bromide; Organon, Oss, Netherlands) dissolved in 2.5% glucose solution was used for maintenance of anesthesia.

5.1.4 Cannulation and physiological parameters

The animals were secured in supine position and were tracheactomized and mechanically ventilated with 30% O_2 and air mixture (Servo-iV3.1, Siemens Medical, Solna, Sweden). Volume controlled mechanical ventilation was calibrated to maintain $PaCO_2$ at about 5 to 5.5 kPa. A 5 cm H_2O positive end-expiratory pressure (PEEP) is used (Sharma et al., 2011). The Capnogram, saturation and electrocardiogram (CO2SMP Plus-8100, Novametrix, Wallingford, CT, USA) were continuously monitored during the experiments. Fluid replacement was done using acetated Ringer's solution (Ringer Acetat, Fresenius Kabi, Stockholm, Sweden) by an infusion pump (Life Care, Abbott Shaw, Miami, FL, USA) at the rate of 30 mL/kg/h during 1st hour and then 10 mL/kg/h for the rest of the period of the experiments. An arterial catheter (18G) was placed into the right carotid artery retrogradely towards heart for withdrawal of blood samples and measurement of blood pressure. A 14- saline filled double lumen catheter was placed into the right atrium through right external jugular vein for the measurement of the right atrial pressure and drug administration. In addition, a 7 branch of fluid filled pulmonary catheter (CritCath, Ohmeda, Madison, WI, USA) equipped with a thermistor was floated into wedge position in the pulmonary artery to record cardiac output (CO) and capillary wedge pressure (PCWP) as describe earlier (Miclescu et al., 2010; Sharma et al., 2011).

5.1.5 Induction of cardiac arrest

All animals were stabilized for at least 1 h with all preparations and ventilated with a mixture of 30% oxygen and air. Ventricular fibrillation was used to induce cardiac arrest (CA) by an alternating transthoracic current as described earlier (Miclescu et al., 2010; Sharma et al., 2011). Following 12 min of CA closed chest cardiopulmonary resuscitation (CPR) was performed using pneumatically driven automatic piston device for CPR (Lucas, Jolife AB, Lund, Sweden) with mechanical ventilation of 100% O_2. About 1 min after CPR Vasopressin (0.4 U/kg, Arg8-Vasopressin, Sigma Chemical Co., St. Louis, MO, USA) was administered to all animals through the right arterial catheter. About 8 min after external chest compression monophasic counter shock at the energy level of 200 J was delivered through the defibrillation electrode pads (Medtronic Physio-Control Corp. Seattle, WA, USA) followed by another 2 defibrillatory shocks at the energy levels 200 and 360 J (Miclescu et al., 2010). If return of spontaneous circulation (ROSC) does not occur after repeated shocks, a bolus of epinephrine 20 µg/kg was given and DC shocks were applied at 360 J for another 5 min (Miclescu et al., 2010; Sharma et al., 2011).

The arterial pH after ROSC if falls below than 7.20 then it was corrected by administering a Tris buffer mixture (1 mM/kg; Tribonat, Fresenius Kabi, Stockholm, Sweden) and increasing minute ventilation to maintain PaCO2 in 5 to 5.5 kPa range. To maintain systolic blood pressure above 70 Torr dobutamine (Dobutrex, Eli Lilly, Indianapolis, IN, USA) was used (Miclescu et al., 2010; Sharma et al., 2011).

5.1.6 Methylene blue treatment in cardiac arrest

The animals were randomly divided into 3 groups. One group served as untreated control that received CA without CPR and the brains were removed immediately after CA. The other group after CA received CPR with infusion of normal saline (55 mL/kg/h). The 3rd group received methylene blue (10 mg/mL equivalent to 8.56 mg/mL water free Metyltioninklorid, Apoteket, Umeå, Sweden) in same dose of saline infusion (7.5 mg/kg/h). After ROSC the saline infusion was reduced to 16.5 mL/kg/h and methylene blue to 2.25 mg/kg/h (Miclescu et al., 2010; Sharma et al., 2011).

In both saline and methylene blue treated animals, the brains were removed at 30, 60 or 180 min after ROSC for histological and biochemical analysis (Sharma et al., 2011).

5.1.7 Processing of brain for histochemistry and biochemical analysis

After the end of the experiments animals were killed by intravenous administration of 20 mM/mL KCl (10 mL) and brains were rapidly removed for later processing of immunocytochemistry and histopathology as described earlier (Sharma et al., 2011, 2015).

For immunohistochemistry the brains were immersion fixed in a cold 4% buffered paraformaldehyde (PFA 4%, HL96753.1000, Askim, Sweden) for 4 weeks. To facilitate better fixation, the tissue pieces from cerebral cortex, hippocampus and cerebellum were cut and immersed in the same fixative and placed at 4 °C in a refrigerator (Sharma et al., 2011).

For biochemical analysis, the brain tissues were immersed either in liquid nitrogen or stored at −80 °C for later biochemical investigation (Wiklund et al., 2013).

6 Brain pathology in cardiac arrest

6.1 Light microscopy

The brain tissues after fixation were processed for histological analysis using Sakura FineTek automatic tissue processing system (Torrance, CA, USA). About 3-µm thick sections were cut and processed for histological examination of neuronal injury using Nissl or Hematoxylin & Eosin (H&E) stain using commercial protocol (Sharma et al., 2011, 2015).

6.2 Immunohistochemistry

Immunohistochemistry of albumin leakage to detect blood-brain barrier (BBB) breakdown, glial fibrillary acidic protein (GFAP) for astrocytic reaction and the enzyme hemeoxygenase for upregulation of carbon monoxide production (Sharma and Westman, 2004) was evaluated according to standard protocol as described earlier (Sharma et al., 1998b,e, 2003, 2006, 2011, 2016, 2019a,b).

6.3 Albumin immunohistochemistry

Endogenous leakage of albumin into the brain parenchyma, one of the most abundant plasma proteins was examined as a marker of BBB breakdown after CA (Sharma et al., 2011). Albumin immunohistochemistry was examined on 3-μm thick paraffin sections using monoclonal rabbit albumin antibody (DAKOCytomation #A0001, Hamburg, Germany) dilution (1:400) according to commercial protocol (Sharma et al., 2011).

6.4 Glial fibrillary acidic protein (GFAP) immunohistochemistry

The astrocytic activation after CA was evaluated using GFAP immunohistochemistry (DAKOCytomation #Z0334, Hamburg, Germany) on paraffin sections using glial fibrillary acidic protein monoclonal rabbit antibody dilution (1:800) according to the commercial protocol (Sharma et al., 1992, 2011).

6.5 Hemeoxygenase (HO) HO-1 and HO-2 immunochemistry

Activation of hemeoxygenase (HO) enzymes inducible isoform HO-1 and constitutive isoform HO-2 immunohistochemistry was examined on paraffin section using anti-heme oxygenase 1 antibody (HO-1-1; ab 13,248, abcam, Cambridge, MA, USA) monoclonal mouse antibody (dilution 1:500) and HO-2 monoclonal mouse antibody (# MA5-25747, Thermo Fisher Scientific, Waltham, MA; USA) dilution (1:150) according to respective commercial protocol (Sharma et al., 2000b,c, 2020a; Sharma and Westman, 2004).

6.6 Semiquantitative analysis of images

The number of distorted nerve cells in Nissl or H&E stained paraffin section in the desired areas were counted manually by at least 2 independent workers 3 times and the median values were used for statistical analysis. For immunostaining the identical procedures were used to evaluate albumin, GFAP, HO-1 and HO-2 immunoreactivity in porcine brain as described earlier (Sharma and Kiyatkin, 2009).

6.7 Transmission electron microscopy

To examine ultrastructural changes in the porcine brain after CA transmission electron microscopy (TEM) analysis was done using standard procedures (Sharma et al., 2011). In brief, small tissues pieces from different brain regions were post-fixed in osmium tetraoxide (OsO4) and embedded in plastic (Epon 812, Sigma Aldrich, St. Louis, MO, USA). About 1 μn thick sections were cut and stained with Toluidine blue and examined under a bench microscope to identify the area for ultrathin sections (Sharma and Olsson, 1990). After trimming the blocks serial ultrathin sections (50 nm) were cut on an Ultramicrotome (LKB, Stockholm, Sweden) using a diamond knife (Ultra 45°; Diatome, Nidau, Switzerland). The sections were collected on a one hole copper grid (600 nm) and counterstained with uranyl acetate and lead citrate. Some sections were examined unstained. The grids were examined under a Phillips 400 (Eindhoven, Netherlands) and images were captured at 4 k to 12 k magnifications on the attached digital camera (K3 Gatan Imaging system, Pleasanton, CA, USA). The images are processed in an Apple Macintosh Power Book (Os X El Capitan 10.11.6) using commercial software Adobe Photoshop 12.4. using identical filters for control, experimental and drug treated groups (Sharma and Sjoquist, 2002; Sharma et al., 2011).

6.8 Biochemical analysis

The following biochemical analysis on porcine brain after CA was assessed using standard procedures.

6.9 Hemeoxygenase HO-1 and HO-2 ELISA

Hemeoxygenase HO-1 and HO-2 was measured in porcine brain using commercial ELISA kit according to described protocol. For HO-1 pig hemeoxygenase-1 ELISA kit (#MBS914287 HMOX1, My BioSource, San Diego, CA, USA) was used. For HO-2 ELISA Kit (#MBS2506874; My BioSource, San Diego, CA, USA) was used. The ELISA kit sensitivity of HO-1 was 0.937 ng/mL to 60 ng/mL and HO-2 detection range was 0.375 ng/mL to 40 ng/mL (Ewing and Maines, 1992; Kitchin et al., 2001).

6.10 Brain water content

Small tissue pieces of porcine brain from cerebral cortex, hippocampus and cerebellum were dissected immediately after removal and weighed immediately on a pre-weighed filter paper on an electronic analytical balance (sensitivity 0.1 mg; Mettler Toledo AB, Stockholm, Sweden). The brain samples were then placed in a laboratory incubator (Thermo Fischer Scientific, Waltham, MA, USA) maintained at 90 °C for 72 h to obtain dry weight of the tissue samples. Brain water content was calculated from the differences between wet and dry weight of the sample as described earlier (Sharma and Olsson, 1990).

6.11 Statistical analysis of the data

ANOVA following Dunnett's test for multiple group comparison from one control was used to analyze quantitative data obtained. Semiquantitative data were analyzed using Chi Square test. The statistical analysis was done using commercial software StatView 5 (Abacus concepts Ltd., Oklahoma city, OK, USA) using Macintosh computer in classical environment. A P-value <0.05 is considered significant.

7 Results obtained

7.1 Cardiac arrest and blood-brain barrier permeability

Subjection of CA followed by ROSC induced breakdown of the BBB as seen using increased albumin immunoreactivity in different regions of porcine brain as described earlier (Sharma et al., 2011, 2015). Thus, the albumin leakage was seen in the cerebral cortex, hippocampus, cerebellum, thalamus, and hypothalamus of after 180 min of ROSC following CA (Fig. 1). The albumin deposits was present in the neurons as well as in some astrocytes and around microvessels showing profound neuropathological damage (Fig. 1A).

Semiquantitative analysis of albumin immunohistochemistry showed significant increase in albumin positive cells present in the cerebral cortex (98 ± 6 cells, $P < 0.05$) as compared to the control group (1 ± 2 cells). In the hippocampus, the number of albumin positive cells was also significantly higher (80 ± 4 cells, $P < 0.05$) as compared to the control group (2 ± 1 cells). While in cerebellum albumin positive cells were about 89 ± 6 ($P < 0.05$) as compared to the control group (1 ± 1 cells) (Fig. 2A). These semiquantitative albumin positive cells after 180 min of ROSC following CA show a global increase in the BBB breakdown (Sharma et al., 2011, 2015).

7.2 Cardiac arrest and brain edema formation

Leakage of serum proteins after CA and ROSC induces vasogenic edema formation (Sharma et al., 2011, 2015). Measurement of brain water content in the cerebral cortex, hippocampus and cerebellum exhibited profound increase in the brain water content (Fig. 3). The brain water content in the cerebral cortex after 180 min of ROSC following CA showed $80.34\% \pm 0.65\%$ ($P < 0.05$) as compared to the control group ($76.32\% \pm 0.45\%$). This amounts to about 16% increase in volume swelling in the cerebral cortex (Fig. 3). In the hippocampus the brain water content rose to $81.34\% \pm 0.31\%$ ($P < 0.05$) from the control value ($78.54\% \pm 0.83\%$) amounting to about 12% increase in volume swelling (Fig. 3) after 180 min of ROSC following CA. In the cerebellum the brain water content increased to $80.56\% \pm 0.31\%$

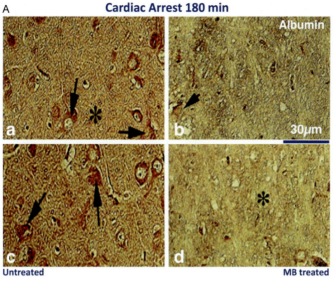

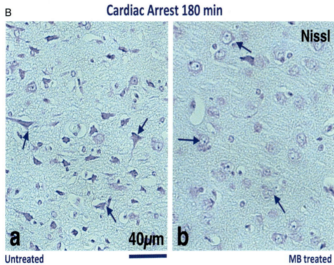

FIG. 1

Cardiac arrest and return of spontaneous circulation (ROSC) 180 min and neuroprotection of albumin leakage (A) and neuronal injury (B) by methylene blue (MB) treatment. MB (10 mg/mL equivalent to 8.56 mg/mL after ROSC was administered 2.25 mg/kg/h (i.v.). (A) Leakage of albumin (*arrows*) seen in the neuropil of parietal cerebral cortex around neurons (a) and microvessels (c) in CA that was markedly reduced by MB (b, d). (B) Neuronal damage (*arrows*, a) is frequently seen in CA in parietal cerebral cortex that was reduced by MB treatment (b). Several neurons look normal in appearance (*arrow*, b). Paraffin sections 3 μm, Bar A = 30 μm, B = 40 μm. For details see text.

Data modified after Sharma, H.S., Miclescu, A., Wiklund, L., 2011. Cardiac arrest-induced regional blood-brain barrier breakdown, edema formation and brain pathology: a light and electron microscopic study on a new model for neurodegeneration and neuroprotection in porcine brain. J. Neural Transm. (Vienna) 118 (1), 87–114. doi:10.1007/s00702-010-0486-4; Sharma, H.S., Patnaik, R., Sharma, A., Lafuente, J.V., Miclescu, A., Wiklund, L., 2015. Cardiac arrest alters regional ubiquitin levels in association with the blood-brain barrier breakdown and neuronal damages in the porcine brain. Mol. Neurobiol. 52 (2), 1043–1053. doi:10.1007/s12035-015-9254-7.

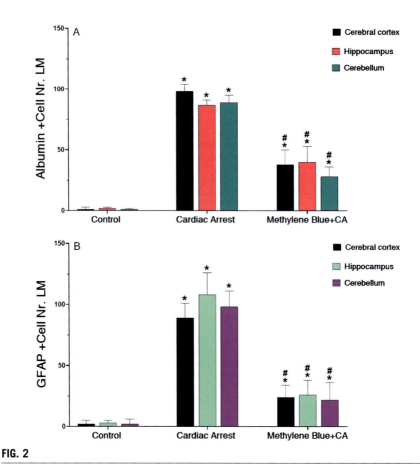

FIG. 2

Semiquantitative data on Cardiac arrest & return of spontaneous circulation (ROSC) 180 min and neuroprotection of albumin leakage (A) and glial fibrillary acidic protein (GFAP) positive cells (B) by methylene blue (MB) treatment. MB (10 mg/mL equivalent to 8.56 mg/mL after ROSC was administered 2.25 mg/kg/h (i.v.). Values are Mean ± SD of 5 to 6 animals at each point. Non-parametric Chi-Square test, * $P<0.05$ from control group; # $P<0.05$ from Cardiac Arrest (CA).

$(P<0.05)$ from the control value (76.42% ± 0.23%) corresponding to about 18% increase in volume swelling after 180 min of ROSC following CA (Fig. 3). This indicates a global brain edema following 180 min of ROSC after CA.

7.3 Cardiac arrest and HO-1 and HO-2 biochemistry

We measured HO-1 and HO-2 enzymes in the cerebral cortex, hippocampus and in cerebellum. The results are shown in Fig. 4. The HO-1 level significantly increased to 22.67 ± 0.13 ng/mL $(P<0.05)$ in the cerebral cortex after 180 min of ROSC

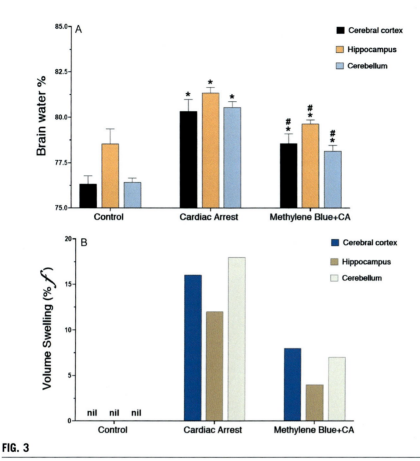

FIG. 3

Cardiac arrest & return of spontaneous circulation (ROSC) 180 min and measurement of brain edema formation as seen using measurement of brain water content (A) and volume swelling (%f) calculated from differences between brain water in control and experimental group (B) by methylene blue (MB) treatment. MB (10 mg/mL equivalent to 8.56 mg/mL after ROSC was administered 2.25 mg/kg/h (i.v.). About 1% difference between control and experimental group is roughly equal to 4% increase in volume swelling (%f) Sharma et al., 2011, 2015. Values are Mean ± SD of 5 to 6 animals at each point. * $P<0.05$ from control group; # $P<0.05$ from Cardiac Arrest (CA). ANOVA followed by Dunnett's test for multiple group comparison from one control.

following CA as compared to the control value (1.80 ± 0.02 ng/mL). In hippocampus the HO.1 value increased by 18.98 ± 0.16 ng/mL ($P<0.05$) from the control value (2.03 ± 0.01 ng/mL) (Fig. 4). In cerebellum the HO-1 was increased to 14.32 ± 0.18 ng/mL ($P<0.05$) as compared to the control group (1.45 ± 0.01 ng/mL) after 180 min ROSC following CA (Fig. 4).

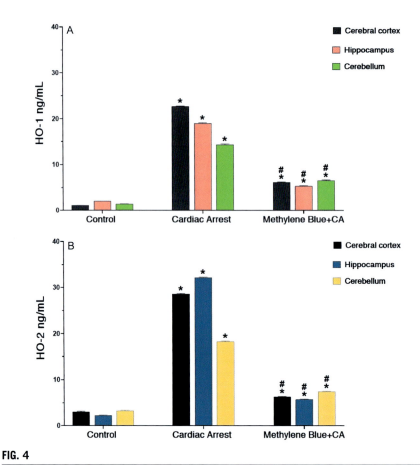

FIG. 4

Cardiac arrest & return of spontaneous circulation (ROSC) 180 min and measurement of hemeoxygenese-1 (HO-1) (A) hemeoxugenase-2 (HO-2) (B) and their modification with methylene blue (MB) treatment. MB (10 mg/mL equivalent to 8.56 mg/mL after ROSC was administered 2.25 mg/kg/h (i.v.). Values are Mean ± SD of 5 to 6 animals at each point. * $P < 0.05$ from control group; # $P < 0.05$ from Cardiac Arrest (CA). ANOVA followed by Dunnett's test for multiple group comparison from one control.

The HO-2 level was also significantly increased in the cerebral cortex, hippocampus and the cerebellum after 180 min ROSC following CA (Fig. 4). Thus, the cerebral cortex revealed HO-2 values (28.56 ± 0.14 ng/mL, $P < 0.05$) from the control group (3.02 ± 0.14 ng/mL). In the hippocampus the HO-2 values increased up to 32.08 ± 0.17 ($P < 0.05$) as compared to the control group (2.18 ± 0.16 ng/mL). The cerebellum values of HO-2 showed 18.23 ± 0.16 ng/mL ($P < 0.05$) from the control group $(3.23 \pm 0.14 \, \text{ng/mL})$ after 180 min of ROSC following CA (Fig. 4). These observations show that both HO-1 and HO-2 are significantly increased in the brain after ROSC following CA.

7.4 Brain pathology

The brain pathology was examined using neuronal, glial and myelin changes using light and electron microscopy.

7.5 Neuronal changes

Profound neuronal changes were observed after 180 min of ROSC following CA (Fig. 5). Thus, several neurons are shrunken, and some are swollen with perineuronal edema present in the expanded and spongy neuropil. In a semiquantittative study about 92 ± 10 neuronal cells $(P < 0.05)$ seem damaged in the cerebral cortex as compared to the control group $(2 \pm 1$ cells$)$. In hippocampus about 104 ± 9 cells $(P < 0.05)$ look abnormal showing neuropathological changes and are located within the CA-1 to CA-3 sectors and in dentate gyrus as compared to the controls $(2 \pm 3$ cells$)$ (Fig. 5). In the cerebellum both Purkinje cells and granule cells were damaged and present in the cerebellar vermis or cerebellar cortices. In the cerebellum about 89 ± 8 cells $(P < 0.05)$ were abnormal or damaged as compared to the control group $(2 \pm 2$ cells$)$ (Fig. 5). These observations suggest that neuronal damages are widespread following ROSC after CA.

7.6 Astrocytic changes

Using GFAP immunohistochemistry, several astrocytes were immunoreactive the in cerebral cortex, hippocampus and in cerebellum after 180 min of ROSC following CA (Fig. 2). Thus, about 89 ± 2 cells $(P < 005)$ appear GFAP positive (Fig. 4) as compared to the control group (2 ± 3) in the cerebral cortex (Fig. 2). In hippocampus 108 ± 18 cells $(P < 0.05)$ located within the CA-1 to CA-3 sector and in dentate gyrus were GFAP positive as compared to the control group $(3 \pm 2$ cells$)$. In cerebellum GFAP positive cells were present among Purkinje and granule cells in high numbers $(98 \pm 13$ cells, $P < 0.05)$ as compared to the control group $(2 \pm 4$ cells$)$ (Fig. 2). These observations suggest that ROSC after CA activated GFAP positive astrocytes throughout the brain.

7.7 HO-1 and HO-2 immunoreactivity

Immunoreactivity of HO-1 and HO-2 were examined in the cerebral cortex, hippocampus and in cerebellum after 180 min of ROSC following CA (Fig. 6). Our results showed significant increase in HO-1 immunoreactivity in the cerebral cortex

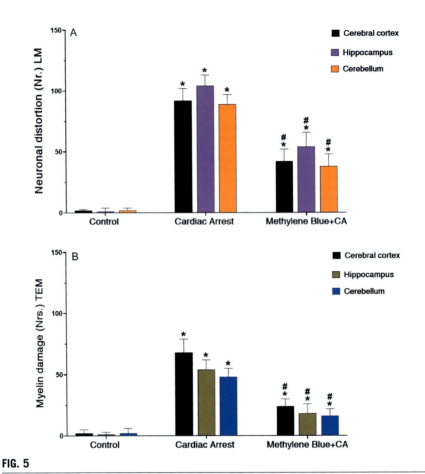

FIG. 5

Semiquantitative data on Cardiac arrest & return of spontaneous circulation (ROSC) 180 min and neuroprotection of neuronal distortion or injury at light microscopy (LM) (A) and myelin vesiculation seen at Transmission electron microscopy (TEM) (B) by methylene blue (MB) treatment. MB (10 mg/mL equivalent to 8.56 mg/mL after ROSC was administered 2.25 mg/kg/h (i.v.). Values are Mean ± SD of 5 to 6 animals at each point. Non-parametric Chi-Square test, * $P < 0.05$ from control group; # $P < 0.05$ from Cardiac Arrest (CA).

(18 ± 2 cells, $P < 0.05$) as compared to the controls (2 ± 1 cells). These HO-1 positive cells were located within the layers II to V in the cerebral cortex. Both small and large neurons were HO-1 positive (Fig. 5). In the hippocampus the HO-1 positive cells were seen within the CA-1 to CA-3 regions along with the dentate gyrus areas. In hippocampus about 22 ± 4 cells ($P < 0.05$) appear HO-1 positive as compared to the control groups (1 ± 1 cell). In the cerebellum both Purkinje and grannie cells

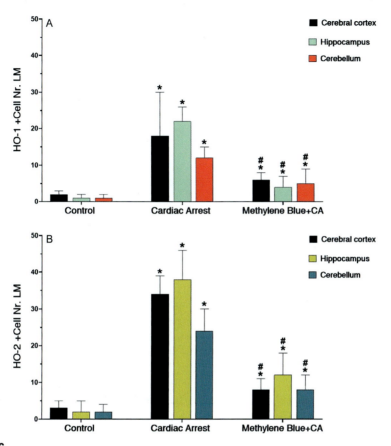

FIG. 6

Semiquantitative data on Cardiac arrest and return of spontaneous circulation (ROSC) 180 min and neuroprotection of hemeoxygenase-1 (HO-1 (A) and hemeoxygenase-2 (HO-2) (B) by methylene blue (MB) treatment. MB (10 mg/mL equivalent to 8.56 mg/mL after ROSC was administered 2.25 mg/kg/h (i.v.). HO-1 ad HO-2 immunohistochemistry was performed on paraffin sections (3-μm thick) using standard protocol and respective antibodies. For details see text. Values are Mean ± SD of 5 to 6 animals at each point. Non-parametric Chi-Square test, * $P < 0.05$ from control group; # $P < 0.05$ from Cardiac Arrest (CA).

exhibited HO-1 positive immunoreactivity following ROSC after CA (Fig. 6). Thus, about 12 ± 3 cells ($P < 0.05$) exhibited HO-1 immunoreactivity as compared to the control group (1 ± 1 cell).

The HO-2 that is constitutive enzyme shows large number of HO-2 immunore-active cells in the cerebral cortex, hippocampus and in cerebellum after 180 min of

ROSC following CA (Fig. 6). Thus, the cerebral cortex exhibited about 34 ± 5 cells ($P < 0.05$) as compared to the controls (3 ± 2 cells). These cells were seen in the areas of layer II to V in the cerebral cortex (Fig. 5). In hippocampus about 38 ± 8 cells ($P < 0.05$) were HO-2 positive as compared to the control group (2 ± 3 cells). These HO-2 positive cells were located within the CA-1 to CA-3 and dentate gyrus sectors of the hippocampus. In cerebellum about 24 ± 6 cells ($P < 0.05$) were HO-2 positive as compared to the control group (2 ± 2 cells) (Fig. 6). These HO-2 positive cells were distributed among Purkinje and granule cells of the cerebellar vermis and cerebellar cortices.

These observations suggest that ROSC after CA activated both HO-1 and HO-2 upregulation in the brain.

7.8 Myelin changes

Myelin changes after 180 min ROSC following CA was examined using transmission electron microscopy (Fig. 5). As evident with the figure, myelin damage and vesiculation was frequent after ROSC following CA. This myelin damage and vesiculation was seen in the cerebral cortex, hippocampus and an in cerebellum. In the cerebral cortex semiquantittative study shows about 68 ± 11 axons ($P < 0.05$) were distorted exhibiting myelin vesiculation as compared to the control group (2 ± 3 axons, Fig. 5). In the hippocampus, several axons show myelin damage and vesiculation (54 ± 8 axons, $P < 0.05$) as compared to the controls (1 ± 2 axons). In the cerebellum there were more than 48 ± 7 axons ($P < 0.05$) showing myelin damage and vesiculation as compared to the control groups (2 ± 4 axons). These observations show that ROSC after CA is associated with myelin damage in several brain regions.

7.9 Methylene blue treatment induces neuroprotection

Treatment with methylene blue (MB) significantly reduced the albumin leakage in the cerebral cortex, hippocampus and in cerebellum together with reduction in brain edema formation (Fig. 1). MB significantly reduced the volume swelling in the cerebral cortex to 8%, in hippocampus (4%) and in cerebellum to 7% (Fig. 2). The HO-1 and HO-2 levels were also significantly reduced (Fig. 3).

The number of GFAP positive astrocytes was also down in MB treated group after 180 min ROSC following CA (Fig. 1). The number of distorted neurons was considerable reduced in MB treated group after ROSC and CA (Fig. 1). The HO-1 and HO-2 immunohistochemistry was significantly reduced by MB (Fig. 6).

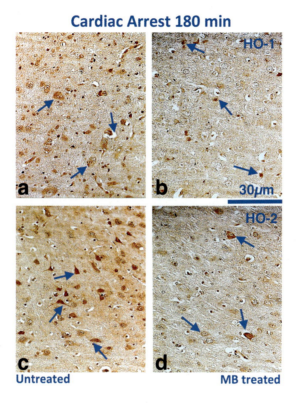

Cardiac Arrest 180 min

FIG. 7

Cardiac arrest & return of spontaneous circulation (ROSC) 180 min and neuroprotection of hemeoxygenase-1 (HO-1 (A) and hemeoxygenase-2 (HO-2) (B) by methylene blue (MB) treatment. MB (10 mg/mL equivalent to 8.56 mg/mL after ROSC was administered 2.25 mg/kg/h (i.v.). HO-1 ad HO-2 immunohistochemistry was performed on paraffin sections (3-μm thick) using standard protocol and respective antibodies. For details see text. In CA several HO-1 and HO-2 positive cells (*arrows*) are seen in the parietal cerebral cortex (A, C) of porcine brain and MB treatment markedly attenuated the HO-1 and HO-2 positive cells in the neuropil (B, D). Bar = 30 μm.

The number of axons showing myelin damage or vesiculation was also reduced significantly in MB treated CA group after ROSC (Fig. 5).

Morphological examination showed significant neuroprotection of neurons (Figs. 9 & 11) after MB treatment in CA. Activation of astrocytes as seen using GFAP immunoreactivity in CA (Fig. 10) was also reduced markedly by MB treatment. Immunoreactivity of HO-1 and HO-2 in parietal cerebral cortex (Fig. 7) and in temporal

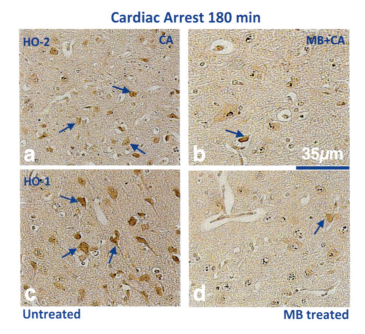

FIG. 8

Cardiac arrest & return of spontaneous circulation (ROSC) 180 min and neuroprotection of hemeoxygenase-1 (HO-1 (A) and hemeoxygenase-2 (HO-2) (B) by methylene blue (MB) treatment. MB (10 mg/mL equivalent to 8.56 mg/mL after ROSC was administered 2.25 mg/kg/h (i.v.). HO-1 ad HO-2 immunohistochemistry was performed on paraffin sections (3-μm thick) using standard protocol and respective antibodies. For details see text. In CA several HO-1 and HO-2 positive cells (*arrows*) are seen in the temporal cerebral cortex (A, C) of porcine brain and MB treatment markedly attenuated the HO-1 and HO-2 positive cells in the neuropil (B, D). Interestingly, HO-1 positive cells in temporal cortex appear to be greater in numbers than HO-2 positive cells in CA. However, both the HO-1 and HO-2 positive cells were markedly reduced by MB treatment in the temporal cerebral cortex. Bar = 35 μm.

cerebral cortex (Fig. 8) was also considerable reduced by MB treatment in CA. These light and electron microscopic data clearly show that MB is a powerful neuroprotective agent in CA on brain pathology (Figs. 9–11).

Cardiac Arrest and Neuronal Damage

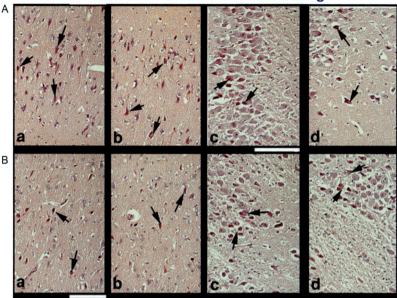

FIG. 9

Cardiac arrest & return of spontaneous circulation (ROSC) induced neuronal damages in the cortex A (a, b) and B (a, b) and in hippocampal areas A (c, d) and B (c, d). Damaged neurons show perineuronal edema, distorted nerve cells in the neuropil (*arrows*). Sponginess and edematous expansion is clearly seen. Bar = 30 μm.

Data modified from Sharma, H.S., Miclescu, A., Wiklund, L., 2011. Cardiac arrest-induced regional blood-brain barrier breakdown, edema formation and brain pathology: a light and electron microscopic study on a new model for neurodegeneration and neuroprotection in porcine brain. J. Neural Transm. (Vienna) 118 (1), 87–114. doi:10.1007/s00702-010-0486-4.

Cardiac Arrest 180 min

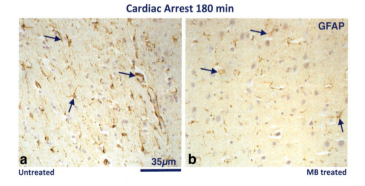

FIG. 10

Cardiac arrest and return of spontaneous circulation (ROSC) induced activation of astrocytes as seen using glial fibrillary acidic protein (GFAP) immunoreactivity in the occipital cortex of porcine brain (A) and its attenuation with methylene blue (MB) treatment (B). MB (10 mg/mL equivalent to 8.56 mg/mL) after ROSC was administered 2.25 mg/kg/h (i.v.). Paraffin sections 3-μm thick, Bar = 35 μm.

Cardiac Arrest 180 min

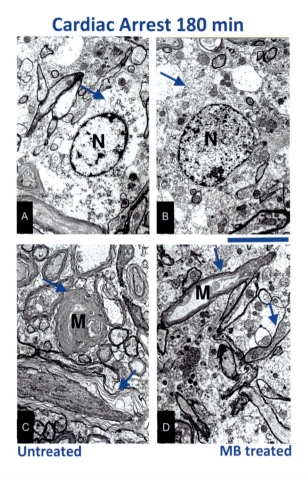

Untreated   **MB treated**

FIG. 11

Transmission electron micrograph after Cardiac arrest and return of spontaneous circulation (ROSC) 180 min and neuroprotection of nerve cells (N) (B) and myelin (M) vesiculation (D) by methylene blue (MB) treatment as compared to CA alone (A, C). MB (10 mg/mL equivalent to 8.56 mg/mL) after ROSC was administered 2.25 mg/kg/h (i.v.). In CA neuronal damage (N) (A) and vesiculation of myelin (M) (C) *arrows* are quite prominent and MB treatment markedly reduced these neuronal (N) (B) and myelin (M) (D) *arrows* changes. Bar = 1 μm.

8 Discussion

The salient new findings of this investigation show that CA followed by ROSC results in upregulation of HO-1 and HO-2 in different regions of the brain. This indicates that HO-1 and HO-2 levels are altered in CA following ROSC. Methylene blue treatment significantly attenuated the HO-1 and HO-2 activity in CA. This indicates that HO-1 and HO-2 activity in CA occurs due to oxidative stress, not reported earlier.

CA induced oxidative stress by cessation of blood flow to the brain and generation of reactive oxygen species occurs after ROSC (Diao et al., 2020; Ottolenghi et al., 2020; Secher et al., 2018). Sudden cessation of heart severely affects delivery of oxygen to brain tissues and other vital organs resulting in cell and tissue death. CA patients after ROSC show various organs including myocardial and brain tissue damage due to oxidative stress and generation of reactive oxygen and nitrogen species (Chen et al., 2020; Hackenhaar et al., 2017; Li et al., 2020; Uray et al., 2018).

CA followed by ROSC leads generation of reactive oxygen species that reacts to nitric oxide (NO) to form perioxynitrite also known as reactive nitrogen species and damages cell membranes and other organelles to induce cell death (Dixon and Stockwell, 2014; Nakamura et al., 2019; Radi, 2018; Ramdial et al., 2017; Su et al., 2019). The most important observation of reactive nitrogen species is the measurement of nitrite and nitrates in the brain of CA. Our observations clearly show that CA significantly elevates nitrites and nitrate levels after CA and ROSC (L. Wiklund et al., unpublished observations). Another important landmark of oxidative stress is the generation of lipid peroxidation (Del Rio et al., 2005; Ramana et al., 2017; Tsikas, 2017). The best markers for lipid peroxidation are generation of malondialdehyde. Previously we have shown generation of malondialdehyde in CA in porcine model that clearly suggests that lipid peroxidation is an important factor in generation of oxidative injury to the brain (Wiklund et al., 2018).

Oxidative stress is one of the important determinants of upregulation of hemeoxygenase enzymes (Elmarakby et al., 2012; Hung et al., 2011). In the present study we observed upregulation of HO-1 and HO-2 in the porcine brain after CA. This indicates that HO system is activated following CA and ROSC (Warenits et al., 2020). After ROSC, several mechanisms are activated including cellular stress, alterations in redox homeostasis, neuroinflammation signaling and cell death (Rohnert et al., 2012; Sairanen et al., 2001; Yakubov et al., 2004; Yin et al., 2004). In these situations, the HO system supports neuronal functions and provides defense to oxidative stress (Dennery, 2014; Maines et al., 1993). The two isoforms of HO are identified that are HO-1 the inducible isoform and the HO-2 the constitutive isoform in the central nervous system plays key roles in the brain pathology (Chen-Roetling and Regan, 2006; Intagliata et al., 2019; Neis et al., 2018; Seta et al., 2006). In brain pathology both HO-1 and HO-2 upregulation occurs (Chen-Roetling and Regan, 2006; Seta et al., 2006). The constitutive isoform HO-2 upregulation maintains homeostasis and function in neuronal cells (Parfenova and Leffler, 2008). Whereas HO-1, the inducible isoform increased in brain following cellular stress associated with hypoxia and ischemia (Zhang et al., 2020; Zhao et al., 2020). Several hours after ROSC ischemia leads to induction of HO-1 in neurons and astrocytes in the brain and upregulates HO-2 isoform in several brain regions (Dore et al., 1999; Garnier et al., 2001; Geddes et al., 1996; Sutherland et al., 2009; Wang et al., 2016).

In the brain, HO-1 and HO-2 are differentially expressed following oxidative stress. HO-1 is largely expressed in brain following hypoxic and ischemic episodes in the hippocampus, cerebral cortex, hypothalamus, cerebellum and brain stem. On the other hand, HO-2 is abundantly expressed in the brain mainly in neurons in the cerebral cortex, cerebellum, hippocampus, basal ganglia, thalamus and brain stem (Ewing and Maines, 1997, 2006; Liu et al., 2018).

HO-1 and HO-2 are expressed against oxidative stress following hypoxia and ischemia has neuroprotective roles in the brain (Chen, 2014; Maines and Panahian, 2001). HO-2 is also responsible for the production of carbon monoxide (CO) for physiological functions in neuronal populations against oxidative damage in the brain (Alm et al., 2000; Sharma et al., 2000a,b,c). HO-2 also acts as oxygen sensors that are related to redox signals and CO metabolism (Maines, 1997; Munoz-Sanchez and Chanez-Cardenas, 2014; Williams et al., 2004). Thus, HO-2 is one of the first responders of oxidative stress and global ischemia in the brain (Kemp et al., 2006; Shibahara et al., 2007). Apart from maintenance of neuronal functions, HO-2 also regulates vascular tone of the cerebral vessels (Chodorowski et al., 2005; He et al., 2010; Kim et al., 2011).

In our investigations we observed breakdown of the BBB after CA. This observation is in line with our previous study in CA (Sharma et al., 2011, 2015). Oxidative stress and generation of lipid peroxidation with oxygen radicle induce breakdown of the cell membranes of microvessels, neurons and astrocytes (Mattson, 1998; McCall et al., 1987). Thus, neuronal astrocytes and endothelial cell injury caused by lipid peroxidation results in BBB leakage and brain pathology in CA (Hall et al., 2016).

Breakdown of the BBB to serum proteins lead to vasogenic edema formation (Hossmann et al., 1983; Kalimo et al., 1986; Klatzo, 1987; Sharma et al., 1998b). After CA and ROSC albumin leakage into the brain fluid environment clearly indicates vasogenic brain edema associated with the BBB breakdown (Sharma et al., 2011, 2015). Breakdown of the BBB and spread of edema fluid is also one of the key mediators of HO upregulation (Sharma et al., 1998a,c,d, 2000a,b,c; Yakovlev and Faden, 2001). Thus, breakdown of the BBB in hyperthermic brain injury or spinal cord injury enhances HO-2 upregulation in the brain and spinal cord (Alm et al., 2000; Sharma et al., 2000a,b,c). When serotonin synthesis inhibitor drug p-chlorophenylalanine (p-CPA) or neurotrophins were administered in spinal cord injury breakdown of the BBB associated with edema formation and HO-2 upregulation in the cord was attenuated (Sharma et al., 2000a,b,c, 2003). Also, in hyperthermic brain injury HO-1 and HO-2 expressions were reduced by prevention of the BBB breakdown, edema formation and cell injury by drugs modifying oxidative stress, serotonin metabolism or nitric oxide synthase (Sharma and Westman, 2004; Sharma et al., 1998d,e,f, 2000a,b,c, 2003). This suggests that BBB breakdown and spread of the edema fluid are key mediators of HO expression in the brain or spinal cord following brain pathology.

Methylene blue (MB) is a potent antioxidant compound and has the ability to reduce pathophysiology of brain following shock, ischemia or traumatic brain injuries (Bozkurt et al., 2015; Cefalu et al., 2020; Jang et al., 2013, 2015;

Rey-Funes et al., 2016; Saha and Burns, 2020; Salaris et al., 1991; Talley Watts et al., 2014). MB was able to reduce the BBB breakdown and edema formation in CA model of porcine brains earlier (Miclescu et al., 2010; Sharma et al., 2011, 2015; Wiklund et al., 2007, 2012, 2013, 2018). MB was also able to attenuate nitric oxide synthase upregulation in porcine cardiac arrest model (Wiklund et al., 2007). These observations suggest a putative role of MB in CA induced brain pathology. In this investigation, our results for the first time show that MB is capable in attenuating HO-1 and HO-2 upregulation both at biochemical level and confirmed this finding using immunohistochemistry in porcine brain after CA. Our results added new dimension in CA followed by ROSC induced pathophysiology with regard to the involvement of HO-1 and HO-2 system, not reported earlier. Its is quite likely that CA induced ischemia and hypoxia stimulated HO-1 and HO-2 upregulation that somehow try to attenuate cell pathology and take art in antioxidant defense mechanisms. When MB treatment was made in CA cases the ROSC induced oxidative stress subsided. Due to counterbalancing oxidative stress with the antioxidant MB treatment the neural and glial cells do not feel enough stimulus to upregulate HO system. As a result, the upregulation of HO-1 and HO-2 were thwarted by MB in CA treatment.

It would be interesting to examine time related events with MB treatment and HO-1 and HO-2 upregulation. Also, it would be very interesting to use nanotechnology to deliver MB in CA to induce superior neuroprotective effects on the BBB permeability, edema formation and cell injury together with HO-1 and HO-2 upregulation. Our earlier studies on nanowired delivery of MB exhibited significant reduction in brain pathology via regulating protein kinase system in brain injury (Sharma et al., 2020a,b; Zhao et al., 2020).

In summary, our results for the first time show that CA followed by ROSC activated hemeoxygenase system as evidenced with upregulation of HO-1 and HO-2 in different brain areas. This is evident with both the biochemical measurement of HO-1 and HO-2 in the cerebral cortex, cerebellum and in hippocampus. These elevated HO-1 and HO-2 levels in the porcine brain were further confirmed by their immunohistochemical localization in the nerve cells. This HO-1 and HO-2 upregulation is significantly reduced in CA followed by ROSC by MB treatment, not reported earlier. It would be interesting to examine nanowired delivery of MB in CA to induce further superior neuroprotection on brain pathology and HO-1 and HO-2 upregulation. It is believed that hemeoxygebase system activated to counteract antioxidant defense mechanism to reduce brain pathology after CA. Further studies are needed to confirm this point, a feature currently being examined in our laboratory.

Acknowledgments

This investigation is supported by grants from the Air Force Office of Scientific Research (EOARD, London, UK), and Air Force Material Command, USAF, under grant number FA8655-05-1-3065; Grants from the Alzheimer's Association (IIRG-09- 132087), the Laerdal

Foundation for Acute Medicine, Stavanger, Norway, the National Institutes of Health (R01 AG028679) and the Dr. Robert M. Kohrman Memorial Fund (RJC); Swedish Medical Research Council (Nr 2710-HSS), Göran Gustafsson Foundation, Stockholm, Sweden (HSS), Astra Zeneca, Mölndal, Sweden (HSS/AS), The University Grants Commission, New Delhi, India (HSS/AS), Ministry of Science & Technology, Govt. of India (HSS/AS), Indian Medical Research Council, New Delhi, India (HSS/AS) and India-EU Co-operation Program (RP/AS/HSS) and IT-901/16 (JVL), Government of Basque Country and PPG 17/51 (JVL), JVL thanks to the support of the University of the Basque Country (UPV/EHU) PPG 17/51 and 14/08, the Basque Government (IT-901/16 and CS-2203) Basque Country, Spain; and Foundation for Nanoneuroscience and Nanoneuroprotection (FSNN), Romania. Technical and human support provided by Dr. Ricardo Andrade from SGIker (UPV/EHU) is gratefully acknowledged. Dr. Seaab Sahib is supported by Research Fellowship at University of Arkansas Fayetteville AR by Department of Community Health; Middle Technical University; Wassit; Iraq, and The Higher Committee for Education Development in Iraq; Baghdad; Iraq. We thank Suraj Sharma, Blekinge Inst. Technology, Karlskrona, Sweden and Dr. Saja Alshafeay, University of Arkansas Fayetteville, Fayetteville AR, USA for computer and graphic support. The U.S. Government is authorized to reproduce and distribute reprints for Government purpose notwithstanding any copyright notation thereon. The views and conclusions contained herein are those of the authors and should not be interpreted as necessarily representing the official policies or endorsements, either expressed or implied, of the Air Force Office of Scientific Research or the U.S. Government.

Conflict of interest

The authors declare no conflict of interest with any entity mentioned here.

References

Abraham, N., Drummond, G., Lutton, J., Kappas, A., 1996. The biological significance and physiological role of heme oxygenase. Cell. Physiol. Biochem. 6 (3), 129–168.

Adibhatla, R.M., Hatcher, J., 2006. Phospholipase A2, reactive oxygen species, and lipid peroxidation in cerebral ischemia. Free Radic. Biol. Med. 40 (3), 376–387.

Alici-Evcimen, Y., Breitbart, W.S., 2007. Ifosfamide neuropsychiatric toxicity in patients with cancer. Psychooncology 16 (10), 956–960. https://doi.org/10.1002/pon.1161.

Alm, P., Sharma, H.S., Sjoquist, P.O., Westman, J., 2000. A new antioxidant compound H-290/51 attenuates nitric oxide synthase and heme oxygenase expression following hyperthermic brain injury. An experimental study using immunohistochemistry in the rat. Amino Acids 19 (1), 383–394. https://doi.org/10.1007/s007260070069.

Alpers, B.J., 1940. Personality and Emotional Disorders Associated with Hypothalamic Lesions. Research Publications of the Association for Research in Nervous & Mental Disease.

Althaus, J., Bernaudin, M., Petit, E., Toutain, J., Touzani, O., Rami, A., 2006. Expression of the gene encoding the pro-apoptotic BNIP3 protein and stimulation of hypoxia-inducible factor-1alpha (HIF-1alpha) protein following focal cerebral ischemia in rats. Neurochem. Int. 48 (8), 687–695. https://doi.org/10.1016/j.neuint.2005.12.008.

Andersen, L.W., Holmberg, M.J., Berg, K.M., Donnino, M.W., Granfeldt, A., 2019. In-hospital cardiac arrest: a review. JAMA 321 (12), 1200–1210. https://doi.org/10.1001/jama.2019.1696.

Anderson, D.K., Demediuk, P., Saunders, R.D., Dugan, L.L., Means, E.D., Horrocks, L.A., 1985. Spinal cord injury and protection. Ann. Emerg. Med. 14 (8), 816–821. https://doi.org/10.1016/s0196-0644(85)80064-0.

Annunziato, L., Amoroso, S., Pannaccione, A., Cataldi, M., Pignataro, G., D'Alessio, A., Sirabella, R., Secondo, A., Sibaud, L., Di Renzo, G.F., 2003. Apoptosis induced in neuronal cells by oxidative stress: role played by caspases and intracellular calcium ions. Toxicol. Lett. 139 (2–3), 125–133. https://doi.org/10.1016/s0378-4274(02)00427-7.

Applegate, L.A., Luscher, P., Tyrrell, R.M., 1991. Induction of heme oxygenase: a general response to oxidant stress in cultured mammalian cells. Cancer Res. 51 (3), 974–978. Retrieved from https://www.ncbi.nlm.nih.gov/pubmed/1988141.

Asahi, M., Hoshimaru, M., Uemura, Y., Tokime, T., Kojima, M., Ohtsuka, T., Matsuura, N., Aoki, T., Shibahara, K., Kikuchi, H., 1997. Expression of interleukin-1 beta converting enzyme gene family and bcl-2 gene family in the rat brain following permanent occlusion of the middle cerebral artery. J. Cereb. Blood Flow Metab. 17 (1), 11–18. https://doi.org/10.1097/00004647-199701000-00003.

Atamna, H., Nguyen, A., Schultz, C., Boyle, K., Newberry, J., Kato, H., Ames, B.N., 2008. Methylene blue delays cellular senescence and enhances key mitochondrial biochemical pathways. FASEB J. 22 (3), 703–712. https://doi.org/10.1096/fj.07-9610com.

Atamna, H., Mackey, J., Dhahbi, J.M., 2012. Mitochondrial pharmacology: electron transport chain bypass as strategies to treat mitochondrial dysfunction. Biofactors 38 (2), 158–166.

Babu, P.P., Yoshida, Y., Su, M., Segura, M., Kawamura, S., Yasui, N., 2000. Immunohistochemical expression of Bcl-2, Bax and cytochrome c following focal cerebral ischemia and effect of hypothermia in rat. Neurosci. Lett. 291 (3), 196–200.

Balentine, J.D., 1985. Hypotheses in spinal cord trauma research. In: Central Nervous System Trauma Status Report. vol. 455, p. 461.

Basu, S., Nozari, A., Liu, X.L., Rubertsson, S., Wiklund, L., 2000. Development of a novel biomarker of free radical damage in reperfusion injury after cardiac arrest. FEBS Lett. 470 (1), 1–6. https://doi.org/10.1016/s0014-5793(00)01279-5.

Basu, S., Liu, X., Nozari, A., Rubertsson, S., Miclescu, A., Wiklund, L., 2003. Evidence for time-dependent maximum increase of free radical damage and eicosanoid formation in the brain as related to duration of cardiac arrest and cardio-pulmonary resuscitation. Free Radic. Res. 37 (3), 251–256.

Biegon, A., Fry, P.A., Paden, C.M., Alexandrovich, A., Tsenter, J., Shohami, E., 2004. Dynamic changes in N-methyl-D-aspartate receptors after closed head injury in mice: implications for treatment of neurological and cognitive deficits. Proc. Natl. Acad. Sci. U. S. A. 101 (14), 5117–5122. https://doi.org/10.1073/pnas.0305741101.

Binks, A., Nolan, J.P., 2010. Post-cardiac arrest syndrome. Minerva Anestesiol. 76 (5), 362–368. Retrieved from https://www.ncbi.nlm.nih.gov/pubmed/20395899.

Bjorklund, E., Lindberg, E., Rundgren, M., Cronberg, T., Friberg, H., Englund, E., 2014. Ischaemic brain damage after cardiac arrest and induced hypothermia—a systematic description of selective eosinophilic neuronal death. A neuropathologic study of 23 patients. Resuscitation 85 (4), 527–532. https://doi.org/10.1016/j.resuscitation.2013.11.022.

Borgers, M., Thone, F., Van Reempts, J., Verheyen, F., 1983. The role of calcium in cellular dysfunction. Am. J. Emerg. Med. 1 (2), 154–161. https://doi.org/10.1016/0735-6757(83)90083-9.

Bozkurt, B., Dumlu, E.G., Tokac, M., Ozkardes, A.B., Ergin, M., Orhun, S., Kilic, M., 2015. Methylene blue as an antioxidant agent in experimentally-induced injury in rat liver. Bratisl. Lek. Listy 116 (3), 157–161. https://doi.org/10.4149/bll_2015_032.

Broughton, B.R., Reutens, D.C., Sobey, C.G., 2009. Apoptotic mechanisms after cerebral ischemia. Stroke 40 (5), e331–e339. https://doi.org/10.1161/STROKEAHA. 108.531632.

Bruchey, A.K., Gonzalez-Lima, F., 2008. Behavioral, physiological and biochemical hormetic responses to the autoxidizable dye methylene blue. Am. J. Pharmacol. Toxicol. 3 (1), 72–79. https://doi.org/10.3844/ajptsp.2008.72.79.

Butterfield, D.A., Lauderback, C.M., 2002. Lipid peroxidation and protein oxidation in Alzheimer's disease brain: potential causes and consequences involving amyloid beta-peptide-associated free radical oxidative stress. Free Radic. Biol. Med. 32 (11), 1050–1060. https://doi.org/10.1016/s0891-5849(02)00794-3.

Callaway, N.L., Riha, P.D., Bruchey, A.K., Munshi, Z., Gonzalez-Lima, F., 2004. Methylene blue improves brain oxidative metabolism and memory retention in rats. Pharmacol. Biochem. Behav. 77 (1), 175–181. https://doi.org/10.1016/j.pbb.2003.10.007.

Cantor, J.B., Ashman, T., Gordon, W., Ginsberg, A., Engmann, C., Egan, M., Spielman, L., Dijkers, M., Flanagan, S., 2008. Fatigue after traumatic brain injury and its impact on participation and quality of life. J. Head Trauma Rehabil. 23 (1), 41–51. https://doi.org/10.1097/01.HTR.0000308720.70288.af.

Cantu-Medellin, N., Kelley, E.E., 2013. Xanthine oxidoreductase-catalyzed reactive species generation: a process in critical need of reevaluation. Redox Biol. 1 (1), 353–358. https://doi.org/10.1016/j.redox.2013.05.002.

Cao, G., Xiao, M., Sun, F., Xiao, X., Pei, W., Li, J., Graham, S.H., Simon, R.P., Chen, J., 2004. Cloning of a novel Apaf-1-interacting protein: a potent suppressor of apoptosis and ischemic neuronal cell death. J. Neurosci. 24 (27), 6189–6201. https://doi.org/10.1523/JNEUROSCI.1426-04.2004.

Cao, Z., Zhao, M., Xu, C., Zhang, T., Jia, Y., Wang, T., Zhu, B., 2020. Evaluation of agonal cardiac function for sudden cardiac death in forensic medicine with postmortem brain natriuretic peptide (BNP) and NT-proBNP: a meta-analysis. J. Forensic Sci. 65 (3), 686–691. https://doi.org/10.1111/1556-4029.14232.

Castejón, O.J., 2009. The extracellular space in the edematous human cerebral cortex: an electron microscopic study using cortical biopsies. Ultrastruct. Pathol. 33 (3), 102–111.

Castejon, O.J., Valero, C., Diaz, M., 1995. Synaptic degenerative changes in human traumatic brain edema. An electron microscopic study of cerebral cortical biopsies. J. Neurosurg. Sci. 39 (1), 47–65. Retrieved from https://www.ncbi.nlm.nih.gov/pubmed/8568555.

Castejon, O.J., Castejon, H.V., Zavala, M., Sanchez, M.E., Diaz, M., 2002. A light and electron microscopic study of oedematous human cerebral cortex in two patients with post-traumatic seizures. Brain Inj. 16 (4), 331–346. https://doi.org/10.1080/02699050110088209.

Cefalu, J.N., Joshi, T.V., Spalitta, M.J., Kadi, C.J., Diaz, J.H., Eskander, J.P., Cornett, E.M., Kaye, A.D., 2020. Methemoglobinemia in the operating room and intensive care unit: early recognition, pathophysiology, and management. Adv. Ther. 37 (5), 1714–1723. https://doi.org/10.1007/s12325-020-01282-5.

Chambers, D.J., Fallouh, H.B., 2010. Cardioplegia and cardiac surgery: pharmacological arrest and cardioprotection during global ischemia and reperfusion. Pharmacol. Ther. 127 (1), 41–52. https://doi.org/10.1016/j.pharmthera.2010.04.001.

Chan, P.H., 1996. Role of oxidants in ischemic brain damage. Stroke 27 (6), 1124–1129. https://doi.org/10.1161/01.str.27.6.1124.

Chan, P.H., 2004. Mitochondria and neuronal death/survival signaling pathways in cerebral ischemia. Neurochem. Res. 29 (11), 1943–1949. https://doi.org/10.1007/s11064-004-6869-x.

Chan, P.H., Fishman, R.A., 1980. Transient formation of superoxide radicals in polyunsaturated fatty acid-induced brain swelling. J. Neurochem. 35 (4), 1004–1007. https://doi.org/10.1111/j.1471-4159.1980.tb07100.x.

Chen, J., 2014. Heme oxygenase in neuroprotection: from mechanisms to therapeutic implications. Rev. Neurosci. 25 (2), 269–280. https://doi.org/10.1515/revneuro-2013-0046.

Chen, K., Rose, C.L., Clowes, G., 1933. Methylene blue, nitrites, and sodium thiosulphate against cyanide poisoning. Proc. Soc. Exp. Biol. Med. 31 (2), 250–251.

Chen, G., Ray, R., Dubik, D., Shi, L., Cizeau, J., Bleackley, R.C., Saxena, S., Gietz, R.D., Greenberg, A.H., 1997. The E1B 19K/Bcl-2–binding protein Nip3 is a dimeric mitochondrial protein that activates apoptosis. J. Exp. Med. 186 (12), 1975–1983.

Chen, X., Guo, C., Kong, J., 2012. Oxidative stress in neurodegenerative diseases. Neural Regen. Res. 7 (5), 376–385. https://doi.org/10.3969/j.issn.1673-5374.2012.05.009.

Chen, W., Wang, H., Wang, Z., Zhao, C., Xu, J., Chen, Q., 2020. Resolvin D1 improves Post-resuscitation cardiac and cerebral outcomes in a porcine model of cardiac arrest. Shock 54 (4), 548–554. https://doi.org/10.1097/SHK.0000000000001528.

Chen-Roetling, J., Regan, R.F., 2006. Effect of heme oxygenase-1 on the vulnerability of astrocytes and neurons to hemoglobin. Biochem. Biophys. Res. Commun. 350 (1), 233–237. https://doi.org/10.1016/j.bbrc.2006.09.036.

Cho, B.B., Toledo-Pereyra, L.H., 2008. Caspase-independent programmed cell death following ischemic stroke. J. Invest. Surg. 21 (3), 141–147. https://doi.org/10.1080/08941930802029945.

Cho, D.H., Hong, Y.M., Lee, H.J., Woo, H.N., Pyo, J.O., Mak, T.W., Jung, Y.K., 2004. Induced inhibition of ischemic/hypoxic injury by APIP, a novel Apaf-1-interacting protein. J. Biol. Chem. 279 (38), 39942–39950. https://doi.org/10.1074/jbc.M405747200.

Chodorowski, Z., Sein Anand, J., Nowak-Banasik, L., Szydlowska, M., Klimek, J., Kaletha, K., 2005. Carbon monoxide—a regulator of vascular tone in hypoxia? Przegl. Lek. 62 (6), 438–440. Retrieved from https://www.ncbi.nlm.nih.gov/pubmed/16225090.

Clemens, J.A., 2000. Cerebral ischemia: gene activation, neuronal injury, and the protective role of antioxidants. Free Radic. Biol. Med. 28 (10), 1526–1531. https://doi.org/10.1016/s0891-5849(00)00258-6.

Clifton, I., Leikin, J.B., 2003. Methylene blue. Am. J. Ther. 10, 289–291.

Cocco, T., Di Paola, M., Papa, S., Lorusso, M., 1999. Arachidonic acid interaction with the mitochondrial electron transport chain promotes reactive oxygen species generation. Free Radic. Biol. Med. 27 (1–2), 51–59. https://doi.org/10.1016/s0891-5849(99)00034-9.

Cooksey, C.J., 2017. Quirks of dye nomenclature. 8. Methylene blue, azure and violet. Biotech. Histochem. 92 (5), 347–356. https://doi.org/10.1080/10520295.2017.1315775.

Costantini, V., Lenti, M., 2002. Treatment of acute occlusion of peripheral arteries. Thromb. Res. 106 (6), V285–V294. https://doi.org/10.1016/s0049-3848(02)00104-4.

Culmsee, C., Zhu, C., Landshamer, S., Becattini, B., Wagner, E., Pellecchia, M., Blomgren, K., Plesnila, N., 2005. Apoptosis-inducing factor triggered by poly(ADP-ribose) polymerase and Bid mediates neuronal cell death after oxygen-glucose deprivation and focal cerebral ischemia. J. Neurosci. 25 (44), 10262–10272. https://doi.org/10.1523/JNEUROSCI.2818-05.2005.

Das, N.K., Mandal, B.M., 1982. Methylene blue as a retarder of free radical polymerization: 1. Polymerization of acrylonitrile, methyl methacrylate and styrene. Polymer 23 (11), 1653–1658.

Daugas, E., Susin, S.A., Zamzami, N., Ferri, K.F., Irinopoulou, T., Larochette, N., Prevost, M.C., Leber, B., Andrews, D., Penninger, J., Kroemer, G., 2000. Mitochondrio-nuclear translocation of AIF in apoptosis and necrosis. FASEB J. 14 (5), 729–739. Retrieved from https://www.ncbi.nlm.nih.gov/pubmed/10744629.

de Baaij, J.H., Hoenderop, J.G., Bindels, R.J., 2015. Magnesium in man: implications for health and disease. Physiol. Rev. 95 (1), 1–46. https://doi.org/10.1152/physrev.00012.2014.

Dean, R.T., 1987. Free radicals, membrane damage and cell-mediated cytolysis. Br. J. Cancer Suppl. 8, 39–45. Retrieved from https://www.ncbi.nlm.nih.gov/pubmed/3307872.

Deby, C., Goutier, R., 1990. New perspectives on the biochemistry of superoxide anion and the efficiency of superoxide dismutases. Biochem. Pharmacol. 39 (3), 399–405. https://doi.org/10.1016/0006-2952(90)90043-k.

Del Rio, D., Stewart, A.J., Pellegrini, N., 2005. A review of recent studies on malondialdehyde as toxic molecule and biological marker of oxidative stress. Nutr. Metab. Cardiovasc. Dis. 15 (4), 316–328. https://doi.org/10.1016/j.numecd.2005.05.003.

Dennery, P.A., 2014. Signaling function of heme oxygenase proteins. Antioxid. Redox Signal. 20 (11), 1743–1753. https://doi.org/10.1089/ars.2013.5674.

Deutsch, S.I., Rosse, R.B., Schwartz, B.L., Fay-McCarthy, M., Rosenberg, P.B., Fearing, K., 1997. Methylene blue adjuvant therapy of schizophrenia. Clin. Neuropharmacol. 20 (4), 357–363. https://doi.org/10.1097/00002826-199708000-00008.

Dey, S., DeMazumder, D., Sidor, A., Foster, D.B., O'Rourke, B., 2018. Mitochondrial ROS drive sudden cardiac death and chronic proteome remodeling in heart failure. Circ. Res. 123 (3), 356–371. https://doi.org/10.1161/CIRCRESAHA.118.312708.

Diao, M.Y., Zheng, J., Shan, Y., Xi, S., Zhu, Y., Hu, W., Lin, Z., 2020. Hypothermia prevents hippocampal oxidative stress and apoptosis via the GSK-3beta/Nrf2/HO-1 signaling pathway in a rat model of cardiac arrest-induced brain damage. Neurol. Res. 42 (9), 773–782. https://doi.org/10.1080/01616412.2020.1774210.

Dirnagl, U., Iadecola, C., Moskowitz, M.A., 1999. Pathobiology of ischaemic stroke: an integrated view. Trends Neurosci. 22 (9), 391–397. https://doi.org/10.1016/s0166-2236(99)01401-0.

DiSanto, A.R., Wagner, J.G., 1972. Pharmacokinetics of highly ionized drugs. II. Methylene blue—absorption, metabolism, and excretion in man and dog after oral administration. J. Pharm. Sci. 61 (7), 1086–1090. https://doi.org/10.1002/jps.2600610710.

Dixon, S.J., Stockwell, B.R., 2014. The role of iron and reactive oxygen species in cell death. Nat. Chem. Biol. 10 (1), 9–17. https://doi.org/10.1038/nchembio.1416.

Dodel, R.C., Du, Y., Bales, K.R., Ling, Z., Carvey, P.M., Paul, S.M., 1999. Caspase-3-like proteases and 6-hydroxydopamine induced neuronal cell death. Brain Res. Mol. Brain Res. 64 (1), 141–148. https://doi.org/10.1016/s0169-328x(98)00318-0.

Dore, S., Sampei, K., Goto, S., Alkayed, N.J., Guastella, D., Blackshaw, S., Gallagher, M., Traystman, R.J., Hurn, P.D., Koehler, R.C., Snyder, S.H., 1999. Heme oxygenase-2 is neuroprotective in cerebral ischemia. Mol. Med. 5 (10), 656–663. Retrieved from https://www.ncbi.nlm.nih.gov/pubmed/10602774.

Elmarakby, A.A., Faulkner, J., Baban, B., Saleh, M.A., Sullivan, J.C., 2012. Induction of hemeoxygenase-1 reduces glomerular injury and apoptosis in diabetic spontaneously hypertensive rats. Am. J. Physiol. Renal Physiol. 302 (7), F791–F800. https://doi.org/10.1152/ajprenal.00472.2011.

Elmer, J., Callaway, C.W., 2017. The brain after cardiac arrest. Semin. Neurol. 37 (1), 19–24. https://doi.org/10.1055/s-0036-1597833.

Elmore, S., 2007. Apoptosis: a review of programmed cell death. Toxicol. Pathol. 35 (4), 495–516. https://doi.org/10.1080/01926230701320337.

Endisch, C., Westhall, E., Kenda, M., Streitberger, K.J., Kirkegaard, H., Stenzel, W., Storm, C., Ploner, C.J., Cronberg, T., Friberg, H., Englund, E., Leithner, C., 2020. Hypoxic-ischemic encephalopathy evaluated by brain autopsy and neuroprognostication after cardiac arrest. JAMA Neurol. 77 (11), 1–10. https://doi.org/10.1001/jamaneurol. 2020.2340.

Endres, M., Namura, S., Shimizu-Sasamata, M., Waeber, C., Zhang, L., Gomez-Isla, T., Hyman, B.T., Moskowitz, M.A., 1998. Attenuation of delayed neuronal death after mild focal ischemia in mice by inhibition of the caspase family. J. Cereb. Blood Flow Metab. 18 (3), 238–247. https://doi.org/10.1097/00004647-199803000-00002.

Esdaille, C.J., Coppler, P.J., Faro, J.W., Weisner, Z.M., Condle, J.P., Elmer, J., Callaway, C.-W., Pittsburgh Post Cardiac Arrest, S., 2020. Duration and clinical features of cardiac arrest predict early severe cerebral edema. Resuscitation 153, 111–118. https://doi.org/10.1016/j.resuscitation.2020.05.049.

Ewing, J.F., Maines, M.D., 1992. In situ hybridization and immunohistochemical localization of heme oxygenase-2 mRNA and protein in normal rat brain: differential distribution of isozyme 1 and 2. Mol. Cell. Neurosci. 3 (6), 559–570. https://doi.org/10.1016/1044-7431 (92)90068-d.

Ewing, J.F., Maines, M.D., 1997. Histochemical localization of heme oxygenase-2 protein and mRNA expression in rat brain. Brain Res. Brain Res. Protoc. 1 (2), 165–174. https://doi. org/10.1016/s1385-299x(96)00027-x.

Ewing, J.F., Maines, M.D., 2006. Regulation and expression of heme oxygenase enzymes in aged-rat brain: age related depression in HO-1 and HO-2 expression and altered stress-response. J. Neural Transm. (Vienna) 113 (4), 439–454. https://doi.org/10.1007/s00702-005-0408-z.

Farber, J.L., 1982. Biology of disease: membrane injury and calcium homeostasis in the pathogenesis of coagulative necrosis. Lab. Invest. 47 (2), 114–123. Retrieved from https:// www.ncbi.nlm.nih.gov/pubmed/7109537.

Ferrer, I., Friguls, B., Dalfo, E., Justicia, C., Planas, A.M., 2003. Caspase-dependent and caspase-independent signalling of apoptosis in the penumbra following middle cerebral artery occlusion in the adult rat. Neuropathol. Appl. Neurobiol. 29 (5), 472–481. https://doi.org/10.1046/j.1365-2990.2003.00485.x.

Fichet, J., Dumas, F., Charbonneau, H., Giovanetti, O., Cariou, A., 2010. What is the outcome of cardiac arrest survivors? Presse Med. 39 (6), 694–700.

Fiskum, G., 1983. Involvement of mitochondira in ischemic cell injury and in regulation of intracellular calcium. Am. J. Emerg. Med. 1 (2), 147–153.

Fiskum, G., 1985. Mitochondrial damage during cerebral ischemia. Ann. Emerg. Med. 14 (8), 810–815. https://doi.org/10.1016/s0196-0644(85)80063-9.

Freeman, B.A., Crapo, J.D., 1982. Biology of disease: free radicals and tissue injury. Lab. Invest. 47 (5), 412–426. Retrieved from https://www.ncbi.nlm.nih.gov/pubmed/6290784.

Frisch, S., Thiel, F., Schroeter, M.L., Jentzsch, R.T., 2017. Apathy and cognitive deficits in patients with transient global ischemia after cardiac arrest. Cogn. Behav. Neurol. 30 (4), 172–175. https://doi.org/10.1097/WNN.0000000000000139.

Fujimura, M., Morita-Fujimura, Y., Murakami, K., Kawase, M., Chan, P.H., 1998. Cytosolic redistribution of cytochrome c after transient focal cerebral ischemia in rats. J. Cereb. Blood Flow Metab. 18 (11), 1239–1247. https://doi.org/10.1097/00004647-199811000-00010.

Fujimura, M., Morita-Fujimura, Y., Kawase, M., Copin, J.C., Calagui, B., Epstein, C.J., Chan, P.H., 1999. Manganese superoxide dismutase mediates the early release of mito-chondrial cytochrome C and subsequent DNA fragmentation after permanent focal cerebral ischemia in mice. J. Neurosci. 19 (9), 3414–3422. Retrieved from https://www.ncbi.nlm.nih.gov/pubmed/10212301.

Fukuda, K., Richmon, J.D., Sato, M., Sharp, F.R., Panter, S.S., Noble, L.J., 1996. Induction of heme oxygenase-1 (HO-1) in glia after traumatic brain injury. Brain Res. 736 (1–2), 68–75. https://doi.org/10.1016/0006-8993(96)00680-4.

Garnier, P., Demougeot, C., Bertrand, N., Prigent-Tessier, A., Marie, C., Beley, A., 2001. Stress response to hypoxia in gerbil brain: HO-1 and Mn SOD expression and glial acti-vation. Brain Res. 893 (1–2), 301–309. https://doi.org/10.1016/s0006-8993(01)02009-1.

Gascon, S., Sobrado, M., Roda, J.M., Rodriguez-Pena, A., Diaz-Guerra, M., 2008. Excitotoxi-city and focal cerebral ischemia induce truncation of the NR2A and NR2B subunits of the NMDA receptor and cleavage of the scaffolding protein PSD-95. Mol. Psychiatry 13 (1), 99–114. https://doi.org/10.1038/sj.mp.4002017.

Geddes, J.W., Pettigrew, L.C., Holtz, M.L., Craddock, S.D., Maines, M.D., 1996. Permanent focal and transient global cerebral ischemia increase glial and neuronal expression of heme oxygenase-1, but not heme oxygenase-2, protein in rat brain. Neurosci. Lett. 210 (3), 205–208. https://doi.org/10.1016/0304-3940(96)12703-8.

Geocadin, R.G., Koenig, M.A., Jia, X., Stevens, R.D., Peberdy, M.A., 2008. Management of brain injury after resuscitation from cardiac arrest. Neurol. Clin. 26 (2), 487–506. ix https://doi.org/10.1016/j.ncl.2008.03.015.

Geocadin, R.G, Wijdicks, E., Armstrong, M.J., Damian, M., Mayer, S.A., Ornato, J.P., Rabinstein, A., Suarez, J.I., Torbey, M.T., Dubinsky, R.M., Lazarou, J., 2017. Practice guideline summary: reducing brain injury following cardiopulmonary resuscitation: Re-port of the Guideline Development, Dissemination, and Implementation Subcommittee of the American Academy of Neurology. Pract. Guidel. Neurol. 88 (22), 2141–2149. https://doi.org/10.1212/WNL.0000000000003966.

George, J., Struthers, A.D., 2009. Role of urate, xanthine oxidase and the effects of allopurinol in vascular oxidative stress. Vasc. Health Risk Manag. 5 (1), 265–272. https://doi.org/10.2147/vhrm.s4265.

Globus, M.Y., Ginsberg, M.D., Busto, R., 1991. Excitotoxic index—a biochemical marker of selective vulnerability. Neurosci. Lett. 127 (1), 39–42. https://doi.org/10.1016/0304-3940(91)90889-2.

Go, K., 1980. Changes of brain extracellular space as reflected by the composition of brain edema fluid. Adv. Neurol. 28, 9–13.

Gonzalez-Lima, F., Auchter, A., 2015. Protection against neurodegeneration with low-dose methylene blue and near-infrared light. Front. Cell. Neurosci. 9, 179. https://doi.org/10.3389/fncel.2015.00179.

Greijer, A.E., van der Wall, E., 2004. The role of hypoxia inducible factor 1 (HIF-1) in hypoxia induced apoptosis. J. Clin. Pathol. 57 (10), 1009–1014. https://doi.org/10.1136/jcp.2003.015032.

Guerrero, A.D., Chen, M., Wang, J., 2008. Delineation of the caspase-9 signaling cascade. Apoptosis 13 (1), 177–186. https://doi.org/10.1007/s10495-007-0139-8.

Guo, K., Searfoss, G., Krolikowski, D., Pagnoni, M., Franks, C., Clark, K., Yu, K.T., Jaye, M., Ivashchenko, Y., 2001. Hypoxia induces the expression of the pro-apoptotic gene BNIP3. Cell Death Differ. 8 (4), 367–376. https://doi.org/10.1038/sj.cdd.4400810.

Hackenhaar, F.S., Medeiros, T.M., Heemann, F.M., Behling, C.S., Putti, J.S., Mahl, C.D., Verona, C., da Silva, A.C.A., Guerra, M.C., Goncalves, C.A.S., Oliveira, V.M.,

Riveiro, D.F.M., Vieira, S.R.R., Benfato, M.S., 2017. Therapeutic hypothermia reduces oxidative damage and alters antioxidant defenses after cardiac arrest. Oxid. Med. Cell. Longev. 2017, 8704352. https://doi.org/10.1155/2017/8704352.

Haggard, H.W., Greenberg, L.A., 1933. Methylene blue: a synergist, not an antidote, for carbon monoxide. JAMA 100 (25), 2001–2003.

Haglund, M., Lindberg, E., Englund, E., 2019. Hippocampus and basal ganglia as potential sentinel sites for ischemic pathology after resuscitated cardiac arrest. Resuscitation 139, 230–233. https://doi.org/10.1016/j.resuscitation.2019.04.012.

Hall, E.D., Braughler, J.M., 1986. Role of lipid peroxidation in post-traumatic spinal cord degeneration: a review. Cent. Nerv. Syst. Trauma 3 (4), 281–294. https://doi.org/10.1089/cns.1986.3.281.

Hall, E.D., Yonkers, P.A., Andrus, P.K., Cox, J.W., Anderson, D.K., 1992. Biochemistry and pharmacology of lipid antioxidants in acute brain and spinal cord injury. J. Neurotrauma 9 (Suppl. 2), S425–S442. Retrieved from https://www.ncbi.nlm.nih.gov/pubmed/1613805.

Hall, E.D., Wang, J.A., Bosken, J.M., Singh, I.N., 2016. Lipid peroxidation in brain or spinal cord mitochondria after injury. J. Bioenerg. Biomembr. 48 (2), 169–174. https://doi.org/10.1007/s10863-015-9600-5.

Hamacher-Brady, A., Brady, N.R., Logue, S.E., Sayen, M.R., Jinno, M., Kirshenbaum, L.A., Gottlieb, R.A., Gustafsson, A.B., 2007. Response to myocardial ischemia/reperfusion injury involves Bnip3 and autophagy. Cell Death Differ. 14 (1), 146–157. https://doi.org/10.1038/sj.cdd.4401936.

Hamud, F., Fiskum, G., 1985. Loss of maximal capacities for Ca-2+ accumulation and oxidative-phosphorylation by rat-brain mitochondria during cerebral-ischemia. Paper presented at the Biophys. J. 47, A414.

Hartmann, A., Hunot, S., Michel, P.P., Muriel, M.P., Vyas, S., Faucheux, B.A., Mouatt-Prigent, A., Turmel, H., Srinivasan, A., Ruberg, M., Evan, G.I., Agid, Y., Hirsch, E.C., 2000. Caspase-3: a vulnerability factor and final effector in apoptotic death of dopaminergic neurons in Parkinson's disease. Proc. Natl. Acad. Sci. U. S. A. 97 (6), 2875–2880. https://doi.org/10.1073/pnas.040556597.

Harvel, C., Arrich, J., Losert, H., Stoneham, M., Thompson, J., Delinger, R., Levy, M., Rhodes, A., Mietto, C., Pinciroli, R., 2015. Trial of early, goal-directed resuscitation for septic shock. J. Intensive Care Soc. 16 (Suppl. 4), 129–136.

Hawkes, M.A., Rabinstein, A.A., 2019. Neurological prognostication after cardiac arrest in the era of target temperature management. Curr. Neurol. Neurosci. Rep. 19 (2), 10. https://doi.org/10.1007/s11910-019-0922-2.

Hayman, E.G., Patel, A.P., Kimberly, W.T., Sheth, K.N., Simard, J.M., 2018. Cerebral edema after cardiopulmonary resuscitation: a therapeutic target following cardiac arrest? Neurocrit. Care 28 (3), 276–287. https://doi.org/10.1007/s12028-017-0474-8.

He, J.Z., Ho, J.J.D., Gingerich, S., Courtman, D.W., Marsden, P.A., Ward, M.E., 2010. Enhanced translation of heme oxygenase-2 preserves human endothelial cell viability during hypoxia. J. Biol. Chem. 285 (13), 9452–9461. https://doi.org/10.1074/jbc.M109.077230.

Heydrick, S.J., Reed, K.L., Cohen, P.A., Aarons, C.B., Gower, A.C., Becker, J.M., Stucchi, A.F., 2007. Intraperitoneal administration of methylene blue attenuates oxidative stress, increases peritoneal fibrinolysis, and inhibits intraabdominal adhesion formation. J. Surg. Res. 143 (2), 311–319. https://doi.org/10.1016/j.jss.2006.11.012.

Higgins, G.C., Beart, P.M., Shin, Y.S., Chen, M.J., Cheung, N.S., Nagley, P., 2010. Oxidative stress: emerging mitochondrial and cellular themes and variations in neuronal injury. J. Alzheimers Dis. 20 (Suppl. 2(s2)), S453–S473. https://doi.org/10.3233/JAD-2010-100321.

Hill-Kapturczak, N., Chang, S.H., Agarwal, A., 2002. Heme oxygenase and the kidney. DNA Cell Biol. 21 (4), 307–321. https://doi.org/10.1089/104454902753759726.

Hochstenbach, J.B., Anderson, P.G., van Limbeek, J., Mulder, T.T., 2001. Is there a relation between neuropsychologic variables and quality of life after stroke? Arch. Phys. Med. Rehabil. 82 (10), 1360–1366.

Hoek, T.L.V., Shao, Z., Li, C., Schumacker, P.T., Becker, L.B., 1997. Mitochondrial electron transport can become a significant source of oxidative injury in cardiomyocytes. J. Mol. Cell. Cardiol. 29 (9), 2441–2450.

Hong, S.J., Dawson, T.M., Dawson, V.L., 2004. Nuclear and mitochondrial conversations in cell death: PARP-1 and AIF signaling. Trends Pharmacol. Sci. 25 (5), 259–264. https://doi.org/10.1016/j.tips.2004.03.005.

Hopper, C.P., Meinel, L., Steiger, C., Otterbein, L.E., 2018. Where is the clinical breakthrough of heme oxygenase-1/carbon monoxide therapeutics? Curr. Pharm. Des. 24 (20), 2264–2282.

Horton, J.W., 2003. Free radicals and lipid peroxidation mediated injury in burn trauma: the role of antioxidant therapy. Toxicology 189 (1–2), 75–88. https://doi.org/10.1016/s0300-483x(03)00154-9.

Hosokawa, M., Arai, T., Masuda-Suzukake, M., Nonaka, T., Yamashita, M., Akiyama, H., Hasegawa, M., 2012. Methylene blue reduced abnormal tau accumulation in P301L tau transgenic mice. PLoS One 7 (12). https://doi.org/10.1371/journal.pone.0052389, e52389.

Hossmann, K.A., Hürter, T., Oschlies, U., 1983. The effect of dexamethasone on serum protein extravasation and edema development in experimental brain tumors of cat. Acta Neuropathol. 60 (3–4), 223–231. https://doi.org/10.1007/bf00691870.

Huang, S., Du, F., Shih, Y.Y., Shen, Q., Gonzalez-Lima, F., Duong, T.Q., 2013. Methylene blue potentiates stimulus-evoked fMRI responses and cerebral oxygen consumption during normoxia and hypoxia. Neuroimage 72, 237–242. https://doi.org/10.1016/j.neuroimage.2013.01.027.

Hung, T.C., Huang, L.W., Su, S.J., Hsieh, B.S., Cheng, H.L., Hu, Y.C., Chen, Y.H., Hwang, C.-C., Chang, K.L., 2011. Hemeoxygenase-1 expression in response to arecoline-induced oxidative stress in human umbilical vein endothelial cells. Int. J. Cardiol. 151 (2), 187–194. https://doi.org/10.1016/j.ijcard.2010.05.015.

Hunt, J.V., Dean, R.T., Wolff, S.P., 1988. Hydroxyl radical production and autoxidative glycosylation. Glucose autoxidation as the cause of protein damage in the experimental glycation model of diabetes mellitus and ageing. Biochem. J. 256 (1), 205–212.

Idris, A.H., Roberts 2nd, L.J., Caruso, L., Showstark, M., Layon, A.J., Becker, L.B., Vanden Hoek, T., Gabrielli, A., 2005. Oxidant injury occurs rapidly after cardiac arrest, cardiopulmonary resuscitation, and reperfusion. Crit. Care Med. 33 (9), 2043–2048. https://doi.org/10.1097/01.ccm.0000174104.50799.bd.

Intagliata, S., Salerno, L., Ciaffaglione, V., Leonardi, C., Fallica, A.N., Carota, G., Amata, E., Marrazzo, A., Pittala, V., Romeo, G., 2019. Heme oxygenase-2 (HO-2) as a therapeutic target: activators and inhibitors. Eur. J. Med. Chem. 183, 111703. https://doi.org/10.1016/j.ejmech.2019.111703.

Jadon, N., Jain, R., Aribam, N.G., Chauhan, P., 2017. Monitoring of endogenous antioxidants: an electroanalytical approach. J. Electrochem. Soc. 164 (4), H266–H277.

Jakkula, P., Pettila, V., Skrifvars, M.B., Hastbacka, J., Loisa, P., Tiainen, M., Wilkman, E., Toppila, J., Koskue, T., Bendel, S., Birkelund, T., Laru-Sompa, R., Valkonen, M., Reinikainen, M., COMACARE study group, 2018. Targeting low-normal or high-normal mean arterial pressure after cardiac arrest and resuscitation: a randomised pilot trial. Intensive Care Med. 44 (12), 2091–2101. https://doi.org/10.1007/s00134-018-5446-8.

Jang, D.H., Nelson, L.S., Hoffman, R.S., 2013. Methylene blue for distributive shock: a potential new use of an old antidote. J. Med. Toxicol. 9 (3), 242–249. https://doi.org/10.1007/s13181-013-0298-7.

Jang, D.H., Donovan, S., Nelson, L.S., Bania, T.C., Hoffman, R.S., Chu, J., 2015. Efficacy of methylene blue in an experimental model of calcium channel blocker-induced shock. Ann. Emerg. Med. 65 (4), 410–415. https://doi.org/10.1016/j.annemergmed.2014.09.015.

Jones, E.G., 2009. Synchrony in the interconnected circuitry of the thalamus and cerebral cortex. Ann. N. Y. Acad. Sci. 1157 (1), 10–23. https://doi.org/10.1111/j.1749-6632.2009.04534.x.

Josephson, R.A., Silverman, H.S., Lakatta, E.G., Stern, M.D., Zweier, J.L., 1991. Study of the mechanisms of hydrogen peroxide and hydroxyl free radical-induced cellular injury and calcium overload in cardiac myocytes. J. Biol. Chem. 266 (4), 2354–2361. Retrieved from https://www.ncbi.nlm.nih.gov/pubmed/1846625.

Kajimoto, M., Nuri, M., Sleasman, J.R., Charette, K.A., Nelson, B.R., Portman, M.A., 2020. Inhaled nitric oxide reduces injury and microglia activation in porcine hippocampus after deep hypothermic circulatory arrest. J. Thorac. Cardiovasc. Surg. 161, e485–e498. https://doi.org/10.1016/j.jtcvs.2019.12.075.

Kalimo, H., Fredriksson, K., Nordborg, C., Auer, R.N., Olsson, Y., Johansson, B., 1986. The spread of brain oedema in hypertensive brain injury. Med. Biol. 64 (2–3), 133–137.

Kaur, C., Ling, E.A., 2008. Blood brain barrier in hypoxic-ischemic conditions. Curr. Neurovasc. Res. 5 (1), 71–81. https://doi.org/10.2174/156720208783565645.

Keijzer, H.M., Hoedemaekers, C.W.E., Meijer, F.J.A., Tonino, B.A.R., Klijn, C.J.M., Hofmeijer, J., 2018. Brain imaging in comatose survivors of cardiac arrest: pathophysiological correlates and prognostic properties. Resuscitation 133, 124–136. https://doi.org/10.1016/j.resuscitation.2018.09.012.

Kelner, M.J., Bagnell, R., Hale, B., Alexander, N.M., 1988. Potential of methylene blue to block oxygen radical generation in reperfusion injury. In: Oxygen Radicals in Biology and Medicine. Springer, pp. 895–898.

Kemp, P.J., Williams, S.E., Mason, H.S., Wootton, P., Iles, D.E., Riccardi, D., Peers, C., 2006. Functional proteomics of BK potassium channels: defining the acute oxygen sensor. Novartis Found. Symp. 272, 141–151. discussion 151-146, 214-147. Retrieved from https://www.ncbi.nlm.nih.gov/pubmed/16686434.

Khalid, M.A., Ashraf, M., 1993. Direct detection of endogenous hydroxyl radical production in cultured adult cardiomyocytes during anoxia and reoxygenation. Is the hydroxyl radical really the most damaging radical species? Circ. Res. 72 (4), 725–736.

Kikuchi, G., Yoshida, T., Noguchi, M., 2005. Heme oxygenase and heme degradation. Biochem. Biophys. Res. Commun. 338 (1), 558–567. https://doi.org/10.1016/j.bbrc.2005.08.020.

Kim, S.D., Moon, C.K., Eun, S.-Y., Ryu, P.D., Jo, S.A., 2005. Identification of ASK1, MKK4, JNK, c-Jun, and caspase-3 as a signaling cascade involved in cadmium-induced neuronal cell apoptosis. Biochem. Biophys. Res. Commun. 328 (1), 326–334.

Kim, Y.M., Pae, H.O., Park, J.E., Lee, Y.C., Woo, J.M., Kim, N.H., Choi, Y.K., Lee, B.S., Kim, S.R., Chung, H.T., 2011. Heme oxygenase in the regulation of vascular biology: from molecular mechanisms to therapeutic opportunities. Antioxid. Redox Signal. 14 (1), 137–167. https://doi.org/10.1089/ars.2010.3153.

Kim, H.I., Lee, I.H., Park, J.S., Kim, D.M., You, Y., Min, J.H., Cho, Y.C., Jeong, W.J., Ahn, H.J., Kang, C., Lee, B.K., 2020. The usefulness of quantitative analysis of blood-brain barrier disruption measured using contrast-enhanced magnetic resonance imaging to predict neurological prognosis in out-of-hospital cardiac arrest survivors: a preliminary study. J. Clin. Med. 9 (9), 3013. https://doi.org/10.3390/jcm9093013.

Kirkland, R.A., Windelborn, J.A., Kasprzak, J.M., Franklin, J.L., 2002. A bax-induced pro-oxidant state is critical for cytochromec release during programmed neuronal death. J. Neurosci. 22 (15), 6480–6490.

Kitchin, K.T., Anderson, W.L., Suematsu, M., 2001. An ELISA assay for heme oxygenase (HO-1). J. Immunol. Methods 247 (1–2), 153–161. https://doi.org/10.1016/s0022-1759 (00)00325-2.

Kiyatkin, E.A., Sharma, H.S., 2009. Permeability of the blood-brain barrier depends on brain temperature. Neuroscience 161 (3), 926–939. https://doi.org/10.1016/j.neuroscience. 2009.04.004.

Klatzo, I., 1987. Pathophysiological aspects of brain edema. Acta Neuropathol. 72 (3), 236–239. https://doi.org/10.1007/bf00691095.

Kontos, H.A., Wei, E.P., Povlishock, J.T., Christman, C.W., 1984. Oxygen radicals mediate the cerebral arteriolar dilation from arachidonate and bradykinin in cats. Circ. Res. 55 (3), 295–303. https://doi.org/10.1161/01.res.55.3.295.

Kontos, H.A., Wei, E.P., Ellis, E.F., Jenkins, L.W., Povlishock, J.T., Rowe, G.T., Hess, M.L., 1985. Appearance of superoxide anion radical in cerebral extracellular space during increased prostaglandin synthesis in cats. Circ. Res. 57 (1), 142–151. https://doi.org/ 10.1161/01.res.57.1.142.

Kubes, P., Granger, D.N., 1992. Nitric oxide modulates microvascular permeability. Am. J. Physiol. 262 (2 Pt. 2), H611–H615. https://doi.org/10.1152/ajpheart.1992.262.2.H611.

Kuppusamy, P., Zweier, J.L., 1989. Characterization of free radical generation by xanthine oxidase. Evidence for hydroxyl radical generation. J. Biol. Chem. 264 (17), 9880–9884. Retrieved from https://www.ncbi.nlm.nih.gov/pubmed/2542334.

Kvietys, P.R., Inauen, W., Bacon, B.R., Grisham, M.B., 1989. Xanthine oxidase-induced injury to endothelium: role of intracellular iron and hydroxyl radical. Am. J. Physiol. 257 (5 Pt. 2), H1640–H1646. https://doi.org/10.1152/ajpheart.1989.257.5.H1640.

Le Greves, P., Sharma, H.S., Westman, J., Alm, P., Nyberg, F., 1997. Acute heat stress induces edema and nitric oxide synthase upregulation and down-regulates mRNA levels of the NMDAR1, NMDAR2A and NMDAR2B subunits in the rat hippocampus. In: Brain Edema X. Springer, pp. 275–278.

Le, D.A., Wu, Y., Huang, Z., Matsushita, K., Plesnila, N., Augustinack, J.C., Hyman, B.T., Yuan, J., Kuida, K., Flavell, R.A., Moskowitz, M.A., 2002. Caspase activation and neuroprotection in caspase-3- deficient mice after in vivo cerebral ischemia and in vitro oxygen glucose deprivation. Proc. Natl. Acad. Sci. U. S. A. 99 (23), 15188–15193. https://doi.org/10.1073/pnas.232473399.

Lee, B.I., Lee, D.J., Cho, K.J., Kim, G.W., 2005. Early nuclear translocation of endonuclease G and subsequent DNA fragmentation after transient focal cerebral ischemia in mice. Neurosci. Lett. 386 (1), 23–27. https://doi.org/10.1016/j.neulet.2005.05.058.

Li, H., Zhu, H., Xu, C.J., Yuan, J., 1998. Cleavage of BID by caspase 8 mediates the mitochondrial damage in the Fas pathway of apoptosis. Cell 94 (4), 491–501. https://doi.org/10.1016/s0092-8674(00)81590-1.

Li, Y., Tang, Q., Wang, P., Qin, J., Wu, H., Lin, J., Huang, Z., 2017. Dynamic changes of mitochondrial fusion and fission in brain injury after cardiac arrest in rats. Biomed. Res. Int. 2017, 1948070. https://doi.org/10.1155/2017/1948070.

Li, J., Wang, J., Shen, Y., Dai, C., Chen, B., Huang, Y., Xu, S., Wu, Y., Li, Y., 2020. Hyper-oxygenation with cardiopulmonary resuscitation and targeted temperature management improves post-cardiac arrest outcomes in rats. J. Am. Heart Assoc. 9 (19). https://doi.org/10.1161/JAHA.120.016730, e016730.

Lin, A.-L., Poteet, E., Du, F., Gourav, R.C., Liu, R., Wen, Y., Bresnen, A., Huang, S., Fox, P.T., Yang, S.-H., 2012a. Methylene blue as a cerebral metabolic and hemodynamic enhancer. PLoS One 7 (10), e46585.

Lin, C.L., Kong, Q., Cuny, G.D., Glicksman, M.A., 2012b. Glutamate transporter EAAT2: a new target for the treatment of neurodegenerative diseases. Future Med. Chem. 4 (13), 1689–1700. https://doi.org/10.4155/fmc.12.122.

Little, J.R., 1978. Modification of acute focal ischemia by treatment with mannitol. Stroke 9 (1), 4–9.

Liu, Y., Belayev, L., Zhao, W., Busto, R., Belayev, A., Ginsberg, M.D., 2001. Neuroprotective effect of treatment with human albumin in permanent focal cerebral ischemia: histopathology and cortical perfusion studies. Eur. J. Pharmacol. 428 (2), 193–201. https://doi.org/10.1016/s0014-2999(01)01255-9.

Liu, X.L., Wiklund, L., Nozari, A., Rubertsson, S., Basu, S., 2003. Differences in cerebral reperfusion and oxidative injury after cardiac arrest in pigs. Acta Anaesthesiol. Scand. 47 (8), 958–967. https://doi.org/10.1034/j.1399-6576.2003.00189.x.

Liu, D.Z., Ander, B.P., Xu, H., Shen, Y., Kaur, P., Deng, W., Sharp, F.R., 2010. Blood-brain barrier breakdown and repair by Src after thrombin-induced injury. Ann. Neurol. 67 (4), 526–533. https://doi.org/10.1002/ana.21924.

Liu, C., Zhong, L., Tian, X.L., Han, Y.C., 2018. Protective effects of 8-MOP on blood-brain barrier via the Nrf-2/HO-1 pathway in mice model of cerebral infarction. Eur. Rev. Med. Pharmacol. Sci. 22 (13), 4278–4287. https://doi.org/10.26355/eurrev_201807_15424.

Logue, S.E., Martin, S.J., 2008. Caspase activation cascades in apoptosis. Biochem. Soc. Trans. 36 (Pt. 1), 1–9. https://doi.org/10.1042/BST0360001.

Lopez Soto, C., Dragoi, L., Heyn, C.C., Kramer, A., Pinto, R., Adhikari, N.K.J., Scales, D.C., 2020. Imaging for neuroprognostication after cardiac arrest: systematic review and meta-analysis. Neurocrit. Care 32 (1), 206–216. https://doi.org/10.1007/s12028-019-00842-0.

Love, S., 2003. Apoptosis and brain ischaemia. Prog. Neuropsychopharmacol. Biol. Psychiatry 27 (2), 267–282. https://doi.org/10.1016/S0278-5846(03)00022-8.

Ma, J., Endres, M., Moskowitz, M.A., 1998. Synergistic effects of caspase inhibitors and MK-801 in brain injury after transient focal cerebral ischaemia in mice. Br. J. Pharmacol. 124 (4), 756–762. https://doi.org/10.1038/sj.bjp.0701871.

Madder, R.D., Reynolds, J.C., 2018. Multidisciplinary management of the post-cardiac arrest patient. Cardiol. Clin. 36 (1), 85–101. https://doi.org/10.1016/j.ccl.2017.08.005.

Maines, M.D., 1997. The heme oxygenase system: a regulator of second messenger gases. Annu. Rev. Pharmacol. Toxicol. 37, 517–554. https://doi.org/10.1146/annurev.pharmtox.37.1.517.

Maines, M.D., Panahian, N., 2001. The heme oxygenase system and cellular defense mechanisms. Do HO-1 and HO-2 have different functions? Adv. Exp. Med. Biol. 502, 249–272. https://doi.org/10.1007/978-1-4757-3401-0_17.

Maines, M.D., Mark, J.A., Ewing, J.F., 1993. Heme oxygenase, a likely regulator of cGMP production in the brain: induction in vivo of HO-1 compensates for depression in NO synthase activity. Mol. Cell. Neurosci. 4 (5), 396–405. https://doi.org/10.1006/mcne.1993.1050.

Mattson, M.P., 1998. Modification of ion homeostasis by lipid peroxidation: roles in neuronal degeneration and adaptive plasticity. Trends Neurosci. 21 (2), 53–57. https://doi.org/10.1016/s0166-2236(97)01188-0.

Maulik, N., Sharma, H.S., Das, D.K., 1996. Induction of the haem oxygenase gene expression during the reperfusion of ischemic rat myocardium. J. Mol. Cell. Cardiol. 28 (6), 1261–1270. https://doi.org/10.1006/jmcc.1996.0116.

Mayer, B., Brunner, F., Schmidt, K., 1993. Inhibition of nitric oxide synthesis by methylene blue. Biochem. Pharmacol. 45 (2), 367–374. https://doi.org/10.1016/0006-2952(93)90072-5.

McCall, J.M., Braughler, J.M., Hall, E.D., 1987. Lipid peroxidation and the role of oxygen radicals in CNS injury. Acta Anaesthesiol. Belg. 38 (4), 373–379. Retrieved from https://www.ncbi.nlm.nih.gov/pubmed/3126621.

McCord, J.M., 1985. Oxygen-derived free radicals in postischemic tissue injury. N. Engl. J. Med. 312 (3), 159–163. https://doi.org/10.1056/NEJM198501173120305.

Medina, D.X., Caccamo, A., Oddo, S., 2011. Methylene blue reduces abeta levels and rescues early cognitive deficit by increasing proteasome activity. Brain Pathol. 21 (2), 140–149. https://doi.org/10.1111/j.1750-3639.2010.00430.x.

Mehta, A., Prabhakar, M., Kumar, P., Deshmukh, R., Sharma, P., 2013. Excitotoxicity: bridge to various triggers in neurodegenerative disorders. Eur. J. Pharmacol. 698 (1–3), 6–18.

Meissner, P.E., Mandi, G., Coulibaly, B., Witte, S., Tapsoba, T., Mansmann, U., Rengelshausen, J., Schiek, W., Jahn, A., Walter-Sack, I., Mikus, G., Burhenne, J., Riedel, K.D., Schirmer, R.H., Kouyate, B., Muller, O., 2006. Methylene blue for malaria in Africa: results from a dose-finding study in combination with chloroquine. Malar. J. 5 (1), 84. https://doi.org/10.1186/1475-2875-5-84.

Mela, L., 1979. Reversibility of mitochondrial metabolic response to circulatory shock and tissue ischemia. Circ. Shock Suppl. 1, 61–67. Retrieved from https://www.ncbi.nlm.nih.gov/pubmed/288534.

Mergenthaler, P., Dirnagl, U., Meisel, A., 2004. Pathophysiology of stroke: lessons from animal models. Metab. Brain Dis. 19 (3–4), 151–167. https://doi.org/10.1023/b:mebr.0000043966.46964.e6.

Miclescu, A., Basu, S., Wiklund, L., 2006. Methylene blue added to a hypertonic-hyperoncotic solution increases short-term survival in experimental cardiac arrest. Crit. Care Med. 34 (11), 2806–2813. https://doi.org/10.1097/01.CCM.0000242517.23324.27.

Miclescu, A., Basu, S., Wiklund, L., 2007. Cardio-cerebral and metabolic effects of methylene blue in hypertonic sodium lactate during experimental cardiopulmonary resuscitation. Resuscitation 75 (1), 88–97. https://doi.org/10.1016/j.resuscitation.2007.03.014.

Miclescu, A., Sharma, H.S., Martijn, C., Wiklund, L., 2010. Methylene blue protects the cortical blood-brain barrier against ischemia/reperfusion-induced disruptions. Crit. Care Med. 38 (11), 2199–2206. https://doi.org/10.1097/CCM.0b013e3181f26b0c.

Middelkamp, W., Moulaert, V.R., Verbunt, J.A., van Heugten, C.M., Bakx, W.G., Wade, D.T., 2007. Life after survival: long-term daily life functioning and quality of life of patients with hypoxic brain injury as a result of a cardiac arrest. Clin. Rehabil. 21 (5), 425–431.

Miyazaki, Y., Ichinose, F., 2020. Nitric oxide in post-cardiac arrest syndrome. J. Cardiovasc. Pharmacol. 75 (6), 508–515. https://doi.org/10.1097/FJC.0000000000000765.

Morano, K.A., Grant, C.M., Moye-Rowley, W.S., 2012. The response to heat shock and oxidative stress in *Saccharomyces cerevisiae*. Genetics 190 (4), 1157–1195. https://doi.org/10.1534/genetics.111.128033.

Mortberg, E., Cumming, P., Wiklund, L., Rubertsson, S., 2009. Cerebral metabolic rate of oxygen (CMRO2) in pig brain determined by PET after resuscitation from cardiac arrest. Resuscitation 80 (6), 701–706. https://doi.org/10.1016/j.resuscitation.2009.03.005.

Moulaert, V.R., Verbunt, J.A., van Heugten, C.M., Wade, D.T., 2009. Cognitive impairments in survivors of out-of-hospital cardiac arrest: a systematic review. Resuscitation 80 (3), 297–305. https://doi.org/10.1016/j.resuscitation.2008.10.034.

Moulaert, V.R., Wachelder, E.M., Verbunt, J.A., Wade, D.T., van Heugten, C.M., 2010. Determinants of quality of life in survivors of cardiac arrest. J. Rehabil. Med. 42 (6), 553–558. https://doi.org/10.2340/16501977-0547.

Mouw, G., Zechel, J.L., Zhou, Y., Lust, W.D., Selman, W.R., Ratcheson, R.A., 2002. Caspase-9 inhibition after focal cerebral ischemia improves outcome following reversible focal ischemia. Metab. Brain Dis. 17 (3), 143–151. https://doi.org/10.1023/a:1019921904378.

Mukherjee, A., Wong, T.M., Templeton, G., Buja, L.M., Willerson, J.T., 1979. Influence of volume dilution, lactate, phosphate, and calcium on mitochondrial functions. Am. J. Physiol. 237 (2), H224–H238. https://doi.org/10.1152/ajpheart.1979.237.2.H224.

Munoz-Sanchez, J., Chanez-Cardenas, M.E., 2014. A review on hemeoxygenase-2: focus on cellular protection and oxygen response. Oxid. Med. Cell. Longev. 2014, 604981. https://doi.org/10.1155/2014/604981.

Muresanu, D.F., Sharma, A., Sharma, H.S., 2010. Diabetes aggravates heat stress-induced blood-brain barrier breakdown, reduction in cerebral blood flow, edema formation, and brain pathology: possible neuroprotection with growth hormone. Ann. N. Y. Acad. Sci. 1199 (1), 15–26. https://doi.org/10.1111/j.1749-6632.2009.05328.x.

Nakamura, T., Naguro, I., Ichijo, H., 2019. Iron homeostasis and iron-regulated ROS in cell death, senescence and human diseases. Biochim. Biophys. Acta, Gen. Subj. 1863 (9), 1398–1409. https://doi.org/10.1016/j.bbagen.2019.06.010.

Nakase, T., Sohl, G., Theis, M., Willecke, K., Naus, C.C., 2004. Increased apoptosis and inflammation after focal brain ischemia in mice lacking connexin43 in astrocytes. Am. J. Pathol. 164 (6), 2067–2075. https://doi.org/10.1016/S0002-9440(10)63765-0.

Namura, S., Zhu, J., Fink, K., Endres, M., Srinivasan, A., Tomaselli, K.J., Yuan, J., Moskowitz, M.A., 1998. Activation and cleavage of caspase-3 in apoptosis induced by experimental cerebral ischemia. J. Neurosci. 18 (10), 3659–3668. Retrieved from https://www.ncbi.nlm.nih.gov/pubmed/9570797.

Nath, K.A., Balla, G., Vercellotti, G.M., Balla, J., Jacob, H.S., Levitt, M., Rosenberg, M.E., 1992. Induction of heme oxygenase is a rapid, protective response in rhabdomyolysis in the rat. J. Clin. Invest. 90 (1), 267–270.

Naylor, G.J., Martin, B., Hopwood, S., Watson, Y., 1986. A two-year double-blind crossover trial of the prophylactic effect of methylene blue in manicdepressive psychosis. Biol. Psychiatry 21 (10), 915–920.

Neis, V.B., Rosa, P.B., Moretti, M., Rodrigues, A.L.S., 2018. Involvement of heme oxygenase-1 in neuropsychiatric and neurodegenerative diseases. Curr. Pharm. Des. 24 (20), 2283–2302. https://doi.org/10.2174/1381612824666180717160623.

Niizuma, K., Endo, H., Chan, P.H., 2009. Oxidative stress and mitochondrial dysfunction as determinants of ischemic neuronal death and survival. J. Neurochem. 109, 133–138.

Norberg, E., Gogvadze, V., Ott, M., Horn, M., Uhlen, P., Orrenius, S., Zhivotovsky, B., 2008. An increase in intracellular Ca 2+ is required for the activation of mitochondrial calpain to release AIF during cell death. Cell Death Differ. 15 (12), 1857.

Ohlow, M.J., Moosmann, B., 2011. Phenothiazine: the seven lives of pharmacology's first lead structure. Drug Discov. Today 16 (3–4), 119–131. https://doi.org/10.1016/j.drudis.2011.01.001.

Ohomoto, T., Miura, I., Konno, S., 1976. A new method of external cardiac massage to improve diastolic augmentation and prolong survival time. Ann. Thorac. Surg. 21 (4), 284–290. https://doi.org/10.1016/s0003-4975(10)64312-3.

Olmez, I., Ozyurt, H., 2012. Reactive oxygen species and ischemic cerebrovascular disease. Neurochem. Int. 60 (2), 208–212. https://doi.org/10.1016/j.neuint.2011.11.009.

Oshitari, T., Yamamoto, S., Hata, N., Roy, S., 2008. Mitochondria- and caspase-dependent cell death pathway involved in neuronal degeneration in diabetic retinopathy. Br. J. Ophthalmol. 92 (4), 552–556. https://doi.org/10.1136/bjo.2007.132308.

Ottolenghi, S., Sabbatini, G., Brizzolari, A., Samaja, M., Chiumello, D., 2020. Hyperoxia and oxidative stress in anesthesia and critical care medicine. Minerva Anestesiol. 86 (1), 64–75. https://doi.org/10.23736/S0375-9393.19.13906-5.

Oz, M., Lorke, D.E., Petroianu, G.A., 2009. Methylene blue and Alzheimer's disease. Biochem. Pharmacol. 78 (8), 927–932. https://doi.org/10.1016/j.bcp.2009.04.034.

Panchal, A.R., Berg, K.M., Hirsch, K.G., Kudenchuk, P.J., Del Rios, M., Cabanas, J.G., Link, M.S., Kurz, M.C., Chan, P.S., Morley, P.T., Hazinski, M.F., Donnino, M.W., 2019. American heart association focused update on advanced cardiovascular life support: use of advanced airways, vasopressors, and extracorporeal cardiopulmonary resuscitation during cardiac arrest: an update to the American heart association guidelines for cardiopulmonary resuscitation and emergency cardiovascular care. Circulation 140 (24), e881–e894. https://doi.org/10.1161/CIR.0000000000000732.

Pandey, A.K., Patnaik, R., Muresanu, D.F., Sharma, A., Sharma, H.S., 2012. Quercetin in hypoxia-induced oxidative stress: novel target for neuroprotection. Int. Rev. Neurobiol. 102, 107–146. https://doi.org/10.1016/B978-0-12-386986-9.00005-3.

Parfenova, H., Leffler, C.W., 2008. Cerebroprotective functions of HO-2. Curr. Pharm. Des. 14 (5), 443–453. https://doi.org/10.2174/138161208783597380.

Park, J.S., You, Y., Min, J.H., Yoo, I., Jeong, W., Cho, Y., Ryu, S., Lee, J., Kim, S.W., Cho, S.U., Oh, S.K., Ahn, H.J., Lee, J., Lee, I.H., 2019. Study on the timing of severe blood-brain barrier disruption using cerebrospinal fluid-serum albumin quotient in post cardiac arrest patients treated with targeted temperature management. Resuscitation 135, 118–123. https://doi.org/10.1016/j.resuscitation.2018.10.026.

Pasalic, D., Marinkovic, N., Feher-Turkovic, L., 2012. Uric acid as one of the important factors in multifactorial disorders–facts and controversies. Biochem. Med. (Zagreb) 22 (1), 63–75.

Pelgrims, J., De Vos, F., Van den Brande, J., Schrijvers, D., Prove, A., Vermorken, J.B., 2000. Methylene blue in the treatment and prevention of ifosfamide-induced encephalopathy: report of 12 cases and a review of the literature. Br. J. Cancer 82 (2), 291–294. https://doi.org/10.1054/bjoc.1999.0917.

Pérez-Pinzón, M.A., Xu, G.P., Born, J., Lorenzo, J., Busto, R., Rosenthal, M., Sick, T.J., 1999. Cytochrome C is released from mitochondria into the cytosol after cerebral anoxia or ischemia. J. Cereb. Blood Flow Metab. 19 (1), 39–43.

Peter, C., Hongwan, D., Kupfer, A., Lauterburg, B.H., 2000. Pharmacokinetics and organ distribution of intravenous and oral methylene blue. Eur. J. Clin. Pharmacol. 56 (3), 247–250. https://doi.org/10.1007/s002280000124.

Piantadosi, C.A., 2008. Carbon monoxide, reactive oxygen signaling, and oxidative stress. Free Radic. Biol. Med. 45 (5), 562–569. https://doi.org/10.1016/j.freeradbiomed.2008.05.013.

Plesnila, N., Zinkel, S., Amin-Hanjani, S., Qiu, J., Korsmeyer, S.J., Moskowitz, M.A., 2002. Function of BID—a molecule of the bcl-2 family—in ischemic cell death in the brain. Eur. Surg. Res. 34 (1–2), 37–41. https://doi.org/10.1159/000048885.

Poh Loh, K., Hong Huang, S., De Silva, R., Tan, H., Benny, K., Zhun Zhu, Y., 2006. Oxidative stress: apoptosis in neuronal injury. Curr. Alzheimer Res. 3 (4), 327–337.

Poteet, E., Winters, A., Yan, L.J., Shufelt, K., Green, K.N., Simpkins, J.W., Wen, Y., Yang, S.-H., 2012. Neuroprotective actions of methylene blue and its derivatives. PLoS One 7 (10). https://doi.org/10.1371/journal.pone.0048279, e48279.

Prinsze, C., Dubbelman, T.M., Van Steveninck, J., 1990. Protein damage, induced by small amounts of photodynamically generated singlet oxygen or hydroxyl radicals. Biochim. Biophys. Acta 1038 (2), 152–157. https://doi.org/10.1016/0167-4838(90)90198-o.

Quinsay, M.N., Lee, Y., Rikka, S., Sayen, M.R., Molkentin, J.D., Gottlieb, R.A., Gustafsson, A.B., 2010. Bnip3 mediates permeabilization of mitochondria and release of cytochrome c via a novel mechanism. J. Mol. Cell. Cardiol. 48 (6), 1146–1156. https://doi.org/10.1016/j.yjmcc.2009.12.004.

Radi, R., 2018. Oxygen radicals, nitric oxide, and peroxynitrite: redox pathways in molecular medicine. Proc. Natl. Acad. Sci. U. S. A. 115 (23), 5839–5848. https://doi.org/10.1073/pnas.1804932115.

Ralston, S.H., Babbs, C.F., Niebauer, M.J., 1982. Cardiopulmonary resuscitation with interposed abdominal compression in dogs. Anesth. Analg. 61 (8), 645–651. Retrieved from https://www.ncbi.nlm.nih.gov/pubmed/7201267.

Ramana, K.V., Srivastava, S., Singhal, S.S., 2017. Lipid peroxidation products in human health and disease 2016. Oxid. Med. Cell. Longev. 2017, 2163285. https://doi.org/10.1155/2017/2163285.

Ramdial, K., Franco, M.C., Estevez, A.G., 2017. Cellular mechanisms of peroxynitrite-induced neuronal death. Brain Res. Bull. 133, 4–11. https://doi.org/10.1016/j.brainresbull.2017.05.008.

Rami, A., Kogel, D., 2008. Apoptosis meets autophagy-like cell death in the ischemic penumbra: two sides of the same coin? Autophagy 4 (4), 422–426. https://doi.org/10.4161/auto.5778.

Rami, A., Sims, J., Botez, G., Winckler, J., 2003. Spatial resolution of phospholipid scramblase 1 (PLSCR1), caspase-3 activation and DNA-fragmentation in the human hippocampus after cerebral ischemia. Neurochem. Int. 43 (1), 79–87. https://doi.org/10.1016/s0197-0186(02)00194-8.

Rao, P.S., Cohen, M.V., Mueller, H.S., 1983. Production of free radicals and lipid peroxides in early experimental myocardial ischemia. J. Mol. Cell. Cardiol. 15 (10), 713–716. https://doi.org/10.1016/0022-2828(83)90260-2.

Ray, R., Chen, G., Velde, C.V., Cizeau, J., Park, J.H., Reed, J.C., Gietz, R.D., Greenberg, A.H., 2000. BNIP3 heterodimerizes with Bcl-2/Bcl-XL and induces cell death independent of a Bcl-2 homology 3 (BH3) domain at both mitochondrial and nonmitochondrial sites. J. Biol. Chem. 275 (2), 1439–1448.

Reed, D.J., Woodbury, D.M., Holtzer, R.I., 1964. Brain edema, electrolytes, and extracellular space. Effect of triethyl tin or brain and skeletal muscle. Arch. Neurol. 10 (6), 604–616. https://doi.org/10.1001/archneur.1964.00460180070007.

Rehncrona, S., Mela, L., Siesjo, B.K., 1979. Recovery of brain mitochondrial function in the rat after complete and incomplete cerebral ischemia. Stroke 10 (4), 437–446. https://doi.org/10.1161/01.str.10.4.437.

Reis, C., Akyol, O., Araujo, C., Huang, L., Enkhjargal, B., Malaguit, J., Gospodarev, V., Zhang, J.H., 2017. Pathophysiology and the monitoring methods for cardiac arrest associated brain injury. Int. J. Mol. Sci. 18 (1), 129. https://doi.org/10.3390/ijms18010129.

Rengelshausen, J., Burhenne, J., Frohlich, M., Tayrouz, Y., Singh, S.K., Riedel, K.D., Muller, O., Hoppe-Tichy, T., Haefeli, W.E., Mikus, G., Walter-Sack, I., 2004. Pharmacokinetic interaction of chloroquine and methylene blue combination against malaria. Eur. J. Clin. Pharmacol. 60 (10), 709–715. https://doi.org/10.1007/s00228-004-0818-0.

Rey-Funes, M., Larrayoz, I.M., Fernández, J.C., Contartese, D.S., Rolón, F., Inserra, P.I., Martínez-Murillo, R., López-Costa, J.J., Dorfman, V.B., Martínez, A., Loidl, C.F., 2016. Methylene blue prevents retinal damage in an experimental model of ischemic proliferative retinopathy. Am. J. Physiol. Regul. Integr. Comp. Physiol. 310 (11), R1011–R1019. https://doi.org/10.1152/ajpregu.00266.2015.

Riha, P.D., Bruchey, A.K., Echevarria, D.J., Gonzalez-Lima, F., 2005. Memory facilitation by methylene blue: dose-dependent effect on behavior and brain oxygen consumption. Eur. J. Pharmacol. 511 (2–3), 151–158. https://doi.org/10.1016/j.ejphar.2005.02.001.

Riley, P.A., 1994. Free radicals in biology: oxidative stress and the effects of ionizing radiation. Int. J. Radiat. Biol. 65 (1), 27–33. https://doi.org/10.1080/09553009414550041.

Rizzo, A.M., Berselli, P., Zava, S., Montorfano, G., Negroni, M., Corsetto, P., Berra, B., 2010. Endogenous antioxidants and radical scavengers. In: Bio-Farms for Nutraceuticals. Springer, pp. 52–67.

Rizzuto, R., Pinton, P., Ferrari, D., Chami, M., Szabadkai, G., Magalhaes, P.J., Di Virgilio, F., Pozzan, T., 2003. Calcium and apoptosis: facts and hypotheses. Oncogene 22 (53), 8619–8627. https://doi.org/10.1038/sj.onc.1207105.

Roberts, B.W., Kilgannon, J.H., Hunter, B.R., Puskarich, M.A., Pierce, L., Donnino, M., Leary, M., Kline, J.A., Jones, A.E., Shapiro, N.I., Abella, B.S., Trzeciak, S., 2018. Association between early hyperoxia exposure after resuscitation from cardiac arrest and neurological disability: prospective multicenter protocol-directed cohort study. Circulation 137 (20), 2114–2124. https://doi.org/10.1161/CIRCULATIONAHA.117.032054.

Rohnert, P., Schroder, U.H., Ziabreva, I., Tager, M., Reymann, K.G., Striggow, F., 2012. Insufficient endogenous redox buffer capacity may underlie neuronal vulnerability to cerebral ischemia and reperfusion. J. Neurosci. Res. 90 (1), 193–202. https://doi.org/10.1002/jnr.22754.

Rojas, J.C., Bruchey, A.K., Gonzalez-Lima, F., 2012. Neurometabolic mechanisms for memory enhancement and neuroprotection of methylene blue. Prog. Neurobiol. 96 (1), 32–45. https://doi.org/10.1016/j.pneurobio.2011.10.007.

Rokyta, R., Racek, J., Holecek, V., 1996. Free radicals in the central nervous system. Cesk. Fysiol. 45 (1), 4–12. Retrieved from https://www.ncbi.nlm.nih.gov/pubmed/8665612.

Rosenthal, R.E., Hamud, F., Fiskum, G., Varghese, P.J., Sharpe, S., 1987. Cerebral ischemia and reperfusion: prevention of brain mitochondrial injury by lidoflazine. J. Cereb. Blood Flow Metab. 7 (6), 752–758. https://doi.org/10.1038/jcbfm.1987.130.

Rossetti, A.O., Rabinstein, A.A., Oddo, M., 2016. Neurological prognostication of outcome in patients in coma after cardiac arrest. Lancet Neurol. 15 (6), 597–609. https://doi.org/10.1016/S1474-4422(16)00015-6.

Rud, J., May, T.L., Riker, R.R., Seder, D.B., 2020. Early cerebral edema after cardiac arrest and its ramifications. Resuscitation 154, 112–114. https://doi.org/10.1016/j.resuscitation.2020.06.025.

Ryter, S.W., Alam, J., Choi, A.M., 2006. Heme oxygenase-1/carbon monoxide: from basic science to therapeutic applications. Physiol. Rev. 86 (2), 583–650. https://doi.org/10.1152/physrev.00011.2005.

Sagi, M., Fluhr, R., 2006. Production of reactive oxygen species by plant NADPH oxidases. Plant Physiol. 141 (2), 336–340. https://doi.org/10.1104/pp.106.078089.

Saha, B.K., Burns, S.L., 2020. The story of nitric oxide, sepsis and methylene blue: a comprehensive pathophysiologic review. Am. J. Med. Sci. 360 (4), 329–337. https://doi.org/10.1016/j.amjms.2020.06.007.

Sairanen, T.R., Lindsberg, P.J., Brenner, M., Carpen, O., Siren, A., 2001. Differential cellular expression of tumor necrosis factor-alpha and type I tumor necrosis factor receptor after transient global forebrain ischemia. J. Neurol. Sci. 186 (1–2), 87–99. https://doi.org/10.1016/s0022-510x(01)00508-1.

Salaris, S.C., Babbs, C.F., Voorhees III, W.D., 1991. Methylene blue as an inhibitor of superoxide generation by xanthine oxidase: a potential new drug for the attenuation of ischemia/reperfusion injury. Biochem. Pharmacol. 42 (3), 499–506.

Sato, M., Paschen, W., Pawlik, G., Heiss, W.D., 1984. Neurologic deficit and cerebral ATP depletion after temporary focal ischemia in cats. J. Cereb. Blood Flow Metab. 4 (2), 173–177. https://doi.org/10.1038/jcbfm.1984.25.

Sato, K., Akaike, T., Kohno, M., Ando, M., Maeda, H., 1992. Hydroxyl radical production by H2O2 plus Cu,Zn-superoxide dismutase reflects the activity of free copper released from the oxidatively damaged enzyme. J. Biol. Chem. 267 (35), 25371–25377. Retrieved from https://www.ncbi.nlm.nih.gov/pubmed/1334093.

Saugstad, O.D., Hallman, M., Abraham, J.L., Epstein, B., Cochrane, C., Gluck, L., 1984. Hypoxanthine and oxygen induced lung injury: a possible basic mechanism of tissue damage? Pediatr. Res. 18 (6), 501–504. https://doi.org/10.1203/00006450-198406000-00002.

Schaller, B., Graf, R., 2004. Cerebral ischemia and reperfusion: the pathophysiologic concept as a basis for clinical therapy. J. Cereb. Blood Flow Metab. 24 (4), 351–371.

Schleien, C.L., Kuluz, J.W., Gelman, B., 1998. Hemodynamic effects of nitric oxide synthase inhibition before and after cardiac arrest in infant piglets. Am. J. Physiol. Heart Circ. Physiol. 274 (4), H1378–H1385.

Schmidley, J.W., 1990. Free radicals in central nervous system ischemia. Stroke 21 (7), 1086–1090. https://doi.org/10.1161/01.str.21.7.1086.

Schwab, M.E., Bartholdi, D., 1996. Degeneration and regeneration of axons in the lesioned spinal cord. Physiol. Rev. 76 (2), 319–370. https://doi.org/10.1152/physrev.1996.76.2.319.

Scott, A., Hunter, F.E., 1966. Support of thyroxine-induced swelling of liver mitochondria by generation of high energy intermediates at any one of three sites in electron transport. J. Biol. Chem. 241 (5), 1060–1066.

Secher, N., Ostergaard, L., Tonnesen, E., Hansen, F.B., Granfeldt, A., 2018. Impact of age on cardiovascular function, inflammation, and oxidative stress in experimental asphyxial cardiac arrest. Acta Anaesthesiol. Scand. 62 (1), 49–62. https://doi.org/10.1111/aas.13014.

Sekhon, M.S., Ainslie, P.N., Griesdale, D.E., 2017. Clinical pathophysiology of hypoxic ischemic brain injury after cardiac arrest: a "two-hit" model. Crit. Care 21 (1), 90. https://doi.org/10.1186/s13054-017-1670-9.

Semenas, E., Sharma, H.S., Wiklund, L., 2014. Adrenaline increases blood-brain-barrier permeability after haemorrhagic cardiac arrest in immature pigs. Acta Anaesthesiol. Scand. 58 (5), 620–629. https://doi.org/10.1111/aas.12293.

Seta, F., Bellner, L., Rezzani, R., Regan, R.F., Dunn, M.W., Abraham, N.G., Gronert, K., Laniado-Schwartzman, M., 2006. Heme oxygenase-2 is a critical determinant for execution of an acute inflammatory and reparative response. Am. J. Pathol. 169 (5), 1612–1623. https://doi.org/10.2353/ajpath.2006.060555.

Sharma, H.S., Ali, S.F., 2008. Acute administration of 3, 4-methylenedioxymethamphetamine induces profound hyperthermia, blood–brain barrier disruption, brain edema formation, and cell injury: an experimental study in rats and mice using biochemical and morphologic approaches. Ann. N. Y. Acad. Sci. 1139 (1), 242–258.

Sharma, H.S., Kiyatkin, E.A., 2009. Rapid morphological brain abnormalities during acute methamphetamine intoxication in the rat: an experimental study using light and electron microscopy. J. Chem. Neuroanat. 37 (1), 18–32.

Sharma, H.S., Olsson, Y., 1990. Edema formation and cellular alterations following spinal cord injury in the rat and their modification with p-chlorophenylalanine. Acta Neuropathol. 79 (6), 604–610. https://doi.org/10.1007/BF00294237.

Sharma, H.S., Sharma, A., 2010. Breakdown of the blood-brain barrier in stress alters cognitive dysfunction and induces brain pathology: new perspectives for neuroprotective strategies. In: Brain Protection in Schizophrenia, Mood and Cognitive Disorders. Springer, pp. 243–303.

Sharma, H.S., Sjoquist, P.O., 2002. A new antioxidant compound H-290/51 modulates glutamate and GABA immunoreactivity in the rat spinal cord following trauma. Amino Acids 23 (1–3), 261–272. https://doi.org/10.1007/s00726-001-0137-z.

Sharma, H.S., Westman, J., 2004. The heat shock proteins and hemeoxygenase response in central nervous system injuries. In: Blood-Spinal Cord and Brain Barriers in Health and Disease. Elsevier, Academic Press, pp. 329–360.

Sharma, H.S., Zimmer, C., Westman, J., Cervos-Navarro, J., 1992. Acute systemic heat stress increases glial fibrillary acidic protein immunoreactivity in brain: experimental observations in conscious normotensive young rats. Neuroscience 48 (4), 889–901. https://doi.org/10.1016/0306-4522(92)90277-9.

Sharma, H., Nyberg, F., Gordh, T., Alm, P., Westman, J., 1998a. Neurotrophic factors attenuate neuronal nitric oxide synthase upregulation, microvascular permeability disturbances, edema formation and cell injury in the spinal cord following trauma: an experimental study using topical application of BDNF and IGF-1 in the rat. In: Spinal Cord Monitoring. Springer, pp. 181–210.

Sharma, H.S., Alm, P., Westman, J., 1998b. Nitric oxide and carbon monoxide in the brain pathology of heat stress. In: Progress in Brain Research. vol. 115. Elsevier, pp. 297–333.

Sharma, H.S., Alm, P., Westman, J., 1998c. Nitric oxide and carbon monoxide in the brain pathology of heat stress. Prog. Brain Res. 115, 297–333. https://doi.org/10.1016/s0079-6123(08)62041-5.

Sharma, H.S., Nyberg, F., Westman, J., Alm, P., Gordh, T., Lindholm, D., 1998d. Brain derived neurotrophic factor and insulin like growth factor-1 attenuate upregulation of nitric oxide synthase and cell injury following trauma to the spinal cord. An immunohistochemical study in the rat. Amino Acids 14 (1–3), 121–129. https://doi.org/10.1007/BF01345252.

Sharma, H.S., Westman, J., Nyberg, F., 1998e. Pathophysiology of brain edema and cell changes following hyperthermic brain injury. In: Progress in Brain Research. vol. 115. Elsevier, pp. 351–412.

Sharma, H.S., Westman, J., Nyberg, F., 1998f. Pathophysiology of brain edema and cell changes following hyperthermic brain injury. Prog. Brain Res. 115, 351–412. https://doi.org/10.1016/s0079-6123(08)62043-9.

Sharma, H.S., Das, D.K., Verdouw, P.D., 1999. Enhanced expression and localization of heme oxygenase-1 during recovery phase of porcine stunned myocardium. In: Stress Adaptation, Prophylaxis and Treatment. Springer, pp. 133–139.

Sharma, H.S., Alm, P., Sjoquist, P.O., Westman, J., 2000a. A new antioxidant compound H-290/51 attenuates upregulation of constitutive isoform of heme oxygenase (HO-2) following trauma to the rat spinal cord. Acta Neurochir. Suppl. 76, 153–157. https://doi.org/10.1007/978-3-7091-6346-7_31.

Sharma, H.S., Nyberg, F., Gordh, T., Alm, P., Westman, J., 2000b. Neurotrophic factors influence upregulation of constitutive isoform of heme oxygenase and cellular stress response in the spinal cord following trauma. An experimental study using immunohistochemistry in the rat. Amino Acids 19 (1), 351–361. Retrieved from https://www.ncbi.nlm.nih.gov/pubmed/11026506.

Sharma, H.S., Westman, J., Gordh, T., Alm, P., 2000c. Topical application of brain derived neurotrophic factor influences upregulation of constitutive isoform of heme oxygenase in the spinal cord following trauma an experimental study using immunohistochemistry in the rat. Acta Neurochir. Suppl. 76, 365–369. https://doi.org/10.1007/978-3-7091-6346-7_76.

Sharma, H.S., Drieu, K., Westman, J., 2003. Antioxidant compounds EGB-761 and BN-52021 attenuate brain edema formation and hemeoxygenase expression following hyperthermic brain injury in the rat. Acta Neurochir. Suppl. 86, 313–319. https://doi.org/10.1007/978-3-7091-0651-8_68.

Sharma, H.S., Nyberg, F., Gordh, T., Alm, P., 2006. Topical application of dynorphin a (1-17) antibodies attenuates neuronal nitric oxide synthase up-regulation, edema formation, and cell injury following focal trauma to the rat spinal cord. Acta Neurochir. Suppl. 96, 309–315. https://doi.org/10.1007/3-211-30714-1_66.

Sharma, H.S., Sjoquist, P.O., Ali, S.F., 2007. Drugs of abuse-induced hyperthermia, blood-brain barrier dysfunction and neurotoxicity: neuroprotective effects of a new antioxidant compound H-290/51. Curr. Pharm. Des. 13 (18), 1903–1923. https://doi.org/10.2174/138161207780858375.

Sharma, H.S., Muresanu, D., Sharma, A., Patnaik, R., 2009. Cocaine-induced breakdown of the blood-brain barrier and neurotoxicity. Int. Rev. Neurobiol. 88, 297–334. https://doi.org/10.1016/S0074-7742(09)88011-2.

Sharma, H.S., Miclescu, A., Wiklund, L., 2011. Cardiac arrest-induced regional blood-brain barrier breakdown, edema formation and brain pathology: a light and electron microscopic study on a new model for neurodegeneration and neuroprotection in porcine brain. J. Neural Transm. (Vienna) 118 (1), 87–114. https://doi.org/10.1007/s00702-010-0486-4.

Sharma, H.S., Patnaik, R., Sharma, A., Lafuente, J.V., Miclescu, A., Wiklund, L., 2015. Cardiac arrest alters regional ubiquitin levels in association with the blood-brain barrier breakdown and neuronal damages in the porcine brain. Mol. Neurobiol. 52 (2), 1043–1053. https://doi.org/10.1007/s12035-015-9254-7.

Sharma, A., Menon, P., Muresanu, D.F., Ozkizilcik, A., Tian, Z.R., Lafuente, J.V., Sharma, H.-S., 2016. Nanowired drug delivery across the blood-brain barrier in central nervous system injury and repair. CNS Neurol. Disord. Drug Targets 15 (9), 1092–1117. https://doi.org/10.2174/1871527315666160819123059.

Sharma, A., Castellani, R.J., Smith, M.A., Muresanu, D.F., Dey, P.K., Sharma, H.S., 2019a. 5-Hydroxytryptophan: a precursor of serotonin influences regional blood-brain barrier breakdown, cerebral blood flow, brain edema formation, and neuropathology. Int. Rev. Neurobiol. 146, 1–44. https://doi.org/10.1016/bs.irn.2019.06.005.

Sharma, A., Muresanu, D.F., Ozkizilcik, A., Tian, Z.R., Lafuente, J.V., Manzhulo, I., Mossler, H., Sharma, H.S., 2019b. Sleep deprivation exacerbates concussive head injury induced brain pathology: neuroprotective effects of nanowired delivery of cerebrolysin with alpha-melanocyte-stimulating hormone. Prog. Brain Res. 245, 1–55. https://doi.org/10.1016/bs.pbr.2019.03.002.

Sharma, A., Muresanu, D.F., Castellani, R.J., Nozari, A., Lafuente, J.V., Sahib, S., Tian, Z.R., Buzoianu, A.D., Patnaik, R., Wiklund, L., Sharma, H.S., 2020a. Mild traumatic brain injury exacerbates Parkinson's disease induced hemeoxygenase-2 expression and brain pathology: neuroprotective effects of co-administration of TiO2 nanowired mesenchymal stem cells and cerebrolysin. Prog. Brain Res. 258, 157–231. https://doi.org/10.1016/bs.pbr.2020.09.010.

Sharma, H.S., Sahib, S., Tian, Z.R., Muresanu, D.F., Nozari, A., Castellani, R.J., Lafuente, J.-V., Wiklund, L., Sharma, A., 2020b. Protein kinase inhibitors in traumatic brain injury and repair: new roles of nanomedicine. Prog. Brain Res. 258, 233–283. https://doi.org/10.1016/bs.pbr.2020.09.009.

Shen, Q., Du, F., Huang, S., Rodriguez, P., Watts, L.T., Duong, T.Q., 2013. Neuroprotective efficacy of methylene blue in ischemic stroke: an MRI study. PLoS One 8 (11), e79833. https://doi.org/10.1371/journal.pone.0079833.

Shibahara, S., Han, F., Li, B., Takeda, K., 2007. Hypoxia and heme oxygenases: oxygen sensing and regulation of expression. Antioxid. Redox Signal. 9 (12), 2209–2225. https://doi.org/10.1089/ars.2007.1784.

Shin, M.H., Moon, Y.J., Seo, J.E., Lee, Y., Kim, K.H., Chung, J.H., 2008. Reactive oxygen species produced by NADPH oxidase, xanthine oxidase, and mitochondrial electron transport system mediate heat shock-induced MMP-1 and MMP-9 expression. Free Radic. Biol. Med. 44 (4), 635–645. https://doi.org/10.1016/j.freeradbiomed.2007.10.053.

Shlosberg, D., Benifla, M., Kaufer, D., Friedman, A., 2010. Blood–brain barrier breakdown as a therapeutic target in traumatic brain injury. Nat. Rev. Neurol. 6 (7), 393.

Sies, H., 2014. Role of metabolic H2O2 generation: redox signaling and oxidative stress. J. Biol. Chem. 289 (13), 8735–8741. https://doi.org/10.1074/jbc.R113.544635.

Siesjo, B.K., 1981. Cell damage in the brain: a speculative synthesis. J. Cereb. Blood Flow Metab. 1 (2), 155–185. https://doi.org/10.1038/jcbfm.1981.18.

Sisein, E.A., 2014. Biochemistry of free radicals and antioxidants. Sch. Acad. J. Biosci. 2 (2), 110–118.

Sowter, H.M., Ratcliffe, P.J., Watson, P., Greenberg, A.H., Harris, A.L., 2001. HIF-1-dependent regulation of hypoxic induction of the cell death factors BNIP3 and NIX in human tumors. Cancer Res. 61 (18), 6669–6673. Retrieved from https://www.ncbi.nlm.nih.gov/pubmed/11559532.

Stack, C., Jainuddin, S., Elipenahli, C., Gerges, M., Starkova, N., Starkov, A.A., Jove, M., Portero-Otin, M., Launay, N., Pujol, A., Kaidery, N.A., Thomas, B., Tampellini, D., Beal, M.F., Dumont, M., 2014. Methylene blue upregulates Nrf2/ARE genes and prevents tau-related neurotoxicity. Hum. Mol. Genet. 23 (14), 3716–3732. https://doi.org/10.1093/hmg/ddu080.

Su, L.J., Zhang, J.H., Gomez, H., Murugan, R., Hong, X., Xu, D., Jiang, F., Peng, Z.Y., 2019. Reactive oxygen species-induced lipid peroxidation in apoptosis, autophagy, and ferroptosis. Oxid. Med. Cell. Longev. 2019, 5080843. https://doi.org/10.1155/2019/5080843.

Sugawara, T., Fujimura, M., Morita-Fujimura, Y., Kawase, M., Chan, P.H., 1999. Mitochondrial release of cytochrome c corresponds to the selective vulnerability of hippocampal CA1 neurons in rats after transient global cerebral ischemia. J. Neurosci. 19 (22), RC39. Retrieved from https://www.ncbi.nlm.nih.gov/pubmed/10559429.

Sugawara, T., Lewen, A., Gasche, Y., Yu, F., Chan, P.H., 2002. Overexpression of SOD1 protects vulnerable motor neurons after spinal cord injury by attenuating mitochondrial cytochrome c release. FASEB J. 16 (14), 1997–1999. https://doi.org/10.1096/fj.02-0251fje.

Sugawara, T., Fujimura, M., Noshita, N., Kim, G.W., Saito, A., Hayashi, T., Narasimhan, P., Maier, C.M., Chan, P.H., 2004. Neuronal death/survival signaling pathways in cerebral ischemia. NeuroRx 1 (1), 17–25. https://doi.org/10.1602/neurorx.1.1.17.

Sutherland, B.A., Rahman, R.M., Clarkson, A.N., Shaw, O.M., Nair, S.M., Appleton, I., 2009. Cerebral heme oxygenase 1 and 2 spatial distribution is modulated following injury from hypoxia-ischemia and middle cerebral artery occlusion in rats. Neurosci. Res. 65 (4), 326–334. https://doi.org/10.1016/j.neures.2009.08.007.

Talley Watts, L., Long, J.A., Chemello, J., Van Koughnet, S., Fernandez, A., Huang, S., Shen, Q., Duong, T.Q., 2014. Methylene blue is neuroprotective against mild traumatic brain injury. J. Neurotrauma 31 (11), 1063–1071.

Tamura, T., Suzuki, M., Hayashida, K., Kobayashi, Y., Yoshizawa, J., Shibusawa, T., Sano, M., Hori, S., Sasaki, J., 2020. Hydrogen gas inhalation alleviates oxidative stress in patients with post-cardiac arrest syndrome. J. Clin. Biochem. Nutr. 67 (2), 214–221. https://doi.org/10.3164/jcbn.19-101.

Tator, C.H., Fehlings, M.G., 1991. Review of the secondary injury theory of acute spinal cord trauma with emphasis on vascular mechanisms. J. Neurosurg. 75 (1), 15–26. https://doi.org/10.3171/jns.1991.75.1.0015.

Teschendorf, P., Albertsmeier, M., Vogel, P., Padosch, S.A., Spohr, F., Kirschfink, M., Schwaninger, M., Bottiger, B.W., Popp, E., 2008. Neurological outcome and inflammation after cardiac arrest- -effects of protein C in rats. Resuscitation 79 (2), 316–324. https://doi.org/10.1016/j.resuscitation.2008.05.005.

Tesoriere, L., D'arpa, D., Conti, S., Giaccone, V., Pintaudi, A., Livrea, M., 1999. Melatonin protects human red blood cells from oxidative hemolysis: new insights into the radical-scavenging activity. J. Pineal Res. 27 (2), 95–105.

Torbey, M.T., Geocadin, R., Bhardwaj, A., 2004. Brain arrest neurological outcome scale (BrANOS): predicting mortality and severe disability following cardiac arrest. Resuscitation 63 (1), 55–63. https://doi.org/10.1016/j.resuscitation.2004.03.021.

Tsai, R.M., Boxer, A.L., 2014. Treatment of frontotemporal dementia. Curr. Treat. Options. Neurol. 16 (11), 319. https://doi.org/10.1007/s11940-014-0319-0.

Tsikas, D., 2017. Assessment of lipid peroxidation by measuring malondialdehyde (MDA) and relatives in biological samples: analytical and biological challenges. Anal. Biochem. 524, 13–30. https://doi.org/10.1016/j.ab.2016.10.021.

Uray, T., Lamade, A., Elmer, J., Drabek, T., Stezoski, J.P., Misse, A., Janesko-Feldman, K., Garman, R.H., Chen, N., Kochanek, P.M., Dezfulian, C., Callaway, C.W., Doshi, A.A., Frisch, A., Guyette, F.X., Reynolds, J.C., Rittenberger, J.C., University of Pittsburgh Post-Cardiac Arrest Service, 2018. Phenotyping cardiac arrest: bench and bedside characterization of brain and heart injury based on etiology. Crit. Care Med. 46 (6), e508–e515. https://doi.org/10.1097/CCM.0000000000003070.

Vaagenes, P., Ginsberg, M., Ebmeyer, U., Ernster, L., Fischer, M., Gisvold, S.E., Gurvitch, A., Hossmann, K.A., Nemoto, E.M., Radovsky, A., Severinghaus, J.W., Safar, P., Schlichtig, R., Sterz, F., Tonnessen, T., White, R.J., Xiao, F., Zhou, Y., 1996. Cerebral resuscitation from cardiac arrest: pathophysiologic mechanisms. Crit. Care Med. 24 (Suppl. 2), S57–S68. Retrieved from https://www.ncbi.nlm.nih.gov/pubmed/8608707.

van de Port, I.G., Kwakkel, G., Schepers, V.P., Heinemans, C.T., Lindeman, E., 2007. Is fatigue an independent factor associated with activities of daily living, instrumental activities of daily living and health-related quality of life in chronic stroke? Cerebrovasc. Dis. 23 (1), 40–45.

Velde, C.V., Cizeau, J., Dubik, D., Alimonti, J., Brown, T., Israels, S., Hakem, R., Greenberg, A., 2000. BNIP3 and genetic control of necrosis-like cell death through the mitochondrial permeability transition pore. Mol. Cell. Biol. 20 (15), 5454–5468.

Verma, A., Hirsch, D.J., Glatt, C.E., Ronnett, G.V., Snyder, S.H., 1993. Carbon monoxide: a putative neural messenger. Science 259 (5093), 381–384. https://doi.org/10.1126/science.7678352.

Visser-Meily, J.M., Rhebergen, M.L., Rinkel, G.J., van Zandvoort, M.J., Post, M.W., 2009. Long-term health-related quality of life after aneurysmal subarachnoid hemorrhage: relationship with psychological symptoms and personality characteristics. Stroke 40 (4), 1526–1529. https://doi.org/10.1161/STROKEAHA.108.531277.

Voorhees 3rd, W.D., Ralston, S.H., Babbs, C.F., 1984. Regional blood flow during cardiopulmonary resuscitation with abdominal counterpulsation in dogs. Am. J. Emerg. Med. 2 (2), 123–128. https://doi.org/10.1016/S0735-6757(84)80003-0.

Voorhees, W.D., Niebauer, M.J., Babbs, C.F., 1983. Improved oxygen delivery during cardiopulmonary resuscitation with interposed abdominal compressions. Ann. Emerg. Med. 12 (3), 128–135. https://doi.org/10.1016/s0196-0644(83)80550-2.

Wagner, S.J., Skripchenko, A., Robinette, D., Foley, J.W., Cincotta, L., 1998. Factors affecting virus photoinactivation by a series of phenothiazine dyes. Photochem. Photobiol. 67 (3), 343–349. Retrieved from https://www.ncbi.nlm.nih.gov/pubmed/9523534.

Wagner, K.R., Dean, C., Beiler, S., Bryan, D.W., Packard, B.A., Smulian, A.G., Linke, M.J., de Courten-Myers, G.M., 2005. Plasma infusions into porcine cerebral white matter induce early edema, oxidative stress, pro-inflammatory cytokine gene expression and DNA fragmentation: implications for white matter injury with increased blood-brain-barrier permeability. Curr. Neurovasc. Res. 2 (2), 149–155.

Wainwright, M., Crossley, K.B., 2002. Methylene blue-a therapeutic dye for all seasons? J. Chemother. 14 (5), 431–443.

Walker, J.W., Bruestle, J.C., White, B.C., Evans, A.T., Indreri, R., Bialek, H., 1984. Perfusion of the cerebral cortex by use of abdominal counterpulsation during cardiopulmonary resuscitation. Am. J. Emerg. Med. 2 (5), 391–393.

Wan, J., Martinvalet, D., Ji, X., Lois, C., Kaech, S.M., Von Andrian, U.H., Lieberman, J., Ahmed, R., Manjunath, N., 2003. The Bcl-2 family pro-apoptotic molecule, BNIP3 regulates activation-induced cell death of effector cytotoxic T lymphocytes. Immunology 110 (1), 10–17. https://doi.org/10.1046/j.1365-2567.2003.01710.x.

Wang, R., Wang, S.T., Wang, Y.D., Wu, G., Du, Y., Qian, M.Q., Liang, X.G., Elbatreek, M.H., Yang, H.Y., Liu, Z.R., Fukunaga, K., Liu, J.X., Lu, Y.M., 2016. Stress-responsive heme oxygenase-1 isoenzyme participates in toll-like receptor 4-induced inflammation during brain ischemia. Neuroreport 27 (6), 445–454. https://doi.org/10.1097/WNR.0000000000000561.

Wang, X., Sun, D., Hu, Y., Xu, X., Jiang, W., Shang, H., Cui, D., 2019. The roles of oxidative stress and Beclin-1 in the autophagosome clearance impairment triggered by cardiac arrest. Free Radic. Biol. Med. 136, 87–95. https://doi.org/10.1016/j.freeradbiomed.2018.12.039.

Ward, P.A., Till, G.O., Kunkel, R., Beauchamp, C., 1983. Evidence for role of hydroxyl radical in complement and neutrophil-dependent tissue injury. J. Clin. Invest. 72 (3), 789–801. https://doi.org/10.1172/JCI111050.

Warenits, A.M., Hatami, J., Mullebner, A., Ettl, F., Teubenbacher, U., Magnet, I.A.M., Bauder, B., Janata, A., Miller, I., Moldzio, R., Kramer, A.M., Sterz, F., Holzer, M., Hogler, S., Weihs, W., Duvigneau, J.C., 2020. Motor cortex and hippocampus display decreased heme oxygenase activity 2 weeks after ventricular fibrillation cardiac arrest in rats. Front. Med. (Lausanne) 7, 513. https://doi.org/10.3389/fmed.2020.00513.

Weil, M.H., von Planta, M., Gazmuri, R.J., Rackow, E.C., 1988. Incomplete global myocardial ischemia during cardiac arrest and resuscitation. Crit. Care Med. 16 (10), 997–1001. https://doi.org/10.1097/00003246-198810000-00011.

Wen, Y., Li, W., Poteet, E.C., Xie, L., Tan, C., Yan, L.-J., Ju, X., Liu, R., Qian, H., Marvin, M.-A., 2011. Alternative mitochondrial electron transfer as a novel strategy for neuroprotection. J. Biol. Chem. 286 (18), 16504–16515.

Wendel, W.B., 1939. The control of methemoglobinemia with methylene blue. J. Clin. Invest. 18 (2), 179–185. https://doi.org/10.1172/JCI101033.

White, B.C., Winegar, C.D., Jackson, R.E., Joyce, K.M., Vigor, D.N., Hoehner, T.J., Krause, G.S., Wilson, R.F., 1983. Cerebral cortical perfusion during and following resuscitation from cardiac arrest in dogs. Am. J. Emerg. Med. 1 (2), 128–138. https://doi.org/10.1016/0735-6757(83)90080-3.

White, B.C., Hildebrandt, J.F., Evans, A.T., Aronson, L., Indrieri, R.J., Hoehner, T., Fox, L., Huang, R., Johns, D., 1985. Prolonged cardiac arrest and resuscitation in dogs: brain mitochondrial function with different artificial perfusion methods. Ann. Emerg. Med. 14 (5), 383–388. https://doi.org/10.1016/s0196-0644(85)80278-x.

Wiklund, L., Basu, S., Miclescu, A., Wiklund, P., Ronquist, G., Sharma, H.S., 2007. Neuro- and cardioprotective effects of blockade of nitric oxide action by administration of methylene blue. Ann. N. Y. Acad. Sci. 1122, 231–244. https://doi.org/10.1196/annals.1403.016.

Wiklund, L., Martijn, C., Miclescu, A., Semenas, E., Rubertsson, S., Sharma, H.S., 2012. Central nervous tissue damage after hypoxia and reperfusion in conjunction with cardiac arrest and cardiopulmonary resuscitation: mechanisms of action and possibilities for mitigation. Int. Rev. Neurobiol. 102, 173–187. https://doi.org/10.1016/B978-0-12-386986-9.00007-7.

Wiklund, L., Zoerner, F., Semenas, E., Miclescu, A., Basu, S., Sharma, H.S., 2013. Improved neuroprotective effect of methylene blue with hypothermia after porcine cardiac arrest. Acta Anaesthesiol. Scand. 57 (8), 1073–1082. https://doi.org/10.1111/aas.12106.

Wiklund, L., Patnaik, R., Sharma, A., Miclescu, A., Sharma, H.S., 2018. Cerebral tissue oxidative ischemia-reperfusion injury in connection with experimental cardiac arrest and cardiopulmonary resuscitation: effect of mild hypothermia and methylene blue. Mol. Neurobiol. 55 (1), 115–121. https://doi.org/10.1007/s12035-017-0723-z.

Williams, S.E., Wootton, P., Mason, H.S., Bould, J., Iles, D.E., Riccardi, D., Peers, C., Kemp, P.J., 2004. Hemoxygenase-2 is an oxygen sensor for a calcium-sensitive potassium channel. Science 306 (5704), 2093–2097. https://doi.org/10.1126/science.1105010.

Wiltgen, B.J., Zhou, M., Cai, Y., Balaji, J., Karlsson, M.G., Parivash, S.N., Li, W., Silva, A.J., 2010. The hippocampus plays a selective role in the retrieval of detailed contextual memories. Curr. Biol. 20 (15), 1336–1344. https://doi.org/10.1016/j.cub.2010.06.068.

Winkler, T., Sharma, H., Stålberg, E., Westman, J., 1998. Spinal cord bioelectrical activity, edema and cell injury following a focal trauma to the rat spinal cord. An experimental study using pharmacological and morphological approaches. In: Spinal Cord Monitoring. Springer, pp. 283–363.

Wolff, S.P., Garner, A., Dean, R.T., 1986. Free radicals, lipids and protein degradation. Trends Biochem. Sci. 11 (1), 27–31.

Wollborn, J., Schlueter, B., Steiger, C., Hermann, C., Wunder, C., Schmidt, J., Diel, P., Meinel, L., Buerkle, H., Goebel, U., Schick, M.A., 2019. Extracorporeal resuscitation with carbon monoxide improves renal function by targeting inflammatory pathways in cardiac arrest in pigs. Am. J. Physiol. Renal Physiol. 317 (6), F1572–F1581. https://doi.org/10.1152/ajprenal.00241.2019.

Wollborn, J., Steiger, C., Doostkam, S., Schallner, N., Schroeter, N., Kari, F.A., Meinel, L., Buerkle, H., Schick, M.A., Goebel, U., 2020. Carbon monoxide exerts functional neuroprotection after cardiac arrest using extracorporeal resuscitation in pigs. Crit. Care Med. 48 (4), e299–e307. https://doi.org/10.1097/CCM.0000000000004242.

Wright, R.O., Lewander, W.J., Woolf, A.D., 1999. Methemoglobinemia: etiology, pharmacology, and clinical management. Ann. Emerg. Med. 34 (5), 646–656. https://doi.org/10.1016/s0196-0644(99)70167-8.

Wrubel, K.M., Riha, P.D., Maldonado, M.A., McCollum, D., Gonzalez-Lima, F., 2007. The brain metabolic enhancer methylene blue improves discrimination learning in rats. Pharmacol. Biochem. Behav. 86 (4), 712–717. https://doi.org/10.1016/j.pbb.2007.02.018.

Wu, J., Li, Y., Yang, P., Huang, Y., Lu, S., Xu, F., 2019. Novel role of carbon monoxide in improving neurological outcome after cardiac arrest in aged rats: involvement of inducing mitochondrial autophagy. J. Am. Heart Assoc. 8 (9). https://doi.org/10.1161/JAHA.118.011851, e011851.

Wu, J., Li, Z., Yuan, W., Zhao, Y., Li, J., Li, Z., Li, J., Li, C., 2020. Changes of endothelin-1 and nitric oxide systems in brain tissue during mild hypothermia in a porcine model of cardiac arrest. Neurocrit. Care 33 (1), 73–81. https://doi.org/10.1007/s12028-019-00855-9.

Xiao, D., Zikopoulos, B., Barbas, H., 2009. Laminar and modular organization of prefrontal projections to multiple thalamic nuclei. Neuroscience 161 (4), 1067–1081. https://doi.org/10.1016/j.neuroscience.2009.04.034.

Yakovlev, A.G., Faden, A.I., 2001. Caspase-dependent apoptotic pathways in CNS injury. Mol. Neurobiol. 24 (1–3), 131–144. https://doi.org/10.1385/MN:24:1-3:131.

Yakubov, E., Gottlieb, M., Gil, S., Dinerman, P., Fuchs, P., Yavin, E., 2004. Overexpression of genes in the CA1 hippocampus region of adult rat following episodes of global ischemia. Brain Res. Mol. Brain Res. 127 (1–2), 10–26. https://doi.org/10.1016/j.molbrainres.2004.05.010.

Yet, S.F., Perrella, M.A., Layne, M.D., Hsieh, C.M., Maemura, K., Kobzik, L., Wiesel, P., Christou, H., Kourembanas, S., Lee, M.E., 1999. Hypoxia induces severe right ventricular dilatation and infarction in heme oxygenase-1 null mice. J. Clin. Invest. 103 (8), R23–R29. https://doi.org/10.1172/JCI6163.

Yin, L., Ohtaki, H., Nakamachi, T., Kudo, Y., Makino, R., Shioda, S., 2004. Delayed expressed TNFR1 co-localize with ICAM-1 in astrocyte in mice brain after transient focal ischemia. Neurosci. Lett. 370 (1), 30–35. https://doi.org/10.1016/j.neulet.2004.07.083.

Yoo, A.J., Sheth, K.N., Kimberly, W.T., Chaudhry, Z.A., Elm, J.J., Jacobson, S., Davis, S.M., Donnan, G.A., Albers, G.W., Stern, B.J., Gonzalez, R.G., 2013. Validating imaging biomarkers of cerebral edema in patients with severe ischemic stroke. J. Stroke Cerebrovasc. Dis. 22 (6), 742–749. https://doi.org/10.1016/j.jstrokecerebrovasdis.2012.01.002.

Young, W., 1988. Session 2: secondary CNS injury. J. Neurotrauma 5 (3), 219–221.

Yu, S.P., Canzoniero, L.M., Choi, D.W., 2001. Ion homeostasis and apoptosis. Curr. Opin. Cell Biol. 13 (4), 405–411. https://doi.org/10.1016/s0955-0674(00)00228-3.

Yuan, Y., Kitts, D., 1997. Endogenous antioxidants: role of antioxidant enzymes in biological systems. In: Natural Antioxidants: Chemistry, Health Effects, and Applications. 15. AOCS Press, Champaign, Illinois, pp. 258–270.

Zador, Z., Stiver, S., Wang, V., Manley, G.T., 2009. Role of aquaporin-4 in cerebral edema and stroke. In: Aquaporins. Springer, pp. 159–170.

Zhang, J., Ney, P.A., 2009. Role of BNIP3 and NIX in cell death, autophagy, and mitophagy. Cell Death Differ. 16 (7), 939–946. https://doi.org/10.1038/cdd.2009.16.

Zhang, X., Rojas, J.C., Gonzalez-Lima, F., 2006. Methylene blue prevents neurodegeneration caused by rotenone in the retina. Neurotox. Res. 9 (1), 47–57. https://doi.org/10.1007/BF03033307.

Zhang, D., Li, Y., Heims-Waldron, D., Bezzerides, V., Guatimosim, S., Guo, Y., Gu, F., Zhou, P., Lin, Z., Ma, Q., Liu, J., Wang, D.Z., Pu, W.T., 2018. Mitochondrial cardiomyopathy caused by elevated reactive oxygen species and impaired cardiomyocyte proliferation. Circ. Res. 122 (1), 74–87. https://doi.org/10.1161/CIRCRESAHA.117.311349.

Zhang, Y., Zhang, J.J., Liu, X.H., Wang, L., 2020. CBX7 suppression prevents ischemia-reperfusion injury-induced endoplasmic reticulum stress through the Nrf-2/HO-1 pathway. Am. J. Physiol. Renal Physiol. 318 (6), F1531–F1538. https://doi.org/10.1152/ajprenal.00088.2020.

Zhao, T., Chen, S., Wang, B., Cai, D., 2020. L-carnitine reduces myocardial oxidative stress and alleviates myocardial ischemia-reperfusion injury by activating nuclear transcription-related factor 2 (Nrf2)/heme oxygenase-1 (HO-1) signaling pathway. Med. Sci. Monit. 26, e923251. https://doi.org/10.12659/MSM.923251.

Multimodal imaging in the differential diagnosis of glioma recurrence from treatment-related effects: A protocol for systematic review and network meta-analysis

Huijing Chen[a,†], Yanwen Luo[b,†], Cong Li[c,†], Wengang Zhan[c], Qijia Tan[c], Caijun Xie[c], Aruna Sharma[d,*], Hari Shanker Sharma[d,*], and Zhiqiang Zhang[c,*]

[a]*Guangzhou University of Chinese Medicine, Guangzhou, China*
[b]*Qionghai Hospital of traditional Chinese medicine, Qionghai, Hainan Province, China*
[c]*The Second Affiliated Hospital of Guangzhou University of Chinese Medicine, Guangdong Province Hospital of Chinese Medical, Guangzhou, China*
[d]*International Experimental Central Nervous System Injury & Repair (IECNSIR), Department of Surgical Sciences, Anesthesiology & Intensive Care Medicine, Uppsala University Hospital, Uppsala University, Uppsala, Sweden*
*Corresponding authors: Tel.: +46-70-21-95-963; Fax: +46-18-24-38-99 (Aruna Sharma); Tel.: +46-70-20-11-801; Fax: +46-18-24-38-99 (Hari Shanker Sharma; Aruna Sharma); Tel.: +86-13826296218; Fax: +86-02081887233 (Zhiqiang Zhang), e-mail address: aruna.sharma@surgsci.uu.se; sharma@surgsci.uu.se; doctorzzq@163.com

Abstract

Background: Glioma is the most common malignant primary brain tumor and it will always recur. To date, various multimodal imaging including magnetic resonance imaging (MRI) and positron emission tomography computed tomography (PET/CT) was used to differentiate the diagnosis of true tumor recurrent (TuR) and treatment-related effects (TrE) in glioma patient but with no overall conclusion. In this study, SROC curve and Bayesian network meta-analysis

---

[†]Contributed equally to this article.

Progress in Brain Research, Volume 265, ISSN 0079-6123, https://doi.org/10.1016/bs.pbr.2021.06.011

377

will be used to conduct a comprehensive analysis of the results of different clinical reports, and assess the efficacy of multimodal imaging in difference TuR and TrE.

Methods: To find more comprehensive information about the application of multimodal imaging in glioma patients, we searched the EMBASE, Pubmed, and Cochrane Central Register of Controlled Trials for relevant clinical trials. We also reviewed their reference lists to avoid omissions. QUADAS-2, RevMan software, Stata, and R software will be used.

Results: This study will provide reliable evidence for the efficacy of multimodal imaging in the differential diagnosis of TuR and TrE in glioma patients.

Conclusion: We will evaluate the effectiveness of different and rank each imaging method in glioma patients to provide a decision-making reference on which method to choose for clinicians.

Protocol registration number: CRD42020217861.

Keywords

Glioma, Treatment-related effects, Pseudoprogression, Protocol, Network meta-analysis

1 Introduction

Glioma is a tumor that originates from the glial cells of the brain or spinal cord, which is the most common brain tumor with a high recurrence rate and fatality rate (Ostrom et al., 2020). The current standard therapeutic strategy for high-grade glioma involves maximal safe surgical resection, followed by radiation or chemoradiation therapy (Wen and Kesari, 2008). Treatment-related effects (TrE) include early pseudoprogression (PsP) and late radiation necrosis (RN) which are mainly caused by radiation and related to the destruction of the blood-brain barrier (BBB) (Delgado-Lopez et al., 2018; Sharma et al., 2020). PsP often occurs in glioma-treated patients (between 10% and 30%) in around several weeks up to 4 months after radiotherapy (Zikou et al., 2018). During this time, no further biopsy or surgery is reasonable, imaging is believed to be a reliably detect progression. However, differentiating tumor recurrence (TuR) from TrE remains challenging due to overlapping imaging features in conventional structural magnetic resonance imaging (MRI). For example, newly developed or increased size of enhancing lesions with surrounding edema may occur in both TuR and TrE. Nowadays, there are various imaging methods were widely used in glioma, but no single modality can provide a reliable diagnosis between TuR from TrE, causing uncertain clinical judgment and high inspection fees. MRI is the most commonly used imaging method in glioma diagnosis and treatment follow-up because of it can well show brain tumor lesions. In recent years, advanced MR imaging such as diffusion MRI sequences, amide proton transfer MRI sequences, perfusion MRI sequences, and MRI spectroscopy imaging were providing more structural and metabolic information (Li et al., 2020). In addition to MRI examination, clinicians also often use positron emission tomography-computed tomography (PET/CT) to assistance in the diagnosis when encountering suspected recurrent gliomas. TuR is usually hypermetabolic in PET/CT which can provide more metabolic information. The most commonly used PET/CT tracers for imaging of gliomas are ^{18}F-fluorodeoxyglucose (^{18}F-FDG), ^{13}N-ammonia (^{13}N-NH$_3$), ^{11}C-methylmethionine (^{11}C-MET), ^{18}F-fluoroethyl-L-tyrosine (^{18}F-FET), and ^{18}F-fluoro-L-dihydroxyphenylalanine (^{18}F-FDOPA) (Li et al., 2020). The Response Assessment in Neuro-Oncology working group and European Association for Neuro-Oncology have recommended the use of PET/CT imaging in glioma and they emphasize that PET/CT has higher diagnostic accuracy than MR when differentiation of TuR from TrE (Albert et al., 2016). However, there is currently no standard imaging technology to achieve a differential diagnosis and the evaluation results of different multimodal imaging including MR and PET/CT used in this critical issue of glioma follow-up process are also mixed. What's more, the imaging diagnosis of recurrent gliomas is particularly important. If necessary, surgery should be performed again (Salvati et al., 2019; Zhong et al., 2019). We are going to a systematic review and network meta-analysis the clinical potential of multimodal imaging including MRI and PET/CT in the differential diagnosis of TuR and TrE in glioma patients, to improve the efficiency of clinical diagnosis.

2 Methods

2.1 Literature search

A thorough and exhaustive searching method was conducted for relevant articles published before August 18th, 2019 in the PubMed, Embase and Cochrane Library with ("glioblastoma" or "glioma" or "gliomas" or "astrocytoma" or "glial cell tumors" or "glial cell tumor" or "tumor, glial cell" or "tumors, glial cell" or "mixed glioma" or "glioma, mixed" or "glioma, mixed" or "mixed gliomas" or "malignant glioma" or "glioma, malignant" or "gliomas, malignant" or "malignant gliomas") and ("DWI" or "ADC" or "PWI" or "MRS" or "DOPA" or "NH3" or "FDG" or (PET) or (PET/CT)) and ("malignant gliomas" or "treatment-induced lesions" or "treatment-induced changes" or "treatment-induced changes" or "radiation-induced brain injury" or "treatment-related effects" or "treatment effects" or "post-treatment related changes" or "pseudoprogression" or "radionecrosis" or "radiation necrosis" or "radiation-induced cerebral necrosis"). Articles in English or Chinese were selected. We also will extensively scan the references cited in the retrieved articles to find other articles that may meet the criteria. Before the conclusion is made, if any valuable imaging diagnosis method is discovered, we will include it in the systematic review.

2.2 Selection of studies and data collection process

After excluding duplicates using Endnote, two authors will screen the titles and abstracts of all records separately, choosing those meeting the eligibility criteria. Case reports, reviews, letters and irrelevant articles will be excluded. The third author as an arbitrator will make the final reasonable decision between the two authors in disagreement.

Included data will be selected based on the following criteria: (i) patients are histopathologically diagnosed with glioma, and has a history of treatment with surgery and radiotherapy or chemotherapy; (ii) diagnostic equipment is performed by DWI or PWI or MRS or PET; (iii) histological confirmation or follow-up by MRI is applied as a golden standard; (iv) the number of valid cases is greater than 10; (v) retrospective and prospective studies are allowed; (vi) extracted data contained sensitivity, specificity, AUC, and general characteristics. For more potential results, true positives, false positives, true negatives, and false negatives are welcomed to be recorded; (vii) general characteristics included total first author, published time, number of patients, gender, study design, age, tumor histology, cut-off (if provided in the paper), used reference standard.

2.3 Quality evaluation

Study quality was assessed according to the quality assessment of diagnostic accuracy studies (QUADAS-2) (Whiting et al., 2011). All data will be reviewed and separately extracted by two independent investigators, and the third author will act as an arbitrator. The flow chart is shown in Fig. 1.

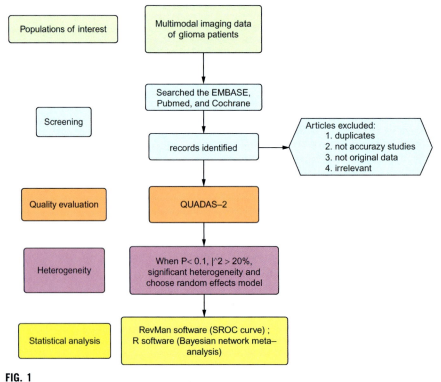

FIG. 1

Flow diagram on this systematic review and network meta-analysis.

2.4 **Geometry of the network**

Stata SE 15.0 will be used to draw network plots, where the size of nodes represents the number of studies evaluating each treatment, and the thickness of the lines between the nodes represents the number of direct comparisons between tests.

2.5 **Statistical analysis**

For heterogeneity, we will use the Q value and the inconsistency index ($P > 0.05$ and $I^2 < 50\%$ will be considered low heterogeneity, a fixed model will be adopted). The "multiple test analysis" of the diagnostic test module in RevMan software can display multiple diagnostics in the same graph. The SROC curve takes into account both sensitivity and specificity, and comprehensively reflects the diagnostic value of the diagnostic test for the target disease. The test SROC curve chart compares the accuracy of each diagnostic test by observing the area under the different SROC curve. We use the ANOVA model in R software to realize the Bayesian network meta-analysis of the diagnostic test accuracy and calculate the diagnostic accuracy of the diagnostic technology. This method can provide relative sensitivity, relative specificity, and diagnostic test ranking results.

3 **Protocol and registration**

For accuracy, we followed the preferred reporting items for systematic review and meta-analysis protocols (PRISMA-P 2015) (Moher et al., 2015). We've registered this protocol on the international prospective register of Systematic Review (PROSPERO). The website is https://www.crd.york.ac.uk/PROSPERO/ and the registration number is CRD42020217861.

4 **Discussion**

Glioma is the most common human primary brain tumor (Ostrom et al., 2020). Despite the first-line treatments including surgical resection, radiation, and chemotherapy, the survival time for the most malignant glioblastoma patients is still only 12–15 months (Wen and Kesari, 2008). Recurrence is an important cause of poor prognosis, early detection and diagnosis of recurrent glioma may improve the prognosis of patients. But the diagnosis of recurrent glioma is difficult through imaging because of radiotherapy is associated with a high incidence of TrE, including pseudoprogression and radiation necrosis. This phenomenon is believed to be related to the destruction of the blood-brain barrier after radiation therapy and chemotherapy leading to increased leakiness of contrast agent showing enhancement in the area on MR imaging which is similar to the imaging findings of TuR (Hygino et al., 2011; Sharma et al., 2020). Even though advanced multimodal MR imaging plays an important role in the post-treatment follow-up of glioma patient's care in recent years

(Hygino et al., 2011; Liu et al., 2020) and the PET/CT play as a molecular imaging technique allowing quantitative measurement of biological processes non-invasively, which has become an integral supplemental imaging tool for differential diagnosis beyond MRI (Albert et al., 2016). The diagnosis of TuR requires a change in the therapeutic approach to a more active anti-tumor strategy, while a diagnosis of TrE supports the effectiveness of current treatment. Therefore, accurate and reliable imaging evaluation is very important for making clinical decisions and some studies are also exploring these issues. In 2015, Wang et al. compared magnetic resonance spectroscopy (MRS) and PET/CT in the detection of tumor recurrence in post-treatment of glioma, concluded that MRS is highly sensitive in the detection of tumor recurrence in glioma (Wang et al., 2015). And they found that ^{18}F-FDG PET/CT was highly specific in recurrence diagnosis and ^{11}C-MET PET/CT did not have a noticeable advantage over ^{18}F-FDG PET/CT. Bart R. J. van Dijken et al. performed a systematic meta-analysis showing that advanced techniques had higher diagnostic accuracy than anatomical MRI (Abbasi et al., 2018). Besides, there was a report that ^{11}C-choline PET has high diagnostic accuracy for the identification of tumor relapse from radiation induced necrosis in glioma (Gao et al., 2018). There are still many advanced MRI to detecting the TuR from TrE (Strauss et al., 2021). Because of the vague diagnosis, clinicians cannot make accurate judgments and corresponding clinical decisions. How to distinguish treatment-related necrosis and recurrent glioma has become a clinical research hotspot of neurology. In this systematic review and network meta-analysis, we will analyze recent applications of advanced MRI imaging and advanced PET/CT with multiple nuclides in the evaluation of differential diagnosis of TrE from TuR, to provide an important reference for clinical application.

Funding

Scientific research project of Guangdong Bureau of traditional Chinese Medicine (No.20211159).

References

Abbasi, A.W., Westerlaan, H.E., Holtman, G.A., Aden, K.M., van Laar, P.J., van der Hoorn, A., 2018. Incidence of tumour progression and pseudoprogression in high-grade gliomas: a systematic review and meta-analysis. Clin. Neuroradiol. 28, 401–411.

Albert, N.L., Weller, M., Suchorska, B., Galldiks, N., Soffietti, R., Kim, M.M., la Fougere, C., Pope, W., Law, I., Arbizu, J., Chamberlain, M.C., Vogelbaum, M., Ellingson, B.M., Tonn, J.C., 2016. Response Assessment in Neuro-Oncology working group and European Association for Neuro-Oncology recommendations for the clinical use of PET imaging in gliomas. Neuro Oncol. 18, 1199–1208.

Delgado-Lopez, P.D., Rinones-Mena, E., Corrales-Garcia, E.M., 2018. Treatment-related changes in glioblastoma: a review on the controversies in response assessment criteria and the concepts of true progression, pseudoprogression, pseudoresponse and radionecrosis. Clin. Transl. Oncol. 20, 939–953.

Gao, L., Xu, W., Li, T., Zheng, J., Chen, G., 2018. Accuracy of 11C-choline positron emission tomography in differentiating glioma recurrence from radiation necrosis: a systematic review and meta-analysis. Medicine (Baltimore) 97, e11556.

Hygino, D.C.L.J., Rodriguez, I., Domingues, R.C., Gasparetto, E.L., Sorensen, A.G., 2011. Pseudoprogression and pseudoresponse: imaging challenges in the assessment of post-treatment glioma. AJNR Am. J. Neuroradiol. 32, 1978–1985.

Li, C., Gan, Y., Chen, H., Chen, Y., Deng, Y., Zhan, W., Tan, Q., Xie, C., Sharma, H.S., Zhang, Z., 2020. Advanced multimodal imaging in differentiating glioma recurrence from post-radiotherapy changes. Int. Rev. Neurobiol. 151, 281–297.

Liu, J., Li, C., Chen, Y., Lv, X., Lv, Y., Zhou, J., Xi, S., Dou, W., Qian, L., Zheng, H., Wu, Y., Chen, Z., 2020. Diagnostic performance of multiparametric MRI in the evaluation of treatment response in glioma patients at 3T. J. Magn. Reson. Imaging 51, 1154–1161.

Moher, D., Shamseer, L., Clarke, M., Ghersi, D., Liberati, A., Petticrew, M., Shekelle, P., Stewart, L.A., 2015. Preferred reporting items for systematic review and meta-analysis protocols (PRISMA-P) 2015 statement. Syst. Rev. 4, 1.

Ostrom, Q.T., Patil, N., Cioffi, G., Waite, K., Kruchko, C., Barnholtz-Sloan, J.S., 2020. CBTRUS statistical report: primary brain and other central nervous system tumors diagnosed in the United States in 2013-2017. Neuro Oncol. 22, v1–v96.

Salvati, M., Pesce, A., Palmieri, M., Floriana Brunetto, G.M., Santoro, A., Frati, A., 2019. The role and real effect of an iterative surgical approach for the management of recurrent high-grade glioma: an observational analytic cohort study. World Neurosurg. 124, e480–e488.

Sharma, H.S., Muresanu, D.F., Castellani, R.J., Nozari, A., Lafuente, J.V., Tian, Z.R., Sahib, S., Bryukhovetskiy, I., Bryukhovetskiy, A., Buzoianu, A.D., Patnaik, R., Wiklund, L., Sharma, A., 2020. Pathophysiology of blood-brain barrier in brain tumor. Novel therapeutic advances using nanomedicine. Int. Rev. Neurobiol. 151, 1–66.

Strauss, S.B., Meng, A., Ebani, E.J., Chiang, G.C., 2021. Imaging glioblastoma posttreatment: progression, pseudoprogression, pseudoresponse, radiation necrosis. Neuroimaging Clin. N. Am. 31, 103–120.

Wang, X., Hu, X., Xie, P., Li, W., Li, X., Ma, L., 2015. Comparison of magnetic resonance spectroscopy and positron emission tomography in detection of tumor recurrence in post-treatment of glioma: a diagnostic meta-analysis. Asia Pac. J. Clin. Oncol. 11, 97–105.

Wen, P.Y., Kesari, S., 2008. Malignant gliomas in adults. N. Engl. J. Med. 359, 492–507.

Whiting, P.F., Rutjes, A.W., Westwood, M.E., Mallett, S., Deeks, J.J., Reitsma, J.B., Leeflang, M.M., Sterne, J.A., Bossuyt, P.M., 2011. QUADAS-2: a revised tool for the quality assessment of diagnostic accuracy studies. Ann. Intern. Med. 155, 529–536.

Zhong, D., Hai, Y., Ning, W., Wahap, A., Jia, W., Tuo, W., Changwang, D., Maode, W., 2019. Impact of preoperative Karnofsky Performance Scale (KPS) and American Society of Anesthesiologists (ASA) scores on perioperative complications in patients with recurrent glioma undergoing repeated operation. J. Neurorestoratology 07, 143–152.

Zikou, A., Sioka, C., Alexiou, G.A., Fotopoulos, A., Voulgaris, S., Argyropoulou, M.I., 2018. Radiation necrosis, pseudoprogression, pseudoresponse, and tumor recurrence: imaging challenges for the evaluation of treated gliomas. Contrast Media Mol. Imaging 2018, 6828396.

Manganese nanoparticles induce blood-brain barrier disruption, cerebral blood flow reduction, edema formation and brain pathology associated with cognitive and motor dysfunctions

Aruna Sharma[a],*, Lianyuan Feng[b], Dafin F. Muresanu[c,d], Seaab Sahib[e], Z. Ryan Tian[e], José Vicente Lafuente[f], Anca D. Buzoianu[g], Rudy J. Castellani[h], Ala Nozari[i], Lars Wiklund[a], and Hari Shanker Sharma[a],*

[a]*International Experimental Central Nervous System Injury & Repair (IECNSIR), Department of Surgical Sciences, Anesthesiology & Intensive Care Medicine, Uppsala University Hospital, Uppsala University, Uppsala, Sweden*
[b]*Department of Neurology, Bethune International Peace Hospital, Shijiazhuang, China*
[c]*Department of Clinical Neurosciences, University of Medicine & Pharmacy, Cluj-Napoca, Romania*
[d]*"RoNeuro" Institute for Neurological Research and Diagnostic, Cluj-Napoca, Romania*
[e]*Department of Chemistry & Biochemistry, University of Arkansas, Fayetteville, AR, United States*
[f]*LaNCE, Department of Neuroscience, University of the Basque Country (UPV/EHU), Leioa, Bizkaia, Spain*
[g]*Department of Clinical Pharmacology and Toxicology, "Iuliu Hatieganu" University of Medicine and Pharmacy, Cluj-Napoca, Romania*
[h]*Department of Pathology, University of Maryland, Baltimore, MD, United States*
[i]*Anesthesiology & Intensive Care, Massachusetts General Hospital, Boston, MA, United States*
*Corresponding authors: Tel.: +46 70 21 95 963; Fax: +46 18 24 38 99 (Aruna Sharma); Tel.: +46 70 2011 801; Fax: +46 18 24 38 99 (Hari Shanker Sharma), e-mail address: aruna.sharma@surgsci.uu.se; sharma@surgsci.uu.se; harishanker_sharma55@icloud.com; hssharma@aol.com

Abstract

Nanoparticles affect blood-brain barrier (BBB) and brain edema formation resulting in sensory-motor dysfunction. Exposure of Mn nanoparticles from industrial sources in humans could target basal ganglia resulting in Parkinson's disease. In present investigation, Mn exposure on brain pathology in a rat model was examined. Rats received Mn nanoparticles (30–40 nm size) in a dose of 10 or 20 mg/kg, i.p. once daily for 7 days and behavioral dysfunctions on Rota Rod performance, inclined plane angle and grid-walking tests as well as gait performances were examined. In addition, BBB breakdown to Evans blue and radioiodine, brain edema formation and neural injuries were also evaluated.

Mn nanoparticles treated rats exhibited cognitive and motor dysfunction on the 8th day. At this time, BBB disruption, reduction in cerebral blood flow (CBF), brain edema formation and brain pathology were most marked in the sensory-motor cortex, hippocampus, caudate putamen, cerebellum and thalamus followed by hypothalamus, pons, medulla and spinal cord. In these brain areas, neuronal injuries using Nissl staining was clearly seen. These effects of Mn nanoparticle are dose dependent. These results are the first to demonstrate that Mn nanoparticles induce selective brain pathology resulting in cognitive and motor dysfunction, not reported earlier.

Keywords

Mn nanoparticles, Blood-brain barrier, Cerebral blood flow, Brain edema, Brain pathology

1 Introduction

Manganese (Mn) is an essential component for human brain function (Balachandran et al., 2020; Blanc, 2018; Erikson and Aschner 2019). However, excessive exposure is injurious to health and could induce neurotoxicity (Dobson et al., 2004; Ke et al., 2019; Lucchini et al., 2018; Normandin et al., 2002; Nyarko-Danquah et al., 2020; Peres et al., 2016; Tjalkens et al., 2017). Mn affects Globus pallidus especially and degeneration of this brain region leads to Parkinson's disease (PD) (Ajsuvakova et al., 2020; Guilarte and Gonzales, 2015; Martins et al., 2019; Mehkari et al., 2020; Olanow, 2004; Pajarillo et al., 2020). However the detail mechanisms of Mn toxicity in the central nervous system (CNS) is not well understood.

Military personnel are often prone to Mn exposure during burn pits that may lead to serious health hazards (Butler et al., 2017; Mallon et al., 2016; Powell et al., 2012; Rohrbeck et al., 2016). Repeated or long-term exposure of Mn could develop Parkinson's disease syndrome and related motor function abnormalities. Thus, effects of Mn exposure on sensory motor functions, together with neurological abnormalities needs further investigations.

Deficiency of Mn could also result in neurological dysfunction and mental abnormality (Jenkitkasemwong et al., 2018; Soares et al., 2020; Takeda, 2003). However, excess exposure of Mn from industrial sources, welding machine, battery assembly, hazardous wastes and or deep mines could results in physical disability and mental abnormalities (Blanc, 2018; Cortez-Lugo et al., 2018; O'Neal and Zheng, 2015). Workers exposed to such activities often show high levels of blood Mn that is often correlated with their neurological dysfunctions (Aschner, 2000; de Water et al., 2018; Jiang et al., 2007). However, the basic mechanisms of Mn induced neurodegeneration or brain dysfunction requires further investigation in great details.

Available evidences suggest that oxidative stress caused by Mn could result in selective brain damage (Erikson et al., 2004; Farina et al., 2013; Neely et al., 2017). Specific brain damage leads to functional disability and impairment of motor functions (Kinder et al., 2019; Caeyenberghs et al., 2011; Fujimoto et al., 2004). It is quite likely that Mn induced oxidative stress induces disruption of the blood-brain barrier (BBB) function causing edema formation (Yang et al., 2019; Banks et al., 2015; Ronaldson and Davis 2020). Breakdown of the BBB, reduction in cerebral blood flow (CBF) and associated edema formation appears to be crucial for neurological impairments and neurodegeneration (Erdő et al., 2017; Katsnelson et al., 2016; Ransohoff, 2016; Sharma, 2009; Sharma et al., 1998; Sweeney et al., 2019). Thus, a possibility exists that Mn could exert these neurological dysfunctions through opening of the BBB, reducing CBF and inducing brain edema formation. Swelling of the brain regions and neurodegeneration in specific area would affect sensory motor functions as well (Sharma et al., 2006, 2009a,b).

In present investigation, effects of Mn nanoparticles on BBB breakdown, CBF, edema formation and cell injury was examined in a rat model. In addition, the functional disturbances in sensory motor dysfunction were also evaluated in Mn nanoparticle treated rats.

2 Materials and methods

2.1 Animals

Experiments were carried out on Male Wistar Rats (200–250 g body weight) housed at controlled room temperature ($21 \pm 1\,°C$) with 12 h light and dark schedule. The rat feed and tap water were supplied ad libitum. All experiments were conducted according to the National Institute of Health (NIH) Guidelines and Care for experimental animals and approved by the Institutional Ethics Committee.

2.2 Exposure to manganese nanoparticles

MnO2 nanoparticles (NPs, 30–40 nm) were purchased from SkySpring Nanomaterials, Inc. (Houston, TX, USA) and administered intraperitoneally in a dose of either 10 or 20 mg/kg in a suspension of Tween 80 once daily for 7 days (Sadeghi et al., 2018; Sharma and Sharma, 2007; Sharma et al., 2009a,b). On the 8th day various parameters of brain function and behavioral tests were carried out. Control rats received saline instead of MnO2 NPs.

2.3 Blood-brain barrier permeability

On the 8th day, animals were anesthetized with Equithesin and Evans blue albumin (2%, 3 mL/kg) or [131]-Iodine (100 µCi/kg) was administered together into the right femoral vein (Sharma 1987; Sharma and Dey, 1986a,b, 1987). After allowing the tracer to circulate for 5 min, the intravascular tracer was washed out with 0.9% saline perfusion through heart. After that brain and spinal cord was removed, dissected into the desired regions, weighed immediately and the radioactivity determined in a Gamma counter (Sharma 1987). Before saline perfusion 1 mL whole blood was taken out from the left cardiac ventricle for total blood radioactivity determination. After radioactivity counted in brain samples the tissue pieces were homogenized in a mixture of sodium sulfate and acetone to extract Evans blue due entered into these brain samples and the supernatant was placed in spectrophotometer at 620 nm to determine dye concentration against standard graph of Evans blue dye (Sharma and Dey, 1987).

2.4 Cerebral blood flow

Cerebral blood flow (CBF) was measured using carbonized microspheres (15 ± 0.6 µm in diameter) labeled with [125]-Iodine as described (Sharma, 1987; Sharma and Dey, 1986a,b, 1987; Sharma et al., 1990). In brief, the microspheres were injected into the heart through common carotid artery and the reference blood flow was obtained from the right femoral artery using standard procedures (Sharma, 1987; Sharma et al., 1990).

2.5 Brain edema formation and volume swelling

For this purpose, the animals were anesthetized on the 8th day and the brains were removed after decapitation. Large blood vessels and blood clots if any over the surface of the brain or spinal cord were removed and the desired pieces were dissected out, weighed immediately and placed in a an incubator maintained at 90 °C for 72 h to obtain their dry weights. Brain water content was calculated form the differences between wet and dry weight of the samples (Sharma, 2007; Sharma and Cervós-Navarro 1990; Sharma and Olsson, 1990) and volume swelling was calculated according to Elliott and Jasper (1949).

2.6 Sensory motor functions

The sensory motor functions were determined in control and MnO2 treated animals on the 8th day using Rota Rod performances, inclined angle test, walking on a mesh grid and foot print analyses using standard protocol as described earlier (Sharma, 2006; Sharma et al., 2009a,b).

2.7 Morphological analyses

Alterations in neuronal structure were examined using standard morphological methods using Hematoxylin Eosin or Nissl staining on paraffin sections (Sharma and Cervós-Navarro, 1990; Sharma and Olsson, 1990; Sharma et al., 1991). For this purpose, the animals were perfused in situ through cardiac puncture with 4% buffered paraformaldehyde preceded with a brief saline rinse. After perfusion the brains were taken out and coronals sections passing through hippocampus was embedded in paraffin. About 3-μm thick sections were cut and stained with HE or Nissl staining according to standard protocol. Also, using glial fibrillary acidic protein (GFAP) immunoreactivity, astrocytic activation is also analyzed using standard procedures as described earlier (Sharma et al., 1992a,b). The sections were then examine under an Inverted Zeiss Microscope and photographed using a digital camera. The images were processed in a Macintosh computer Mac OS 10.11.6 using commercial Photoshop Abacus Inc., USA, 12.4 software (Sharma and Sjöquist, 2002).

2.8 Statistical analyses of the data

Quantitative data obtained from these investigations were analyzed using ANOVA followed by Dunnett's test for multiple group comparison using one control group. Semiquantitative data from morphological investigations were analyzed using non-parametric Chi-Square test. A P-value < 0.05 was considered significant.

3 Results

3.1 Effect of MnO2 nanoparticles on blood-brain barrier permeability

MnO2 nanoparticles exposure for 7 days resulted in profound leakage of Evans blue and radioiodine across several brain regions on the 8th day in a dose dependent manner. The blue staining was seen on the dorsal surface as well as in the ventral surfaces of the brains indicating BBB breakdown to Evans blue in Mn NPs treated rats. The magnitude and intensity of the blue staining was related to the dose of Mn NPs used (Table 1). The blue staining of the ventricular walls and dorsal surfaces of the hippocampus, caudate-Putamen clearly shows that the blood-cerebrospinal fluid barrier (BCSFB) was also broken down by the Mn NPs intoxication. Thus, cerebral cortex, hippocampus showed moderate to deep blue staining and about 3–5 fold increase radioiodine extravasation with 10 mg and 20 mg dose of MnO2 intoxication respectively (Fig. 1). The magnitude of BBB leakage in thalamus hypothalamus and spinal cord were 2–3 fold higher depending on the dose of MnO2 used (Fig. 1 and Table 1).

Table 1 MnO2 induced blood-brain barrier leakage of Evans blue albumin (EBA) in the rat brain.

Brain regions	Evans blue albumin leakage			
	MnO2 10mg	EBA mg %	MnO2 20mg	EBA mg %
1. Cingulate cortex	+++	0.54±0.02*	++++	0.75±0.06*#
2. Frontal cortex	++	0.63±0.05*	++++	0.84±0.04*#
3. Parietal cortex	+++	0.64±0.08*	++++	0.88±0.07*#
4. Temporal cortex	++	n.d.	++++	n.d.
5. Occipital cortex	+++	0.67±0.08*	++++	0.88±0.06*#
6. Hippocampus	++	0.57±0.03*	++++	0.76±0.05*#
7. Caudate nucleus	++	0.68±0.04*	+++	0.89±0.02*#
8. Cerebellar vermis	+++	0.54±0.002*	++++	0.67±0.004*#
9. Cerebellar cortex	++	0.48±0.004*	++++	0.58±0.003*#
10. Massa intermedia	+++	n.d.	++++	n.d.
11. Dorsal thalamus	+++	0.48±0.05*	++++	0.87±0.07*#
12. Hypothalamus	++	0.54±0.04*	++++	0.89±0.05*#
13. Superior colliculi	++	0.52±0.03*	+++	0.76±0.04*#
14. Inferior colliculi	++	n.d.	++++	n.d.
15. Brain stem	++	0.38±0.05*	+++	0.52±0.04*#
16. Cervical spinal cord	++	0.37±0.03*	+++	0.64±0.05*#
17. Lateral ventricle	+++	n.d.	++++	n.d.
18. IV ventricular	+++	n.d.	++++	n.d.

MnO2 nanoparticles (NPs, 30–40nm) administered intraperitoneally either 10 or 20mg/kg in Tween 80 once daily for 7 days. Spread of Evans blue dye on brain surface was evaluated using a magnifying glass and scored by 2 independent investigators on the 8th day and the dye entered into the brain were measured calorimetrically (Sharma, 1987). Control rats received saline instead of MnO2 NPs. EBA staining of brain tissues: ++=Mild, +++=Moderate, ++++=Deep Blue, n.d. = not done; Measurement of EBA. Values Mean±SD of 5–6 rats at each data point. *P<0.05 from control group (results not shown); #P<0.05 from Mn 10mg; ANOVA followed by Dunnett's test for multiple group comparison from one control.

3.2 Effect of MnO2 nanoparticles on cerebral blood flow

MnO2 nanoparticles administration reduced the CBF in dose dependent manner. Thus, reduction in the CBF was seen in almost all 8-brain regions examined. The cerebral cortex, cerebellum, hippocampus, caudate-Putamen areas exhibited greater CBF reductions after Mn NPs intoxication in a dose dependent manner (Fig. 2).

3.3 Effect of MnO2 nanoparticles on brain edema formation and volume swelling

Exposure of rats to MnO2 nanoparticles for 7 days resulted in profound brain edema formation on the 8th day that is about 2–3% higher in the cerebral cortex depending on the dose of the nanoparticles used as compared to saline treated control group. Significant increase in brain edema and volume swelling is seen in almost all brain regions examined in a dose dependent manner (Fig. 3).

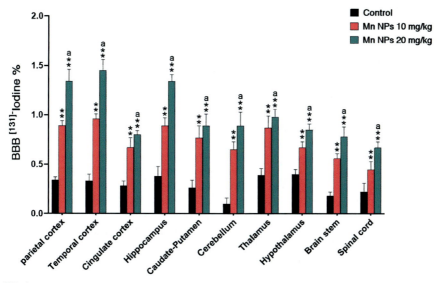

FIG. 1

Regional blood-brain barrier leakage of radioiodine following MnO2 nanoparticles (NPs) intoxication in a dose dependent manner. MnO2 nanoparticles (NPs, 30–40 nm) administered intraperitoneally either 10 or 20 mg/kg in Tween 80 once daily for 7 days. Control rats received saline instead of MnO2 NPs. Values are Mean ± SD of 5–6 rats at each data point. **$P < 0.01$ from control group; [a]$P < 0.05$ from Mn 10 mg; ANOVA followed by Dunnett's test for multiple group comparison from one control.

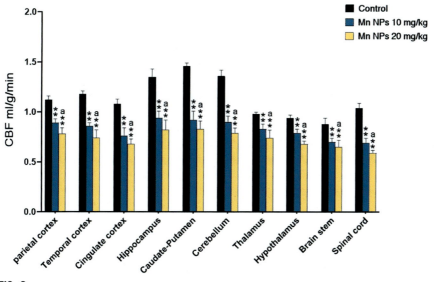

FIG. 2

Regional cerebral blood flow following MnO2 nanoparticles (NPs) intoxication in a dose dependent manner. MnO2 nanoparticles (NPs, 30–40 nm) administered intraperitoneally either 10 or 20 mg/kg in Tween 80 once daily for 7 days. Control rats received saline instead of MnO2 NPs. Values are Mean ± SD of 5–6 rats at each data point. **$P < 0.01$ from control group; [a]$P < 0.05$ from Mn 10 mg; ANOVA followed by Dunnett's test for multiple group comparison from one control.

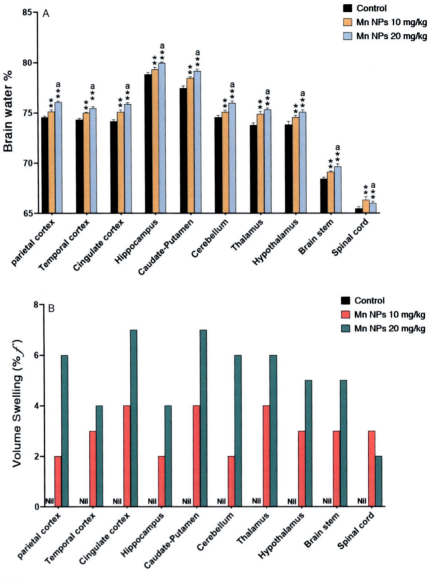

FIG. 3

Regional brain edema formation as measured using brain water content (A) and volume swelling (B) following MnO2 nanoparticles (NPs) intoxication in a dose dependent manner. MnO2 nanoparticles (NPs, 30–40 nm) administered intraperitoneally either 10 or 20 mg/kg in Tween 80 once daily for 7 days. Control rats received saline instead of MnO2 NPs. Volume swelling (% f) was calculated from differences between control and experimental brain water content according to the formula of Elliott and Jasper (1949). About 1% increase in brain water content from the control group equal to approximately 4% increase in volume swelling. Values are Mean ± SD of 5 to 6 rats at each data point. **$P < 0.01$ from control group; [a]$P < 0.05$ from Mn 10 mg; ANOVA followed by Dunnett's test for multiple group comparison from one control.

3.4 Effect of MnO2 nanoparticles on neuronal injury

MnO2 nanoparticles markedly induced neuronal injuries in a dose dependent manner (Figs. 4 and 5). These neuronal damages were most marked in cerebral cortex and hippocampus followed by thalamus, hypothalamus and spinal cord (Fig. 4). The

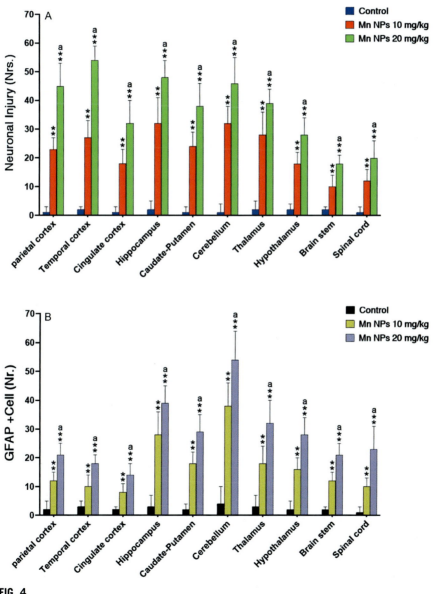

FIG. 4

Semiquantitative analysis of regional neuronal damage (A) and glial fibrillary acidic protein (GFAP) reactivity (B) following MnO2 nanoparticles (NPs) intoxication in a dose

(Continued)

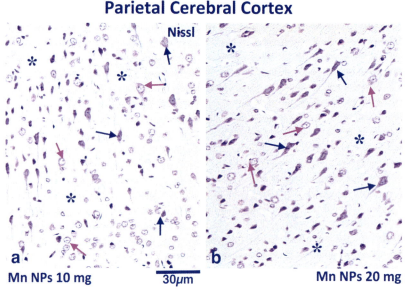

FIG. 5

Representative example of Nissl Stained nerve cells on paraffin sections (3 μm thick) from the parietal cerebral cortex of the rat following MnO2 nanoparticles (NPs) intoxication 10 mg (A) and 20 mg (B) for 7 days. Several dark and distorted neurons (blue arrows) are seen following Mn NPs intoxication in a dose dependent manner. Expansion of neuropil and sponginess (*) is clearly evident. However, some neurons appear quite normal (pink arrows) in Mn NPs intoxicated rats. This indicates that Mn NPs neurotoxicity in the rat brain is dose dependent. MnO2 nanoparticles (NPs, 30–40 nm) administered intraperitoneally either 10 or 20 mg/kg in Tween 80 once daily for 7 days. Bar = 30 μm.

magnitude and intensity of neuronal injuries was dose dependent. The neuronal damage includes vacuolation, perineuronal edema, chromatolysis, and loss of Nissl substance, eccentric nucleolus, sponginess and edematous swelling of neuropil (Fig. 5). These pathological changes were clearly dependent on the dose of MnO2 nanoparticles given.

Using GFAP immunoreactivity, marked activation of astrocytes were seen in the brain areas exhibiting BBB breakdown and edematous swellings (Figs. 4 and 6).

FIG. 4—Cont'd   dependent manner. MnO2 nanoparticles (NPs, 30–40 nm) administered intraperitoneally either 10 or 20 mg/kg in Tween 80 once daily for 7 days. Control rats received saline instead of MnO2 NPs. Values are Mean ± SD of 5–6 rats at each data point. **$P < 0.01$ from control group; [a]$P < 0.05$ from Mn 10 mg; non-parametric test Chi-Square test.

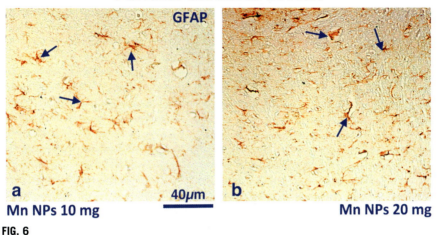

FIG. 6

Representative example of glial fibrillary acidic protein (GFAP) immunoreactive on paraffin sections (3 μm thick) from the parietal cerebral cortex of the rat following MnO2 nanoparticles (NPs) intoxication 10 mg (A) and 20 mg (B) for 7 days. GFAP positive cells (blue arrows) are seen following Mn NPs intoxication in a dose dependent manner. GFAP positive cells are seen around microvessels and astrocytes as well as in the perineuronal areas of the neuropil. This indicates that Mn NPs neurotoxicity in the rat brain is dose dependent. MnO2 nanoparticles (NPs, 30–40 nm) administered intraperitoneally either 10 or 20 mg/kg in Tween 80 once daily for 7 days. Bar = 40 μm.

Activation of astrocytes is seen around microvessels, nerve cells and neuropil in the areas of BBB breakdown (Fig. 6). The magnitude and intensity of GFAP activation is also Mn NPs dose related.

3.5 Effect of MnO2 nanoparticles on sensory motor functions

MnO2 nanoparticles treated animals showed profound disturbances in sensory motor performances on the Rota Rod, inclined plane angle test; waking on a mesh grid and foot print analyses (Figs. 7–10). These changes were dose dependent. Thus, depending on the doses the MnO2 treated animals showed profound diminution on the duration of stay on the Rota Rod without falling (Fig. 7). Also these animals were able to stay only if the inclined plane angle was reduced considerably (Fig. 8). The number of errors on fore paw placement on the mesh grid was enhanced (Fig. 9) and the distances between hid feet increased and the stride length decreased in MnO2 treated animals (Fig. 10).

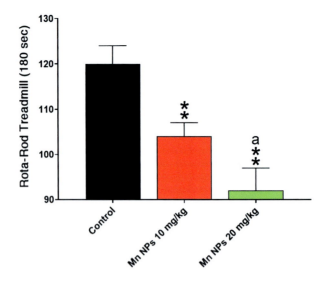

FIG. 7

Rota-Rod treadmill performances of the rat following MnO2 nanoparticles (NPs) intoxication in a dose dependent manner. Mn NPs reduced Rota-Rod performances in the rat in a dose dependent manner. MnO2 nanoparticles (NPs, 30–40nm) administered intraperitoneally either 10 or 20mg/kg in Tween 80 once daily for 7 days. Control rats received saline instead of MnO2 NPs. Values are Mean±SD of 5–6 rats at each data point. **$P < 0.01$ from control group; [a]$P < 0.05$ from Mn 10mg; ANOVA followed by Dunnett's test for multiple group comparison from one control.

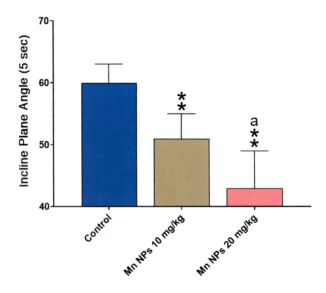

FIG. 8

Inclined plane angle (°) performances of the rat following MnO2 nanoparticles (NPs) intoxication in a dose dependent manner. Mn NPs reduced Inclined plane angle (°) in the rat in a dose dependent manner. MnO2 nanoparticles (NPs, 30–40nm) administered intraperitoneally either 10 or 20mg/kg in Tween 80 once daily for 7 days. Control rats received saline instead of MnO2 NPs. Values are Mean±SD of 5–6 rats at each data point. **$P < 0.01$ from control group; [a]$P < 0.05$ from Mn 10mg; ANOVA followed by Dunnett's test for multiple group comparison from one control.

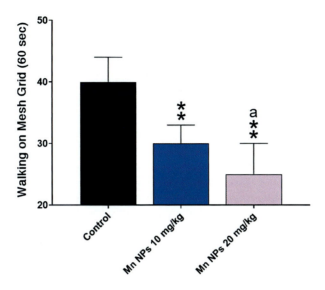

FIG. 9

Walking on an inclined mesh grid performance of the rat following MnO2 nanoparticles (NPs) intoxication in a dose dependent manner. Mn NPs reduced performances on walking on an inclined mesh grid in the rat in a dose dependent manner. MnO2 nanoparticles (NPs, 30–40 nm) administered intraperitoneally either 10 or 20 mg/kg in Tween 80 once daily for 7 days. Control rats received saline instead of MnO2 NPs. Values are Mean ± SD of 5–6 rats at each data point. **$P < 0.01$ from control group; [a]$P < 0.05$ from Mn 10 mg; ANOVA followed by Dunnett's test for multiple group comparison from one control.

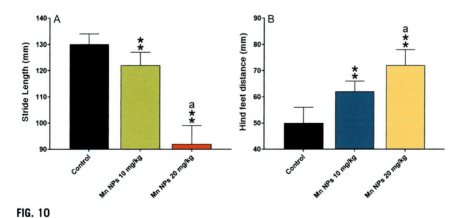

FIG. 10

Gait performance of the rat following MnO2 nanoparticles (NPs) intoxication in a dose dependent manner. Mn NPs reduced gait performances of the rat by reducing stride length (A) and increased distances between hind legs (B) in a dose dependent manner. MnO2 nanoparticles (NPs, 30–40 nm) administered intraperitoneally either 10 or 20 mg/kg in Tween 80 once daily for 7 days. Control rats received saline instead of MnO2 NPs. Values are Mean ± SD of 5–6 rats at each data point. **$P < 0.01$ from control group; [a]$P < 0.05$ from Mn 10 mg; ANOVA followed by Dunnett's test for multiple group comparison from one control.

4 Discussion

The present results are the first to demonstrate that MnO2 nanoparticles when administered for 1 week are able to induce profound neurotoxicity by disruption of the BBB function and reduction in the CBF in a dose dependent manner. Furthermore MnO2 intoxication induces brain edema formation and sensory motor alterations, the magnitude of which closely corresponds to the severity of neuronal injury and BBB disturbances. This suggests that MnO2 induced neurotoxicity is mediated via breakdown of the BBB function.

Another interesting observation in this investigation is a clear dose dependent response of MnO2 nanoparticles in inducing brain damage. This is evident from the neuronal and astrocytic reactions seen in the areas of BBB disruption and swelling of the neuropil. This indicates that alterations in the microfluid environment in the brain are crucial for neuronal and glial cells abnormal reactions and injuries (Sharma, 2009; Sharma and Sharma, 2007; Sharma et al., 1998, 2009a,b). This suggests that MnO2 induced cellular and molecular changes in the brain or blood could partly be responsible for such cell and tissue injuries (Alarifi et al., 2017; Huang et al., 2020; Sadeghi et al., 2018). MnO2 is known to induce oxidative stress and it appears that in this investigation also, an increased oxidative stress could be instrumental in inducing neuronal injuries (Khatri et al., 2018; Mo et al., 2019; Yin et al., 2017). Obviously, oxidative stress production following MnO2 intoxication could be dose dependent.

Our observations further show that MnO2 intoxication resulted in profound sensory motor disturbances (Sadeghi et al., 2018). It appears that these sensory motor disturbances following MnO2 administration could be due to specific neurotoxicity caused by the nanoparticles in different brain regions (Sharma et al., 2009a,b, 2015, 2018). Since massive neuronal injuries could be seen in the sensory motor cortex, hippocampus, thalamus, hypothalamus and spinal cord it appears that neuronal damage and associated sensory motor pathways could contribute to functional disturbance in MnO2 treated animals (Sharma, 2006). A dose dependent disturbance in sensory motor function and neuronal injuries further supports this idea.

In addition, breakdown of the BBB could results in entry of serum proteins into the brain causing vasogenic edema formation (Blixt et al., 2015; Sadeghian et al., 2019; Sharma et al., 2019; Yang et al., 2019). Reduction in the CBF further contributes to ischemic insults leading to edema and cell injuries (Dobrivojević et al., 2015; Rosenberg, 1999; von Kummer and Dzialowski, 2017). Marked swelling of the brain into a closed cranial compartment will result in severe compression of the brain causing disturbances in sensory motor functions (Feng et al., 2018; Niu et al., 2020; Sharma et al., 2007). This is further supported from our previous studies on heat stress induced hyperthermia in which pronounced brain swelling resulted in similar disturbances of sensory motor function (Sharma, 2006; Sharma et al., 2009a,b). This suggests that brain edema formation and neuronal injuries could partly be responsible for sensory motor disturbances in MnO2 treated rats.

The possible link between MnO2 intoxication and breakdown of the BBB is unclear form this study. However, available evidences suggest that reduction in the CBF associated with oxidative stress induced generation of free radicals could contribute to cell membrane damage in the brain and spinal cord, e.g., endothelium, neurons and glial cells (Gu et al., 2011; Sharma and Sharma 2012a,b; Sharma et al., 2011a,b; Siesjö et al., 1995). As a result, breakdown of the BBB occurs due to damage of cerebral endothelium. Leakage of serum proteins across the BBB could further initiate the immunological processes and adverse reaction within the brain microfluid environments in which neurons and glial cells are suspended leading to brain pathology (Sharma, 2004; Sharma and Westman, 2004; Sharma et al., 2011a,b). Alterations in the brain fluid microenvironment could result in neuronal and glial cells dysfunction causing abnormal brain function and sensory motor disturbances (Sharma, 2006).

To further test this hypothesis in MnO2 intoxicated rats, pretreatment with antioxidant compound is needed to see whether blockade of oxidative stress could induce neuroprotection and attenuate sensory motor disturbances. This aspect of investigation is in progress in our laboratory. Furthermore, in MnO2 treated animals generation of oxidative stress is also being examined using biochemical approaches.

In conclusion, our results are the first to show that MnO2 neurotoxicity is associated with profound sensory motor dysfunction and profound neurological abnormalities. It appears that these neurological changes could be largely due to MnO2 induced BBB breakdown resulting in vasogenic edema formation, not reported earlier.

Acknowledgments

This investigation is supported by grants from the Ministry of Science & Technology, People Republic of China, the Air Force Office of Scientific Research (EOARD, London, UK), and Air Force Material Command, USAF, under grant number FA8655-05-1-3065; Grants from the Alzheimer's Association (IIRG-09-132087), the National Institutes of Health (R01 AG028679) and the Dr. Robert M. Kohrman Memorial Fund (R.J.C.); Swedish Medical Research Council (Nr 2710-HSS), Göran Gustafsson Foundation, Stockholm, Sweden (H.S.S.), Astra Zeneca, Mölndal, Sweden (H.S.S./A.S.), The University Grants Commission, New Delhi, India (H.S.S./A.S.), Ministry of Science & Technology, Govt. of India (H.S.S./A.S.), Indian Medical Research Council, New Delhi, India (H.S.S./A.S.) and India-EU Co-operation Program (R.P./A.S./H.S.S.) and IT-901/16 (J.V.L.), Government of Basque Country and PPG 17/51 (J.V.L.), J.V.L. thanks to the support of the University of the Basque Country (UPV/ EHU) PPG 17/51 and 14/08, the Basque Government (IT-901/16 and CS-2203) Basque Country, Spain; and Foundation for Nanoneuroscience and Nanoneuroprotection (FSNN), Romania. Technical and human support provided by Dr. Ricardo Andrade from SGIker (UPV/EHU) is gratefully acknowledged. Dr. Seaab Sahib is supported by Research Fellowship at the University of Arkansas, Fayetteville, AR, by Department of Community Health; Middle Technical University, Wassit, Iraq, and The Higher Committee for Education Development in Iraq;

Baghdad; Iraq. We thank Suraj Sharma, Blekinge Inst. Technology, Karlskrona, Sweden and Dr. Saja Alshafeay, University of Arkansas, Fayetteville, AR, USA, for computer and graphic support. The U.S. Government is authorized to reproduce and distribute reprints for Government purpose notwithstanding any copyright notation thereon. The views and conclusions contained herein are those of the authors and should not be interpreted as necessarily representing the official policies or endorsements, either expressed or implied, of the Air Force Office of Scientific Research or the U.S. Government.

Conflict of interest

There is no conflict of interest between any entity and/or organization mentioned here.

References

Ajsuvakova, O.P., Tinkov, A.A., Willkommen, D., Skalnaya, A.A., Danilov, A.B., Pilipovich, A.A., Aschner, M., Skalny, A.V., Michalke, B., Skalnaya, M.G., 2020. Assessment of copper, iron, zinc and manganese status and speciation in patients with Parkinson's disease: a pilot study. J. Trace Elem. Med. Biol. 59, 126423. https://doi.org/10.1016/j.jtemb.2019.126423. (Epub 2019 Oct 28).

Alarifi, S., Ali, D., Alkahtani, S., 2017. Oxidative stress-induced DNA damage by manganese dioxide nanoparticles in human neuronal cells. Biomed. Res. Int. 2017, 5478790. https://doi.org/10.1155/2017/5478790. (Epub 2017 May 17).

Aschner, M., 2000. Manganese: brain transport and emerging research needs. Environ. Health Perspect. 108 (Suppl. 3), 429–432. https://doi.org/10.1289/ehp.00108s3429.

Balachandran, R.C., Mukhopadhyay, S., McBride, D., Veevers, J., Harrison, F.E., Aschner, M., Haynes, E.N., Bowman, A.B., 2020. Brain manganese and the balance between essential roles and neurotoxicity. J. Biol. Chem. 295 (19), 6312–6329. https://doi.org/10.1074/jbc.REV119.009453. (Epub 2020 Mar 18).

Banks, W.A., Gray, A.M., Erickson, M.A., Salameh, T.S., Damodarasamy, M., Sheibani, N., Meabon, J.S., Wing, E.E., Morofuji, Y., Cook, D.G., Reed, M.J., 2015. Lipopolysaccharide-induced blood-brain barrier disruption: roles of cyclooxygenase, oxidative stress, neuroinflammation, and elements of the neurovascular unit. J. Neuroinflammation 12, 223. https://doi.org/10.1186/s12974-015-0434-1.

Blanc, P.D., 2018. The early history of manganese and the recognition of its neurotoxicity, 1837-1936. Neurotoxicology 64, 5–11. https://doi.org/10.1016/j.neuro.2017.04.006. (Epub 2017 Apr 14).

Blixt, J., Svensson, M., Gunnarson, E., Wanecek, M., 2015. Aquaporins and blood-brain barrier permeability in early edema development after traumatic brain injury. Brain Res. 1611, 18–28. https://doi.org/10.1016/j.brainres.2015.03.004. (Epub 2015 Mar 11).

Butler, D.A., Styka, A.N., Savitz, D.A. (Eds.), 2017. Assessment of the Department of Veterans Affairs Airborne Hazards and Open Burn Pit Registry. National Academies of Sciences, Engineering, and Medicine; Health and Medicine Division; Board on Population Health and Public Health Practice; Board on the Health of Select Populations; Committee on the Assessment of the Department of Veterans Affairs Airborne Hazards and Open Burn Pit Registry; National Academies Press (US), Washington (DC).

Caeyenberghs, K., Leemans, A., Geurts, M., Linden, C.V., Smits-Engelsman, B.C., Sunaert, S., Swinnen, S.P., 2011. Correlations between white matter integrity and motor function in traumatic brain injury patients. Neurorehabil. Neural Repair 25 (6), 492–502. https://doi.org/10.1177/1545968310394870. (Epub 2011 Mar 22).

Cortez-Lugo, M., Riojas-Rodríguez, H., Moreno-Macías, H., Montes, S., Rodríguez-Agudelo, Y., Hernández-Bonilla, D., Catalán-Vázquez, M., Díaz-Godoy, R., Rodríguez-Dozal, S., 2018. Evaluation of the effect of an environmental management program on exposure to manganese in a mining zone in Mexico. Neurotoxicology 64, 142–151. https://doi.org/10.1016/j.neuro.2017.08.014. (Epub 2017 Sep 10).

de Water, E., Proal, E., Wang, V., Medina, S.M., Schnaas, L., Téllez-Rojo, M.M., Wright, R.O., Tang, C.Y., Horton, M.K., 2018. Prenatal manganese exposure and intrinsic functional connectivity of emotional brain areas in children. Neurotoxicology 64, 85–93. https://doi.org/10.1016/j.neuro.2017.06.006. (Epub 2017 Jun 10).

Dobrivojević, M., Špiranec, K., Sinđić, A., 2015. Involvement of bradykinin in brain edema development after ischemic stroke. Pflugers Arch. 467 (2), 201–212. https://doi.org/10.1007/s00424-014-1519-x. (Epub 2014 Apr 23).

Dobson, A.W., Erikson, K.M., Aschner, M., 2004. Manganese neurotoxicity. Ann. N. Y. Acad. Sci. 1012, 115–128. https://doi.org/10.1196/annals.1306.009.

Elliott, K.A., Jasper, H., 1949. Measurement of experimentally induced brain swelling and shrinkage. Am. J. Physiol. 157 (1), 122–129. https://doi.org/10.1152/ajplegacy.1949.157.1.122.

Erdő, F., Denes, L., de Lange, E., 2017. Age-associated physiological and pathological changes at the blood-brain barrier: a review. J. Cereb. Blood Flow Metab. 37 (1), 4–24. https://doi.org/10.1177/0271678X16679420. (Epub 2016 Nov 11).

Erikson, K.M., Aschner, M., 2019. Manganese: its role in disease and health. In: Essential Metals in Medicine: Therapeutic Use and Toxicity of Metal Ions in the Clinic. De Gruyter. https://doi.org/10.1515/9783110527872-010 (Chapter 10).

Erikson, K.M., Dobson, A.W., Dorman, D.C., Aschner, M., 2004. Manganese exposure and induced oxidative stress in the rat brain. Sci. Total Environ. 334–335, 409–416. https://doi.org/10.1016/j.scitotenv.2004.04.044.

Farina, M., Avila, D.S., da Rocha, J.B., Aschner, M., 2013. Metals, oxidative stress and neurodegeneration: a focus on iron, manganese and mercury. Neurochem. Int. 62 (5), 575–594. https://doi.org/10.1016/j.neuint.2012.12.006. (Epub 2012 Dec 21).

Feng, L., Sharma, A., Niu, F., Huang, Y., Lafuente, J.V., Muresanu, D.F., Ozkizilcik, A., Tian, Z.R., Sharma, H.S., 2018. TiO(2)-nanowired delivery of DL-3-n-butylphthalide (DL-NBP) attenuates blood-brain barrier disruption, brain edema formation, and neuronal damages following concussive head injury. Mol. Neurobiol. 55 (1), 350–358. https://doi.org/10.1007/s12035-017-0746-5.

Fujimoto, S.T., Longhi, L., Saatman, K.E., Conte, V., Stocchetti, N., McIntosh, T.K., 2004. Motor and cognitive function evaluation following experimental traumatic brain injury. Neurosci. Biobehav. Rev. 28 (4), 365–378. https://doi.org/10.1016/j.neubiorev.2004.06.002.

Gu, Y., Dee, C.M., Shen, J., 2011. Interaction of free radicals, matrix metalloproteinases and caveolin-1 impacts blood-brain barrier permeability. Front. Biosci. (Schol. Ed.) 3, 1216–1231. https://doi.org/10.2741/222.

Guilarte, T.R., Gonzales, K.K., 2015. Manganese-induced parkinsonism is not idiopathic Parkinson's disease: environmental and genetic evidence. Toxicol. Sci. 146 (2), 204–212. https://doi.org/10.1093/toxsci/kfv099.

Huang, W., Wu, G., Xiao, H., Song, H., Gan, S., Ruan, S., Gao, Z., Song, J., 2020. Transformation of m-aminophenol by birnessite (delta-MnO(2)) mediated oxidative processes: reaction kinetics, pathways and toxicity assessment. Environ. Pollut. 256, 113408. https://doi.org/10.1016/j.envpol.2019.113408. (Epub 2019 Oct 19).

Jenkitkasemwong, S., Akinyode, A., Paulus, E., Weiskirchen, R., Hojyo, S., Fukada, T., Giraldo, G., Schrier, J., Garcia, A., Janus, C., Giasson, B., Knutson, M.D., 2018. SLC39A14 deficiency alters manganese homeostasis and excretion resulting in brain manganese accumulation and motor deficits in mice. Proc. Natl. Acad. Sci. U. S. A. 115 (8), E1769–E1778. https://doi.org/10.1073/pnas.1720739115. Epub.

Jiang, Y., Zheng, W., Long, L., Zhao, W., Li, X., Mo, X., Lu, J., Fu, X., Li, W., Liu, S., Long, Q., Huang, J., Pira, E., 2007. Brain magnetic resonance imaging and manganese concentrations in red blood cells of smelting workers: search for biomarkers of manganese exposure. Neurotoxicology 28 (1), 126–135. https://doi.org/10.1016/j.neuro.2006.08.005. (Epub 2006 Aug 22).

Katsnelson, A., De Strooper, B., Zoghbi, H.Y., 2016. Neurodegeneration: from cellular concepts to clinical applications. Sci. Transl. Med. 8 (364), 364ps18. https://doi.org/10.1126/scitranslmed.aal2074.

Ke, T., Sidoryk-Wegrzynowicz, M., Pajarillo, E., Rizor, A., Soares, F.A.A., Lee, E., Aschner, M., 2019. Role of astrocytes in manganese neurotoxicity revisited. Neurochem. Res. 44 (11), 2449–2459. https://doi.org/10.1007/s11064-019-02881-7. (Epub 2019 Sep 30).

Khatri, N., Thakur, M., Pareek, V., Kumar, S., Sharma, S., Datusalia, A.K., 2018. Oxidative stress: major threat in traumatic brain injury. CNS Neurol. Disord. Drug Targets 17 (9), 689–695. https://doi.org/10.2174/1871527317666180627120501.

Kinder, H.A., Baker, E.W., Wang, S., Fleischer, C.C., Howerth, E.W., Duberstein, K.J., Mao, H., Platt, S.R., West, F.D., 2019. Traumatic brain injury results in dynamic brain structure changes leading to acute and chronic motor function deficits in a pediatric piglet model. J. Neurotrauma 36 (20), 2930–2942. https://doi.org/10.1089/neu.2018.6303. (Epub 2019 Jun 17).

Lucchini, R.G., Aschner, M., Landrigan, P.J., Cranmer, J.M., 2018. Neurotoxicity of manganese: indications for future research and public health intervention from the manganese 2016 conference. Neurotoxicology 64, 1–4. https://doi.org/10.1016/j.neuro.2018.01.002. (Epub 2018 Feb 3).

Mallon, C.T., Rohrbeck, M.P., Haines, M.K., Jones, D.P., Utell, M., Hopke, P.K., Phipps, R.P., Walker, D.I., Thatcher, T., Woeller, C.F., Baird, C.P., Pollard, H.B., Dalgard, C.L., Gaydos, J.C., 2016. Introduction to Department of Defense research on burn pits, biomarkers, and health outcomes related to deployment in Iraq and Afghanistan. J. Occup. Environ. Med. 58 (8 Suppl. 1), S3–S11. https://doi.org/10.1097/JOM.0000000000000775.

Martins, A.C., Morcillo, P., Ijomone, O.M., Venkataramani, V., Harrison, F.E., Lee, E., Bowman, A.B., Aschner, M., 2019. New insights on the role of manganese in Alzheimer's disease and Parkinson's disease. Int. J. Environ. Res. Public Health 16 (19), 3546. https://doi.org/10.3390/ijerph16193546.

Mehkari, Z., Mohammed, L., Javed, M., Althwanay, A., Ahsan, F., Oliveri, F., Goud, H.K., Rutkofsky, I.H., 2020. Manganese, a likely cause of 'Parkinson's in Cirrhosis', a unique clinical entity of acquired hepatocerebral degeneration. Cureus 12 (9), e10448. https://doi.org/10.7759/cureus.10448.

Mo, J., Enkhjargal, B., Travis, Z.D., Zhou, K., Wu, P., Zhang, G., Zhu, Q., Zhang, T., Peng, J., Xu, W., Ocak, U., Chen, Y., Tang, J., Zhang, J., Zhang, J.H., 2019. AVE 0991 attenuates oxidative stress and neuronal apoptosis via Mas/PKA/CREB/UCP-2 pathway after subarachnoid hemorrhage in rats. Redox Biol. 20, 75–86. https://doi.org/10.1016/j.redox.2018.09.022. (Epub 2018 Sep 28).

Neely, M.D., Davison, C.A., Aschner, M., Bowman, A.B., 2017. From the cover: manganese and rotenone-induced oxidative stress signatures differ in iPSC-derived human dopamine neurons. Toxicol. Sci. 159 (2), 366–379. https://doi.org/10.1093/toxsci/kfx145.

Niu, F., Sharma, A., Wang, Z., Feng, L., Muresanu, D.F., Sahib, S., Tian, Z.R., Lafuente, J.V., Buzoianu, A.D., Castellani, R.J., Nozari, A., Patnaik, R., Wiklund, L., Sharma, H.S., 2020. Co-administration of TiO(2)-nanowired dl-3-n-butylphthalide (dl-NBP) and mesenchymal stem cells enhanced neuroprotection in Parkinson's disease exacerbated by concussive head injury. Prog. Brain Res. 258, 101–155. https://doi.org/10.1016/bs.pbr.2020.09.011. (Epub 2020 Nov 9).

Normandin, L., Panisset, M., Zayed, J., 2002. Manganese neurotoxicity: behavioral, pathological, and biochemical effects following various routes of exposure. Rev. Environ. Health 17 (3), 189–217. https://doi.org/10.1515/reveh.2002.17.3.189.

Nyarko-Danquah, I., Pajarillo, E., Digman, A., Soliman, K.F.A., Aschner, M., Lee, E., 2020. Manganese accumulation in the brain via various transporters and its neurotoxicity mechanisms. Molecules 25 (24), 5880. https://doi.org/10.3390/molecules25245880.

Olanow, C.W., 2004. Manganese-induced parkinsonism and Parkinson's disease. Ann. N. Y. Acad. Sci. 1012, 209–223. https://doi.org/10.1196/annals.1306.018.

O'Neal, S.L., Zheng, W., 2015. Manganese toxicity upon overexposure: a decade in review. Curr. Environ. Health Rep. 2 (3), 315–328. https://doi.org/10.1007/s40572-015-0056-x.

Pajarillo, E., Rizor, A., Son, D.S., Aschner, M., Lee, E., 2020. The transcription factor REST up-regulates tyrosine hydroxylase and antiapoptotic genes and protects dopaminergic neurons against manganese toxicity. J. Biol. Chem. 295 (10), 3040–3054. https://doi.org/10.1074/jbc.RA119.011446. (Epub 2020 Jan 30).

Peres, T.V., Schettinger, M.R., Chen, P., Carvalho, F., Avila, D.S., Bowman, A.B., Aschner, M., 2016. Manganese-induced neurotoxicity: a review of its behavioral consequences and neuroprotective strategies. BMC Pharmacol. Toxicol. 17 (1), 57. https://doi.org/10.1186/s40360-016-0099-0.

Powell, T.M., Smith, T.C., Jacobson, I.G., Boyko, E.J., Hooper, T.I., Gackstetter, G.D., Phillips, C.J., Smith, B., 2012. Millennium cohort study team. Prospective assessment of chronic multisymptom illness reporting possibly associated with open-air burn pit smoke exposure in Iraq. J. Occup. Environ. Med. 54 (6), 682–688. https://doi.org/10.1097/JOM.0b013e318255ba39.

Ransohoff, R.M., 2016. How neuroinflammation contributes to neurodegeneration. Science 353 (6301), 777–783. https://doi.org/10.1126/science.aag2590.

Rohrbeck, P., Hu, Z., Mallon, C.T., 2016. Assessing health outcomes after environmental exposures associated with open pit burning in deployed US service members. J. Occup. Environ. Med. 58 (8 Suppl. 1), S104–S110. https://doi.org/10.1097/JOM.0000000000000802.

Ronaldson, P.T., Davis, T.P., 2020. Regulation of blood-brain barrier integrity by microglia in health and disease: a therapeutic opportunity. J. Cereb. Blood Flow Metab. 40 (Suppl. 1), S6–S24. https://doi.org/10.1177/0271678X20951995. (Epub 2020 Sep 14).

Rosenberg, G.A., 1999. Ischemic brain edema. Prog. Cardiovasc. Dis. 42 (3), 209–216. https://doi.org/10.1016/s0033-0620(99)70003-4.

Sadeghi, L., Babadi, V.Y., Tanwir, F., 2018. Manganese dioxide nanoparticle induces Parkinson like neurobehavioral abnormalities in rats. Bratisl. Lek. Listy 119 (6), 379–384. https://doi.org/10.4149/BLL_2018_070.

Sadeghian, N., Shadman, J., Moradi, A., Ghasem Golmohammadi, M., Panahpour, H., 2019. Calcitriol protects the blood-brain barrier integrity against ischemic stroke and reduces vasogenic brain edema via antioxidant and antiapoptotic actions in rats. Brain Res. Bull. 150, 281–289. https://doi.org/10.1016/j.brainresbull.2019.06.010. (Epub 2019 Jun 18).

Sharma, H.S., 1987. Effect of captopril (a converting enzyme inhibitor) on blood-brain barrier permeability and cerebral blood flow in normotensive rats. Neuropharmacology 26 (1), 85–92. https://doi.org/10.1016/0028-3908(87)90049-9.

Sharma, H.S., 2004. Histamine influences the blood-spinal cord and brain barriers following injuries to the central nervous system. In: Sharma, H.S., Westman, J. (Eds.), The Blood-Spinal Cord and Brain Barriers in Health and Disease. Elsevier Academic Press, San Diego, pp. 159–190.

Sharma, H.S., 2006. Hyperthermia influences excitatory and inhibitory amino acid neurotransmitters in the central nervous system. An experimental study in the rat using behavioural, biochemical, pharmacological, and morphological approaches. J. Neural Transm. (Vienna) 113 (4), 497–519. https://doi.org/10.1007/s00702-005-0406-1.

Sharma, H.S., 2007. Methods to produce hyperthermia-induced brain dysfunction. Prog. Brain Res. 162, 173–199. https://doi.org/10.1016/S0079-6123(06)62010-4.

Sharma, H.S., 2009. Blood–central nervous system barriers: the gateway to neurodegeneration, neuroprotection and neuroregeneration. In: Lajtha, A., Banik, N., Ray, S.K. (Eds.), Handbook of Neurochemistry and Molecular Neurobiology: Brain and Spinal Cord Trauma. Springer Verlag, Berlin, Heidelberg, New York, pp. 363–457.

Sharma, H.S., Cervós-Navarro, J., 1990. Brain oedema and cellular changes induced by acute heat stress in young rats. Acta Neurochir. Suppl. (Wien) 51, 383–386. https://doi.org/10.1007/978-3-7091-9115-6_129.

Sharma, H.S., Dey, P.K., 1986a. Influence of long-term immobilization stress on regional blood-brain barrier permeability, cerebral blood flow and 5-HT level in conscious normotensive young rats. J. Neurol. Sci. 72 (1), 61–76. https://doi.org/10.1016/0022-510x(86)90036-5.

Sharma, H.S., Dey, P.K., 1986b. Probable involvement of 5-hydroxytryptamine in increased permeability of blood-brain barrier under heat stress in young rats. Neuropharmacology 25 (2), 161–167. https://doi.org/10.1016/0028-3908(86)90037-7.

Sharma, H.S., Dey, P.K., 1987. Influence of long-term acute heat exposure on regional blood-brain barrier permeability, cerebral blood flow and 5-HT level in conscious normotensive young rats. Brain Res. 424 (1), 153–162. https://doi.org/10.1016/0006-8993(87)91205-4.

Sharma, H.S., Olsson, Y., 1990. Edema formation and cellular alterations following spinal cord injury in the rat and their modification with p-chlorophenylalanine. Acta Neuropathol. 79 (6), 604–610. https://doi.org/10.1007/BF00294237.

Sharma, H.S., Sharma, A., 2007. Nanoparticles aggravate heat stress induced cognitive deficits, blood-brain barrier disruption, edema formation and brain pathology. Prog. Brain Res. 162, 245–273. https://doi.org/10.1016/S0079-6123(06)62013-X.

Sharma, H.S., Sjöquist, P.O., 2002. A new antioxidant compound H-290/51 modulates glutamate and GABA immunoreactivity in the rat spinal cord following trauma. Amino Acids 23 (1–3), 261–272. https://doi.org/10.1007/s00726-001-0137-z.

Sharma, H.S., Westman, J., 2004. The Blood-Spinal Cord and Brain Barriers in Health and Disease. Academic Press, San Diego, pp. 1–617. Release date: Nov. 9, 2003.

Sharma, H.S., Olsson, Y., Dey, P.K., 1990. Changes in blood-brain barrier and cerebral blood flow following elevation of circulating serotonin level in anesthetized rats. Brain Res. 517 (1–2), 215–223. https://doi.org/10.1016/0006-8993(90)91029-g.

Sharma, H.S., Cervós-Navarro, J., Dey, P.K., 1991. Acute heat exposure causes cellular alteration in cerebral cortex of young rats. Neuroreport 2 (3), 155–158. https://doi.org/10.1097/00001756-199103000-00012.

Sharma, H.S., Kretzschmar, R., Cervós-Navarro, J., Ermisch, A., Rühle, H.J., Dey, P.K., 1992a. Age-related pathophysiology of the blood-brain barrier in heat stress. Prog. Brain Res. 91, 189–196. https://doi.org/10.1016/s0079-6123(08)62334-1.

Sharma, H.S., Zimmer, C., Westman, J., Cervós-Navarro, J., 1992b. Acute systemic heat stress increases glial fibrillary acidic protein immunoreactivity in brain: experimental observations in conscious normotensive young rats. Neuroscience 48 (4), 889–901. https://doi.org/10.1016/0306-4522(92)90277-9.

Sharma, H.S., Westman, J., Cervós-Navarro, J., Dey, P.K., Nyberg, F., 1998. Blood-brain barrier in stress: a gateway to various brain diseases. In: Levy, A., Grauer, E., Ben-Nathan, D., de Kloet, E.R. (Eds.), New Frontiers of Stress Research: Modulation of Brain Function. Harwood Academic Publishers Inc, Amsterdam, pp. 259–276.

Sharma, H.S., Wiklund, L., Badgaiyan, R.D., Mohanty, S., Alm, P., 2006. Intracerebral administration of neuronal nitric oxide synthase antiserum attenuates traumatic brain injury-induced blood-brain barrier permeability, brain edema formation, and sensory motor disturbances in the rat. Acta Neurochir. Suppl. 96, 288–294. https://doi.org/10.1007/3-211-30714-1_62.

Sharma, H.S., Patnaik, R., Patnaik, S., Mohanty, S., Sharma, A., Vannemreddy, P., 2007. Antibodies to serotonin attenuate closed head injury induced blood brain barrier disruption and brain pathology. Ann. N. Y. Acad. Sci. 1122, 295–312. https://doi.org/10.1196/annals.1403.022.

Sharma, H.S., Sharma, A., 2012a. Nanowired drug delivery for neuroprotection in central nervous system injuries: modulation by environmental temperature, intoxication of nanoparticles, and comorbidity factors. Wiley Interdiscip. Rev. Nanomed. Nanobiotechnol. 4 (2), 184–203. https://doi.org/10.1002/wnan.172. (Epub 2011 Dec 8).

Sharma, H.S., Sharma, A., 2012b. Neurotoxicity of engineered nanoparticles from metals. CNS Neurol. Disord. Drug Targets 11 (1), 65–80. https://doi.org/10.2174/187152712799960817.

Sharma, H.S., Ali, S.F., Hussain, S.M., Schlager, J.J., Sharma, A., 2009a. Influence of engineered nanoparticles from metals on the blood-brain barrier permeability, cerebral blood flow, brain edema and neurotoxicity. An experimental study in the rat and mice using biochemical and morphological approaches. J. Nanosci. Nanotechnol. 9 (8), 5055–5072. https://doi.org/10.1166/jnn.2009.gr09.

Sharma, H.S., Ali, S.F., Tian, Z.R., Hussain, S.M., Schlager, J.J., Sjöquist, P.O., Sharma, A., Muresanu, D.F., 2009b. Chronic treatment with nanoparticles exacerbate hyperthermia induced blood-brain barrier breakdown, cognitive dysfunction and brain pathology in the rat. Neuroprotective effects of nanowired-antioxidant compound H-290/51. J. Nanosci. Nanotechnol. 9 (8), 5073–5090. https://doi.org/10.1166/jnn.2009.gr10.

Sharma, H.S., Miclescu, A., Wiklund, L., 2011a. Cardiac arrest-induced regional blood-brain barrier breakdown, edema formation and brain pathology: a light and electron microscopic study on a new model for neurodegeneration and neuroprotection in porcine brain. J. Neural Transm. (Vienna) 118 (1), 87–114. https://doi.org/10.1007/s00702-010-0486-4. (Epub 2010 Oct 21).

Sharma, H.S., Muresanu, D.F., Patnaik, R., Stan, A.D., Vacaras, V., Perju-Dumbrav, L., Alexandru, B., Buzoianu, A., Opincariu, I., Menon, P.K., Sharma, A., 2011b. Superior neuroprotective effects of cerebrolysin in heat stroke following chronic intoxication of Cu or Ag engineered nanoparticles. A comparative study with other neuroprotective agents using biochemical and morphological approaches in the rat. J. Nanosci. Nanotechnol. 11 (9), 7549–7569. https://doi.org/10.1166/jnn.2011.5114.

Sharma, H.S., Muresanu, D.F., Lafuente, J.V., Sjöquist, P.O., Patnaik, R., Sharma, A., 2015. Nanoparticles exacerbate both ubiquitin and heat shock protein expressions in spinal cord injury: neuroprotective effects of the proteasome inhibitor carfilzomib and the antioxidant compound H-290/51. Mol. Neurobiol. 52 (2), 882–898. https://doi.org/10.1007/s12035-015-9297-9. (Epub 2015 Jul 1).

Sharma, A., Muresanu, D.F., Lafuente, J.V., Sjöquist, P.O., Patnaik, R., Ryan Tian, Z., Ozkizilcik, A., Sharma, H.S., 2018. Cold environment exacerbates brain pathology and oxidative stress following traumatic brain injuries: potential therapeutic effects of nano-wired antioxidant compound H-290/51. Mol. Neurobiol. 55 (1), 276–285. https://doi.org/10.1007/s12035-017-0740-y.

Sharma, A., Castellani, R.J., Smith, M.A., Muresanu, D.F., Dey, P.K., Sharma, H.S., 2019. 5-Hydroxytryptophan: a precursor of serotonin influences regional blood-brain barrier breakdown, cerebral blood flow, brain edema formation, and neuropathology. Int. Rev. Neurobiol. 146, 1–44. https://doi.org/10.1016/bs.irn.2019.06.005. (Epub 2019 Jul 18).

Siesjö, B.K., Zhao, Q., Pahlmark, K., Siesjö, P., Katsura, K., Folbergrová, J., 1995 May. Glutamate, calcium, and free radicals as mediators of ischemic brain damage. Ann. Thorac. Surg. 59 (5), 1316–1320. https://doi.org/10.1016/0003-4975(95)00077-x.

Soares, A.T.G., Silva, A.C., Tinkov, A.A., Khan, H., Santamaría, A., Skalnaya, M.G., Skalny, A.V., Tsatsakis, A., Bowman, A.B., Aschner, M., Ávila, D.S., 2020. The impact of manganese on neurotransmitter systems. J. Trace Elem. Med. Biol. 61, 126554. https://doi.org/10.1016/j.jtemb.2020.126554. Online ahead of print.

Sweeney, M.D., Zhao, Z., Montagne, A., Nelson, A.R., Zlokovic, B.V., 2019. Blood-brain barrier: from physiology to disease and back. Physiol. Rev. 99 (1), 21–78. https://doi.org/10.1152/physrev.00050.2017.

Takeda, A., 2003. Manganese action in brain function. Brain Res. Brain Res. Rev. 41 (1), 79–87. https://doi.org/10.1016/s0165-0173(02)00234-5.

Tjalkens, R.B., Popichak, K.A., Kirkley, K.A., 2017. Inflammatory activation of microglia and astrocytes in manganese neurotoxicity. Adv. Neurobiol. 18, 159–181. https://doi.org/10.1007/978-3-319-60189-2_8.

von Kummer, R., Dzialowski, I., 2017. Imaging of cerebral ischemic edema and neuronal death. Neuroradiology 59 (6), 545–553. https://doi.org/10.1007/s00234-017-1847-6. (Epub 2017 May 24).

Yang, C., Hawkins, K.E., Doré, S., Candelario-Jalil, E., 2019. Neuroinflammatory mechanisms of blood-brain barrier damage in ischemic stroke. Am. J. Physiol. Cell Physiol. 316 (2), C135–C153. https://doi.org/10.1152/ajpcell.00136.2018. (Epub 2018 Oct 31).

Yin, Y., Sun, G., Li, E., Kiselyov, K., Sun, D., 2017. ER stress and impaired autophagy flux in neuronal degeneration and brain injury. Ageing Res. Rev. 34, 3–14. https://doi.org/10.1016/j.arr.2016.08.008. (Epub 2016 Sep 1).

Targeted therapy with anlotinib for a leptomeningeal spread recurrent glioblastoma patient

10

Cong Li[a,†], Wenyu Li[b,†], Shuang Dai[c], Aruna Sharma[d], Hari Shanker Sharma[d,]*, and Youliang Wu[a,]*

[a]*The Second Affiliated Hospital of Guangzhou University of Chinese Medicine, Guangdong Province Hospital of Chinese Medical, Guangzhou, China*
[b]*Department of Oncology, The First Affiliated Hospital, Sun Yat-sen University, Guangzhou, China*
[c]*Department of Oncology, The Seventh Affiliated Hospital, Sun Yat-sen University, Shenzhen, China*
[d]*International Experimental Central Nervous System Injury & Repair (IECNSIR), Department of Surgical Sciences, Anesthesiology & Intensive Care Medicine, Uppsala University Hospital, Uppsala University, Uppsala, Sweden*
**Corresponding authors: Tel.: +46-70-20-11-801; Fax: +46-18-24-38-99 (Hari Shanker Sharma); Tel.: +86-13632311724; Fax: +86-02081887233 (Youliang Wu), e-mail address: harishanker_sharma55@icloud.com; sharma@surgsci.uu.se; wuyouliang@gzucm.edu.cn*

Abstract

Glioblastoma (GBM) is the most common and the most aggressive primary malignant brain tumor in adults. Although tumor recurrence is inevitable, leptomeningeal spread is relatively rare. We describe a case of leptomeningeal spread recurrent GBM treated with anlotinib in this report. When the recurrent GBM patient had leptomeningeal spread was administered anlotinib 10 mg p.o. once every day and added oral temozolomide chemotherapy 100 mg/m^2 (days 1–7, days 15–21, 28-day cycle) after 3 months. The patient's overall survival time was more than 5 months and developed oral ulcer and acute cerebral infarction during his oral administration of anlotinib. This patient showed a favorable clinic outcome for treatment of leptomeningeal spread recurrent GBM with anlotinib and didn't show serious side effects.

[†]Contributed equally to this article.

Progress in Brain Research, Volume 265, ISSN 0079-6123, https://doi.org/10.1016/bs.pbr.2021.06.018

Keywords

Glioblastoma, Anlotinib, Leptomeningeal spread

1 Introduction

Glioma is the most common and fatal primary brain tumor. Despite surgery, radiotherapy, and chemotherapy, the median survival time of glioblastoma (GBM) patients which is the highest degree of malignancy is 14–16 months (Stupp et al., 2005). Glioma will recur in nearly all patients, but the follow-up treatment opinions of recurrent glioma are still inconsistent (Tan et al., 2020). The molecular level research of recurrent glioma has made some progress which also improves personalized diagnosis and treatment including some targeted drugs that have been used for the treatment of patients with GBM. The common targeted drugs are bevacizumab, thalidomide, cetuximab, etc. However, these drugs have limited effectiveness for patients with recurrent glioma (Olson et al., 2014) what's more, as the prognosis of GBM patients improves, leptomeningeal spread (LMS) becomes a more frequent clinical issue in neuro-oncology. Interestingly, given the advances in glioma therapeutics including molecular targeted therapies and immunotherapies. However, investigations of these innovative treatments remain limited in the setting of LMS and need further studies.

Anlotinib is a novel multitarget tyrosine kinase inhibitor that targets angiogenesis-related kinases such as vascular endothelial growth factor receptor (VEGFR)1/2/3, fibroblast growth factor receptors (FGFR)1/2/3, and other tumor-associated kinases such as c-Kit and Ret (Xie et al., 2018). Anlotinib can inhibit not only tumor angiogenesis but also tumor cell growth. Preclinical results showed that anlotinib significantly inhibited FGFR1–4, especially FGFR2 (Sun et al., 2016). Anlotinib has been reported for the treatment of nonsmall cell lung cancer, metastatic renal cell carcinoma, and sarcoma, with good effect and mild side effect (Shen et al., 2018; Xie et al., 2018; Zhang et al., 2019). There were also multiple brain metastases from other parts of the body, which can disappear or did not continue to increase after taking anlotinib. However, there is a little study that reported the effect of the anlotinib on patients with recurrent GBM, especially in LMS recurrent glioma. Here, we reported the efficacy and security of a case with LMS recurrent GBM after taking anlotinib.

2 Case presentation

A 25-year-old man began to have a headache in September 2017, presenting with whole head moderate degree pain, no nausea, vomiting, limb fatigue, and other discomforts. The patient's headache broke out intermittently without further treatment. Until April 2018, the patient had a seizure, which was characterized by sudden crying and limbs twitched with unconsciousness, and his eyes turned up. The twitch lasted for about 2 min and he woke up about 4 min later. After waking up, he felt headache, fatigue, and soreness. Then he received emergency treatment. The brain computed tomography (CT) and magnetic resonance imaging (MRI) showed that there was a lesion in the left frontal lobe (Fig. 1). The long-term electroencephalogram (EEG) showed limited slow activity in the left anterior head, especially the θ waves. The patient underwent left frontal brain lesion resection under the navigation and intra-operative EEG monitoring on April 12, 2018. The patient's postoperative recovery was good. The pathological results showed glioblastoma, IDH1 wild type, WHO Grade IV, TERT wild type, and no deletion of 1p19q. The patient received concurrent chemo-radiotherapy and begun temozolomide (TMZ) chemotherapy after operation. Six courses of TMZ adjuvant chemotherapy (standard Stupp regimen) were carried out and end in January 2019. During this period, the patient took anti-epileptic drugs regularly, and had limb convulsions twice, but did not relapse after adjusting the drug dose. On September 10, 2019, MRI showed that there was a new small nodule located in the left temporal lobe. Then gamma knife treatment was performed because of considering tumor recurrence. On December 31, 2019, the brain MRI showed that the masses and nodules in the right lateral fissure cistern, left temporal lobe and the tumor in the temporal occipital cortex near the sulcus was larger than before, considering the possibility of tumor spreading in cerebrospinal fluid. Then the LMS recurrent GBM patient was treated with TMZ 80 mg p.o. QD has begun January 6, 2020. However, on February 4, 2020, MRI showed that the lesion was still in progress. Then the administration of Anlotinib 10 mg p.o. QD was started and TMZ was stopped on February 11. Then MRI on February 22 showed that the

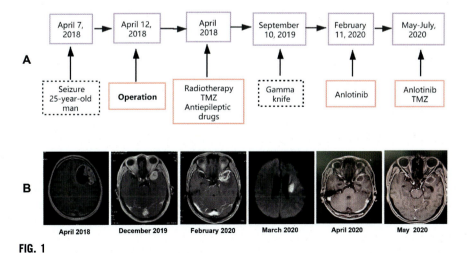

FIG. 1

The treatment flow chart (A) of this LMS recurrent GBM patient and the regular MRI image (B). In addition to the MR diffusion sequence in March 2020, the rest are MR with enhancement sequence.

lesions were generally smaller. On March 28 and April 10, the second and third course of Anlotinib were given and the patient was accompanied by I degree of oral ulcer. The symptoms of limb weakness and aphasia were improved. However, during this period, the patient developed acute cerebral infarction in the left radial crown area (Fig. 1), and the drug was discontinued for about a week. On April 29, MRI suggested multiple intracranial lesions were stable disease (SD). While on May 8, MRI showed a new lesion in the brain column and was evaluated as progressive disease (PD). Then $100\,mg/m^2$ (days 1–7, days 15–21, 28-day cycle) was added to Anlotinib. The patient's MRI was evaluated as SD on May and July 2020. The 7-day regimen of TMZ combined with Anlotinib was continued. After telephone follow-up in September, the patient's symptoms worsened without further follow-up anti-tumor treatment, just have palliative treatment.

3 Discussion

The current standard treatment consisting of maximal tumor resection, followed by radiation and TMZ-based chemotherapy, the survival of gliomas patient is still not optimistic, the 5-year survival rate of GBM patients is less than 10% (Ostrom et al., 2020). Glioma will recur in nearly all patients, but the diagnosis and follow-up treatment opinions of recurrent glioma are still inconsistent (Li et al., 2020). Many methods including repeat surgery, chemotherapy, re-radiotherapy, methods of Neurorestoratology, traditional Chinese Medicine, molecular targeted therapy, and immunotherapy can be used in recurrent glioma (Tan et al., 2020; Xinyu et al., 2020; Zhong et al., 2019). what's more, the GBM can metastasize to cerebrospinal fluid (CSF) flow stream inducing LMS GBM. LMS results from spreading of tumor cells from brain parenchyma to leptomeninges and CSF and is one of the most severe complications of GBM. The median overall survival of GBM patients with LMS varies between 2 and 5 months (Birzu et al., 2020; Noh et al., 2015).

Given the difficulty to treat recurrent glioma, there is an urgent need to develop effective therapies. Molecular effective targeted drugs are one of the key research directions. GBM is one of the most vascularized tumors and glioma cells can produce VEGF. The levels of VEGF usually induce a worse prognosis in glioma patients (Louis et al., 2007). Bevacizumab is a monoclonal antibody directed against VEGF, has become the standard treatment for patients with recurrent GBM. However, the bevacizumab can improve progression-free survival (PFS), did not prolong over survival (OS) of recurrent GBM patients (Chen et al., 2018). Following on from bevacizumab, there have been several trials of VEGF or multikinase TKIs to target the tumor microenvironment, and they showed mixed results. Cediranib is an oral tyrosine kinase inhibitor (TKI), which failed to show a survival benefit in a randomized phase 3 trial, either as monotherapy or in combination with lomustine, in recurrent GBM (Batchelor et al., 2013). More recently, a phase 2 trial of regorafenib in the relapsed setting showed an efficacy signal with an OS benefit compared with lomustine (Lombardi et al., 2019). Trials of other drugs, such as tivozanib (Kalpathy-Cramer et al., 2017), pazopanib (Iwamoto et al., 2010), and sunitinib

(Hutterer et al., 2014), have shown minimal activity, and suggest that VEGF mono-therapy may have a limited role in an unselected glioma patient. These anti-VEGF drugs including cediranib, imatinib, sunitinib, and sorafenib have mild toxicities, but their efficacy of anti-tumor was not satisfactory.

Anlotinib is a multitarget TKI that can inhibit not only tumor angiogenesis but also tumor cell proliferation. Anlotinib can inhibit more targets than some other target drugs can, including sorafenib, sunitinib, and pazopanib (Shen et al., 2018). Bevacizumab was recommended for recurrent glioma, but this drug was given to this patient because it has an anti-angiogenic effect similar to bevacizumab. The same dosage and cycle of anlotinib as lung cancer treatment were applied to glioma pa-tients, and this dosage has been confirmed to be safe by clinical trial, and toxicities were well tolerated (Shen et al., 2018).

This patient started to use anlotinib after recurrent GBM had LMS, and the final prognosis was great. The OS from starting to use anlotinib was more than 9 months. This is a well prognosis in LMS recurrent GBM patients and also better than the recurrent GBM case had anlotinib Yajuan Lv et al. reported, whose PFS was about 2 months and OS was just 110 days (Zhang et al., 2019). As far as we know, this is the first reported case of anlotinib used in LMS GBM patients and achieved good performance.

Of course, the safety of this drug is also important. This patient developed an oral ulcer and acute infarction during the use of anlotinib. With the continuous expansion of the clinical application of anlotinib, the monitoring of adverse drug reactions (ADR) cannot be ignored. Zhou et al. (2020) summarized a total of 1240 cases of anlotinib, and 4596 adverse reactions were reported (summarized in Table 1). The most common ADRs were hypertension, fatigue, and dyslipidemia. The most com-mon ADRs (grade ≥ 3) were hypertension, dyslipidemia, and hyponatremia (Zhou et al., 2020). The oral ulcer may be related to anlotinib, but whether the acute infarc-tion has anything to do with anlotinib is unknown. It has been reported that gamma knife can increases the risk of cerebral artery stenosis or occlusion. However, the risk of ischemic complications caused by gamma knife is very low (Graffeo et al., 2019). The appearance of acute brain ischemic in this young patient suggests that anlotinib may accelerate the process of cerebral artery stenosis or occlusion, but the exact mechanism of anlotinib in this process is unknown. But we need to pay attention to the effect of this drug on blood vessels when treatment of glioma.

Above all, the use of anlotinib in glioma can be treated as a single drug, convenient oral administration, relatively good patient compliance, and well clinical applications. Clinicians should strengthen the follow-up in the process of medica-tion, reduce and avoid the occurrence of ADR, to achieve better clinical treatment effect and improve the quality of life of patients. But there are still some limitations to this case. The survival benefits of anlotinib in LMS recurrent GBM patients are not so sure, because the patient also had a gamma knife and taken TMZ, especially the effect of gamma knife cannot be ignored. Therefore, the efficacy of combination therapy of anlotinib with radiotherapy and other drugs needs to be further evaluated. Besides, the patient's acute infarction may also be accidental, whether it is related to gamma knife and anlotinib is hard to know.

Table 1 Organ/system involvement, main clinical manifestations and incidence of ADR.

Organs/systems involved	Constituent ratio (%)	Main clinical manifestations
Cardiovascular and cerebrovascular system damage	11.25	Hypertension; abnormal electrocardiogram; tachycardia
Skin and its appendages damage	9.05	Hand foot syndrome; rash
Gastrointestinal system damage	16.12	Anorexia; diarrhea; oral mucositis; abdominal pain; vomiting; nausea; toothache; serum amylase abnormality
Systemic damage	9.31	Fatigue, headache, and pain, fever
Respiratory system damage	11.31	Cough, sore throat, hemoptysis, vocal cord paralysis, upper respiratory tract infection, dyspnea, pleural effusion, pneumothorax Chest
Urinary system damage	6.74	Proteinuria, hematuria, urinary tract infection and creatinine clearance rate decreased
Metabolic and nutritional disorders	17.21	Hypertriglyceridemia, hypercholesterolemia, low density lipoprotein increased, weight loss, hyponatremia, hypochloremia, hypophosphatemia, Alkaline phosphatase increased, hypoproteinemia, lactate dehydrogenase increased, myocardial enzyme abnorma
Hepatobiliary system damage	7.88	Glutamyl transpeptidase, hyperbilirubinemia, glutamic pyruvic transaminase, glutamic oxaloacetic transaminase (AST) increased, abnormal liver function
Endocrine disorders	5.94	Increased thyroid stimulating hormone, hypothyroidism, hyperthyroidism, thrombocytopenia, gastrointestinal bleeding, bleeding
Others	2.02	Dysphonia, arthralgia, dizziness, taste disorder

4 Conclusion

This patient showed a favorable clinical outcome for treatment of LMS recurrent GBM with anlotinib and didn't show serious side effects. Anlotinib provides a new option for LMS recurrent GBM patients. However, the efficacy and side effects of this drug still require further clinical trials to confirm.

References

Batchelor, T.T., Mulholland, P., Neyns, B., Nabors, L.B., Campone, M., Wick, A., Mason, W., Mikkelsen, T., Phuphanich, S., Ashby, L.S., Degroot, J., Gattamaneni, R., Cher, L., Rosenthal, M., Payer, F., Jurgensmeier, J.M., Jain, R.K., Sorensen, A.G., Xu, J., Liu, Q., van den Bent, M., 2013. Phase III randomized trial comparing the efficacy of cediranib as monotherapy, and in combination with lomustine, versus lomustine alone in patients with recurrent glioblastoma. J. Clin. Oncol. 31, 3212–3218.

Birzu, C., Tran, S., Bielle, F., Touat, M., Mokhtari, K., Younan, N., Psimaras, D., Hoang-Xuan, K., Sanson, M., Delattre, J.Y., Idbaih, A., 2020. Leptomeningeal spread in glioblastoma: diagnostic and therapeutic challenges. Oncologist 25, e1763–e1776.

Chen, Z., Xu, N., Zhao, C., Xue, T., Wu, X., Wang, Z., 2018. Bevacizumab combined with chemotherapy vs single-agent therapy in recurrent glioblastoma: evidence from randomized controlled trials. Cancer Manag. Res. 10, 2193–2205.

Graffeo, C.S., Link, M.J., Stafford, S.L., Parney, I.F., Foote, R.L., Pollock, B.E., 2019. Risk of internal carotid artery stenosis or occlusion after single-fraction radiosurgery for benign parasellar tumors. J. Neurosurg. 133, 1–8.

Hutterer, M., Nowosielski, M., Haybaeck, J., Embacher, S., Stockhammer, F., Gotwald, T., Holzner, B., Capper, D., Preusser, M., Marosi, C., Oberndorfer, S., Moik, M., Buchroithner, J., Seiz, M., Tuettenberg, J., Herrlinger, U., Wick, A., Vajkoczy, P., Stockhammer, G., 2014. A single-arm phase II Austrian/German multicenter trial on continuous daily sunitinib in primary glioblastoma at first recurrence (SURGE 01-07). Neuro Oncol. 16, 92–102.

Iwamoto, F.M., Lamborn, K.R., Robins, H.I., Mehta, M.P., Chang, S.M., Butowski, N.A., Deangelis, L.M., Abrey, L.E., Zhang, W.T., Prados, M.D., Fine, H.A., 2010. Phase II trial of pazopanib (GW786034), an oral multi-targeted angiogenesis inhibitor, for adults with recurrent glioblastoma (North American Brain Tumor Consortium Study 06-02). Neuro Oncol. 12, 855–861.

Kalpathy-Cramer, J., Chandra, V., Da, X., Ou, Y., Emblem, K.E., Muzikansky, A., Cai, X., Douw, L., Evans, J.G., Dietrich, J., Chi, A.S., Wen, P.Y., Stufflebeam, S., Rosen, B., Duda, D.G., Jain, R.K., Batchelor, T.T., Gerstner, E.R., 2017. Phase II study of tivozanib, an oral VEGFR inhibitor, in patients with recurrent glioblastoma. J. Neurooncol 131, 603–610.

Li, C., Gan, Y., Chen, H., Chen, Y., Deng, Y., Zhan, W., Tan, Q., Xie, C., Sharma, H.S., Zhang, Z., 2020. Advanced multimodal imaging in differentiating glioma recurrence from post-radiotherapy changes. Int. Rev. Neurobiol. 151, 281–297.

Lombardi, G., De Salvo, G.L., Brandes, A.A., Eoli, M., Ruda, R., Faedi, M., Lolli, I., Pace, A., Daniele, B., Pasqualetti, F., Rizzato, S., Bellu, L., Pambuku, A., Farina, M., Magni, G., Indraccolo, S., Gardiman, M.P., Soffietti, R., Zagonel, V., 2019. Regorafenib compared with lomustine in patients with relapsed glioblastoma (REGOMA): a multicentre, open-label, randomised, controlled, phase 2 trial. Lancet Oncol. 20, 110–119.

Louis, D.N., Ohgaki, H., Wiestler, O.D., Cavenee, W.K., Burger, P.C., Jouvet, A., Scheithauer, B.W., Kleihues, P., 2007. The 2007 WHO classification of tumours of the central nervous system. Acta Neuropathol. 114, 97–109.

Noh, J.H., Lee, M.H., Kim, W.S., Lim, D.H., Kim, S.T., Kong, D.S., Nam, D.H., Lee, J.I., Seol, H.J., 2015. Optimal treatment of leptomeningeal spread in glioblastoma: analysis of risk factors and outcome. Acta Neurochir. 157, 569–576.

Olson, J.J., Nayak, L., Ormond, D.R., Wen, P.Y., Kalkanis, S.N., 2014. The role of cytotoxic chemotherapy in the management of progressive glioblastoma: a systematic review and evidence-based clinical practice guideline. J. Neurooncol 118, 501–555.

Ostrom, Q.T., Patil, N., Cioffi, G., Waite, K., Kruchko, C., Barnholtz-Sloan, J.S., 2020. CBTRUS statistical report: primary brain and other central nervous system tumors diagnosed in the United States in 2013-2017. Neuro Oncol. 22, v1–v96.

Shen, G., Zheng, F., Ren, D., Du, F., Dong, Q., Wang, Z., Zhao, F., Ahmad, R., Zhao, J., 2018. Anlotinib: a novel multi-targeting tyrosine kinase inhibitor in clinical development. J. Hematol. Oncol. 11, 120.

Stupp, R., Mason, W.P., van den Bent, M.J., Weller, M., Fisher, B., Taphoorn, M.J., Belanger, K., Brandes, A.A., Marosi, C., Bogdahn, U., Curschmann, J., Janzer, R.C., Ludwin, S.K., Gorlia, T., Allgeier, A., Lacombe, D., Cairncross, J.G., Eisenhauer, E., Mirimanoff, R.O., 2005. Radiotherapy plus concomitant and adjuvant temozolomide for glioblastoma. N. Engl. J. Med. 352, 987–996.

Sun, Y., Niu, W., Du, F., Du, C., Li, S., Wang, J., Li, L., Wang, F., Hao, Y., Li, C., Chi, Y., 2016. Safety, pharmacokinetics, and antitumor properties of anlotinib, an oral multi-target tyrosine kinase inhibitor, in patients with advanced refractory solid tumors. J. Hematol. Oncol. 9, 105.

Tan, A.C., Ashley, D.M., Lopez, G.Y., Malinzak, M., Friedman, H.S., Khasraw, M., 2020. Management of glioblastoma: state of the art and future directions. CA Cancer J. Clin. 70, 299–312.

Xie, C., Wan, X., Quan, H., Zheng, M., Fu, L., Li, Y., Lou, L., 2018. Preclinical characterization of anlotinib, a highly potent and selective vascular endothelial growth factor receptor-2 inhibitor. Cancer Sci. 109, 1207–1219.

Xinyu, W., Nan, S., Xiangqi, M., Meng, C., Chuanlu, J., Jinquan, C., 2020. Review of clinical nerve repair strategies for neurorestoration of central nervous system tumor damage. J. Neurorestoratol. 08, 172–181.

Zhang, A., Liu, B., Xu, D., Sun, Y., 2019. Advanced intrahepatic cholangiocarcinoma treated using anlotinib and microwave ablation: a case report. Medicine (Baltimore) 98, e18435.

Zhong, D., Hai, Y., Ning, W., Wahap, A., Jia, W., Tuo, W., Changwang, D., Maode, W., 2019. Impact of preoperative Karnofsky Performance Scale (KPS) and American Society of Anesthesiologists (ASA) scores on perioperative complications in patients with recurrent glioma undergoing repeated operation. J. Neurorestoratol. 07, 143–152.

Zhou, W., Yang, S., Feng, L., Chen, M., Wang, W., 2020. Bibliometric analysis of adverse drug reactions induced by Anlotinib. Chin. Pharm. Aff. 34, 1467–1474.

Printed in the United States
by Baker & Taylor Publisher Services